THE NUMERICAL JORDAN FORM

THE NUMERICAL JORDAN FORM

Petko Petkov
Bulgarian Academy of Sciences, Sofia, Bulgaria

World Scientific
NEW JERSEY • LONDON • SINGAPORE • BEIJING • SHANGHAI • HONG KONG • TAIPEI • CHENNAI • TOKYO

Published by

World Scientific Publishing Co. Pte. Ltd.
5 Toh Tuck Link, Singapore 596224
USA office: 27 Warren Street, Suite 401-402, Hackensack, NJ 07601
UK office: 57 Shelton Street, Covent Garden, London WC2H 9HE

Library of Congress Cataloging-in-Publication Data
Names: Petkov, P. Hr. (Petko Hr.), 1948– author.
Title: The numerical Jordan form / Petko Petkov, Bulgarian Academy of Sciences, Bulgaria.
Description: New Jersey : World Scientific, [2024] | Includes bibliographical references and index.
Identifiers: LCCN 2024002522 | ISBN 9789811286445 (hardcover) |
ISBN 9789811286452 (ebook for institutions) | ISBN 9789811286469 (ebook for individuals)
Subjects: LCSH: Jordan matrix. | Eigenvalues.
Classification: LCC QA188 P455 2024 | DDC 512/.48--dc23/eng/20240412
LC record available at https://lccn.loc.gov/2024002522

British Library Cataloguing-in-Publication Data
A catalogue record for this book is available from the British Library.

For any available supplementary material, please visit
https://www.worldscientific.com/worldscibooks/10.1142/13682#t=suppl

Desk Editors: Soundararajan Raghuraman/Nijia Liu

Typeset by Stallion Press
Email: enquiries@stallionpress.com

To the memory of my parents

Preface

... as a mathematical probe the Jordan canonical form is still useful, and reports of its dead are greatly exaggerated.

G.W. Stewart, *Matrix Algorithms*, Vol. II: Eigensystems, SIAM, 2001.

There has been a need for a book of this kind for a long time. However, time passed, the book did not appear and in the end I decided to write it myself, although I did not at all think that I could do it better than others.

A.A. Kharkevich, *Spectra and Analysis*, GITTL, Moskow, 1952.

The Jordan form of a matrix is a fundamental concept in the matrix analysis, playing a vital role in solving various problems in applied mathematics, including solution of differential equations, matrix exponential computation, eigenvalue sensitivity analysis, matrix equations solution and determining the stability of dynamic systems described by differential equations.

The purpose of this book is to describe the numerical difficulties in determining the Jordan form of a matrix and to present the available state-of-the art algorithms for reducing a square matrix into numerical Jordan form. In contrast to the "pure" Jordan form whose computation is ill-posed, the numerical Jordan form preserves its structure under small perturbations of the matrix elements, making its determination a well-posed computational problem. If the structure is well-conditioned, it can be determined reliably in the presence of uncertainties and rounding errors.

This book focuses on the algorithmic and computational aspects of finding the Jordan form of a matrix. It offers a comprehensive overview of a class of algorithms designed to transform a complex matrix into numerical Jordan form. Additionally, it delves into the practical application of this form in solving significant problems, such as estimating eigenvalue sensitivity and computing the matrix exponential. A particular emphasis is placed on the Jordan–Schur form of a matrix, which, in the author's perspective, remains underutilized in the field of matrix computations. Since the mathematical objects under consideration can be sensitive to changes in the elements of the given matrix, this book extensively covers perturbation analysis pertaining to eigenvalues, invariant subspaces, Schur form and Jordan form. A large number of numerical examples and figures are included to clarify mathematical concepts.

The text may serve as a comprehensive, self-contained reference for the numerical determination of the Jordan form of a matrix and its practical applications. It covers essential material from matrix analysis, numerical computations and linear algebra. The computational algorithms for finding the Jordan, Weyr and Jordan–Schur forms are introduced step by step, with detailed examples illustrating the concepts. A prerequisite for readers is some familiarity with matrices and basic knowledge of analysis. The intended audience comprises specialists from various scientific disciplines who engage in matrix computations but may not be experts in numerical analysis. Consequently, efforts have been made to minimize theoretical rigor, avoiding detailed proofs of certain results. Instead, readers are directed to relevant literature for in-depth proof of mathematical facts. This book can also benefit advanced master's students studying topics like Matrix Computations, Numerical Analysis, Control Theory and related fields.

It's worth noting that this book serves as an introductory resource to the numerical determination of the Jordan form. A deeper exploration of related issues, which involve multiple mathematical disciplines like differential geometry, topology, singularity theory and bifurcation theory, requires more extensive knowledge. This book references appropriate literature sources for those interested in further study.

In total, this book presents 40 algorithms for computing the Jordan, Weyr and Jordan–Schur forms of matrices, along with various mathematical objects related to eigenvalue computations. Given the increased complexity in determining these forms for real matrices, the algorithms are derived and described within the complex domain. This book employs a MATLAB®-like notation, and it comes with a collection of MATLAB®M-files implementing these algorithms. It's important to note that the included files in this book serve an illustrative purpose and aren't optimized for computational efficiency and accuracy. Developing reliable and efficient software for determining the Jordan form requires additional expertise and effort. The computations presented in this book were carried out using MATLAB Version 9.9 (R2020b) with IEEE double-precision arithmetic. Readers should be mindful that when using different computing machines and versions of MATLAB®, there may be variations in the numerical results compared to those provided in this book.

Regarding the numerical experiments featured in this book, most of the examples have been intentionally designed in a way that their analytical solutions are known and can be computed with high precision. This deliberate design aids in identifying the reasons behind rounding errors in intermediate and final results, as well as pinpointing potential sources of numerical instability. The author strongly encourages readers to perform these experiments themselves, as they offer valuable insights into the numerical behavior of the methods under investigation. Engaging in these experiments also forms an integral part of comprehending this book's content and gaining practical computational experience.

A note on the naming scheme in the LAPACK linear algebra package which is frequently cited in this book: As usual, the LAPACK routines are referred generically, regardless of data type, replacing the first letter by "x". Thus `xGEESX` refers to any of the routines `SGEESX`, `DGEESX`, `CGEESX` and `ZGEESX`.

This book serves as a tribute to the ideas and dedicated efforts of numerous scientists, including Vera Kublanovskaya, Axel Ruhe, Bo Kågström and his research group, who have tirelessly worked to develop reliable numerical algorithms for determining the numerical Jordan form.

I express my deep gratitude to Prof. Stefano Serra-Capizano from the Università dell'Insubria, Italy, to Prof. Daniel Kressner from École Polytechnique Fédérale de Lausanne, Switzerland, and to Prof. Yimin Wei from Fudan University, Shanghai, China, who diligently reviewed the initial version of this book and provided a multitude of valuable comments and suggestions. I extend my thank to Dr. Stefan Johansson from the University of Umeå, Sweden, for his assistance in using the GUPTRI program. I'm also indebted to Prof. Zhonggang Zeng from the Northeastern Illinois University, who generously shared his invaluable report on computing the Jordan form before its official publication. I would like to convey my appreciation to the team at World Scientific Publishing, particularly Mathematics Editor Nijia Liu and Production Editor Rosie Williamson for their helpful assistance in this book's preparation. Lastly, but certainly not least, I want to thank Prof. Volker Mehrmann for his support during challenging times.

Contents

List of Algorithms

Notation

$\mathbb{R}$	the set of real numbers;
$\mathbb{C}$	the set of complex numbers;
$j_o = \sqrt{-1}$	the imaginary unit;
$\mathrm{Re}(x)$	the real part of x;
$\mathrm{Im}(x)$	the imaginary part of x;
$\overline{x}$	the conjugate of x;
$\|x\|$	the absolute value of x;
$\mathbb{R}^n$	the space of real n-vectors;
$\mathbb{C}^n$	the space of complex n-vectors;
$\mathbb{R}^{m\times n}$	the space of real $m \times n$ matrices;
$\mathbb{C}^{m\times n}$	the space of complex $m \times n$ complex matrices;
$A = [a_{ij}]$	matrix with elements a_{ij};
A^T	the transpose of A;
$\overline{A}$	the conjugate of A;
$A^H = \overline{A}^T$	the Hermitian transpose (the conjugate transpose) of A;
I_n	the unit $n \times n$ matrix;
$0_{m\times n}$	the zero $m \times n$ matrix;
A^{-1}	the inverse of A;
$\|A\|$	the matrix of absolute values of the elements of A;
ΔA	the perturbation of A;
$\mathrm{Low}(A)$	the strictly lower triangular part of $A \in \mathbb{C}^{n\times n}$;

$\mathrm{Up}(A)$	the strictly upper triangular part of $A \in \mathbb{C}^{n\times n}$;
$A_{i,1:n}$	the ith row of $A \in \mathbb{C}^{n\times n}$;
$A_{i_1:i_2,j_1:j_2}$	the part of A from row i_1 to i_2 and from column j_1 to j_2;
$\mathrm{diag}(a_1, a_2, \ldots, a_n)$	a diagonal matrix;
$A \otimes B$	the Kronecker product of A and B;
e_j	the jth column of I_n;
$\mathrm{trace}(A)$	the trace of A;
$\det(A)$	the determinant of A;
$\mathrm{rank}(A)$	the rank of A;
$\mathrm{null}(A)$	the nullity of A;
$\lambda_i(A)$	the ith eigenvalue of A;
$p(\lambda)$	the characteristic polynomial;
$\mathrm{spect}(A)$	the spectrum of A;
$J_s(\lambda)$	the Jordan block of order s associated with λ;
$\mathcal{S}(A, \lambda)$	the Segre characteristic of A associated with λ;
$\mathcal{W}(A, \lambda)$	the Weyr characteristic of A associated with λ;
$\sigma_{\min}(A)$	the minimum singular value of A;
$\sigma_{\max}(A)$	the maximum singular value of A;
$\mathrm{sv}_A(B)$	the spectral deviation of B with respect to A;
$\mathrm{hd}(A, B)$	the Hausdorff distance between the spectra of A and B;
$\mathrm{dev}(A)$	the deviation of A from normality;
$\mathrm{sep}(A, B)$	the separation between A and B;
$\chi(\lambda, \mu)$	the chordal distance between the generalized eigenvalues λ and μ;
$\mathrm{dif}(A, B; C, D)$	the difference between the spectra of $(A - \lambda B)$ and $(C - \lambda D)$;
$\lVert\cdot\rVert$	the norm of a vector or matrix;
$\lVert x\rVert_1, \lVert x\rVert_2, \lVert x\rVert_\infty$	the vector 1-, 2- and ∞ norms;
$\lVert A\rVert_1, \lVert A\rVert_2, \lVert A\rVert_\infty$	the matrix 1-, 2- and ∞ norms;
$\lVert A\rVert_F$	the Frobenius norm of A;
$A \preceq B$	the elements of A are smaller than or equal to the elements of B;

$h = O(x^p)$	quantity of order x^p;
$h = o(x)$	$\|h\|/x \to 0$ as $x \to 0$;
$\angle(x, y)$	the angle between x and y;
$x \perp y$	the vector x is orthogonal to the vector y;
$\text{span}(X)$	the subspace spanned by the columns of X;
$\dim(\mathcal{X})$	the dimension of the subspace $\mathcal{X}$;
$\mathcal{X} \oplus \mathcal{Y}$	the direct sum of $\mathcal{X}$ and $\mathcal{Y}$;
$\mathcal{R}(A)$	the image of A;
$\mathcal{N}(A)$	the null space of A;
$\text{null}(A)$	the nullity of A;
$\mathcal{X}^\perp$	the orthogonal complement of $\mathcal{X}$;
$\Theta_{\min}(\mathcal{X}, \mathcal{Y})$	the minimum angle between subspaces $\mathcal{X}$ and $\mathcal{Y}$;
$\Theta_{\max}(\mathcal{X}, \mathcal{Y})$	the maximum angle between subspaces $\mathcal{X}$ and $\mathcal{Y}$;
$\text{vec}(A)$	the vec mapping of A;
$P_n \in \mathbb{R}^{n^2 \times n^2}$	the vec-permutation matrix;
$\mathbf{u}$	the unit roundoff unit of the floating point arithmetic;
$\text{fl}(x)$	the rounded value of x;
$\text{fl}(a \circ b)$	the operation $a \circ b$ computed in a floating point arithmetic;
flop	a floating point addition and multiplication.

Chapter 1

Numerical Matrix Computations

The purpose of this chapter is to review the necessary concepts related to the numerical solution of matrix problems that will be needed in the remaining part of this book. First, a brief analysis is done of the errors arising in the representation of numbers and in performing the basic arithmetic operations in the IEEE floating point arithmetic. Error bounds for the most frequently used matrix operations are derived in detail. Next, we consider the conditioning of computational problems and the numerical stability of algorithms for their solution. We present the construction and analyze the errors associated with computing of the elementary matrix transformations which are the building blocks in developing the numerical methods for matrix computations. The implementation of these transformations is illustrated by the computation of the QR decomposition of a matrix. Finally, we define the important notion of numerical rank and show how to compute the fundamental subspaces related to a given matrix.

For a systematics presentation of the subjects covered in this chapter, the reader should consult the references given at the end of each section.

1.1 Basic Rounding Error Analysis

1.1.1 *Floating Point Numbers*

In the computer processor, the set $\mathbb{R}$ of real numbers is approximated by a finite set $\mathcal{F} \subset \mathbb{R}$ of *floating point numbers* which are exactly representable in the *floating point arithmetic*. The floating point arithmetic is characterized by *base* b, *precision* t and *exponent range* $e_{\min}, e_{\max}$. Here b and t are positive integers, $e_{\min}$ is a negative integer and $e_{\max}$ is a positive integer.

Almost all contemporary processors implement the IEEE standard for binary ($b = 2$) floating point arithmetic 754-2019 (IEEE, 2019) or its subset. Further on it will be called *IEEE arithmetic standard* or shortly *IEEE arithmetic.* In the IEEE arithmetic, each binary floating point number x is represented in the *normalized form:*

$$x = (-1)^s 2^e (b_0.b_1 b_2 \ldots b_{t-1}), \tag{1.1}$$

where $s = 0$ or 1 determines the number sign, $e_{\min} \le e \le e_{\max}$, $b_0 = 1$ and $b_i = 0$ or 1 for $i = 1, 2, \ldots, t-1$. The integer e is called *exponent*, the binary numbers $b_0, b_1, \ldots, b_{t-1}$ are called *significant bits* and b_{t-1} is called *the least significant bit.*

The floating point number (1.1) has the value

$$\pm(b_0 + b_1 2^{-1} + \cdots + b_{t-1} 2^{1-t}) \times 2^e \tag{1.2}$$

$$= \pm f \times 2^e. \tag{1.3}$$

The number

$$f = b_0 + b_1 2^{-1} + \cdots + b_{t-1} 2^{1-t}, \quad 1 \le f \le 2 - 2^{1-t}$$

is referred to as *the fractional part, significand* or *mantissa.*

The *normalization* of the floating point numbers is expressed by the fact that for $x \neq 0$, it holds $b_0 = 1$, from where it follows that $f \ge 1$. Since b_0 is always equal to 1, the bit b_0 is not used explicitly.

The minimum representable in the IEEE arithmetic positive number is obtained for $b_0 = 1, b_1 = b_2 = \cdots = b_{t-1} = 0$ and $e = e_{\min}$:

$$r_{\min} = 1 \times 2^{e_{\min}} = 2^{e_{\min}}.$$

The maximum representable positive number is obtained for $b_0 = b_1 = b_2 = \cdots = b_{t-1} = 1$ and $e = e_{\max}$:

$$r_{\max} = (2 - 2^{1-t}) \times 2^{e_{\max}}.$$

The IEEE standard involves two kinds of number systems: *single precision* and *double precision.* These systems are characterized by different values of t, $e_{\min}$ and $e_{\max}$. In the machine representation of a floating point number, instead of e one uses $E = e + \text{bias}$, where bias is the *exponent bias*, used to make the exponent nonnegative. This allows avoiding the presence of an additional bit, corresponding to the exponent sign. In the single precision arithmetic, bias = 127, while in the double precision, bias = 1023.

The parameters of the single precision and double precision IEEE arithmetic are given in Table 1.1.

The 32-bit and 64-bit machine words used in the IEEE single precision and double precision arithmetic, respectively, are organized as shown in Figure 1.1.

In the IEEE arithmetic, the zero is represented as $e = 0$ and $f = 0$, the sign bit ensuring different representation of $+0$ and -0. Since in this case the most significant bit is 0, the zero has exponent equal to $e_{\min} - 1$ and the biased exponent is $E = e_{\min} - 1 + \text{bias} = 0$.

Table 1.1. Parameters of the IEEE arithmetic.

Precision	t	$e_{\min}$	$e_{\max}$	$r_{\min}$	$r_{\max}$	u
Single	24	-126	127	$1.2 \cdot 10^{-38}$	$3.4 \cdot 10^{38}$	$2^{-24} \approx 5.96 \cdot 10^{-8}$
Double	53	-1022	1023	$2.2 \cdot 10^{-308}$	$1.8 \cdot 10^{308}$	$2^{-53} \approx 1.11 \cdot 10^{-16}$

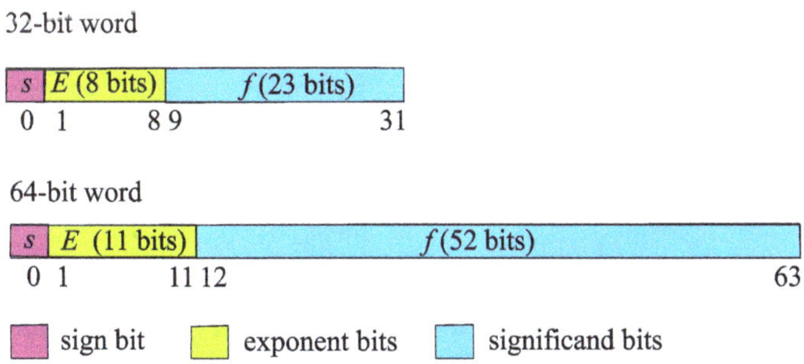

Fig. 1.1. 32-bit and 64-bit words in IEEE arithmetic.

In this arithmetic, there is a special symbol `NaN` ("Not a Number") with $E = e_{\max} + 1 + \text{bias}$ and $f \neq 0$. This symbol is generated in operations as $0/0$, $0 \times \infty$, ∞/∞ and $(+\infty) - (+\infty)$. The arithmetic operations involving `NaN` produce as an answer `NaN`.

The symbol for infinity ∞ is represented by $f = 0$ and the same exponent as `NaN` ($E = e_{\max} + 1 + \text{bias}$), since the sign bit allows making the distinction between $\pm\infty$.

The floating point numbers, represented in the IEEE arithmetic, can be visualized in MATLAB®using the command **`format hex`**. This command shows the internal hexadecimal representation of a single precision or double precision number which can be converted then into binary form.

Example 1.1. The number -2.75 is represented exactly in 2-base arithmetic as

$$
\begin{aligned}
-2.75 &= -(1 \times 2^1 + 0 \times 2^0 + 1 \times 2^{-1} + 1 \times 2^{-2}) \times 2^0 \\
&= -(1 \times 2^0 + 0 \times 2^{-1} + 1 \times 2^{-2} + 1 \times 2^{-3}) \times 2^1.
\end{aligned}
$$

In this way, for $t = 24$ this number is represented as

$$(-1.01100000000000000000000)_2 \times 2^1.$$

The binary representation of this number in the single precision IEEE arithmetic is shown in Figure 1.2. Since the biased exponent in the given case is equal to $E = 1 + \text{bias} = 128$, its equivalent binary representation is 10000000. Note that the most significant bit $b_0 = 1$ is hidden, but it is considered implicitly.

In Figure 1.3, we show the binary representation of some constants and symbols in the single precision IEEE arithmetic. All binary numbers are characterized by a biased exponent. For instance, in the binary representation of the number $r_{\min} = 2^{e_{\min}}$, the exponent is determined from $E = e_{\min} + \text{bias} = -126 + 127 = 1$. In the representation of $r_{\max} = (2 - 2^{1-t}) \times 2^{e_{\max}}$, the biased exponent is $E = e_{\max} + \text{bias} = 127 + 127 = 254$. In case of `NaN` and ∞,

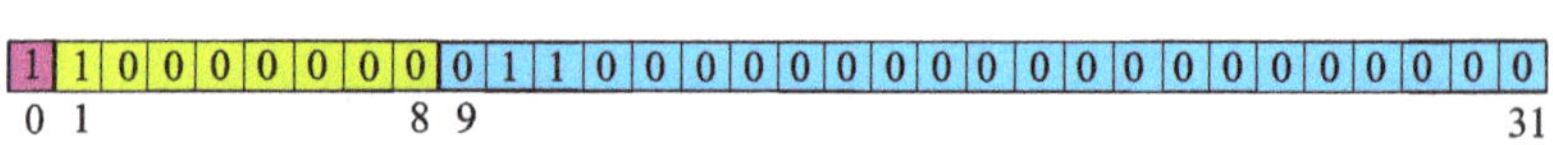

Fig. 1.2. Representation of -2.75 in the single precision IEEE arithmetic.

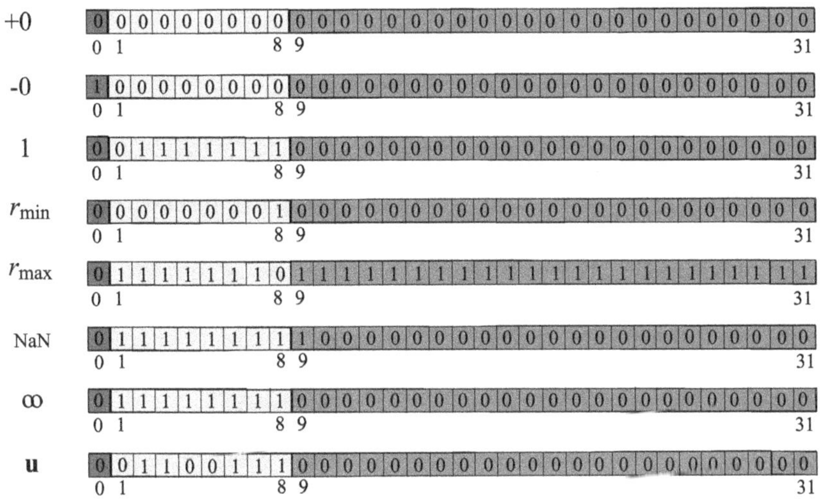

Fig. 1.3 Representation of different constants and symbols in the single precision IEEE arithmetic.

due to the bias the exponent acquires the maximum possible value $E = e_{\max} + 1 + \text{bias} = 127 + 1 + 127 = 255$.

1.1.2 *Approximating Real Numbers*

Let x be an arbitrary real number which can be approximated by a number belonging to the set $\mathcal{F}$ of floating point numbers. *The rounded value* $fl(x)$ of x is defined as the floating point number which is closest to x and the transformation $x \rightarrow \text{fl}(x)$ is called *rounding*. The rounding by default in the IEEE arithmetic is to round to the closest representable number with rounding to the even (zero last digit) in case of equal distance from two representable numbers. The quantity $|\text{fl}(x) - x|$ is called *absolute error* of the floating point representation of x.

It is possible to show that each real number $x \neq 0$ which satisfies

$$r_{\min} \leq |x| \leq r_{\max}$$

can be approximated using rounding to even by an element $\text{fl}(x)$ of $\mathcal{F}$ whose *relative error* is

$$\frac{|\text{fl}(x) - x|}{|x|} \leq \mathbf{u}.$$

The quantity

$$\mathbf{u} = 2^{-t},$$

where t is the precision of the floating point arithmetic, is called *unit roundoff* or *relative machine precision*. The values of $\mathbf{u}$ for computations with single and double precision are given in the last column of Table 1.1.

In MATLAB®, the permanent variable **eps** is equal to the quantity 2^{1-t}, which is two times larger than the unit roundoff $\mathbf{u}$ corresponding to correct rounding.

If x lies in the boundaries of $\mathcal{F}$, one has that

$$\text{fl}(x) = x(1+\delta), \quad |\delta| \le \mathbf{u}. \tag{1.4}$$

If $|\text{fl}(x)| > r_{\text{max}}$, where r_{max} is the maximum representable floating point number, one says that $\text{fl}(x)$ *overflows* and if $0 < |\text{fl}(x)| < r_{\text{min}}$, where r_{min} is the minimum representable floating point number, then $\text{fl}(x)$ *underflows*.

1.1.3 *Arithmetic Floating Point Operations*

Let x, y be floating point numbers and let x op y denote one of the basic arithmetic operations: $+$, $-$, $\times$, $/$.

In accordance with the IEEE standard, all arithmetic operations involving x and y must be done as if they are performed first with infinite accuracy and then are rounded in accordance with the rounding rule. This leads to the following model of the basic arithmetic operations:

$$\text{fl}(x \text{ op } y) = (x \text{ op } y)(1+\delta), \quad |\delta| \le \mathbf{u}, \quad \text{op} = +, -, \times, /. \tag{1.5}$$

This model implies that the computed value of x op y is "as far as accurate" as the rounded exact answer, in the sense that the bound on the relative error is the same in both cases. In a similar way, it is possible to assume that

$$\text{fl}(\sqrt{x}) = \sqrt{x}(1+\delta), \quad |\delta| \le \mathbf{u}. \tag{1.6}$$

The models (1.5) and (1.6) lie in the base of the floating point error analysis. We shall call it the *standard model of the floating point arithmetic.*

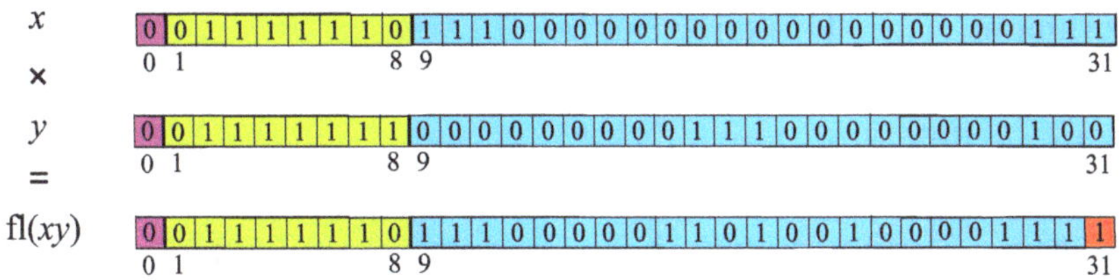

Fig. 1.4. Multiplication of two floating point numbers in single precision IEEE arithmetic.

Example 1.2. Consider the multiplication of the two floating point numbers

$$x = 2^{-1} + 2^{-2} + 2^{-3} + 2^{-4} + 2^{-22} + 2^{-23} + 2^{-24} \approx 0.9375004$$

and

$$y = 2^{0} + 2^{-10} + 2^{-11} + 2^{-12} + 2^{-22} + 2^{-23} + 2^{-24} \approx 1.0017095.$$

The product of these numbers in the single precision IEEE arithmetic in binary form is shown in Figure 1.4. Due to the rounding error in the multiplication, the last bit of the result is wrong. In decimal form, we have that the exact product is (the first seven digits corresponding to the single precision are separated in a box)

$$x \times y = \boxed{0.9391029}81952775,$$

while the computed in the single precision result is

$$\mathrm{fl}(x \times y) = 0.9391031.$$

Note that in the given case the relative error

$$\frac{|fl(x \times y) - x \times y|}{|xy|} = 9.05 \cdot 10^{-8}$$

of the result is almost two times larger than the unit roundoff $\mathbf{u} = 2^{-24} = 5.96 \cdot 10^{-8}$, which is due to the errors in converting the binary numbers x, y and $\mathrm{fl}(x \times y)$ into decimal form.

If x op y does not belong to the set of numbers which can be approximated in the floating point arithmetic, this lead to an arithmetic error. This error can be

overflow, if $|x \text{ op } y| > r_{\max}$ or
underflow, if $0 < |x \text{ op } y| < r_{\min}$.

There exist different ways to handle such arithmetic errors, the overflow and the underflow being avoided in the quality algorithms and software for matrix computations.

1.1.4 *Complex Arithmetic*

The set $\mathbb{C}$ of complex numbers is approximated in the computer processor by ordered pairs of real floating point numbers. That is why the model of the rounding errors in performing the basic arithmetic operations in complex arithmetic is based on the model of the errors in real arithmetic. It is assumed that for the given complex numbers $x = a + \jmath_o b$ and $y = c + \jmath_o d$, one computes

$$x \pm y = a \pm c + \jmath_o(b \pm d), \tag{1.7}$$

$$xy = ac - bd + \jmath_o(ad + bc), \tag{1.8}$$

$$x/y = \frac{ac + bd}{c^2 + d^2} + \jmath_o \frac{bc - ad}{c^2 + d^2}. \tag{1.9}$$

Formula (1.9) is not suitable for practical purposes, since it is possible to happen to overflow if c or d exceeds the overflow threshold, even if x/y does not overflow. That is why if $|c| \geq |d|$, one implements the formula obtained by multiplication of the nominator and denominator by c^{-1}:

$$x/y = \frac{a + \jmath_o b}{c + \jmath_o d} = \frac{a + b(dc^{-1})}{c + d(dc^{-1})} + \jmath_o \frac{b - a(dc^{-1})}{c + d(dc^{-1})} \tag{1.10}$$

and if $|d| \geq |c|$, one uses a similar formula involving d^{-1}.

Lemma 1.3 (Higham (2002)). *For $x, y \in \mathbb{C}$, the basic arithmetic operations performed in accordance with* (1.7), (1.8) *and* (1.10) *in the standard model* (1.5) *satisfy*

$$\mathrm{fl}(x \pm y) = (x \pm y)(1 + \delta), \quad \delta \leq \mathbf{u}, \tag{1.11}$$

$$\mathrm{fl}(xy) = y(1 + \delta), \quad \delta \leq \sqrt{2}\frac{2\mathbf{u}}{1 - 2\mathbf{u}}, \tag{1.12}$$

$$\mathrm{fl}(x/y) = y(1+\delta), \quad \delta \leq \sqrt{2}\frac{7\mathbf{u}}{1-7\mathbf{u}}. \tag{1.13}$$

Since the bounds in Lemma 1.3 have the same form as in the standard model (1.5) of the real arithmetic, most of the results for matrix operations in real arithmetic hold also for the complex arithmetic provided that the constants in the bounds are increased in the corresponding way.

1.1.5 *Notes and References*

The presentation in this and the following sections of this chapter is inspired by the influential books on matrix computations of Stewart (1998), Golub and Van Loan (2013), Björck (2015), Demmel (1997) and Higham (2002).

The computer arithmetic and the floating point computations are described very thoroughly in the books Moler (2004), Overton (2001), Chaitin-Chatelin and Frayssé (1996), Koren (2002) and Muller *et al.* (2010). A very useful paper on this subject is Goldberg (1991). A full description of the IEEE floating point standard can be found in IEEE (2019).

1.2 Error Analysis of Matrix Operations

In this section, we present analysis of the rounding errors arising in the basic matrix operations. The emphasis is put on the cases when these operations can be associated with large relative errors in the result.

In the analysis of rounding errors of matrix operations, we will need some relationships concerning the absolute values of matrix elements.

Definition 1.4. Let $A \in \mathbb{R}^{m\times n}$ with entries a_{ij}. The symbol $|A|$ denotes an $m\times n$ matrix whose ijth element is $|a_{ij}|$. Also, the inequality $|A| \leq |B|$ is equivalent to $|a_{ij}| \leq |b_{ij}|$ for all $i = 1, 2, \ldots, m$ and $j = 1, 2, \ldots, n$.

The following lemma is useful when working with matrix products.

Lemma 1.5. *Let* $A \in \mathbb{R}^{m\times n}$ *and* $B \in \mathbb{R}^{n\times p}$. *Then*

$$|AB| \le |A||B|.$$

Consider first the errors arising when storing a matrix in the computer memory. Assume that $A \in \mathbb{R}^{m\times n}$ and denote the stored matrix by $\mathrm{fl}(A)$. Then

$$[\mathrm{fl}(A)]_{ij} = \mathrm{fl}(a_{ij}) = a_{ij}(1+\delta_{ij}), \quad |\delta_{ij}| \le \mathbf{u}$$

for all i and j, i.e.,

$$|\mathrm{fl}(A) - A| \le \mathbf{u}|A|.$$

This relationship can be easily represented as an inequality involving matrix norm, for instance,

$$\|\mathrm{fl}(A) - A\|_F \le \mathbf{u}\|A\|_F. \tag{1.14}$$

It should be taken into account that in the rounding error analysis of the matrix operations, the expressions involving absolute values sometimes are more useful than those implementing norms since they provide information about the errors of the individual entries (the *componentwise errors*).

Let us now analyze the rounding errors in the multiplication of a matrix by a scalar and in the summation of two matrices. The analysis is elementary and the results can be given in the form of the following two theorems.

Theorem 1.6. *Let* $A \in \mathbb{R}^{m\times n}$ *and* $\alpha \in \mathbb{R}$. *Then* $\mathrm{fl}(\alpha A) = \alpha A + E$, *where* $|E| \le \mathbf{u}|\alpha A|$ *and* $\|E\|_F \le \mathbf{u}|\alpha|\|A\|_F$.

Theorem 1.7. *Let* $A, B \in \mathbb{R}^{m\times n}$. *Then* $\mathrm{fl}(A+B) = (A+B)+E$, *where* $|E| \le \mathbf{u}|A+B|$ *and* $\|E\|_F \le \mathbf{u}\|A+B\|_F$.

1.2.1 *Scalar Product: Strict Error Bounds*

The analysis of rounding errors in matrix products requires first to determine the error in computing the scalar product of two vectors.

Given two real n-vectors x and y, the following algorithm computes the scalar product $s = \sum_{j=1}^{n} x_j y_j$:

Set $s = 0$
for $j = 1, \dots, n$
  $s \leftarrow s + x_j y_j$
end

Denote by $\mathrm{fl}(x^T y)$ the result computed by this algorithm in floating point arithmetic. In the given case, the aim of the rounding error analysis is to find a bound on the quantity

$$|\mathrm{fl}(x^T y) - x^T y|$$

representing the *absolute forward error* of the result, or the quantity

$$\frac{|\mathrm{fl}(x^T y) - x^T y|}{x^T y}$$

representing the *relative forward error* of the result.

Define the kth partial sum s_k as

$$s_k = \mathrm{fl}\left(\sum_{j=1}^{k} x_j y_j\right).$$

Using the standard model of the floating point arithmetic (1.5), (1.6), we obtain that

$$s_1 = x_1 y_1 (1 + \rho_1),$$

where $|\rho_1| \leq \mathbf{u}$ is the rounding error corresponding to the multiplication of x_1 and y_1. In a similar way,

$$\begin{aligned} s_2 &= \mathrm{fl}(s_1 + \mathrm{fl}(x_2 y_2)) \\ &= \mathrm{fl}(s_1 + x_2 y_2 (1 + \rho_2)) \\ &= (s_1 + x_2 y_2 (1 + \rho_2))(1 + \delta_2) \\ &= x_1 y_1 (1 + \rho_1)(1 + \delta_2) + x_2 y_2 (1 + \rho_2)(1 + \delta_2), \end{aligned}$$

where $|\delta_2| \leq \mathbf{u}$ and $|\rho_2| \leq \mathbf{u}$ are rounding errors corresponding to addition and multiplication, respectively.

Continuing in this way, for $k = 3, \dots, n$, we get (Figure 1.5)

$$\begin{aligned} s_n &= \mathrm{fl}(s_{n-1} + \mathrm{fl}(x_n y_n)) \\ &= \mathrm{fl}(s_{n-1} + x_n y_n (1 + \rho_n)) \\ &= (s_{n-1} + x_n y_n (1 + \rho_n))(1 + \delta_n) \end{aligned}$$

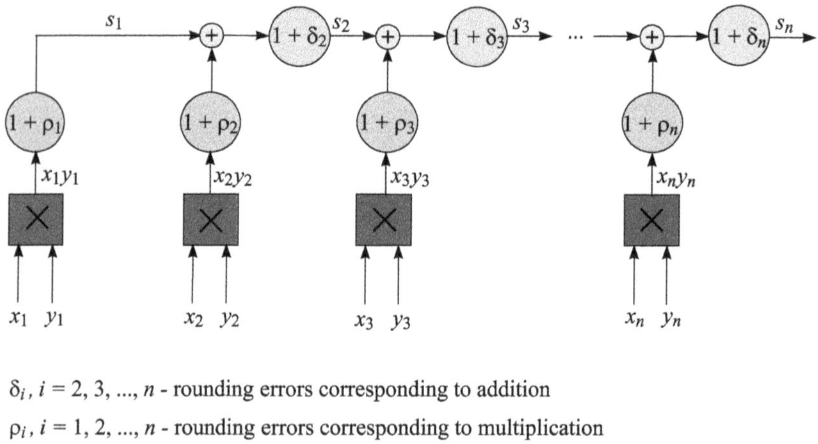

Fig. 1.5. Rounding errors in computing the scalar product.

$$
\begin{aligned}
&= x_1y_1(1+\rho_1)(1+\delta_2)(1+\delta_3)+\cdots+(1+\delta_n)\\
&\quad + x_2y_2(1+\rho_2)(1+\delta_2)(1+\delta_3)+\cdots+(1+\delta_n)\\
&\quad \vdots\\
&\quad + x_{n-1}y_{n-1}(1+\rho_{n-1})(1+\delta_{n-1})(1+\delta_n)\\
&\quad + x_ny_n(1+\rho_n)(1+\delta_n),
\end{aligned}
$$

where $|\delta_k| \leq \mathbf{u}$, $k = 2, \ldots, n$ and $|\rho_k| < \mathbf{u}$, $k = 1, \ldots, n$ are the rounding errors corresponding to addition and multiplication, respectively.

As a result, we obtain that

$$
\mathrm{fl}(x^Ty) = s_n = x_1y_1(1+\gamma_1) + x_2y_2(1+\gamma_2) + \cdots + x_ny_n(1+\gamma_n), \tag{1.15}
$$

where

$$
(1+\gamma_j) = (1+\rho_j)\prod_{i=j}^{n}(1+\delta_i)
$$

with $\delta_1 \equiv 0$.

The numbers γ_i are called *relative backward errors*, since the computed result can be represented by them as the exact result found for a slightly modified initial data.

From (1.15), we find that

$$|\text{fl}(x^T y) - x^T y| \leq \sum_{j=1}^{n} |x_j y_j||\gamma_j|. \tag{1.16}$$

To finish the analysis, we need to use the following lemma.

Lemma 1.8. *Let* $\mathbf{u}$ *be the machine precision and assume that*

$$(1+\gamma) = \prod_{i=1}^{n}(1+\delta_i)$$

with $|\delta_i| < \mathbf{u}$ *and* $n\mathbf{u} < 1$*. Then*

$$|\gamma| < \frac{n\mathbf{u}}{1 - n\mathbf{u}}.$$

Besides this, if $n\mathbf{u} < 0.01$*, then*

$$|\gamma| < 1.01n\mathbf{u}.$$

Further on we shall assume that $n\mathbf{u} < 0.01$ which is always fulfilled in practice. Returning again to equation (1.16), it is easy to prove (taking into account the number of the terms and the identity $\delta_1 = 0$) that

$$|\gamma_j| \leq \begin{cases} 1.01n\mathbf{u} & \text{for } j = 1, \\ 1.01(n-j+2)\mathbf{u} & \text{for } j = 2, \ldots, n. \end{cases} \tag{1.17}$$

In this way, under the "justified" assumption that $n\mathbf{u} < 0.01$, we get

$$\begin{aligned} |\text{fl}(x^T y) - x^T y| &\leq \sum_{j=1}^{n} |x_j y_j| \cdot 1.01n\mathbf{u} \\ &\leq 1.01n\mathbf{u}|x^T||y|, \end{aligned} \tag{1.18}$$

where for $z \in \mathbb{R}^n$ the symbol $|z|$ denotes $[|z_1|, |z_2|, \ldots, |z_n|]^T$. The quantity $|\text{fl}(x^T y) - x^T y|$ represents the absolute forward error of the compute scalar product.

The inequality (1.18) is an example of the so-called *forward error analysis* in which one finds a bound on the error in the result.

The bound (1.18) shows that the error in the computed scalar product $\mathrm{fl}(x^T y)$ can be very large. In fact, the relative error

$$\left|\frac{\mathrm{fl}(x^T y) - x^T y}{x^T y}\right| \leq \frac{1.01n\mathbf{u}|x^T||y|}{|x^T y|} \tag{1.19}$$

is large, when it is fulfilled that

$$|x^T y| \ll |x^T||y|.$$

Example 1.9. Consider the computation of the scalar product of the vectors

$$x = \begin{bmatrix} 0.5382471 \\ 0.3947236 \\ -0.7418354 \end{bmatrix}, \quad y = \begin{bmatrix} 0.2468443 \\ 0.9128155 \\ 0.6648011 \end{bmatrix}$$

whose entries are exactly representable in the single precision arithmetic of MATLAB®. For this aim, the computation of the scalar product is done using the function `single`, which works with accuracy that is approximately equal to seven decimal digits ($\mathbf{u} \approx 5.96 \times 10^{-8}$).

The results of computing the scalar product for $j = 1, 2, 3$ are shown in binary form in Figure 1.6 and in decimal form in Table 1.2. (For convenience in the comparison, the results of the double precision computations are rounded to seven digits after the decimal

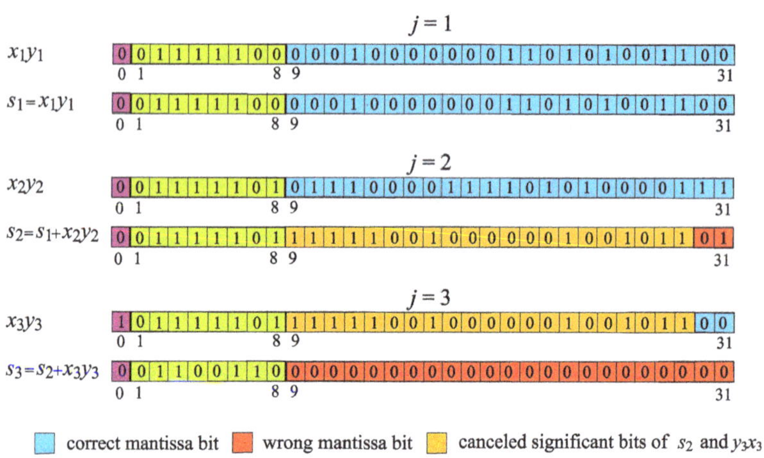

Fig. 1.6. Binary representation of the intermediate results in computing the scalar product in single precision.

Table 1.2. Results from computing the scalar product with single and double precision.

Single precision		
j	$x_j y_j$	s_j
1	1.328632×10^{-1}	1.328632×10^{-1}
2	3.603098×10^{-1}	4.931730×10^{-1}
3	-4.931730×10^{-1}	2.9802322×10^{-8}
Double precision		
j	$x_j y_j$	s_j
1	1.3286323×10^{-1}	1.3286323×10^{-1}
2	3.6030982×10^{-1}	4.9317305×10^{-1}
3	$-4.9317290 \times 10^{-1}$	5.8983390×10^{-8}

point.) The exact value of the scalar product (up to eight decimal digits) is 5.8983390×10^{-8}. In single precision arithmetic, we obtain

$$\mathrm{fl}(x^T y) = 2.9802322 \times 10^{-8},$$

which shows that there is no correct decimal digit in the computed scalar product.

The relative error of the single precision result is

$$\left| \frac{\mathrm{fl}(x^T y) - x^T y}{x^T y} \right| = \left| \frac{2.9802322 \times 10^{-8} - 5.8983390 \times 10^{-8}}{5.8983390 \times 10^{-8}} \right| \approx 0.495,$$

i.e., the error is almost 50%. This large error is due to the fact that $|x^T||y| = 0.9863460$, but $|x^T y| = 5.8983390 \times 10^{-8}$, so that the bound (1.19) on the relative error is very large. As one can see from the table, the large error of the final result is due to a single rounding error done in the addition of the product $\mathrm{fl}(x_3 y_3) = -0.4931730$ to the partial sum $s_2 = 0.4931730$. It is seen in Figure 1.6 that in this addition all 21 leading significant bits in positions from 9 to 29 are canceled producing a result in which even the first significant digit is incorrect. This phenomenon is called *catastrophic cancellation in the floating point arithmetic* and it is one of the causes for obtaining large errors in matrix computations.

1.2.2 Scalar Product — First-Order Bounds

Consider the product

$$1 + \gamma_n = (1 + \rho_n)(1 + \delta_n) = 1 + (\rho_n + \delta_n) + \rho_n\delta_n. \tag{1.20}$$

We have that $|\rho_n + \delta_n| \leq 2\mathbf{u}$ and $|\rho_n\delta_n| \leq \mathbf{u}^2$. If, for instance, $\mathbf{u} = 10^{-16}$, then $2\mathbf{u} = 2 \times 10^{-16}$ and $\mathbf{u}^2 = 10^{-32}$. Hence, the third term in (1.20) is nonsignificant in comparison with the second term and can be neglected. In such case, we obtain

$$\gamma_n \cong \rho_n + \delta_n$$

or

$$|\gamma_n| \lesssim |\rho_n| + |\delta_n| \leq 2\mathbf{u}.$$

Thus,

$$|\gamma_1| \lesssim n\mathbf{u},$$
$$|\gamma_j| \lesssim (n - j + 2)\mathbf{u}, \quad j = 2, \ldots, n.$$

As a result, we have that

$$\begin{aligned} |\mathrm{fl}(x^T y) - x^T y| &\leq \sum_{j=1}^{n} |x_j y_j| n\mathbf{u} \\ &\leq n\mathbf{u}|x^T||y|. \end{aligned} \tag{1.21}$$

The bound on the right-hand side of (1.21) is called *first-order bound* or *linear bound* since it depends linearly on $\mathbf{u}$.

The comparison of the derived bound (1.21) with the strict bound (1.18) shows that the first-order bound yields almost the same result but it is found much easier. That is why the linear bounds are widely applied in the rounding error analysis.

On the basis of (1.18) and (1.21), the bound on the forward rounding error in computing of a scalar product can be written as

$$|\mathrm{fl}(x^T y) - x^T y| \leq n\mathbf{u}|x^T||y| + O(\mathbf{u}^2), \tag{1.22}$$

where the quantity $O(\mathbf{u}^2)$ reflects the effect of second- and higher-order terms in respect to $\mathbf{u}$. The size of this quantity does not exceed 1% of the size of the linear bound $n\mathbf{u}|x^T||y|$.

1.2.3 *Operation Count*

The volume of computational operations in matrix computations is measured in flops, where a *flop* is a floating point addition, subtraction, multiplication or division of numbers. For instance, the computation of the scalar product of two n-vectors requires $2n$ flops. In case of using complex arithmetic, the flops should be interpreted as counting operations on complex numbers.

The following expressions are useful in determining the number of computational operations in working with matrices:

$$\sum_{k=1}^{n} k = \frac{n(n+1)}{2}, \quad \sum_{k=1}^{n} k^2 = \frac{n(n+1)(2n+1)}{6},$$

$$\sum_{k=1}^{n} k^3 = \frac{n^2(n+1)^2}{4}.$$

It is necessary to point out that with the advancing of the computer technologies, the operation count becomes less important than the effect of rounding errors.

1.2.4 *Matrix-Vector Product*

Let $A \in \mathbb{R}^{m\times n}$ and $x \in \mathbb{R}^n$. The matrix-vector product $y = Ax$ can be computed by the algorithm given in Figure 1.7.

This algorithm requires $2mn$ flops.

The analysis of the errors in computing the matrix-vector product can be done by the aid of the analysis of the scalar product presented in the previous section. Let $a^T = [a_1, a_2, \ldots, a_n]^T$ denote a row of A.

```
for i = 1, 2, ..., m
    Set y_i = 0
    for j = 1, 2, ..., n
        y_i = y_i + a_ij x_j
    end
end
```

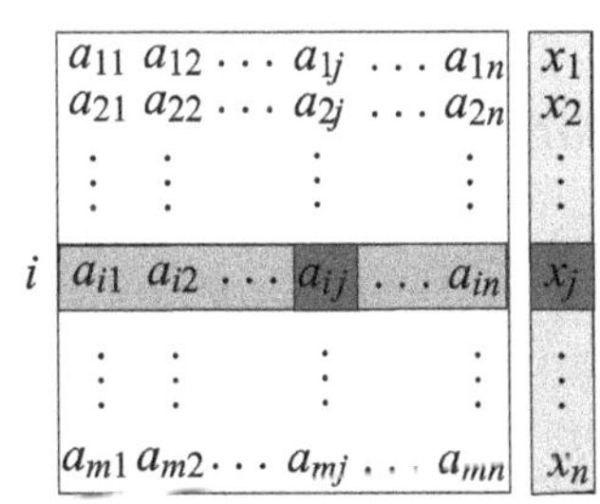

Fig. 1.7. Computing the matrix-vector product.

Then

$$\begin{aligned} \text{fl}(a^T x) &= \sum_{j=1}^{n} a_j x_j (1+\gamma_j) \\ &= (a_j + a_j\gamma_j)x_j, \end{aligned}$$

where γ_j is bounded as in (1.17). Setting $e_j = a_j\gamma_j$, we apply the earlier result to each row of A and write

$$\text{fl}(Ax) = (A+E)x, \tag{1.23}$$

where $|E| \leq |A||D|$ and D is a diagonal matrix whose diagonal elements satisfy

$$|d_{jj}| \leq \begin{cases} 1.01n\mathbf{u} & \text{for } j = 1, \\ 1.01(n-j+2)\mathbf{u} & \text{for } j = 2, \ldots, n. \end{cases} \tag{1.24}$$

This result can be represented in the compact form

$$|Ax - \text{fl}(Ax)| = |Ex| \leq |E||x| \leq |A||D||x| \leq 1.01n\mathbf{u}|A||x|$$

so that the relative error in the computed product is

$$\frac{|Ax - \text{fl}(Ax)|}{|Ax|} \leq 1.01n\mathbf{u}\frac{|A||x|}{|Ax|}.$$

Since $|Ax|$ can be much smaller than $|A||x|$, this bound shows that the multiplication of a matrix by vector in some cases can be accompanied with large errors.

Using the equality $\||A|\|_F = \|A\|_F$, the earlier result can be represented also in the form

$$\frac{\|\text{fl}(Ax) - Ax\|_F}{\|Ax\|_F} \leq 1.01n\mathbf{u}\frac{\|A\|_F\|x\|_F}{\|Ax\|_F}. \tag{1.25}$$

1.2.5 *Matrix–Matrix Product*

Let $A \in \mathbb{R}^{m\times n}$ and $B \in \mathbb{R}^{n\times p}$. The evaluation of the matrix–matrix product $C = AB$ can be done by the algorithm given in Figure 1.8.

This algorithm requires $2mnp$ flops and hence the volume of work in the multiplication of two square matrices of nth order is proportional to $2n^3$.

The errors in obtaining the product $\text{fl}(AB)$ are determined using the error analysis for the matrix–vector product.

```
for i = 1, 2, ..., m
    for j = 1, 2, ..., p
        Set c_ij = 0
        for k = 1, 2, ..., n
            c_ij = c_ij + a_ik b_kj
        end
    end
end
```

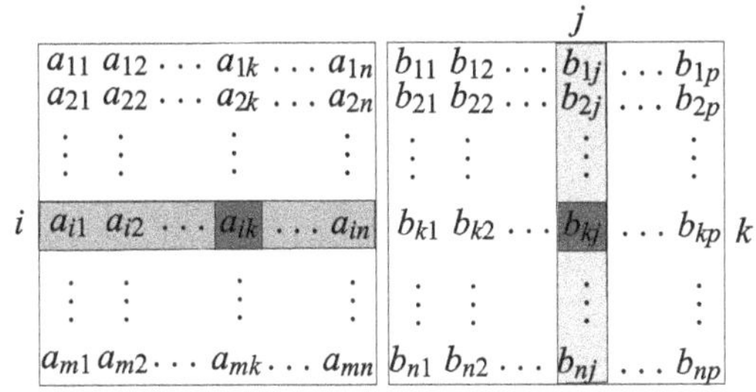

Fig. 1.8. Computing the matrix–matrix product.

Denote by $b^{(j)}$ is the jth column of B and by $c^{(j)}$ is the computed jth column of $C = AB$. Then according to (1.23), we have that

$$c^{(j)} = (A + E^{(j)})b^{(j)},$$

where $|E^{(j)}| \leq |A||D^{(j)}|$ and $D^{(j)}$ is a diagonal matrix whose elements are bounded as in (1.24). It has to be taken into account that the matrices $E^{(j)}$ and $D^{(j)}$ will be different for each j. That is why we obtain that

$$\text{fl}(AB) - AB = [E^{(1)}b^{(1)}, E^{(2)}b^{(2)}, \ldots, E^{(p)}b^{(p)}],$$

from where it follows that

$$\begin{aligned}|\text{fl}(AB) - AB| &\leq [|E^{(1)}||b^{(1)}|, \ldots, |E^{(p)}||b^{(p)}|] \\ &\leq |A||D||B| \\ &\leq 1.01n\mathbf{u}|A||B|.\end{aligned}$$

For the relative error of the result, we find that

$$\frac{|\text{fl}(AB) - AB|}{|AB|} \leq 1.01n\mathbf{u}\frac{|A||B|}{|AB|}.$$

This result can also be represented as

$$\frac{\|\text{fl}(AB) - AB\|_F}{\|AB\|_F} \leq 1.01n\mathbf{u}\frac{\|A\|_F\|B\|_F}{\|AB\|_F}. \tag{1.26}$$

Again, since $\|AB\|_F$ can be much smaller than $\|A\|_F\|B\|_F$, it is not possible to guarantee that the matrix product will be computed with small relative error.

There exists a special case when the matrix multiplication is done certainly with a small relative error. This is the case when we use orthogonal (or unitary in complex transformations) matrices. In fact, let the matrix B in equation (1.26) be orthogonal. In such case, $\|AB\|_F = \|A\|_F$, $\|B\|_F = \sqrt{n}$ and

$$\frac{\|\mathrm{fl}(AB) - AB\|_F}{\|AB\|_F} \leq 1.01 n^{3/2} \mathbf{u} \tag{1.27}$$

so that the relative error of the result is very small.

Equation (1.27) shows the advantage of using orthogonal similarity transformations in the matrix computations.

The following example illustrates the large rounding errors which may appear in the computing of matrix powers which can be considered as a particular case of matrix multiplication.

Example 1.10. Consider the eighth-order matrix

$$A = \begin{bmatrix} -5.478 & -4.402 & 10.368 & -30.564 & 56.288 & 16.440 & -89.156 & -1.824 \\ 0.202 & 0.201 & -0.404 & 1.212 & -2.222 & -0.606 & 3.434 & 0.000 \\ 0.300 & 0.011 & -0.534 & 1.392 & -2.500 & -0.882 & 4.240 & 0.108 \\ 1.470 & 1.449 & -2.856 & 8.588 & -15.870 & -4.428 & 24.750 & 0.372 \\ 2.714 & 2.356 & -5.214 & 15.462 & -28.476 & -8.151 & 44.786 & 0.714 \\ 1.412 & 1.324 & -2.706 & 8.118 & -14.996 & -4.249 & 23.510 & 0.438 \\ 1.838 & 1.501 & -3.504 & 10.332 & -19.010 & -5.514 & 30.034 & 0.528 \\ -0.066 & 0.054 & 0.114 & -0.252 & 0.426 & 0.189 & -0.792 & -0.004 \end{bmatrix}.$$

In Table 1.3, we give the relative errors of the powers A^p, $p = 2, 4, 6, 8$ computed in the single precision arithmetic

Table 1.3. Relative errors in the computed matrix powers.

p	$\|\mathrm{fl}(A^p) - A^p\|_2 / \|A^p\|_2$
1	2.0809×10^{-8}
2	1.5182×10^{-5}
3	7.5369×10^{-3}
4	6.4505×10^{-3}
5	1.0444×10^{-2}
6	1.0853×10^{-2}
7	1.2661×10^{-2}
8	1.1152×10^{-2}

($\mathbf{u} \approx 5.96 \times 10^{-8}$) of MATLAB®, and in Figure 1.9, we show the componentwise relative errors

$$\frac{|\mathrm{fl}(A^p) - A^p|}{\|A^p\|_2}$$

in the computed powers.

It is seen that with the increase in the exponent p the relative error in the computed power A^p increases quickly so that the entries of the eighth power are accurate at most up to the second decimal digit. The cause for this phenomenon is that the norm of the power of A decreases strongly with the increase in p, which is accompanied by large cancellations in obtaining the elements of A^p. As it is shown in Example 1.9, in this cancellations in leading positions come the erroneous significant digits which increases the errors of the computed elements. In the given case, the matrix A is obtained in double precision as $A = QDQ^{-1}$, where

$$Q = \begin{bmatrix} -50 & -26 & 18 & -6 & 0 & -2 & 1 & 2 \\ 2 & 0 & 0 & 0 & 0 & 0 & 0 & -1 \\ 1 & 8 & -6 & 3 & 6 & 4 & -2 & -1 \\ 15 & 0 & 1 & 0 & -6 & -4 & 2 & 1 \\ 26 & 8 & -6 & 2 & -3 & -1 & 1 & -1 \\ 14 & 2 & 0 & 0 & -5 & -2 & 2 & 1 \\ 17 & 8 & -6 & 2 & 0 & 1 & 0 & -1 \\ 0 & -3 & 3 & -1 & -3 & -2 & 1 & 1 \end{bmatrix},$$

$$D = \mathrm{diag}(0.1, -0.03, 0.02, -0.01, 0.005, -0.004, 0.002, -0.001),$$

$$Q^{-1} = \begin{bmatrix} 1 & 1 & -2 & 6 & -11 & -3 & 17 & 0 \\ -1 & 0 & 1 & -3 & 6 & 3 & -12 & -3 \\ 1 & 2 & -2 & 7 & -13 & -3 & 19 & 0 \\ -1 & -2 & 5 & -12 & 21 & 3 & -27 & 8 \\ 2 & 3 & -6 & 18 & -30 & -5 & 40 & -10 \\ 1 & 0 & 0 & 0 & -3 & -3 & 10 & 8 \\ -1 & 0 & 0 & 0 & 4 & 6 & -14 & -14 \\ 2 & 1 & -4 & 12 & -22 & -6 & 34 & 0 \end{bmatrix}.$$

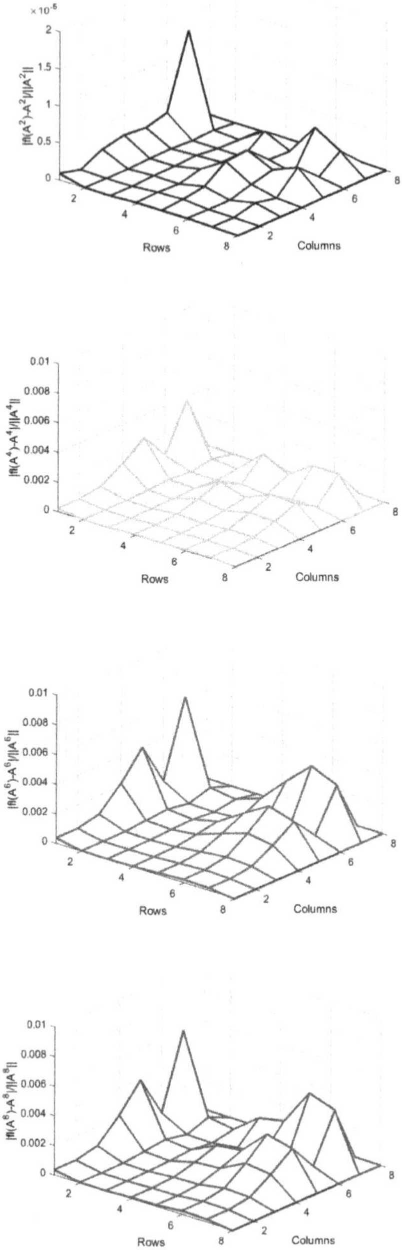

Fig. 1.9. Componentwise relative errors in the matrix powers.

The exact powers of A are determined in double precision as $A^p = QD^pQ^{-1}$ and due to the small diagonal elements of D, the norm of A^p decreases quickly with the increase in p.

1.2.6 *Notes and References*

The foundations of the rounding error analysis are laid out by the English mathematician James H. Wilkinson in his pioneering books (Wilkinson, 2023, 1965). The book of Higham (2002) contains a rigorous presentation of the methods of the rounding error analysis used in matrix computations. The proof of Lemma 1.5 can be found in Laub (2012), Chapter 4, and the proof of Lemma 1.8 is presented in Higham (2002), Chapter 3. It should be noted that the exact error bounds usually are not needed and the purpose of the error analysis is mainly to reveal the sources of large errors in the corresponding algorithm.

In some cases, it can be necessary to use greater precision than the one provided by the double precision arithmetic of IEEE. This may be done by using the variable-precision arithmetic available in the Symbolic Math Toolbox of MATLAB®(Symbolic Toolbox) or implementing some package for multiple precision arithmetic, see for details the rich surveys Abdelfattah *et al.* (2021), Dongarra *et al.* (2020) and Higham and Mary (2022).

1.3 Conditioning and Numerical Stability

The two basic factors leading to large errors in the numerical computations are the ill-conditioning of the solved problem and the effect of the rounding errors on the computed result. The second factor is related to the *numerical stability* of the computational algorithms which is of great importance in matrix computations. In this section, we consider in brief the different essence of the ill-conditioning and numerical instability as well as the ways to reduce the errors caused by these phenomena. As an example, we perform a sensitivity analysis of a square system of linear equations.

1.3.1 *Well-Conditioned and Ill-Conditioned Problems*

The *conditioning* is a property of the computational problem which characterizes the sensitivity of the solution to small changes (*perturbations*) in the data. We say that the problem is *well conditioned*, if small perturbations of the data lead to small changes in the solution. If small changes in the data may eventually cause large changes in the solution, then the problem is called *ill conditioned.* It must be emphasized that the conditioning does not depend on the precision which is used in the solution of the problem and it is exhibited irrelevantly to the properties of the computer arithmetic implemented.

Consider the computational problem as a function f of the input data x. The problem is well conditioned if $f(x)$ is insensitive to small perturbations in x, i.e., if $f(y)$ is close to $f(x)$ when y is close to x. (The exact definition of the notions "insensitive", "small" and "close" depends on the nature of the problem.) On the other hand, if f varies strongly for small changes in x, the problem is ill conditioned. The notions of well-conditioned and ill-conditioned functions are illustrated in Figure 1.10. The sphere on the left represents the domains of values of x and the ellipsoid on the right represents the domains of the values of $f(x)$. The larger ellipsoid on the right illustrates the ill-conditioning of f.

1.3.2 *Backward Error Analysis*

As it is shown in Section 1.2 (see equation (1.15)), the computed scalar product of the vectors $x, y \in \mathbb{R}^n$ in a floating point arithmetic can be represented in the form

$$\mathrm{fl}(x^T y) = x_1 y_1(1+\gamma_1) + x_2 y_2(1+\gamma_2) + \cdots + x_n y_n(1+\gamma_n), \quad (1.28)$$

where the relative errors $|\gamma_j|$, $j = 1, 2, \ldots, n$, are small numbers bounded according to (1.17). The earlier expression can be interpreted schematically in the following way (Figure 1.11). The computed scalar product is a result of the exact addition of the products $x_j y_j$ taken with small relative errors γ_j. The error analysis, in which the computed result is exact for slightly changed data, is called *backward error analysis*. The key moment of this analysis is that the errors

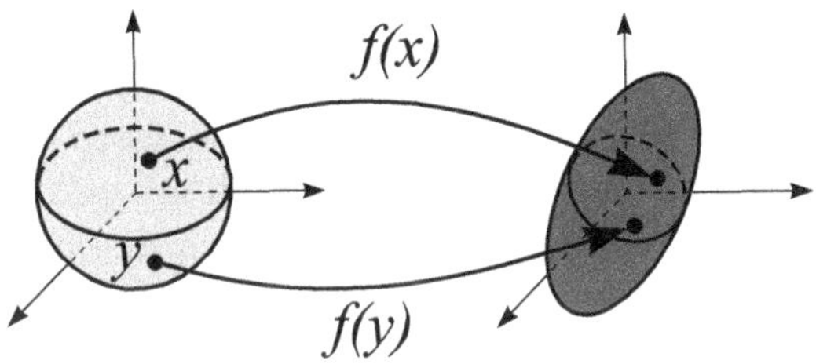

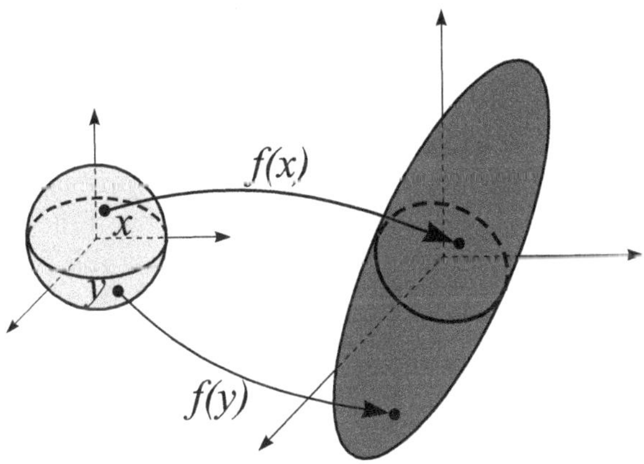

Fig. 1.10. Well-conditioned and ill-conditioned problems.

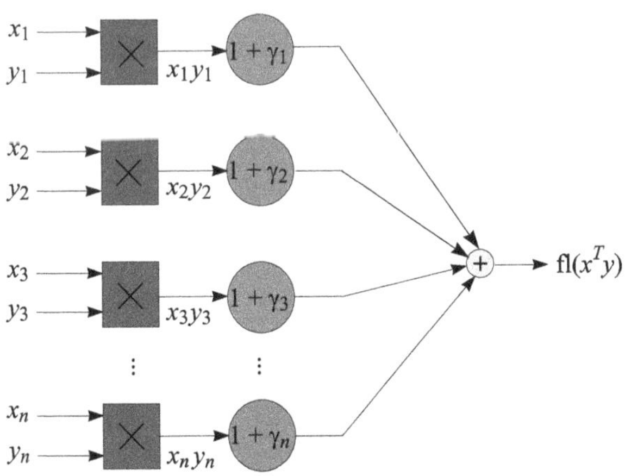

Fig. 1.11. Backward errors in computing the scalar product.

performed during the computations are transferred back to the initial data. These data have an effect on the solution which is equivalent to the computational errors and are called *backward errors.* Algorithms for which the backward errors are small are called *stable* from backward error analysis point of view or *backward stable algorithms.* The implementation of this analysis is justified by the fact that the data used in the solution of the computational problem are never exact since they are obtained by inexact measurements or are result of some inaccurate computations. If the backward errors are smaller than the uncertainty in the initial data, then the algorithm has produced a satisfactory result independently on the size of the errors in the solution.

Let us return to the general case of computing the function $f(x)$. The algorithm for solving the computational problem represented by the function f can be considered as another function g, which approximates f. Assume that the algorithm g is stable from backward error analysis point of view and is applied to the data x. Then $g(x) = f(y)$ for some y which is close to x. If the problem is well conditioned, then $f(x)$ should be close to $f(y)$ and the computed solution $g(x)$ is exact. If, however, the problem is ill conditioned, then the computed result will be generally inexact. Provided that the errors in the data are larger then the backward error, the answer will lie in the domain of the uncertainty generated by the errors in the data. These two situations are illustrated in Figure 1.12.

1.3.3 *Condition Numbers*

Despite of the fact that the backward error analysis characterizes the numerical properties of the algorithm used, it does not give information about the accuracy of the solution obtained. Using the notions introduced earlier, we can only assert that the backward error analysis guarantees that there exists y close to x, such that $g(x) = f(y)$. However, this does not show how far is $f(x)$ from $f(y)$.

The problem for ascertaining the solution accuracy can be stated in a general form which is not related to the rounding errors. This problem can be formulated as follows:

Given are the function f and two arguments x and y, it is necessary to find a bound on the distance between $f(x)$ and $f(y)$ as a function of the distance between x and y.

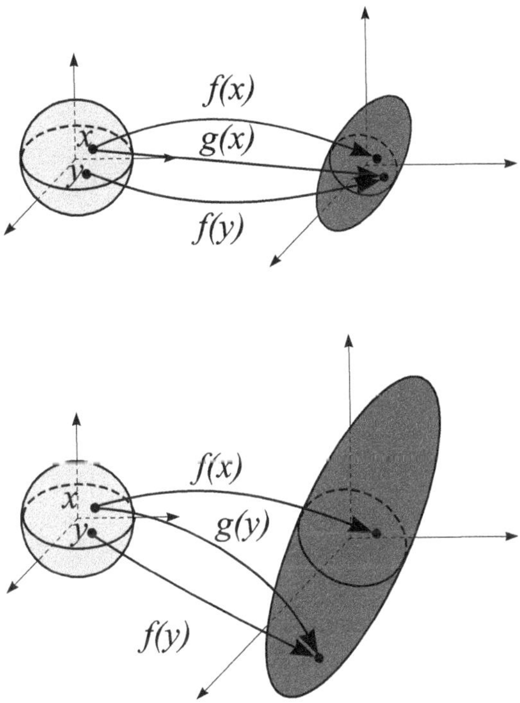

Fig. 1.12. Behavior of a stable algorithm for well-conditioned and ill-conditioned problems.

The solution of this problem is a subject of the mathematical discipline *perturbation theory*. The perturbations which lead to the deviations from x to y can have arbitrary origin, not only be errors due to the rounding in the floating point arithmetic. These perturbations in the data can be a result of measurement errors, errors in the mathematical model and so on.

Consider the perturbations in the sum

$$s = x_1 + x_2 + \cdots + x_n. \tag{1.29}$$

The problem with perturbations is

$$\widehat{s} = x_1(1 + \delta_1) + x_2(1 + \delta_2) + \cdots + x_n(1 + \delta_n), \tag{1.30}$$

where it is assumed that the relative errors δ_i have a common bound

$$|\delta_i| \leq \delta. \tag{1.31}$$

To derive a bound on the perturbation of s, subtract (1.29) from (1.30), which yields

$$\widehat{s} - s = x_1\delta_1 + x_2\delta_2 + \cdots + x_n\delta_n.$$

Taking the absolute values and applying (1.31), we obtain

$$|\widehat{s} - s| \leq (|x_1| + |x_2| + \cdots + |x_n|)\delta.$$

Thus the relative error in the perturbed sum is

$$\frac{|\widehat{s} - s|}{|s|} \leq \text{cond}(s, x)\delta, \tag{1.32}$$

where

$$\text{cond}(s, x) = \frac{|x_1| + |x_2| + \cdots + |x_n|}{|x_1 + x_2 + \cdots + x_n|}. \tag{1.33}$$

The right-hand side of (1.32) gives a bound on the relative error of $\widehat{s}$. The number δ bounds the relative errors in the components $x_1, x_2, \ldots, x_n$. In this way, cond(s, x) is the number, which multiplies the bound on the perturbations of the function arguments to obtain a bound on the perturbation of the function itself. Such number is called *condition number*. Formally, the (relative) condition number of a continuous scalar function $f(x)$ is defined as

$$\text{cond}(f, x) := \text{lim}_{\varepsilon \to 0}\text{max}_{|\Delta x| \leq \delta |x|} \left| \frac{f(x + \Delta x) - f(x)}{\delta f(x)} \right|. \tag{1.34}$$

This definition implies that

$$\begin{aligned} \left| \frac{f(x + \Delta x) - f(x)}{f(x)} \right| &\leq \text{cond}(f, x)\frac{|\Delta x|}{|x|} + o(\Delta x), \qquad (1.35)\\ &\leq \text{cond}(f, x)\delta + o(|\Delta x|), \end{aligned}$$

where the term $h = o(|\Delta x|)$ means that $|h|/|\Delta x| \to 0$ as $|\Delta x| \to 0$. Equation (1.35) represents an estimate of the relative *forward error* in the solution of the computational problem as a function of the relative perturbation $|\Delta x|/|x|$ in the data.

The condition number shows how difficult is the solution of the corresponding computational problem. If the condition number is large, small perturbations in the data can cause a large change in the result. For the problem under consideration, this number is never smaller than one. It is equal to one when the absolute value of the sum is equal to the sum of the absolute values of the arguments. This may happen when all components x_i have equal signs. On the other hand, this number is large when the sum is small in comparison with x_i. Thus, the condition number not only bounds the error in the result but also clarifies the reason for large error in the result.

If the perturbations δ_i in (1.30) reflect the effect of the rounding errors, then the inequality (1.32) allows to estimate the error in the final result. Hence, the combining the backward error analysis with the analysis of the conditioning allows estimating the effect of the rounding errors on the result. Equation (1.32) shows that the error in the result will be small if a numerically stable algorithm is applied on a well-conditioned problem. If the problem is ill conditioned, then the error in the result can be large even if a numerically stable algorithm is used.

In the case of a matrix function $f : \mathbb{C}^{n\times n} \to \mathbb{C}^{n\times n}$, the relative condition number is defined by

$$\operatorname{cond}(f, X) := \lim_{\delta\to 0}\max_{\|\Delta X\|\le\delta\|X\|} \frac{\|f(X+\Delta X) - f(X)\|}{\delta\|f(X)\|}. \quad (1.36)$$

According to this definition, it is fulfilled that

$$\frac{\|f(X+\Delta X) - f(X)\|}{\|f(X)\|} \le \operatorname{cond}(f, X)\frac{\|\Delta X\|}{\|X\|} + o(\|\Delta X\|). \quad (1.37)$$

A schematic interpretation of the forward error estimate is given in Figure 1.13. Geometrically, $r = \operatorname{cond}(f, X)\delta\|f(X)\|$ is the radius

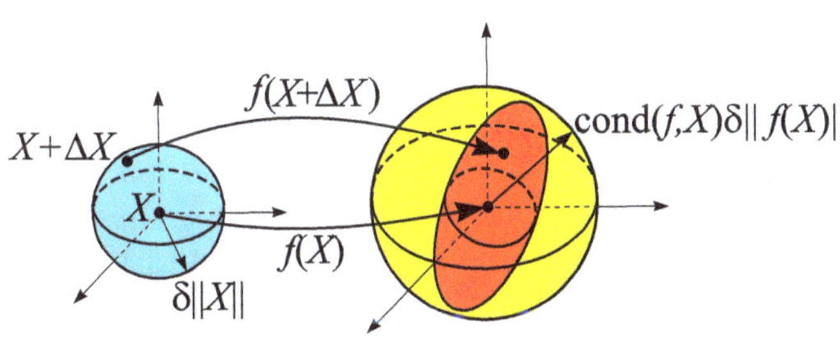

Fig. 1.13. Forward error estimate.

of the smallest sphere in $\mathbb{C}^{n\times n}$ which is centered at $f(X)$ and encloses the image of the set $Y : \|Y - X\| \leq \delta\|X\|$ under the mapping $Y \to f(Y)$.

1.3.4 *Sensitivity of a System of Linear Equations*

As an important application of the perturbation theory, consider the analysis of the sensitivity of a system of linear equations $Ax = b$ with a nonsingular matrix A.

Assume first that the matrix A is known exactly and the vector b — with some uncertainty Δb which leads to an uncertainty Δx in the solution. In this case, we have

$$A(x + \Delta x) = b + \Delta b$$

so that

$$\Delta x = A^{-1}\Delta b.$$

It is fulfilled that

$$\|\Delta x\| \leq \|A^{-1}\| \, \|\Delta b\|, \tag{1.38}$$

where $\|\cdot\|$ denotes any vector norm and the subordinated matrix norm.

Since $b = Ax$, it is fulfilled that

$$\|b\| \leq \|A\|\|x\|. \tag{1.39}$$

Multiplying (1.38) by (1.39) gives

$$\|\Delta x\| \, \|b\| \leq \|A\| \, \|A^{-1}\| \, \|x\| \, \|\Delta b\|. \tag{1.40}$$

Assuming that $b \neq 0$, we obtain from (1.40) that

$$\frac{\|\Delta x\|}{\|x\|} \leq \|A\| \, \|A^{-1}\| \, \frac{\|\Delta b\|}{\|b\|}. \tag{1.41}$$

For arbitrary nonsingular matrix A, the *condition number of the system of linear equations* $Ax = b$ (or *condition number in respect to inversion* of A) is defined as the product $\|A\| \, \|A^{-1}\|$. From the property of the norms $\|AA^{-1}\| \leq \|A\| \, \|A^{-1}\|$, it follows that

$$\text{cond}(A) = \|A\| \, \|A^{-1}\| \geq 1. \tag{1.42}$$

The value of the condition number depends on the norm used. Further on, by $\text{cond}_1(A), \text{cond}_2(A), \text{cond}_F(A)$ and $\text{cond}_\infty(A)$, we shall denote the values of the condition number evaluated using the 1, 2, Frobenius and ∞-norm of A, respectively. Since

$$\|A\|_2 \le \|A\|_F,$$

it follows that

$$\text{cond}_2(A) \le \text{cond}_F(A).$$

If V is an unitary (orthogonal matrix) and $B = VA$, then

$$\|B\|_2\|B^{-1}\|_2 = \|VA\|_2\|A^{-1}V^H\|_2 = \|A\|_2\|A^{-1}\|_2$$

and the same is true if we implement the Frobenius norm. This implies that if we use the 2-norm or the Frobenius norm, the condition number is invariant to unitary (orthogonal in the real case) transformations.

From the inequality (1.41), we obtain that

$$\frac{\|\Delta x\|}{\|x\|} \le \text{cond}(A)\frac{\|\Delta b\|}{\|b\|}. \tag{1.43}$$

The expression (1.43) represents a *sensitivity estimate* of the system of linear equations $Ax = b$.

If $\text{cond}(A) = 1$, then the relative uncertainty in x is exactly equal to the relative uncertainty in b. If, however, $\text{cond}(A) = 10^6$, then $\|\Delta x\|/\|x\|$ can be 10^6 times larger than $\|\Delta b\|/\|b\|$. If $\text{cond}(A)$ is relatively small, then the matrix A is *well conditioned* with respect to the problem of solving linear equations. If $\text{cond}(A)$ is relatively large, then the matrix A is *ill conditioned* with respect to the same problem. Finally, if $\text{cond}(A) \approx 1/\mathbf{u}$, where $\mathbf{u}$ is the unit roundoff, we say that the matrix A is *singular to working precision.* In such a case, we cannot expect accurate decimal digits in the result when solving system of linear equations or when inverting the matrix, although this matrix is theoretically nonsingular. Clearly, this notion depends on the precision used.

The following example shows that if the columns of a matrix tend to become parallel, then the condition number of this matrix increases.

Table 1.4. Angles between the columns of a matrix (in degrees) and its condition number.

$\Theta(a_1, a_2)$	$\Theta(a_2, a_3)$	$\Theta(a_3, a_4)$	$\Theta(a_4, a_5)$	$\text{cond}_2(A)$
70.69	70.48	84.45	86.55	$2.73 \cdot 10^0$
$\Theta(b_1, b_2)$	$\Theta(b_2, b_3)$	$\Theta(b_3, b_4)$	$\Theta(b_4, b_5)$	$\text{cond}_2(B)$
4.43	3.62	4.56	5.31	$4.02 \cdot 10^2$

Example 1.11. Consider the 5×5 matrices

$$A = \begin{bmatrix} 10 & -1 & 3 & 5 & -2 \\ -4 & -12 & 1 & 0 & 7 \\ 2 & -3 & 16 & -1 & 6 \\ -2 & -4 & 1 & -14 & 3 \\ 3 & 4 & -2 & 5 & 15 \end{bmatrix} \quad \text{and} \quad B = \begin{bmatrix} 5 & -6 & 10 & 4 & -5 \\ -2 & 3 & -5 & -2 & 2 \\ -8 & 9 & -15 & -7 & 7 \\ 3 & -4 & 8 & 4 & -4 \\ -1 & 1 & -2 & -1 & 1 \end{bmatrix}.$$

In Table 1.4, we show the angles between the consecutive columns of the matrices A and B, defined as described in Section A.9.3 of the Appendix. In the case of matrix A, the angles between the neighboring columns are sufficiently large (i.e., the columns are far from being parallel) and the condition number of this matrix is very low. (A is not far from a matrix with orthogonal columns.) In contrast to this case, the columns of the matrix B are almost parallel, i.e., almost linearly dependent and its condition number is large. This example demonstrates that if the columns of a matrix are almost linearly dependent, the sensitivity of the corresponding linear system of equations is high.

The earlier considerations indicate that the best conditioned matrices are the unitary (or orthogonal) matrices since their columns are orthogonal. Actually, if U is unitary, we have that $\text{cond}_2(U) = 1$ and $\text{cond}_F(U) = n$.

The condition number plays the same role with respect to the perturbations in A, as to the perturbations in b. Let there be an uncertainty ΔA in the matrix A and the vector b is known exactly. If $x = A^{-1}b$ and

$$x + \Delta x = (A + \Delta A)^{-1} b, \tag{1.44}$$

then

$$\Delta x = [(A + \Delta A)^{-1} - A^{-1}]b. \tag{1.45}$$

The matrix $A + \Delta A$ can be singular, if we do not impose some restrictions on ΔA. Writing

$$A + \Delta A = A(I + A^{-1}\Delta A),$$

we see that $A + \Delta A$ is nonsingular if

$$\|A^{-1}\Delta A\|_2 < 1. \tag{1.46}$$

That is why in (1.44) and (1.45), we assume that (1.46) is fulfilled. Setting $B = A + \Delta A$ in the identity

$$B^{-1} \quad A^{-1} = A^{-1}(A - B)B^{-1},$$

we found from (1.45) that

$$\Delta x = -A^{-1}(\Delta A)(A + \Delta A)^{-1}b = -A^{-1}(\Delta A)(x + \Delta x). \tag{1.47}$$

It follows from (1.47) that

$$\|\Delta x\| \le \|A^{-1}\| \; \|\Delta A\| \; \|x + \Delta x\|.$$

Finally,

$$\frac{\|\Delta x\|}{\|x + \Delta x\|} \le \text{cond}(A)\frac{\|\Delta A\|}{\|A\|}.$$

In this way, the uncertainty in x, relative to $\|x + \Delta x\|$, is bounded by the relative uncertainty in the matrix A, multiplied by cond(A).

Example 1.12. Given is a linear system $Ax = b$, where

$$A = \begin{bmatrix} 0.0050 & -0.9850 & 0.1000 \\ 0.9925 & 0.9775 & -0.1500 \\ 0.0500 & 0.1500 & 1.0000 \end{bmatrix}, \qquad b = \begin{bmatrix} -0.8800 \\ 1.8200 \\ 1.2000 \end{bmatrix}.$$

The exact solution of this system is

$$x = \begin{bmatrix} 1 \\ 1 \\ 1 \end{bmatrix}.$$

The system is well conditioned since the condition number of A with respect to the Frobenius norm is $\text{cond}_F(A) = \|A\|_F \|A^{-1}\|_F = 3.98$.

Consider the behavior of the solution when the right-hand side b of the system is subject to perturbations Δb which satisfy

$$\|\Delta b\|_F / \|b\|_F = \delta = 10^{-6}.$$

The results are interpreted geometrically in Figure 1.14, where the small sphere with radius $\delta = 10^{-6}$ in the center represents the relative perturbations of the right-hand side, the ellipsoid shows the corresponding relative perturbations $\|\widetilde{x} - x\|_2 / \|x\|_2$ of the solution and the large sphere with radius $\text{cond}_F(A)\delta$ exhibits the sensitivity estimate (1.43). Note that the principal axes of the ellipsoid are equal to the singular values $\sigma_1 = 1.601, \sigma_2 = 1.011$ and $\sigma_3 = 0.618$ of A, multiplied by δ.

Obviously, even for well-conditioned system there are perturbations for which the sensitivity estimate exceeds significantly the true change of the solution. If the estimate produce pessimistic estimates for all possible perturbations, we say that it is *conservative*.

1.3.5 *Notes and References*

In the various scientific disciplines, as, for instance, in celestial mechanics, theory of nonlinear oscillations and control theory, there have been developed specific perturbation theories which differ in the types of problems solved and mathematical methods used. All of these theories are based on the idea to investigate an object whose behavior deviates slightly from the behavior of a simple ideal system for which a complete solution of the considered problem is known. For a good introduction to the using of the perturbation theory in Applied Mathematics, see Miller (2015). The most complete source on matrix perturbation analysis at this moment is the book of Stewart and Sun (1990) and a comprehensive presentation of such methods is contained in the survey Li (2014).

The theory of conditioning is developed in the papers Rice (1966) and Geurts (1982). Different kinds of condition numbers are discussed in Gohberg and Koltracht (1993) and the computing of condition estimates is considered in Hager (1984), Higham (2002), Chapter 15, Kenney and Laub (1989) and Mathias (1995). The geometrical interpretation of conditioning and numerical stability in this section is

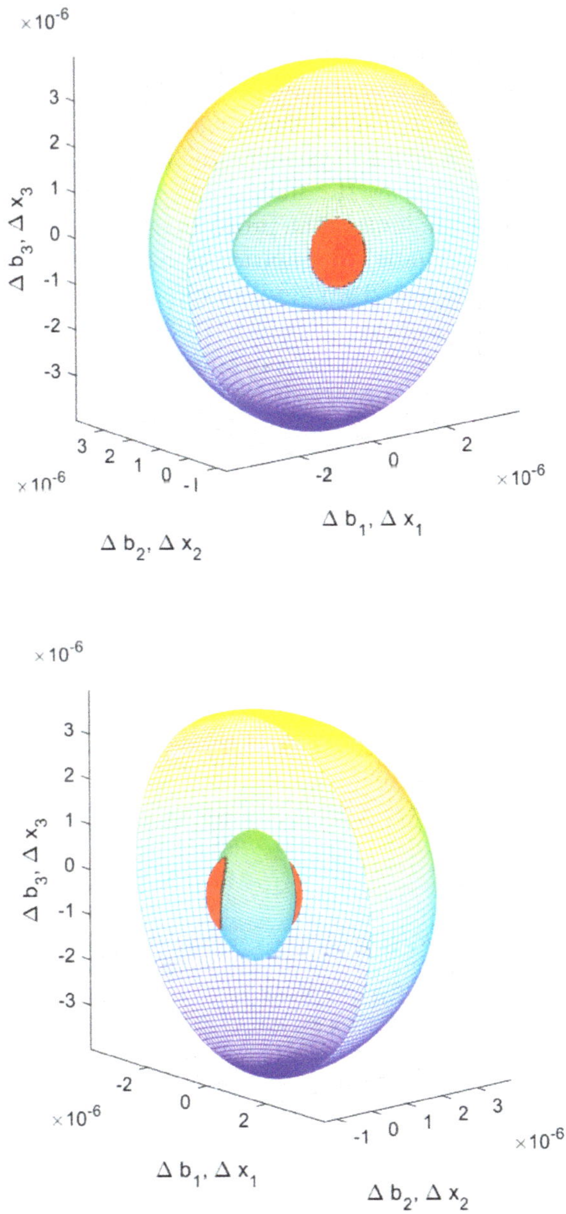

Fig. 1.14. Perturbations in the solution of a linear system shown from two different points of view in $\mathbb{R}^3$. The small sphere (in red) represents the perturbations of the right-hand side, the ellipsoid shows the perturbations of the solution and the large sphere indicates the sensitivity estimate.

modelled after Stewart (1998), Chapter 2. The interpretation of the error in the computed result as a consequence of perturbations in the data is proposed by Givens and is used systematically by Wilkinson in his books (Wilkinson, 2023, 1965). In some cases, it is necessary to use the forward error analysis in which the effect of the rounding errors is transferred to the final result. Since the forward error depends on the conditioning of the problem, the forward analysis cannot always produce reliable information about the numerical properties of the algorithm used. Two kinds of numerical stability in matrix computations are considered in Bunch (1987).

The sensitivity of linear systems of equations is analyzed in several books including Golub and Van Loan (2013), Chapter 2, Higham (2002), Chapter 7, and Stewart (1998), Chapter 3. Efficient estimates of the reciprocal of the condition number $\text{cond}_1(A)$ or $\text{cond}_\infty(A)$ of a nonsymmetric matrix A avoiding the computation of $\|A^{-1}\|$ can be obtained by the LAPACK routines `xGECON` using the matrix norm estimator implementing the Higham's modification (Higham, 1988) of Hager's method (Hager, 1984). More on the condition number estimation in LAPACK can be found in Higham (2002), Chapter 15.

1.4 Solving Triangular Systems of Equations

In this section, we consider the solution of triangular systems of linear equations arising in several computational problems involving matrices.

1.4.1 *Forward Substitution*

Given is the system of linear equations

$$Rx = b, \tag{1.48}$$

where

$$R = \begin{bmatrix} r_{11} & & & \\ r_{21} & r_{22} & & \\ \vdots & \vdots & \ddots & \\ r_{n1} & r_{n2} & \dots & r_{nn} \end{bmatrix}$$

is an nth order real or complex nonsingular lower triangular matrix and

$$b = [b_1, b_2, \ldots, b_n]^T, \quad x = [x_1, x_2, \ldots, x_n]^T.$$

The nonsingularity of R implies that $r_{ii} \neq 0,\ i = 1, 2, \ldots, n$.

The first equation of the system (1.48) is

$$r_{11}x_1 = b_1,$$

which gives

$$x_1 = b_1/r_{11}. \tag{1.49}$$

The second equation of (1.48) is

$$r_{21}x_1 + r_{22}x_2 = b_2$$

which yields

$$x_2 = (b_2 - r_{21}x_1)/r_{22}.$$

Generally, if we know $x_1, x_2, \ldots, x_{i-1}$, the ith equation

$$r_{i1}x_1 + r_{i2}x_2 + \cdots + r_{ii}x_i = b_i,$$

of (1.48) can be solved for x_i yielding

$$x_i = \left(b_i - \sum_{j=1}^{i-1} r_{ij}x_j \right) \Bigg/ r_{ii}. \tag{1.50}$$

The formulas (1.49) and (1.50) define Algorithm 1.1 for solving (1.48).

Algorithm 1.1 is frequently called *forward substitution.* Mathematically, the algorithm cannot produce solution only in the case when some $r_{ii} = 0$, which may happen if the matrix R is singular. However, in computations in floating point arithmetic, it is possible that the algorithm fails due to overflow even when the matrix is nonsingular. For instance, if

$$R = \begin{bmatrix} 10^{-50} & 0 \\ 1 & 10^{-50} \end{bmatrix}$$

Algorithm 1.1. Solving a system of linear equations with a lower triangular matrix

Given a nonsingular $n \times n$ lower triangular matrix R and an n-vector b, this algorithm solves the linear system $Rx = b$.

function $x = \mathrm{tril\_solv}(R, b)$

Input: Matrix $R \in \mathbb{C}^{n\times n}$; vector $b \in \mathbb{C}^n$

Output: $x \in \mathbb{C}^n$ contains the solution vector

for $i = 1,\ 2, \ldots, n$

$$x_i = \left(b_i - \sum_{j=1}^{i-1} r_{ij}x_j\right)\Big/ r_{ii}$$

end

and $b = [1,\ 0]^T$, then $x = [10^{50},\ -10^{100}]^T$. In this way, in IEEE single precision arithmetic, the second element of x cannot be represented as a floating point number and the computations will stop generating a diagnostic message.

The forward substitution requires

$$2\sum_{i=1}^{n}(i-1) = n(n-1)$$

computational operations (flops).

1.4.2 *Back Substitution*

The solution of the system

$$Tx = b,$$

where

$$T = \begin{bmatrix} t_{11} & t_{12} & \ldots & t_{1n} \\ & t_{22} & \ldots & t_{2n} \\ & & \ddots & \vdots \\ & & & t_{nn} \end{bmatrix}$$

is a nonsingular upper triangular matrix, can be done in the sequence $x_n,\ x_{n-1}, \ldots, x_1$ by Algorithm 1.2.

Algorithm 1.2 is called *back substitution*. It also requires $n(n-1)$ flops.

Algorithm 1.2. Solving a system of linear equations with an upper triangular matrix

Given a nonsingular $n \times n$ upper triangular matrix T and an n-vector b, this algorithm solves the linear system $Tx = b$.

function $x = \text{triu\_solv}(T, b)$
Input: Matrix $T \in \mathbb{C}^{n\times n}$, vector $b \in \mathbb{C}^n$
Output: $x \in \mathbb{C}^n$ contains the solution vector
for $i = n,\ n-1, \ldots, 1$

$$x_i = \left(b_i - \sum_{j=i+1}^{n} t_{ij}x_j\right) \Big/ t_{ii}$$

end

1.4.3 *Error Analysis*

Consider the system

$$Rx = b, \tag{1.51}$$

where R is a lower triangular matrix.

Provided that overflows do not occur, the components of x are evaluated by Algorithm 1.1 as

$$x_1 = \text{fl}\left(\frac{b_1}{r_{11}}\right),$$

$$x_2 = \text{fl}\left(\frac{-r_{21}x_1 + b_2}{r_{22}}\right),$$

$$x_i = \text{fl}\left(\frac{-r_{i1}x_1 - r_{i2}x_2 - \cdots - r_{i,i-1}x_{i-1} + b_i}{r_{ii}}\right), \quad i = 3, 4, \ldots, n.$$

Using the expressions for the rounding error bounds in computing the scalar product (Section 1.2.1), we can write that

$$\begin{aligned} x_1 &= \frac{b_1}{r_{11}(1+\delta_{11})},\\ x_2 &= \frac{-r_{21}(1+\delta_{21})x_1 + b_2}{r_{22}(1+\delta_{22})(1+\eta_{22})},\\ x_i &= \frac{-r_{i1}(1+\delta_{i1})x_1 - \cdots - r_{i,i-1}(1+\delta_{i,i-1})x_{i-1} + b_i}{r_{ii}(1+\delta_{ii})(1+\eta_{ii})},\\ i &= 3, 4, \ldots, n, \end{aligned} \tag{1.52}$$

where

$$
\begin{aligned}
&|\delta_{ii}| \le \mathbf{u}, \quad |\eta_{ii}| \le \mathbf{u},\\
&|\delta_{i1}| \le (i-1)\cdot 1.01\mathbf{u}; \quad i = 2,3,\dots,n,\\
&|\delta_{ij}| \le (i-j+1)\cdot 1.01\mathbf{u}; \quad i = 2,3,\dots,n, \quad j = 2,3,\dots,i-1.
\end{aligned}
$$

In (1.52), the factor $(1+\delta_{ii})$ reflects the errors, introduced in the division by r_{ii}, and the factor $(1+\eta_{ii})$ reflects those in the addition of b_i to the scalar product $\sum r_{ij}x_j$. The expressions for the errors are used in such a way that all the factors $1+\delta$ and $1+\eta$ multiply elements of R.

The equations (1.52) can be written in the form

$$
\begin{aligned}
&r_{11}(1+\delta_{11})x_1 = b_1,\\
&r_{21}(1+\delta_{21})x_1 + r_{22}(1+\delta_{22})(1+\eta_{22})x_2 = b_2,\\
&r_{i1}(1+\delta_{i1})x_1 + \cdots + r_{ii}(1+\delta_{ii})(1+\eta_{ii})x_i = b_i, \quad i = 3,\ 4,\dots,n
\end{aligned}
$$

or

$$
(R+\Delta R)x = b,
$$

where

$$
|\Delta R| \le 1.01\mathbf{u}\begin{bmatrix}
|r_{11}| & & & & \\
|r_{21}| & 2|r_{22}| & & & 0\\
2|r_{31}| & 2|r_{32}| & 2|r_{33}| & & \\
\vdots & \vdots & \vdots & \ddots & \\
(n-1)|r_{n1}| & (n-1)|r_{n2}| & (n-2)|r_{n3}| & \dots & 2|r_{nn}|
\end{bmatrix}.
$$

Taking into account the structure of the matrix $|\Delta R|$ and using the fact that $|r_{ij}| \le \max_{i,j}|r_{ij}|$, we obtain that

$$
\begin{aligned}
\|\Delta R\|_\infty &\le \left(\sum_{j=2}^{n-1} j + n - 1\right)\cdot 1.01\mathbf{u}\cdot \max_{i,j}|r_{ij}|\\
&= \frac{1}{2}(n^2+n-4)\cdot 1.01\mathbf{u}\cdot\max_{i,j}|r_{ij}|. \qquad (1.53)
\end{aligned}
$$

Note that a similar expression can be derived for $\|\Delta R\|_1$. From (1.53), we derive an expression for the change of R whose effect is equivalent to the effect of the rounding errors:

$$\|\Delta R\|_\infty \leq \frac{1}{2}(n^2 + n - 4) \cdot 1.01\mathbf{u}\, \|R\|_\infty. \tag{1.54}$$

Consequently, the vector x, computed by (1.52), is the exact solution of the slightly perturbed triangular system

$$(R + \Delta R)x = b, \tag{1.55}$$

where the perturbation ΔR satisfies (1.54). The quantity ΔR is referred to as *backward error* in the solution of the linear system (1.51). From (1.54), it follows that the relative backward error

$$\frac{\|\Delta R\|_\infty}{\|R\|_\infty} \leq \frac{1}{2}(n^2 + n - 4) \cdot 1.01\mathbf{u} \tag{1.56}$$

has the order of the unit roundoff error. This implies that Algorithm 1.1 is numerically stable from backward error analysis point of view and the same holds for Algorithm 1.2.

The analysis performed is an example of a backward error analysis. A forward error analysis can be done using the bound on the backward error and the condition number (Section 1.3.4) of the linear system $Rx = b$. As a result, we obtain that the relative forward error in the computed solution fulfills

$$\begin{aligned} \frac{\|\widehat{x} - x\|_\infty}{\|x\|_\infty} &\leq \text{cond}_\infty(R)\frac{\|\Delta R\|_\infty}{\|R\|_\infty} \\ &\leq \text{cond}_\infty(R)\frac{1}{2}(n^2 + n - 4) \cdot 1.01\mathbf{u}. \end{aligned}$$

High quality subroutines for solving systems of linear equations with different types of matrices are contained in the program packages LINPACK (Dongarra *et al.*, 1979) and LAPACK (Anderson *et al.*, 1999). The solution of triangular systems of equations in LAPACK is done by the routines `xTRTRS` and the solution of the general linear system $Ax = b$ with a nonsymmetric matrix A is obtained by the routines `xGESVX`. `xGESVX` computes a solution $\widehat{x}$ that satisfies $(A + E)\widehat{x} = b + f$, where e_{ij} and f_k are small relative changes in a_{ij}

and b_k, respectively. Along with the solution $\widehat{x}$, these routines also evaluate an estimated forward error bound FERR, such that

$$\frac{\|\widehat{x} - x\|_\infty}{\|x\|_\infty} \leq \text{FERR},$$

and an estimate of the componentwise backward error BERR which is equal to the smallest e_{ij} or f_k that makes $\widehat{x}$ an exact solution of the perturbed system $(A + E)\widehat{x} = b + f$. Both estimates are based on the analysis done in Arioli *et al.* (1989).

Example 1.13. Consider the solution of a family of systems of linear equations $Tx = b$, where T is an upper triangular matrix of varying order n. The components of the exact solution x are chosen as $x_i = i,\ i = 1, 2, \ldots, n$, and the right-hand side is determined as $b = Tx$. The matrix T is obtained as $T = SAS^{-1}$, where A is an upper triangular matrix whose entries are generated by

```
for i = 1 : n
    for j = i : n
        a(i, j) = i + j
    end
end
```

and $S = \text{diag}(1,\ 2, \ldots, s^{n-1})$ is a scaling matrix used to change the conditioning of the system by the change of the positive scalar s. For values of s which are smaller than one and large values of n, the matrix T is ill conditioned with $\text{cond}_\infty(T) = 3.77 \cdot 10^{18}$ for $s = 0.8$ and $n = 100$.

In Figures 1.15 and 1.16, we present the results of the numerical solution of $Tx = b$ for n varying between 1 and 100 and s varying between 0.8 and 1.2. Independently on the large relative forward error

$$\text{rerr} = \frac{\|\widehat{x} - x\|_\infty}{\|x\|_\infty}$$

for small s and large n due to the ill conditioning of the system, the estimated backward relative error BERR remains of the order of the unit machine roundoff $\mathbf{u}$ for all values of n and s. This result

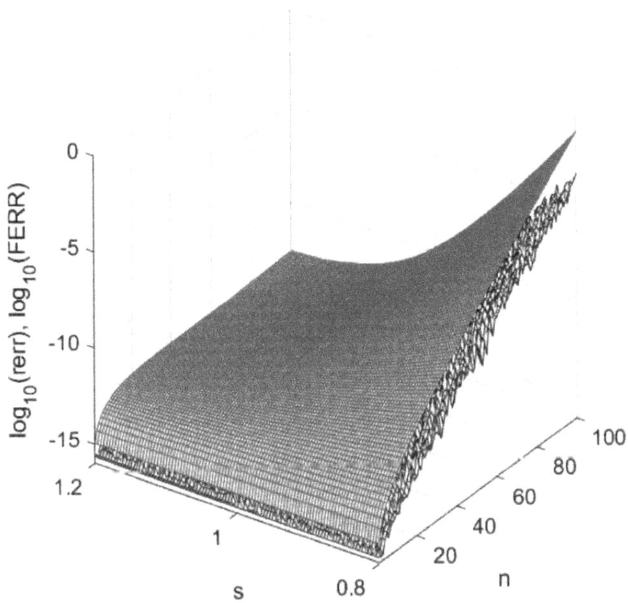

Fig. 1.15. Forward error estimates (the surface above) and the actual errors (the surface beneath) in the solution of systems of linear equations.

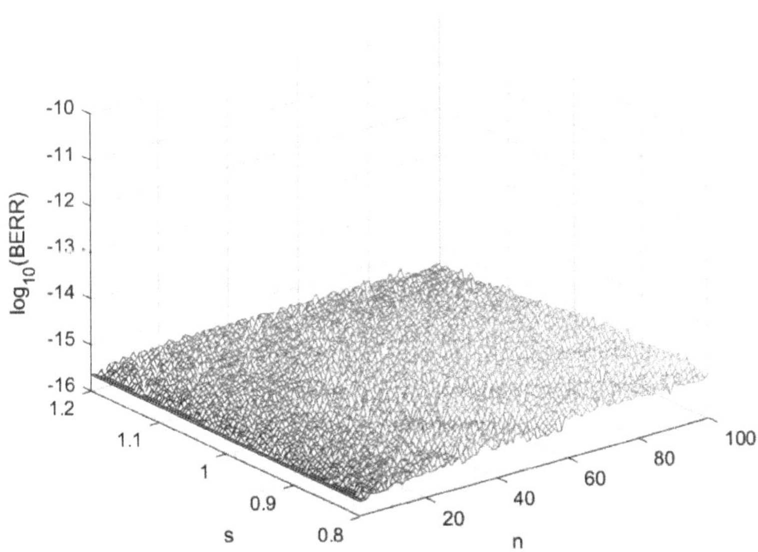

Fig. 1.16. Backward errors in the solution of systems of linear equations.

illustrates the numerical stability of the algorithms for solving linear equations implemented in LAPACK. For $s = 0.8$ and $n = 100$, we obtain that the actual relative forward error is $rerr = 1.45 \cdot 10^{-4}$ with estimated error bound equal to FERR $= 8.26 \cdot 10^{-2}$, while the backward error estimate is BERR $= 6.44 \cdot 10^{-16}$. In the given case, the actual error is severely overestimated by the sensitivity estimate $\text{cond}_\infty(T)\mathbf{u} = 4.19 \cdot 10^2$.

1.4.4 *Notes and References*

The triangular systems of equations usually arise in connection with the Gaussian elimination and LU decomposition in solving linear systems of equations. However, in this book, the triangular systems have different sources that motivate to consider them independently of the solution of linear systems.

The solution of triangular systems of equations and the corresponding rounding error analysis is presented in many books considering the numerical solution of linear systems of equations. A clear presentation of this subject is given in the classic textbooks by George E. Forsythe and Cleve B. Moler (1967) and by Gilbert W. Stewart (1973b) in Chapter 3. Both books still remain useful. The errors in the solution of triangular systems are derived in detail in Higham (2002), Chapter 8, see also in Stewart (1998), Chapter 3. In the last two books, one can find additional information about the practical implementation of Algorithms 1.1 and 1.2 using the Basic Linear Algebra Subroutines (BLAS) (Anderson *et al.*, 1999).

1.5 Elementary Transformations and QR Decomposition

The reduction of matrices into some specific form can be done by a sequence of transformations whose matrices have simple structure. These are the so-called *elementary transformations*, whose matrices are obtained and applied by the aid of a small number of computational operations. From numerical point of view, the most appropriate for this aim are the unitary transformations, but in some problems, one can also apply transformations with elementary triangular matrices. That is why we shall begin the presentation of the elementary transformations with the introduction of the elementary triangular matrices.

1.5.1 Elementary Triangular Matrices

Elementary lower triangular matrix of nth order and kth *index* is a matrix in the form

$$M_k = I_n - me_k^T,$$

where

$$e_i^T m = 0, \quad i = 1, 2, \ldots, k, \tag{1.57}$$

m is the n-vector and e_k is the kth column of the unit $n \times n$ matrix.

The condition (1.57) implies that the first k components of the vector m are zeros, i.e., m has the form

$$m = [0,\ 0, \ldots, 0,\ m_{k+1},\ m_{k+2}, \ldots, m_n]^T.$$

Thus, in the general case, the elementary lower triangular matrix has the form

$$M_k = \begin{bmatrix} 1 & 0 & \ldots & 0 & 0 & \ldots & 0 \\ 0 & 1 & \ldots & 0 & 0 & \ldots & 0 \\ \vdots & \vdots & & \vdots & \vdots & & \vdots \\ 0 & 0 & \ldots & 1 & 0 & \ldots & 0 \\ 0 & 0 & \ldots & -m_{k+1} & 1 & \ldots & 0 \\ \vdots & \vdots & & \vdots & \vdots & & \vdots \\ 0 & 0 & \ldots & -m_n & 0 & \ldots & 1 \end{bmatrix}.$$

The form of M_k shows that the elementary lower triangular matrix is a unit matrix with some additional elements in the kth column below the diagonal. That is why it is called *unit lower triangular matrix.*

These matrices are easily inverted. The inverse matrix of M_k is given by

$$M_k^{-1} = I_n + me_k^T.$$

In fact, let $X = I_n + me_k^T$. Then

$$\begin{aligned} M_k X &= (I_n - me_k^T)(I_n + me_k^T) = \\ &= I_n - me_k^T + me_k^T - me_k^T me_k^T = \\ &= I_n - m(e_k^T m)e_k^T. \end{aligned}$$

It follows from the definition of M_k that $e_k^T m = 0$. Thus $M_k X = I_n$ and $X = M_k^{-1}$, which proves the assertion. This property of the elementary matrices is important in performing similarity transformations since the obtaining of the inverse matrix is not associated with introduction of rounding errors.

It can be shown that the products

$$M_1 M_2 \dots M_{n-1} \quad \text{and} \quad M_{n-1} M_{n-2} \dots M_1$$

are also unit lower triangular matrices.

The computational significance of the elementary lower triangular matrices consists in the fact that they can be used to introduce zero components in a vector.

Let $v = [v_1,\ v_2, \dots, v_n]^T$ be a vector for which

$$e_k^T v = v_k \neq 0.$$

Then there exists an unique elementary lower triangular matrix of kth index, such that

$$M_k v = [v_1,\ v_2, \dots, v_k,\ 0,\ 0, \dots, 0]^T. \tag{1.58}$$

For a proof, M_k is seek in the form $M_k = I - m e_k^T$. Since this matrix should be of index k, then

$$m_i = 0, \quad i = 1,\ 2, \dots, k. \tag{1.59}$$

We have that

$$\begin{aligned} M_k v &= (I - m e_k^T) v = \\ &= v - (e_k^T v) m. \end{aligned}$$

Since the last $n-k$ components of $M_k v$ should be zeros, it follows that

$$v_i = v_k m_i = 0, \quad i = k+1,\ k+2, \dots, n,$$

and, if $v_k \neq 0$, then

$$m_i = \frac{v_i}{v_k}, \quad i = k+1,\ k+2, \dots, n. \tag{1.60}$$

Hence, if M_k exists, it is determined in a unique way from (1.59) and (1.60). On the other hand, when $v_k \neq 0$, the vector m can be

Algorithm 1.3. Multiplication of a vector by an elementary lower triangular matrix

Multiplies a vector a by an elementary lower triangular matrix M_k determined by the vector m

function $b = \text{eltran}(a, m, k)$

Input: Vectors $a \in \mathbb{C}^n$, $m \in \mathbb{C}^n$;
the index k of the elementary transformation M_k

Output: $b \in \mathbb{C}^n$ contains the product $M_k a$

```
for i = 1, 2, ..., n
   if i ≤ k
      b_i = a_i
   else
      b_i = a_i − m_i a_k
   end
end
```

determined from (1.59) and (1.60) and it is easy to prove that there exists a matrix $M_k = I_n - me_k^T$, satisfying (1.58).

So, if $v_k \neq 0$, the vector v can be multiplied by an elementary lower triangular matrix chosen so that the last $n - k$ components of v to become zeros and the rest components to remain unchanged. When $v_k = 0$, such matrix does not exist except when $v_{k+1}, v_{k+2}, \ldots, v_n$ are also zeros. In such case, it is possible to take arbitrary elementary lower triangular matrix of kth index.

Equations (1.59) and (1.60) determine an algorithm for computing of m and hence of M_k.

Let $M_k = I_n - me_k^T$ be an elementary lower triangular matrix of nth order and kth index. Algorithm 1.3 obtains the product $b = M_k a$, where a is a given n-vector.

Algorithm 1.3 requires $2(n - k)$ flops.

Example 1.14. Consider the complex vector

$$a = \begin{bmatrix} 4 - j_o 5 \\ -1 + j_o 7 \\ 5 - j_o 2 \\ 3 + j_o 3 \end{bmatrix}.$$

We want to determine an elementary transformation which annihilates all elements of a, except the first one. This can be done by using a single elementary lower triangular matrix of fourth order and index $k = 1$. Using (1.59) and (1.60), we obtain

$$m = \begin{bmatrix} 0.000000000000000 + j_o 0.000000000000000 \\ -0.951219512195122 + j_o 0.560975609756098 \\ 0.731707317073171 + j_o 0.414634146341463 \\ -0.073170731707317 + j_o 0.658536585365854 \end{bmatrix}$$

and

$$M_1 = \begin{bmatrix} 1.000000000000000 + j_o 0.000000000000000 & 0 & 0 & 0 \\ 0.951219512195122 - j_o 0.560975609756098 & 1 & 0 & 0 \\ -0.731707317073171 - j_o 0.414634146341463 & 0 & 1 & 0 \\ 0.073170731707317 - j_o 0.658536585365854 & 0 & 0 & 1 \end{bmatrix}.$$

Note that the matrix M_1 is well conditioned since

$$\text{cond}_2(M_1) = 4.12.$$

The vector $b = M_1 a$, computed by Algorithm 1.3, is

$$b = \begin{bmatrix} 4.000000000000000 - j_o 5.000000000000000 \\ -0.000000000000000 - j_o 0.000000000000002 \\ 0.000000000000000 + j_o 0.000000000000000 \\ -0.000000000000001 - j_o 0.000000000000000 \end{bmatrix}.$$

The backward error in computing b is

$$\|M_1^{-1} b - a\|_2 = 8.8818 \cdot 10^{-16}.$$

The product of an elementary lower triangular matrix with a given $n \times p$ matrix A can be obtained applying Algorithm 1.3 consecutively to the columns $a_1, a_2, \ldots, a_p$ of the matrix A.

It is seen from (1.60) that the norm of the matrix M_k can be arbitrary large if the element v_k of the vector v is arbitrary small. That is why if such matrices are used to transform a given matrix, it is possible to happen a gradual increase in the size of the elements of the transformed matrix. This leads to numerical instability since at each step the current rounding errors are proportional to the modules of the entries of the transformed matrix.

Generally speaking, increase in the matrix norm during its transformation is one of the main causes for the numerical instability of the matrix algorithms.

In the given case, the numerical instability can be avoided in the following way. Let the maximum of $|v_i|$ $(i = k, \dots, n)$ be attained for $i = k' \geq k$ and let

$$
P_{k,k'} = \begin{array}{cc}
 & \begin{array}{ccccccccc} & & & k & & k' & & & \end{array} \\
\begin{array}{c} \\ \\ \\ k \\ \\ k' \\ \\ \\ \\ \end{array} & \left[\begin{array}{ccccccccc}
1 & & & & & & & & \\
 & \ddots & & \vdots & & \vdots & & & \\
 & & 1 & & & & & & \\
 & \cdots & & 0 & \cdots & 1 & & & \\
 & & & \vdots & & \vdots & & & \\
 & \cdots & & 1 & \cdots & 0 & & & \\
 & & & & & & 1 & & \\
 & & & & & & & \ddots & \\
 & & & & & & & & 1
\end{array}\right]
\end{array}
$$

be an *elementary permutation matrix* obtained from the unit $n \times n$ matrix interchanging columns k and k'. Then the vector $v' = P_{k,k'}v$ is obtained from v by permuting the elements v_k and $v_{k'}$. Hence

$$|v'_k| \geq |v'_i|, \qquad i = k+1,\ k+2, \dots, n$$

and if we choose M_k such that $M_k v'$ has zero components in the position from $k+1$ to n, then

$$m_i = \frac{v'_i}{v'_k}$$

and

$$|m_i| \leq 1.$$

Thus, the matrix M_k will have elements which have modules bounded by 1, and to some degree, we achieve numerical stability since the rounding errors remain bounded. The matrix which transforms v is obtained in this case as a product of a unit lower triangular matrix M_k and the permutation matrix $P_{k,k'}$, since $M_k v' = M_k P_{k,k'} v$.

The process of choosing the maximum element from a given collection is called *pivot choice*. If such choice is performed, M_k is called *stable elementary matrix*.

1.5.2 *Plane Rotations*

The elementary unitary matrices are obtained form the unit matrix by modification with matrices which have rank equal to one.

The *plane rotation* or *Givens rotation* is a 2×2 unitary matrix R in the form

$$R = \begin{bmatrix} c & s \\ -\bar{s} & \bar{c} \end{bmatrix},$$

where

$$|c|^2 + |s|^2 = 1.$$

This transformation is called rotation since in the real case its effect on a two-dimensional vector x is to rotate x clockwise at angle $\Theta = \arccos(c) = \arcsin(s)$ (Figure 1.17).

The plane rotations can be used to introduce zeros in two-dimensional vectors. More specifically, let the complex scalars a and b with $a \neq 0$ be given. If we set

$$g = \sqrt{|a|^2 + |b|^2}$$

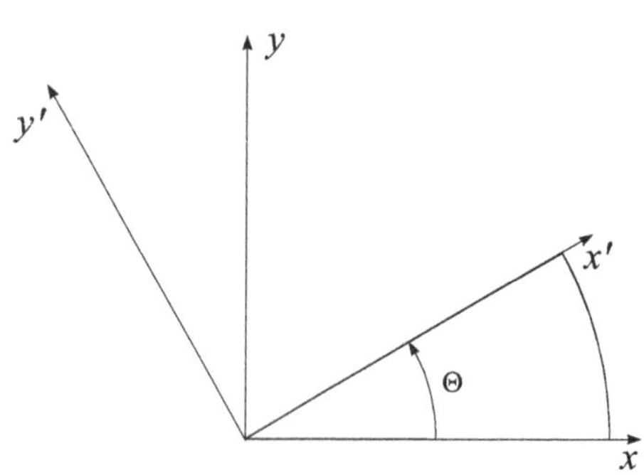

Fig. 1.17. Geometrical interpretation of elementary rotation.

Algorithm 1.4. Computing a plane rotation
For given complex numbers a and b computes a plane rotation satisfying

$$\begin{bmatrix} c & s \\ -\bar{s} & \bar{c} \end{bmatrix} \begin{bmatrix} a \\ b \end{bmatrix} = \begin{bmatrix} ga/|a| \\ 0 \end{bmatrix}$$

function $[a, b, c, s] = \text{rot\_comp}(a, b)$
Input: Scalars $a, b \in \mathbb{C}$
Output: Scalars c and s;
a is overwritten by $ga/|a|$;
b is overwritten by 0.

```
if b = 0
    c = 1, s = 0
    return
end
if a = 0
    c = 0, s = 1, a = b, b = 0
    return
end
```

$\rho = a/|a|$
$t = |a| + |b|$
$g = t\sqrt{|a/t|^2 + |b/t|^2}$
$c = |a|/g$
$s = \rho\bar{b}/g$
$a = g\rho$
$b = 0$

and

$$c = \frac{|a|}{g}, \quad s = \frac{a}{|a|}\frac{\bar{b}}{g}, \tag{1.61}$$

then (note that c is real)

$$\begin{bmatrix} c & s \\ -\bar{s} & \bar{c} \end{bmatrix} \begin{bmatrix} a \\ b \end{bmatrix} = \begin{bmatrix} ga/|a| \\ 0 \end{bmatrix}. \tag{1.62}$$

The computation of a plane rotation is done by Algorithm 1.4.

If b is zero, the algorithm returns the 2×2 unit matrix. If a is zero, the algorithm returns a rotation which exchanges the two vector components. The constant t is a scaling factor which is used to avoid overflow, the underflow being safe. The same algorithm can be used if a and b are real.

In executing Algorithm 1.4 in floating point arithmetic, instead of the exact quantities c, s and R, one obtains their approximations $\widehat{c}$, $\widehat{s}$ and $\widehat{R}$. It may be shown that

$$\widehat{c} = c + \Delta c, \quad \widehat{s} = s + \Delta s, \quad \widehat{R} = R + \Delta R, \quad \Delta R = \begin{bmatrix} \Delta c & \Delta s \\ -\overline{\Delta s} & \overline{\Delta c} \end{bmatrix},$$

where

$$|\Delta c| \leq 6\mathbf{u}, \quad |\Delta s| \leq 9\mathbf{u}, \quad \|\Delta R\|_2 < 11\sqrt{2}\mathbf{u}.$$

This shows that the computed matrix $\widehat{R}$ is close to the exact rotation R, corresponding to the given a and b.

The plane rotations can also be used to introduce zeros in matrices. The plane rotation in the (i, j)th plane is a matrix in the form

$$R_{ij} = \begin{array}{r} \\ \\ \\ i \\ \\ j \\ \\ \\ \\ \end{array} \begin{bmatrix} 1 & \ldots & 0 & \overset{i}{0} & \ldots & \overset{j}{0} & 0 & \ldots & 0 \\ \vdots & \ddots & \vdots & \vdots & & \vdots & \vdots & & \vdots \\ 0 & \ldots & 1 & 0 & \ldots & 0 & 0 & \ldots & 0 \\ 0 & \ldots & 0 & c & \ldots & s & 0 & \ldots & 0 \\ \vdots & & \vdots & \vdots & \ddots & \vdots & \vdots & & \vdots \\ 0 & \ldots & 0 & -\overline{s} & \ldots & \overline{c} & 0 & \ldots & 0 \\ 0 & \ldots & 0 & 0 & \ldots & 0 & 1 & \ldots & 0 \\ \vdots & & \vdots & \vdots & & \vdots & \vdots & \ddots & \vdots \\ 0 & \ldots & 0 & 0 & \ldots & 0 & 0 & \ldots & 1 \end{bmatrix}.$$

Otherwise speaking, the rotation in the (i, j)th plane is a unit matrix in which the plane rotation is build in the submatrix corresponding to the rows and columns i and j.

To see the effect of the rotation in the (i, j)th plane on a matrix, let A be an $n \times n$ matrix and let

$$B = R_{ij} A.$$

If A and B are separated by rows, then

$$
\begin{aligned}
b_i^H &= ca_i^H + sa_j^H, \\
b_j^H &= cx_j^H - \bar{s}a_i^H, \\
b_k^H &= a_k^H, \quad \text{for each } k \neq i,\ j.
\end{aligned}
$$

In this way, the multiplication by a rotation on the left in the (i, j)th plane combines the rows i and j and left the rest unchanged. By a suitable choice of c and s, it is possible to introduce zero everywhere in the ith or jth rows.

The multiplication by R_{ij}^H on the right changes the columns in a similar manner and can introduce zero everywhere in the ith or jth column.

In Figure 1.18, we illustrate the left and right multiplication, respectively, of the matrix A by a plane rotation.

The left multiplication by a plane rotation is done by using Algorithm 1.5 and a similar algorithm can be used to perform right multiplication by a plane rotation. If the matrix A is of nth order, both algorithms require $6n$ flops.

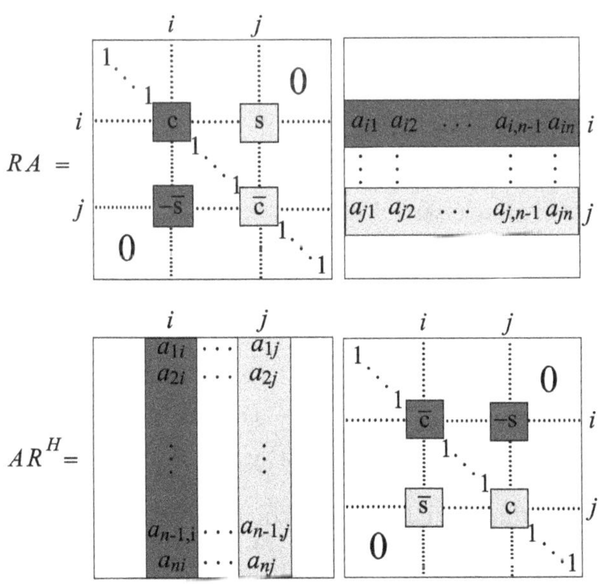

Fig. 1.18. Left and right multiplication of a matrix by plane rotation.

Algorithm 1.5. Left multiplication by a plane rotation
For a plane rotation

$$R = \begin{bmatrix} c & s \\ -\overline{s} & \overline{c} \end{bmatrix}$$

and vectors v and w, computes the product $R\begin{bmatrix} v^T \\ w^T \end{bmatrix}$

function $[v, w] = \text{rot\_mult}(c, s, v, w)$
Input: Plane rotation R, determined by c and s;
vectors v and w

Output: $\begin{bmatrix} v^T \\ w^T \end{bmatrix} \Leftarrow R\begin{bmatrix} v^T \\ w^T \end{bmatrix}$

$t = c \cdot v + s \cdot w$
$w = \overline{c} \cdot w - \overline{s} \cdot v$
$v = t$

Let A be an $n \times p$ matrix. It can be shown that

$$\text{fl}(\widehat{R}A) - RA) = E, \tag{1.63}$$

where

$$\|E\|_F \leq 14\mathbf{u}\|A\|_F.$$

Equation (1.63) can be represented in the form

$$\text{fl}(\widehat{R}A) = R(A + F), \tag{1.64}$$

where $F = R^H E$ and $\|F\|_F < 14\mathbf{u}\|A\|_F$. This result shows that Algorithm 1.5 is backward stable.

Example 1.15. Given is the complex vector

$$a = \begin{bmatrix} 4 - j_o 5 \\ -1 + j_o 7 \\ 5 - j_o 2 \\ 3 + j_o 3 \end{bmatrix},$$

considered in Example 1.14. Using consecutively Algorithms 1.4 and 1.5, we want to zero the elements of a below the first one, implementing plane rotations. This can be done by applying first a plane rotation R_{34} to the 3rd and 4th elements of a, then applying a rotation

R_{23} to the 2nd and 3rd elements of the transformed vector $R_{34}a$ and finally applying a plane rotation R_{12} to the 1st and 2nd elements of $R_{23}R_{34}a$. As a result, we obtain

$$\begin{aligned}\widehat{b} &= \mathrm{fl}(\widehat{R}_{12}\widehat{R}_{23}\widehat{R}_{34}a)\\ &= \begin{bmatrix} 7.338505197694240 - j_o 9.173131497117801 \\ 0.000000000000000 + j_o 0.000000000000000 \\ 0.000000000000000 + j_o 0.000000000000000 \\ 0.000000000000000 + j_o 0.000000000000000 \end{bmatrix}.\end{aligned}$$

For the product of the three rotations, we find (up to four decimal digits)

$$\begin{aligned}\widehat{R} &= \mathrm{fl}(\widehat{R}_{12}\widehat{R}_{23}\widehat{R}_{34})\\ &= \begin{bmatrix} 0.5451 + j_o 0.0000 & -0.5185 - j_o 0.3058 & 0.3988 - j_o 0.2260 \\ 0.7222 - j_o 0.4259 & 0.3913 + j_o 0.0000 & -0.1487 + j_o 0.2583 \\ 0.0000 + j_o 0.0000 & 0.3473 + j_o 0.6032 & 0.5640 + j_o 0.0000 \\ 0.0000 + j_o 0.0000 & 0.0000 + j_o 0.0000 & -0.2438 - j_o 0.5688 \end{bmatrix} \cdots \rightarrow\\ &\leftarrow \cdots \begin{bmatrix} -0.0399 - j_o 0.3589 \\ 0.1409 + j_o 0.1878 \\ 0.1750 - j_o 0.4084 \\ 0.7855 + j_o 0.0000 \end{bmatrix}.\end{aligned}$$

The backward error in computing $\widehat{b} = \mathrm{fl}(\widehat{R}a)$ is

$$\|\widehat{R}^H\widehat{b} - a\|_2 = 1.8611 \cdot 10^{-15}.$$

The deviation of the matrix $\widehat{R}$ from unitarity is characterized by the error

$$\|\widehat{R}^H\widehat{R} - I_4\|_2 = 6.6164 \cdot 10^{-16}.$$

1.5.3 *Householder Transformations*

Consider first an elementary orthogonal matrix in the form

$$H(u) = I_n - \tau u u^T \in \mathbb{R}^{n\times n}, \tag{1.65}$$

where $u \in \mathbb{R}^n$ and $\tau \in \mathbb{R}$. The matrix H is symmetric ($H^T = H$) and for $\tau = 2/u^T u$ it is fulfilled that

$$H^T H = H^2 = I_n - 2\tau u u^T + \tau^2 u(u^T u)u^T = I_n,$$

which implies that H is orthogonal and $H^{-1} = H$. For each vector $a \in \mathbb{R}^n$, we have that

$$Ha = (I_n - \tau u u^T)a = a - \tau(u^T a)u$$

and hence $Ha \in \operatorname{span}(a, u)$.

A matrix of the form (1.65) is called *Householder transformation* or *elementary reflection* and is determined in a unique way by the vector u which is called *Householder vector.* Note that it is not necessary to form the matrix H in explicit form.

The geometrical interpretation of the Householder transformation for $n = 3$ is shown in Figure 1.19. If $H = I - \tau u u^T$ is an elementary

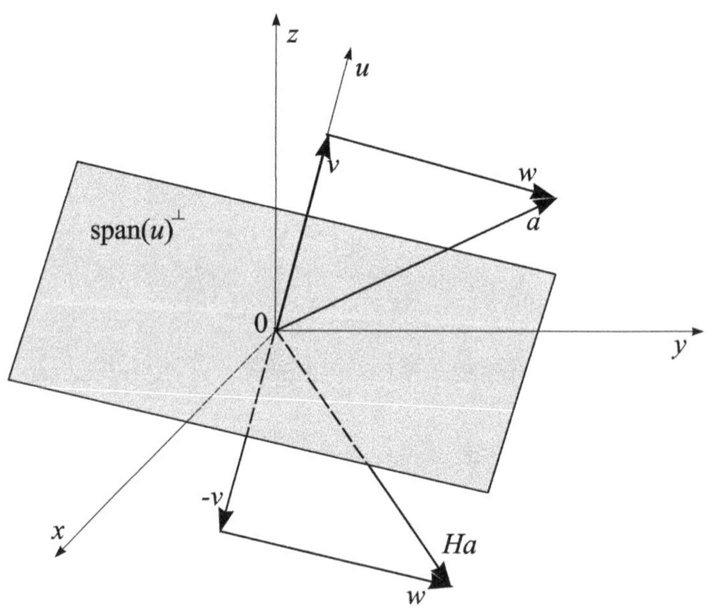

Fig. 1.19. Geometrical interpretation of a Householder transformation.

reflection and a is a given vector represented as $a = v + w$, where v lies on u and w is orthogonal to u, then

$$Ha = -v + w.$$

The effect of the transformation Ha in the space $\mathbb{R}^n$ is to reflect a with respect to the $(n-1)$-dimensional hyperplane $\text{span}(u)^\perp$, which goes through 0 and is normal to the vector u. In particular, $Hu = -u$, i.e., H inverts u and if $a \perp u$, then $Ha = a$.

In matrix computations, the Householder transformations are used for simultaneous introduction of several zeros in a vector or matrix. For a given nonzero vector a, we construct an elementary reflection such that the multiplication by H annihilates all components of a except the first one. This is written as

$$Ha = \rho t e_1, \quad t = \|a\|_2, \quad \rho = \pm 1, \tag{1.66}$$

where e_1 denotes the first column of the unit matrix and the second equation is a consequence of the fact that H is orthogonal. Multiplying (1.66) on the left by H and using the fact that $H^2 = I_n$, we obtain that $He_1 = \pm a/t$. Hence, the first column of H is proportional to a. It is seen that (1.66) is satisfied if we set

$$a = \begin{bmatrix} a_1 \\ a_2 \end{bmatrix}, \quad u = \begin{bmatrix} a_1 - \rho t \\ a_2 \end{bmatrix}. \tag{1.67}$$

Note that u differs from a only in its first component. It may be shown that

$$u^T u = (a - \rho t e_1)^T (a - \rho t e_1) = (t^2 - 2\rho t a_1 + t^2) = 2t(t - \rho a_1).$$

If a is close to the multiple of e_1, then $t \approx |a_1|$ and the cancellation in the above formula can lead to large relative error in $u^T u$. To avoid the cancellation, we take $\rho = -1$ if $a_1 > 0$ and $\rho = +1$, if $a_1 \leq 0$. This yields

$$u_1 = \rho(t + |a_1|), \quad \tau = \frac{1}{t(t + |a_1|)}. \tag{1.68}$$

If $a_1 \neq 0$, then $\rho = -\text{sign}(a_1)$, but this relationship is not accurate when $a_1 = 0$. (Note that in MATLAB®, `sign(0) = 0`, which can lead to errors.)

The Householder transformation described earlier is numerically stable since its computation involves only addition of positive numbers.

The described method for constructing the Householder transformation is not unique. Depending on the way of scaling the vector u, there are possible other versions, for instance, there is a version for which (Stewart, 1998, Chapter 4)

$$H = I - uu^T.$$

In the complex case, the elementary reflection has the form

$$H = I_n - \tau \frac{uu^H}{u^H u}, \quad u \in \mathbb{C}^n. \tag{1.69}$$

It is easy to check that H is Hermitian and unitary matrix ($H = H^H = H^{-1}$). For a given $a \in \mathbb{C}^n$ with $ae_1 = a_1 = e^{j_o\Theta_1}|a_1|$, we want to determine u such that

$$Ha = ge_1, \quad g = \rho t, \quad |\rho| = 1, \quad t = \|a\|_2.$$

Since H is Hermitian matrix, $a^H Ha = \rho t a^H e_1 = \rho \overline{a_1} t$ should be real. Hence, except if a_1 is real, it is not possible to obtain real ρ.

Similar to the real case, the vector u can be taken in the form

$$u = a - \rho t e_1.$$

To avoid cancellation in the first component of u, we take $\rho = -e^{j_o\Theta_1}$ which gives

$$u_1 = a_1 - \rho t = e^{j_o\Theta_1}(|a_1| + t).$$

Since $|\rho|^2 = 1$, we have that

$$u^H u = \left(\|a\|^2 + t(\overline{\rho} a_1 + \rho \overline{a_1}) + |\rho|^2 t^2\right) = 2t(t + |a_1|). \tag{1.70}$$

In this way, the normalizing scalar τ is determined from

$$\tau = \frac{2}{u^H u} = \frac{1}{t(t + |a_1|)} = \frac{1}{\rho\overline{\rho} t(t + |a_1|)} = \frac{1}{\rho t(-\overline{a_1} + \overline{\rho} t)} = -\frac{1}{\rho t \overline{u}_1}.$$

Algorithm 1.6. Computing an elementary reflection
For a given complex vector a, computes the Householder transformation $H = I_n - \tau u u^H$, such that $Ha = ge_1, g = \rho\|a\|_2$.
function $[u, \tau, g] = \text{house\_comp}(a)$
Input: Vector $a \in \mathbb{C}^n$
Output: u contains the Householder vector u;
τ contains the normalizing scalar τ;
g contains the scalar g.

```
s = max_i |a_i|
if s ≠ 0
   t = Σ_{i=1}^n (|a_i/s|)^2
   if a_1 ≠ 0
      ρ = −a_1/|a_1|
   else
      ρ = −1
   end
   g = ρ · √t · s
   u_1 = a_1 − g
   for i = 2, 3, ..., n
      u_i = a_i
   end
   τ = −1/(g · conj(u_1))
else
   g = 0
   τ = 0
end
```

Algorithm 1.6 computes the $n \times n$ Householder transformation $H = I_n - \tau u u^H$ so that for a given complex vector $a \neq 0$, $Ha = ge_1$, $g = \rho\|a\|_2$. In the computation of $\|a\|_2$, the elements of a are scaled by the largest in module element so that to avoid overflow or underflow in the squaring of the elements. If $a = 0$, the algorithm returns $u = 0$, $g = 0$, $\tau = 0$ so that $H = I_n$. The algorithm requires approximately $4n$ flops.

The same algorithm can also be used to find an orthogonal reflection in the case when the vector a is real.

The detailed error analysis shows that in the real case the computed in floating point arithmetic elementary reflection $\widehat{H}$ satisfies

$$\|\widehat{H} - H\|_F \leq (4n + 26)\mathbf{u} + O(\mathbf{u}^2). \qquad (1.71)$$

The bound (1.71) shows that the computed Householder transformation $\widehat{H}$ is close to the exact unitary transformation H.

Consider the implementation of similarity transformations by using Householder reflections. An important peculiarity in the case is that the reflection matrices are not formed explicitly but are represented implicitly. This leads to economy of computational operations and reduces the rounding errors.

When the matrix $A = [a_1, \ldots, a_p] \in \mathbb{C}^{n \times p}$ is multiplied on the left by Householder transformation $H \in \mathbb{C}^{n \times n}$, the product

$$HA = [Ha_1,\ Ha_2, \ldots, Ha_p]$$

can be efficiently computed as (Figure 1.20)

$$Ha_j = (I_n - \tau u u^H) a_j = a_j - \tau (u^H a_j) u, \quad j = 1,\ 2, \ldots,\ p. \qquad (1.72)$$

In a similar way, in the multiplication of A on the right by $H \in \mathbb{C}^{p \times p}$, H acts on the rows of A. Both operations can be represented as

$$HA = A - \tau u (u^H A) = A - u v^H, \quad v^H = \tau (u^H A)$$

and

$$AH = A - \tau (Au) u^H = A - v u^H, \quad v = \tau (Au),$$

respectively. Note that the products uv^H and vu^H are matrices of rank one and the computing of $A - uv^H$ or $A - vu^H$ represents

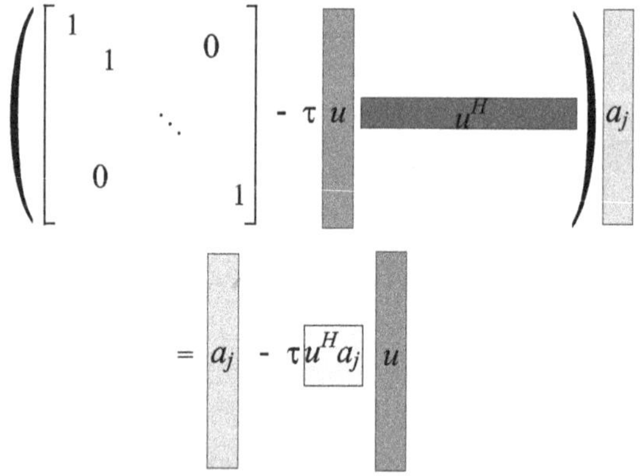

Fig. 1.20. Left multiplication by Householder transformation.

Algorithm 1.7. Left multiplication by an elementary reflection
Multiplies a matrix A on the left by an elementary reflection
$H = I_n - \tau u u^H$, *given by the vector u and the scalar τ.*
function $B = \mathrm{house\_mult}(A, u, \tau)$
Input: Matrix $A \in \mathbb{C}^{n\times p}$;
u contains an n-vector;
τ contains nonzero scalar.
Output: B contains the product HA
for $j = 1, 2, \ldots, p$
$f = \tau u^H a_j$
$b_j = a_j - fu$
end

an *updating of A by a rank one matrix*. Thus, the left and right multiplication by an elementary reflection involve the computing of one matrix-vector product followed by an updating with a matrix of rank one and require $4np + p$ flops.

The left multiplication of a matrix by elementary reflection is done by Algorithm 1.7. A similar algorithm can be used for right multiplication by elementary reflection.

A detailed numerical analysis shows that the computed product

$$\widehat{B} = \mathrm{fl}(\widehat{H}A) \tag{1.73}$$

satisfies

$$\|\widehat{B} - B\|_F \leq (6n + 31)\|A\|_F \mathbf{u} + O(\mathbf{u}^2). \tag{1.74}$$

This result shows that the effect of the rounding errors in the left multiplication by an elementary reflection is proportional to the number of the rows of the matrix and its norm. Since $\|B\|_F = \|A\|_F$, the bound (1.74) reveals that the computed result has small relative error.

From the backward error analysis point of view, we may write that

$$\mathrm{fl}(\widehat{H}A) = H(A + E), \tag{1.75}$$

where according to (1.74) we have the estimate

$$\|E\|_F \leq (6n + 31)\|A\|_F \mathbf{u} + O(\mathbf{u}^2). \tag{1.76}$$

The matrix $\widehat{B} = \mathrm{fl}(\widehat{H}A)$ obtained in this way can be considered as a result of the exact multiplication of H by a slightly perturbed matrix A, i.e., the Algorithm 1.7 is backward stable.

In the complex case the error bounds are larger but qualitatively the results are the same.

Example 1.16. Consider again the complex vector

$$a = \begin{bmatrix} 4 - \jmath_o 5 \\ -1 + \jmath_o 7 \\ 5 - \jmath_o 2 \\ 3 + \jmath_o 3 \end{bmatrix}$$

already used in Examples 1.14 and 1.15. We want to determine a Householder transformation which annihilates all elements of a, except the first one.

Using Algorithm 1.6, we obtain

$$\widehat{u} = \begin{bmatrix} 11.338505197694239 - \jmath_o 14.173131497117799 \\ -1.000000000000000 + \jmath_o 7.000000000000000 \\ 5.000000000000000 - \jmath_o 2.000000000000000 \\ 3.000000000000000 + \jmath_o 3.000000000000000 \end{bmatrix},$$

$$\widehat{\tau} = 4.689998634666 \times 10^{-3},$$

$$\widehat{g} = -7.338505197694239 + \jmath_o 9.173131497117799.$$

For the Householder transformation, we determine (up to four digits)

$$\widehat{H} = \begin{bmatrix} -0.5451 + \jmath_o 0.0000 & 0.5185 + \jmath_o 0.3058 & -0.3988 + \jmath_o 0.2260 \\ 0.5185 - \jmath_o 0.3058 & 0.7655 + \jmath_o 0.0000 & 0.0891 - \jmath_o 0.1548 \\ -0.3988 - \jmath_o 0.2260 & 0.0891 + \jmath_o 0.1548 & 0.8640 + \jmath_o 0.0000 \\ 0.0399 - \jmath_o 0.3589 & -0.0844 + \jmath_o 0.1126 & -0.0422 - \jmath_o 0.0985 \end{bmatrix} \cdots \rightarrow$$

$$\leftarrow \cdots \begin{matrix} 0.0399 + \jmath_o 0.3589 \\ -0.0844 - \jmath_o 0.1126 \\ -0.0422 + \jmath_o 0.0985 \\ 0.9156 + \jmath_o 0.0000 \end{matrix} \Bigg]$$

and

$$\mathrm{fl}(\widehat{H}a) = \widehat{g}e_1.$$

The deviation of the matrix $\widehat{H}$ from unitarity can be characterized by the error

$$\|\widehat{H}^H\widehat{H} - I_4\|_F = 5.6163 \times 10^{-16},$$

which conforms to the bound (1.71).

1.5.4 *Computing the QR Decomposition*

The *QR decomposition* of a rectangular complex $n \times p$, $n \geq p$ matrix A represents a reduction into triangular form:

$$Q^H A = \begin{bmatrix} R \\ 0 \end{bmatrix},$$

where Q is unitary and R is an upper triangular matrix.

In this section, we show how to determine a sequence of Householder transformations $H_1, H_2, \ldots, H_p$ such that

$$H_p H_{p-1} \ldots H_1 A = \begin{bmatrix} R \\ 0 \end{bmatrix},$$

where R is an upper triangular matrix. Since the product $H_p H_{p-1} \ldots H_1$ is unitary, it is equal to the matrix Q^H in the QR decomposition of A. The corresponding process is called *Householder triangularization.*

Let a matrix $A \in \mathbb{C}^{n \times p}$, $n \geq p$, be given and suppose that by the aid of Algorithm 1.6 we computed the Householder transformations $H_1, \ldots, H_{k-1}$ such that (Figure 1.21)

$$H_{k-1} \ldots H_1 A = \begin{bmatrix} R_{11} & r_{1k} & R_{1,k+1} \\ 0 & a_{kk} & A_{k,k+1} \end{bmatrix},$$

where $R_{11} \in \mathbb{C}^{(k-1)\times(k-1)}$ is the upper triangular, $r_{1k} \in \mathbb{C}^{k-1}$, $R_{1,k+1} \in \mathbb{C}^{(k-1)\times(p-k)}$, $a_{kk} \in \mathbb{C}^{n-k+1}$ and $A_{k,k+1} \in \mathbb{C}^{(n-k+1)\times(p-k)}$. Let $\widehat{H}_k \in \mathbb{C}^{(n-k+1)\times(n-k+1)}$ be a Householder transformation such that

$$\widehat{H}_k a_{kk} = g_{kk} e_1.$$

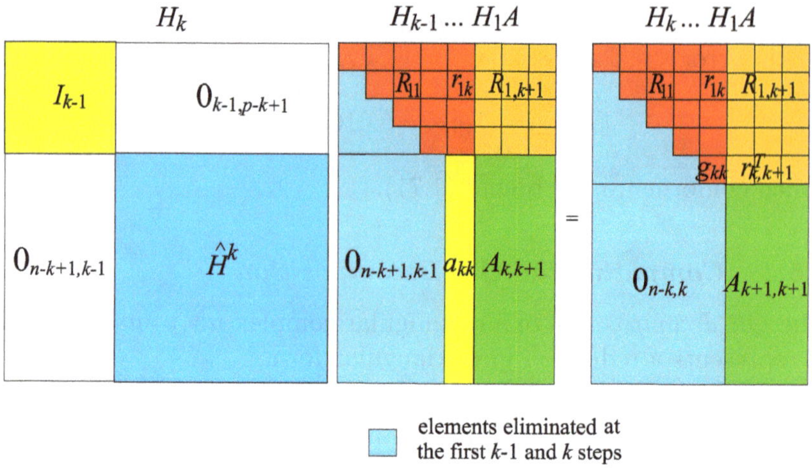

Fig. 1.21. Computing the QR decomposition of a matrix.

Set $H_k = \text{diag}(I_{k-1}, \widehat{H}_k)$. Then

$$H_kH_{k-1}\ldots H_1A = \begin{bmatrix} R_{11} & r_{1k} & R_{1,k+1} \\ 0 & \widehat{H}_k a_{kk} & \widehat{H}_k A_{k,k+1} \end{bmatrix}$$

$$\equiv \begin{bmatrix} R_{11} & r_{1k} & R_{1,k+1} \\ 0 & g_{kk} & r^T_{k,k+1} \\ 0 & 0 & A_{k+1,k+1} \end{bmatrix}.$$

With this process, the triangularization advanced by one step.

The reduction considered is realized by Algorithm 1.8.

Algorithm 1.8 works also for $n < p$ and can be used to find the QR decomposition of a real matrix A as well.

The number of computational operations performed by Algorithm 1.8 can be determined as follows. The work for generating of the transformations can be neglected. At the kth step, the left multiplication by the transformation requires $2(n-k+1)(p-k)$ flops to compute v and the same number of operations to update A. That is why the left multiplication requires

$$4\sum_{k=1}^{n}(n-k+1)(p-k) \approx 2p^2\left(n - \frac{1}{3}p\right) \text{ flops.}$$

Algorithm 1.8. QR decomposition by Householder triangularization

Given a complex $n \times p$ *matrix* A *with* $m = \min\{n, p\}$, *this algorithm computes a sequence of Householder transformations* $H_1, H_2, \ldots, H_m$, $H_k = I_n - \tau_k u_k u_k^H$, *such that*

$$H_m H_{m-1} \ldots H_1 A = \begin{bmatrix} R \\ 0 \end{bmatrix}, \text{ if } n \geq p \text{ or}$$

$$H_m H_{m-1} \ldots H_1 A = \begin{bmatrix} R, S \end{bmatrix}, \text{ if } n < p$$

where R *is an* $m \times m$ *upper triangular matrix. The vectors* u_k, *defining the Householder transformations, are stored in the* $n \times m$ array U and the scalars τ_k — in the m-vector τ.

function $[U, R, \tau] =$ qr_decomp(A)
Input: Matrix $A \in \mathbb{C}^{n \times p}$
Output: U contains the vectors u_k, $k = 1, \ldots, m$
  of the Householder transformations,
  R contains the upper triangular matrix R,
  τ contains the scalars τ_k, $k = 1, \ldots, m$.

for $k = 1, 2, \ldots, m$
  Using Algorithm 1.6, construct the transformation
  $[u, \tau_k, g] =$ house_comp($A_{k:n,k}$)
  $U_{k:n,k} = u$
  $R_{kk} = g$
  Left multiplication by the transformation
  $v^H = u^H A_{k:n,k+1:p}$
  Updating of A
  $A_{k:n,k+1:p} = A_{k:n,k+1:p} - \tau_k u v^H$
  $R_{k,k+1:p} = A_{k,k+1:p}$
end

Thus, Algorithm 1.8 requires about $2p^2(n - \frac{1}{3}p)$ flops for the reduction of A into triangular form. When $n \gg p$, the term $2np^2$ dominates. On the other hand, when $n = p$, the number of operations is reduced to $\frac{4}{3}n^3$.

If $n < p$, the final matrix has the form

$$[R, S],$$

where R is an $n \times n$ upper triangular matrix and S is an $n \times (p - n)$ matrix. In this case, the number of computational operations is $2n^2(p - \frac{1}{3}n)$.

Algorithm 1.9. Accumulating the transformations in the QR decomposition

Computes the product Q of Householder transformations $H_1, H_2, \ldots, H_m$ used in the reduction of an $n \times p$ matrix into triangular form, $m = \min\{n, p\}$.

function $Q = \text{qcomp}(U, \tau)$

Input: $n \times m$ array U, containing the m vectors of the Householder transformations, m-vector τ containing the scalars τ_k.

Output: Matrix Q containing the product of Householder transformations.

$Q = I_n$
for $k = m,\ m-1, \ldots, 1$
  $u = U_{k:n,k}$
  $v^H = u^H Q_{k:n,k:n}$
  $Q_{k:n,k:n} = Q_{k:n,k:n} - \tau_k u v^H$
end

The matrix Q in the QR decomposition of A can be determined as the product $H_1 H_2 \ldots H_m$. The accumulation of the transformations can be done by setting $Q = I_n$ and the transformations are accumulated in the order

$$(\cdots(((QH_1)H_2)H_3)\cdots)H_m.$$

This computation has the advantage that the reduction and the accumulation of the transformations can be done in the same loop. The disadvantage of this method is that since H_1 is a full matrix, each next H_k should be applied to the whole array $Q_{1:n,k:n}$. In Algorithm 1.9, the vectors u_k defining the Householder transformations $\widehat{H}_k = I_{n-k} + \tau_k u_k u_k^H$ are stored during the reduction and after that are accumulated in the order

$$H_1(\cdots(H_{m-2}(H_{m-1}(H_m Q)))\cdots)$$

which improves the efficiency.

The accumulation of Q for each k requires the same number of operations as the Householder triangularization. Hence for $n \geq p$, Algorithm 1.9 requires approximately $2p^2(n - \frac{1}{3}p)$ flops for accumulation of Q.

The QR decomposition of a rectangular matrix A can be obtained by a sequential execution of Algorithms 1.8 and 1.9 and requires $4p^2(n - \frac{1}{3}p)$ flops.

In LAPACK, the QR decomposition is done by the routines `xGEQRF` and in MATLAB®by the function `qr`.

1.5.5 *Error Analysis of the QR Decomposition*

In this section, we present a backward error analysis of the QR decomposition.

Let A be a real $n \times p$ matrix whose entries are floating point numbers. It is necessary to determine the error arising in the implementation of $m = \min\{n, p\}$ consecutive Householder transformations beginning with the matrix A. We have that the exact quantities are

$$A_1 = A,$$
$$A_{k+1} = H_k A_k, \quad k = 1, 2, \ldots, m,$$

where the elementary reflection H_k is constructed so that to introduce zeros in the positions $k+1$ through n of the kth column of $H_k A_k$.

The computed quantities are

$$\widehat{A}_1 = A,$$
$$\widehat{A}_{k+1} = \text{fl}(\widehat{H}_k \widehat{A}_k), \quad k = 1, 2, \ldots, m,$$

where $\widehat{H}_k$ is the computed approximation of the true matrix $\widetilde{H}_k$ which would introduce zeros in the elements from $k+1$ to n of the kth column of the matrix $\widetilde{H}_k \widehat{A}_k$.

Define the error matrix

$$F_k = \text{fl}(\widehat{H}_k \widehat{A}_k) - H_k \widehat{A}_k = \widehat{A}_{k+1} - H_k \widehat{A}_k, \quad k = 1,\ 2, \ldots, m. \tag{1.77}$$

According to the results from Section 1.5.2 (Equations (1.75) and (1.76)),

$$\|F_k\|_F \leq f_{n-k+1} \|\widehat{A}\|_F + O(\mathbf{u}^2), \quad k = 1,\ 2, \ldots, m,$$

where

$$f_j = (6j + 31)\mathbf{u}.$$

Combining equations (1.77) for $k = 1,\ 2, \ldots, m$, we obtain

$$\widehat{A}_{m+1} = H_m \ldots H_1 A + H_m \ldots H_2 F_1 + H_m \ldots H_3 F_2 + \cdots + F_m.$$

It follows from the orthogonality of the matrices H_k that

$$\begin{aligned} \|\widehat{A}_{m+1} - H_m \ldots H_1 A\|_F &\leq \sum_{k=1}^{m} f_{n-k+1} \|\widehat{A}_k\|_F + O(\mathbf{u}^2) \\ &\leq \|A\|_F \sum_{k=1}^{m} f_{n-k+1} + O(\mathbf{u}^2) \qquad (1.78) \\ &\leq (6n - 3m + 34) m\mathbf{u} \|A\|_F + O(\mathbf{u}^2). \end{aligned}$$

It is convenient to define the error matrix

$$E = H_1 \ldots H_m \widehat{A}_{m+1} - A.$$

Then

$$\widehat{A}_{m+1} = H_m \ldots H_1 (A + E) \qquad (1.79)$$

and form the inequality (1.78) we obtain that

$$\|E\|_F \leq (6n - 3m + 34) m\mathbf{u} \|A\|_F. \qquad (1.80)$$

Equation (1.79) and the bound (1.80) show that the computed matrix $\widehat{A}_{m+1}$ is the exact result of an orthogonal transformation of $A + E$, where the backward error $\|E\|_F$ is small in comparison with $\|A\|_F$. A similar result holds in the complex case which shows that Algorithm 1.8 is numerically stable from backward error point of view.

If in (1.78) we take $A = I_n$, then it is possible to obtain the error in the computed product of m approximated Householder transformations. This gives

$$\|fl(\widehat{H}_m \ldots \widehat{U}_1) - H_m \ldots H_1\|_F \leq n^{1/2} (6n - 3m + 34) m\mathbf{u}. \qquad (1.81)$$

The bound obtained shows that the computed product slightly differs from an unitary matrix.

Example 1.17. Consider the 4×3 matrix

$$A = \begin{bmatrix} -1+j_o2 & -2+j_o3 & 5-j_o7 \\ 0-j_o3 & 3-j_o4 & 2+j_o6 \\ 4-j_o1 & -1+j_o5 & -6-j_o3 \\ 6+j_o6 & -4+j_o8 & 1-j_o4 \end{bmatrix}.$$

Using Algorithms 1.8 and 1.9, we obtain the following result for the QR decomposition of A (up to four digits):

$$Q = \begin{bmatrix} -0.2203+j_o0.0000 & 0.5511-j_o0.1688 & 0.6217+j_o0.0070 \\ 0.2644-j_o0.1322 & -0.5094+j_o0.0963 & 0.1134+j_o0.3336 \\ 0.2644+j_o0.3085 & 0.3603+j_o0.5180 & -0.3782-j_o0.3677 \\ -0.2644+j_o0.7932 & 0.0188+j_o0.0237 & -0.0004+j_o0.4594 \end{bmatrix} \cdots \rightarrow$$

$$\leftarrow \cdots \begin{matrix} -0.1772-j_o0.4486 \\ -0.6062-j_o0.3901 \\ -0.2418-j_o0.3165 \\ -0.1936+j_o0.2270 \end{matrix}\Bigg],$$

$$R = \begin{bmatrix} 4.5387-j_o9.0774 & 10.4435+j_o1.3660 & -7.3148+j_o4.7150 \\ 0.0000+j_o0.0000 & -1.1782+j_o5.6285 & -0.2949-j_o4.3344 \\ 0.0000+j_o0.0000 & 0.0000+j_o0.0000 & 6.8223-j_o5.9029 \end{bmatrix}.$$

The backward error of the computed QR decomposition is

$$\left\| Q \begin{bmatrix} R \\ 0 \end{bmatrix} - A \right\|_F = 9.5968 \times 10^{-15}$$

and the error characterizing the deviation from unitarity of the matrix Q is

$$\|Q^H Q - I_4\|_F = 1.5335 \times 10^{-15}.$$

1.5.6 *Notes and References*

The Householder transformations are proposed by the American mathematician Alston Householder in (1958), see also Householder

(1964), Chapter 5. A detailed presentation of the issues related to the computation of the elementary unitary transformations is done in Wilkinson (1965), Chapter 5, Stewart (1973b), Chapter 5, Lawson and Hanson (1995), Chapter 10, Golub and Van Loan (2013), Chapter 5, Stewart (1998), Chapter 4, and Higham (2002), Chapter 19. A comparison of the various methods for constructing Householder transformations can be found in Lehouck (1996). The rounding error analysis of the Householder transformations presented in this section is similar to the analysis done in Lawson and Hanson (1995), Chapter 15. The efficient accumulation of the transformations after the Householder triangularization is proposed in Stewart (2001), Section 2.3.

The QR decomposition of a matrix can also be done by the aid of plane rotations, see Van Loan (1997), Chapter 6, and Stewart (1998), Chapter 4.

The perturbation analysis of the QR decomposition is discussed in Stewart (1977), Sun (1995a), Chang *et al.* (1997), Li and Wei (2015) and Petkov (2022).

1.6 Computing the Rank and Fundamental Subspaces

1.6.1 *Numerical Rank of a Matrix*

An important problem arising in the matrix computations is the determining of the rank of a given matrix A in the presence of uncertainties in its entries. As we see latter on in Chapter 6, this problem is closely related to the numerical determination of the Jordan structure of a matrix. The difficulty of the rank determination is in that the rank does not depend continuously on the elements of the matrix.

Let $A \in \mathbb{C}^{m\times n}$, assuming without loss of generality that $m \geq n$, and let

$$A = U\Sigma V^H$$

be the singular value decomposition of A (Section A.6 of the Appendix), where $U \in \mathbb{C}^{m\times m}$, $V \in \mathbb{C}^{n\times n}$ are unitary matrices and

$$\Sigma = \mathrm{diag}(\sigma_1, \dots, \sigma_k, \dots, \sigma_n)$$

contains the singular values of A, $\sigma_1 \geq \cdots \sigma_k \geq \cdots \geq \sigma_n$. Note that $\sigma_1 = \|A\|_2$. The "theoretical" rank of A is determined as the number of nonzero singular values. In practice, the matrix A is contaminated by measurement, approximation and discretization errors so that instead of the rank of A we have to determine the rank of the matrix $A + \Delta A$, where $\|\Delta A\|_2 \leq tol\|A\|_2$ for some small positive number tol. The number tol can be considered as the relative uncertainty in the matrix A. These considerations lead to the introduction of the notion of *numerical rank* with respect to the tolerance $tol\|A\|_2$, which is defined as

$$k = k(A, \text{tol}) := \min_{\|\Delta A\|_2 \leq \text{tol}\|A\|_2} \text{rank}(A + \Delta A). \tag{1.82}$$

In contrast to the "theoretical" rank, the numerical rank remains stable in the sense that perturbations smaller than the tolerance $tol\|A\|$ will not change the rank of A.

One additional circumstance in the rank determination is that instead of the exact singular values of A, we compute in floating point arithmetic the singular values of the near matrix $\widetilde{A} = A + F$, where $\|F\|_2 \leq \delta\|A\|_2$ for some small number $\delta > 0$, which is of the order of the machine precision $\mathbf{u}$. As a result, we obtain the singular valued decomposition $\widetilde{A} = \widetilde{U}\widetilde{\Sigma}\widetilde{V}^H$, where $\widetilde{U} \in \mathbb{C}^{m\times m}$, $\widetilde{V} \in \mathbb{C}^{n\times n}$ are unitary matrices and

$$\widetilde{\Sigma} = \text{diag}(\widetilde{\sigma}_1, \ldots, \widetilde{\sigma}_k, \ldots, \widetilde{\sigma}_n).$$

Due to the rounding errors, the computed singular values satisfy

$$|\widetilde{\sigma}_i - \sigma_i| = O(\mathbf{u})\|A\|_2.$$

This expression shows that any singular value $\widetilde{\sigma}_i \leq O(\mathbf{u})$ can be treated as zero since it can be a result of rounding errors.

According to Demmel (1997), Section 3.5, the situation described may be illustrated geometrically as shown in Figure 1.22, where each matrix represents a point in the *parameter space* $\mathbb{C}^{m\cdot n}$. (Note that $\mathbb{C}^{m\cdot n}$ is isomorphic to $\mathbb{C}^{m\times n}$.) The coordinates of such point are the entries of the given matrix. In the parameter space, all matrices of incomplete rank (also called *rank deficient matrices*) form a manifold (surface) of a lower dimension $q < m \cdot n$. Let tol be a user supplied tolerance reflecting the relative uncertainty in $\widetilde{A}$, where we assume that $tol > \delta$. Then, in the rank determination, it is reasonable to neglect all singular values which are smaller than $tol\|\widetilde{A}\|_2$.

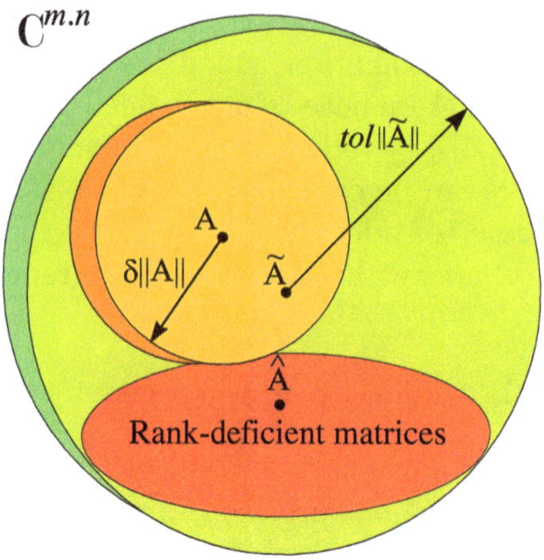

Fig. 1.22. Determining the rank of a matrix.

This results in an approximation matrix $\widehat{A}$ which lies on the surface of the rank-deficient matrices.

In accordance with the Eckart–Young theorem (Eckart and Young, 1936, Golub and Van Loan, 2013, Theorem 2.4.8), the lower rank approximation of $\widetilde{A}$ by a matrix $\widehat{A}$ with $\text{rank}(\widehat{A}) = k$ for $0 \leq k < n$ leads to an error $\|\widehat{A} - \widetilde{A}\|_2$. This error is minimized by taking

$$\widehat{A} = \widetilde{U}\widehat{\Sigma}\widetilde{V}^H,$$

where

$$\widehat{\Sigma} = \text{diag}\left(\widetilde{\sigma}_1, \widetilde{\sigma}_2, \ldots, \widetilde{\sigma}_k, 0_{(m-k)\times(n-k)}\right).$$

In such case, the approximation error is equal to

$$\|\widehat{A} - \widetilde{A}\|_2 = \widetilde{\sigma}_{k+1}, \quad \|\widehat{A} - \widetilde{A}\|_F = \sqrt{\sum_{i=k+1}^{n} \widetilde{\sigma}_i^2}.$$

Based on this theorem, the numerical rank (1.82) can be determined as the largest number k, such that

$$\widetilde{\sigma}_k > tol \cdot \sigma_1 \geq \widetilde{\sigma}_{k+1}.$$

This definition, however, does not guarantee that the rank is stable in respect to small changes in the entries of A, since it can be sensitive to perturbations if $\widetilde{\sigma}_k$ is only slightly larger than $\widetilde{\sigma}_{k+1}$.

Example 1.18. The diagonal matrix Σ with the singular values of a 6×6 matrix A is

$$\Sigma = \text{diag}(1,\ 0.9,\ 0.7,\ 0.6,\ 2 \cdot 10^{-8}, 10^{-10}).$$

Using the tolerance $tol = 10^{-9}$, the given matrix can be approximated by a matrix of rank 5, neglecting the singular value 10^{-10}. As a result, the approximation error will be equal to 10^{-10}. However, the numerical rank determined in this way will be unstable, since as a consequence of perturbations of size $\sigma_2 = 2 \cdot 10^{-8}$ the matrix rank will decrease to 4.

To avoid the rank sensitivity when $\widetilde{\sigma}_k$ is only slightly larger than $\widetilde{\sigma}_{k+1}$, the numerical rank or *pseudorank* of a matrix is defined in the following way.

Definition 1.19. Let the matrix $A \in \mathbb{C}^{m \times n}$ have singular values $\sigma_1 \geq \cdots \sigma_k \geq \cdots \geq \sigma_p$, $p = \min\{m, n\}$. Then the numerical rank of A is the largest number $0 \leq k \leq p$ such that for a given tolerance *tol* and a given constant $gap > 1$, it is fulfilled that (Figure 1.23)

$$\sigma_k > \text{tol} \cdot \sigma_1, \tag{1.83}$$

$$\sigma_k > \text{gap} \cdot \sigma_{k+1}. \tag{1.84}$$

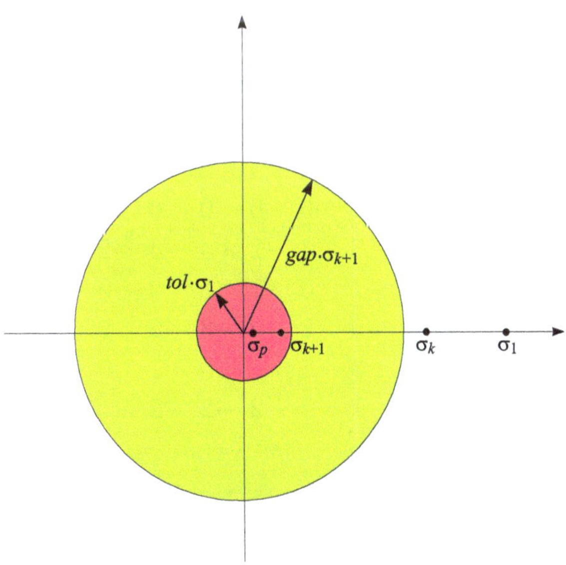

Fig. 1.23. Numerical rank of a matrix.

According to this definition, the numerical rank is preserved under perturbations that are smaller than σ_k. The value of σ_{k+1} determines the size of the approximation error $\|\widehat{A} - A\|_2 = \sigma_{k+1}$. Usually, *gap* is chosen sufficiently large, for instance, equal to 1000. In practice, $tol \cdot \sigma_1 = tol \cdot \|A\|_2$ may be replaced by $tol \cdot \|A\|_F$. In case that $\|A\|_F < 1$, the threshold $tol\|A\|_F$ will be less than the size *tol* of the uncertainty in A so that instead of this threshold it is appropriate to take the quantity $tol \cdot \max\{\|A\|_F, 1\}$.

If the criterion described does not produce an acceptable result in respect to the size of the approximation error $\|\widehat{A} - A\|_2$, the computations should be repeated for other values of *tol* and *gap*.

In the example presented earlier, for $tol = 10^{-7}$ and $gap = 1000$, the numerical rank k is equal to 4 ($\sigma_k = \sigma_4 > tol \cdot \sigma_1$, $\sigma_4 > 1000 \cdot \sigma_5$), and this rank is preserved for perturbations smaller than $\sigma_4 = 0.6$. The approximation error is only $\sigma_5 = 2 \cdot 10^{-8}$. This very good result is due to the fact that the matrix under consideration has neighboring singular values one of which is sufficiently large and the other is very small. If, however, all singular values are sufficiently close, then the numerical rank is not well defined.

Determining of the numerical rank is done by Algorithm 1.10.

Example 1.20. Consider the determining of the numerical rank of a family of matrices whose singular values vary with a parameter. The matrix $A \in \mathbb{C}^{8\times 5}$ is constructed as $A = Q_1 D Q_2$, where

$$Q_1 = \begin{bmatrix} -40 & -11 & 8 & -6 & 2 & -1 & 0 & 2 \\ 24 & 8 & -4 & 4 & 0 & 2 & -1 & -2 \\ 2 & 2 & 0 & 0 & 0 & 0 & 0 & 1 \\ -21 & -13 & 3 & -2 & -4 & -4 & 2 & 1 \\ 1 & -1 & 0 & 1 & 0 & 0 & 0 & -1 \\ -15 & -9 & 2 & -2 & -3 & -2 & 2 & 1 \\ 11 & 3 & -2 & 2 & 0 & 1 & 0 & -1 \\ -14 & -7 & 2 & -2 & -2 & -2 & 1 & 1 \end{bmatrix},$$

Algorithm 1.10. Determining the numerical rank of a matrix
Determines the numerical rank of a matrix using the singular values
function $k = \text{num\_rank}(A, tol, gap)$
Input: The matrix $A \in \mathbb{C}^{m \times n}$; tolerance parameters tol and gap.
Output: The numerical rank k, such that
$\sigma_k > tol \cdot \sigma_1,\ \sigma_k > gap \cdot \sigma_{k+1}$
Set $p = \min\{m, n\}$
$norm_A = \max(\|A\|_F, 1)$
Compute the singular values $\sigma_1 \geq \sigma_2 \geq \cdots \geq \sigma_p$ of A
for $i = 1, 2, \ldots, p$
  if $\sigma_{p-i+1} > tol \cdot norm_A$
    break
  elseif $i = p$
    if $\sigma_1 < tol \cdot norm_A$
      $k = 0$
      return
    end
  end
end
if $i = 1$
  $k = p$
else
  while $i \leq p$
    if $\sigma_{p-i+1} > gap \cdot \sigma_{p-i+2}$
      $k = p - i + 1$
      return
    else
      $i = i + 1$
    end
  end
  The numerical rank can not be determined
  $k = [\,]$
end

Table 1.5. Determining the numerical rank.

j	$tol \cdot \|A\|_F$	σ_2	σ_3	σ_4	k	$gap \cdot \sigma_{k+1}$
1	$5.23 \cdot 10^{-6}$	$2.88 \cdot 10^{1}$	$2.81 \cdot 10^{-1}$	$2.92 \cdot 10^{-4}$	4	$3.28 \cdot 10^{-12}$
2	$5.29 \cdot 10^{-6}$	$2.88 \cdot 10^{1}$	$2.83 \cdot 10^{-3}$	$2.85 \cdot 10^{-7}$	3	$2.85 \cdot 10^{-4}$
3	$5.30 \cdot 10^{-6}$	$2.88 \cdot 10^{1}$	$2.83 \cdot 10^{-4}$	$2.85 \cdot 10^{-10}$	3	$2.85 \cdot 10^{-7}$
4	$5.30 \cdot 10^{-6}$	$2.88 \cdot 10^{1}$	$2.83 \cdot 10^{-5}$	$2.85 \cdot 10^{-13}$	3	$2.85 \cdot 10^{-10}$
5	$5.30 \cdot 10^{-6}$	$2.88 \cdot 10^{1}$	$2.83 \cdot 10^{-6}$	$1.72 \cdot 10^{-14}$	2	$2.83 \cdot 10^{-3}$
6	$5.30 \cdot 10^{-6}$	$2.88 \cdot 10^{1}$	$2.83 \cdot 10^{-7}$	$5.86 \cdot 10^{-15}$	2	$2.83 \cdot 10^{-4}$

$$Q_2 = \begin{bmatrix} 6 & 12 & -5 & 4 & -2 \\ -8 & -12 & 6 & -3 & 2 \\ 2 & 6 & -2 & 2 & -1 \\ -2 & -3 & 1 & 0 & 1 \\ -3 & -5 & 2 & -1 & 1 \end{bmatrix},$$

D is an 8×5 diagonal matrix

$$D = \mathrm{diag}(1, 1, 10^{-j}, 10^{-3j}, 0)$$

and j is a varying parameter. The theoretical rank of this matrix is equal to 4 for all values of j.

The numerical rank k of A is determined for $j = 1, 2, \ldots, 6$ using tolerances $tol = 10^{-8}$ and $gap = 1000$. In Table 1.5, we give the quantities $tol \cdot \|A\|_F$ and $gap \cdot \sigma_{k+1}$, as well as the computed singular values $\sigma_2(A)$, $\sigma_3(A)$, $\sigma_4(A)$, and the numerical rank k as functions of j. The maximum singular value satisfies $\sigma_1 = \|A\|_2 > 500$ and the singular value σ_5 is of order 10^{-15} for each j. The singular value which determines the numerical rank for given j is detached in a box.

According to the results, for $j = 1$, the numerical rank is determined by the singular value σ_4. For $2 \geq j \geq 4$, the singular value σ_4 does no longer satisfy the condition (1.83) and its "role" in determining the numerical rank is taken by σ_3. This leads to the decreasing of the numerical rank by 1. Finally, for $j = 5, 6$, the role in rank determining is taken by σ_2 and the numerical rank drops to 2. Obviously, the numerical rank decreases with the increase in the size of uncertainty in A.

1.6.2 Computing the Fundamental Subspaces

An important application of the numerical rank is the determination of the bases of the fundamental subspaces related to a given matrix A. As shown in Section A.8 of the Appendix, orthonormal bases of all four fundamental subspaces can be found by performing only one singular value decomposition of A. Let $A \in \mathbb{C}^{m\times n}$ be decomposed as

$$A = U\Sigma V^H, \tag{1.85}$$

where $U = [u_1, u_2, \ldots, u_m]$, $V = [v_1, v_2, \ldots, v_n]$ are unitary matrices and

$$\Sigma = \begin{bmatrix} \Sigma_r & 0 \\ 0 & 0 \end{bmatrix}, \quad \Sigma_r = \text{diag}(\sigma_1, \ldots, \sigma_r), \tag{1.86}$$

$$\sigma_1 \geq \cdots \geq \sigma_r > 0.$$

In accordance with the value of the rank of A, $r = \text{rank}(A) \leq \min\{m, n\}$, the matrix A can be represented as

$$A = [\underbrace{U_1}_{r}, \underbrace{U_2}_{m-r}] \begin{bmatrix} \Sigma_r & 0 \\ 0 & 0 \end{bmatrix} \begin{bmatrix} V_1^H \\ V_2^H \end{bmatrix} \begin{matrix} \}\, r \\ \}\, n-r \end{matrix}, \tag{1.87}$$

where

$$\begin{aligned} U_1 &= [u_1, u_2, \ldots, u_r] \in \mathbb{C}^{m\times r}, \\ U_2 &= [u_{r+1}, u_{r+2}, \ldots, u_m] \in \mathbb{C}^{m\times (m-r)}, \\ V_1 &= [v_1, v_2, \ldots, v_r] \in \mathbb{C}^{n\times r}, \\ V_2 &= [v_{r+1}, v_{r+2}, \ldots, v_n] \in \mathbb{C}^{n\times (n-r)}, \end{aligned}$$

$$\|u_i\|_2 = 1, \quad i = 1, 2, \ldots, m; \quad \|v_j\|_2 = 1, 2, \ldots, n.$$

Then the four fundamental subspaces related to A are given by

$$\begin{aligned} \mathcal{R}(A) &= \text{span}(U_1), \\ \mathcal{R}(A)^\perp &= \text{span}(U_2), \\ \mathcal{N}(A) &= \text{span}(V_2), \\ \mathcal{N}(A)^\perp &= \text{span}(V_1). \end{aligned}$$

The accurate determination of the orthonormal bases of these subspaces relies on the proper determination of the matrix rank r.

In the presence of small singular values of A, the orthonormal bases can be very sensitive to perturbations of A. According to a theorem, proved by Stewart (1973a, Theorem 6.4), the perturbations of the orthonormal bases U_1 and V_1 due to a perturbation ΔA of the matrix A are of order $\|\Delta A\|/\sigma_r$. This shows that if σ_r is small, then the perturbations in the bases can be $1/\sigma_r$ times larger than the perturbation of A. Using the numerical rank k of A, the singular value σ_k is obtained sufficiently large and the orthonormal bases of the fundamental subspaces will be less sensitive to the perturbation ΔA.

Example 1.21. Consider a matrix A constructed in the same way as in Example 1.20, for which the diagonal elements of the matrix D are equal to

$$(1, 1, 10^{-5}, 0, 0).$$

Theoretically, the rank of A is equal to 3, but the singular value σ_3 of A is equal to $2.83 \cdot 10^{-6}$. This means that we may expect perturbations in the corresponding subspace which are approximately 10^6 larger than the norm of the perturbation in A. Assume that A is a subject of perturbations in the form

$$\Delta A = 10^{-12+j} \begin{bmatrix} 3 & -2 & 7 & 4 & -6 \\ -1 & 3 & -5 & -8 & 2 \\ 9 & -3 & 6 & 1 & -4 \\ 5 & -8 & -2 & -4 & -6 \\ -2 & 7 & -5 & 2 & 9 \\ -3 & -8 & -4 & -5 & -2 \\ 8 & 7 & 5 & 2 & 7 \\ 4 & -5 & -9 & -3 & 8 \end{bmatrix},$$

where j varies between 1 and 5.

In the last two columns of Table 1.6, we give the values of the maximum angle $\Theta_{\max}(\mathcal{R}(\widetilde{A}), \mathcal{R}(A))$ between the images of the perturbed $\widetilde{A} = A + \Delta A$ and unperturbed matrix A, determined for numerical rank equal to 3 and 2, respectively. The results confirm that the maximum angle between the perturbed and unperturbed images of A in the case of $\text{rank}(A) = 3$ is of order $10^6 \cdot \|\Delta A\|$, as expected. In the case of $\text{rank}(A) = 2$, the singular value σ_2 is equal to $2.88 \cdot 10^2$ and

Table 1.6. Angles between perturbed and unperturbed images of A.

j	$\|\Delta A\|_2$	$k = \text{rank}(A) = 3$ $\Theta_{\max}(\mathcal{R}(\widetilde{A}), \mathcal{R}(A))$ deg	$k = \text{rank}(A) = 2$ $\Theta_{\max}(\mathcal{R}(\widetilde{A}), \mathcal{R}(A))$ deg
1	$2.22 \cdot 10^{-10}$	$1.47 \cdot 10^{-3}$	$2.40 \cdot 10^{-13}$
2	$2.22 \cdot 10^{-9}$	$1.47 \cdot 10^{-2}$	$2.17 \cdot 10^{-12}$
3	$2.22 \cdot 10^{-8}$	$1.47 \cdot 10^{-1}$	$2.16 \cdot 10^{-11}$
4	$2.22 \cdot 10^{-7}$	$1.43 \cdot 10^{0}$	$2.16 \cdot 10^{-10}$
5	$2.22 \cdot 10^{-6}$	$1.25 \cdot 10^{1}$	$2.16 \cdot 10^{-9}$

Table 1.7. Angles between perturbed and unperturbed null spaces of A.

j	$\|\Delta A\|_2$	$k = \text{rank}(A) = 3$ $\Theta_{\max}(\mathcal{N}(A), \mathcal{N}(\widetilde{A}))$ deg	$k = \text{rank}(A) = 2$ $\Theta_{\max}(\mathcal{N}(A), \mathcal{N}(\widetilde{A}))$ deg
1	$2.22 \cdot 10^{-10}$	$8.37 \cdot 10^{-4}$	$2.18 \cdot 10^{-13}$
2	$2.22 \cdot 10^{-9}$	$8.36 \cdot 10^{-3}$	$1.80 \cdot 10^{-12}$
3	$2.22 \cdot 10^{-8}$	$8.31 \cdot 10^{-2}$	$1.80 \cdot 10^{-11}$
4	$2.22 \cdot 10^{-7}$	$7.78 \cdot 10^{-1}$	$1.80 \cdot 10^{-10}$
5	$2.22 \cdot 10^{-6}$	$4.16 \cdot 10^{0}$	$1.80 \cdot 10^{-9}$

the maximum angle between the perturbed and unperturbed images is about 10^3 times smaller than the size of the perturbation in A.

In Table 1.7, we give the perturbations of the null space of A for the same perturbations ΔA and values of the numerical rank k. As in the case of $\mathcal{R}(A)$, the sensitivity of the null space is much lower in the case of numerical rank equal to 2.

The earlier considerations justify the definition of the *numerical image* of A as

$$\mathcal{R}_k(A) = \text{span}(u_1, u_2, \ldots, u_k)$$

and the *numerical null space* of A as

$$\mathcal{N}_k(A) = \text{span}(v_{k+1}, v_{k+2}, \ldots, v_n),$$

where k is the numerical rank of A. Since $u_i^H A = \sigma_i v_i^H$, $i = 1, 2, \ldots, k$, we have that

$$\|u_i^H A\|_2 = \sigma_i \|v_i\|_2 \geq \sigma_k.$$

This inequality shows that any vector u of unit size in $\mathcal{R}_k(A)$ has a large component in the image of A. Similarly, since $Av_{k+i} = \sigma_{k+i}u_{k+i},; i = 1, 2, \ldots, n-k$, it follows that

$$\|Av_{k+i}\|_2 = \sigma_{k+i}\|u_{k+i}\|_2 \leq \sigma_{k+1}.$$

Hence, any vector $v \in \mathcal{N}_k$ of unit size is mapped by the matrix A in a vector whose norm is less than or equal to σ_{k+1}.

In a similar way, the numerical image and the numerical null space of A^H are defined as

$$\mathcal{R}_k(A^H) = \mathcal{N}_k(A)^{\perp} = \operatorname{span}(v_1, v_2, \ldots, v_k),$$
$$\mathcal{N}_k(A^H) = \mathcal{R}_k(A)^{\perp} = \operatorname{span}(u_{k+1}, u_{k+2}, \ldots, u_m).$$

1.6.3 *Notes and References*

The notion of numerical rank of a matrix is discussed in detail in Golub and Van Loan (2013), Section 5.4, Stewart (1998), Chapter 5, Demmel (1997), Section 3.5, and Hansen (1998), Chapter 3, see also Golub *et al.* (1976) and Stewart (1984, 1993). In this book, we use the definition of numerical rank implemented in Kågström and Ruhe (1980a, 1980b), Demmel and Kågström (1993a, 1993b) and Kågström and Wiberg (2000).

The implementation of the QR decomposition for rank determination (the so-called *rank revealing QR decomposition*) presents a cheaper alternative to the using of the singular value decomposition in finding the numerical rank of a matrix. It is studied by several authors, see Chan (1987), Chan and Hansen (1992), Hong and Pan (1992), Chandrasekaran and Ipsen (1994), Gu and Eisenstat (1996), Bischof and Quintana-Ortí (1998b, 1998a), Hansen (1998), Chapter 2, Drmač and Bujanović (2008), Korkmaz *et al.* (2022) and the references therein.

The numerical image and numerical null space are defined in Hansen (1998), Chapter 3.

Chapter 2

The Eigenvalue Problem

In this chapter, we present the basic facts relevant to the eigenvalue problem which are used in the remainder of this book. We consider the properties of the eigenvalues and eigenspaces, the emphasis being put on Schur, Jordan and Weyr canonical forms. Several facts are given without proofs which can be found in the sources referred to at the end of each section.

2.1 Eigenvalues and Eigenvectors

2.1.1 *Basic Definitions*

Definition 2.1. Let $A \in \mathbb{C}^{n\times n}$. The pair (λ, x), where $\lambda \in \mathbb{C}$ and $x \in \mathbb{C}^n$ is called *eigenpair* or *right eigenpair* of A if

1. $x \neq 0$,
2. $Ax = \lambda x$.

The scalar λ is called *eigenvalue* and the vector $x = [\xi_1, \xi_2, \ldots, \xi_n]^T$ is called *eigenvector* or *right eigenvector*. The eigenvector has the special property that it preserves its direction in the space $\mathbb{C}^n$ under multiplication by the matrix A (Figure 2.1). The eigenvalue λ shows how much the magnitude of the eigenvector x changes when it is multiplied by A. Note that the eigenvalue and the corresponding eigenvector may be complex even if the matrix A is real.

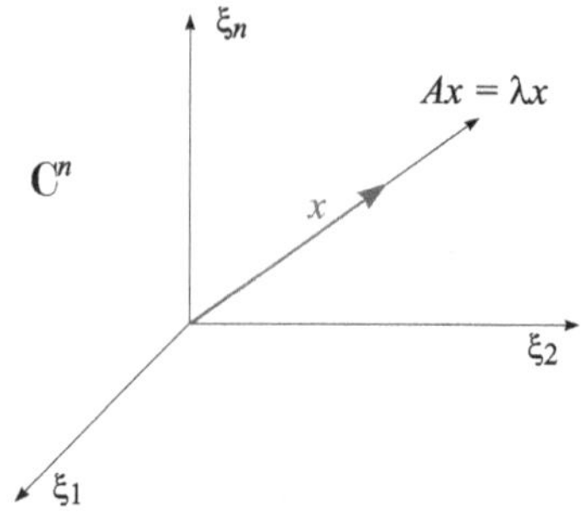

Fig. 2.1. Geometric interpretation of an eigenpair.

The pair (λ, y) is called *left eigenpair*, if $y \neq 0$ and $y^H A = \lambda y^H$. The vector y is called *left eigenvector*.

The *eigenvalue problem* for the square matrix A consists in determining the eigenvalues and eigenvectors of A. We shall also call this problem *ordinary eigenvalue problem* to distinguish it from the *generalized eigenvalue problem* for matrices A and B, considered latter on in Section 2.3.

If (λ, x) is an eigenpair of A, then $\lambda x - Ax = 0$ or equivalently

$$(\lambda I_n - A)x = 0.$$

Since $x \neq 0$, the matrix $\lambda I_n - A$ should be singular. Conversely, if $\lambda I_n - A$ is singular, then it has a zero vector x (i.e., a vector x with at least one nonzero entry that satisfies $(\lambda I_n - A)x = 0$) and (λ, x) is an eigenpair of A. Hence, the eigenvalues of A are those scalars for which the matrix $\lambda I_n - A$ is singular. Since a given matrix is singular if and only if its determinant is equal to zero, the eigenvalues of A are the scalars for which

$$p(\lambda) := \det(\lambda I_n - A) = 0. \tag{2.1}$$

The function

$$p(\lambda) = \lambda^n + p_1\lambda^{n-1} + \cdots + p_{n-1}\lambda + p_n$$

is a polynomial in λ of nth degree which is called *characteristic polynomial* of A. The numbers $p_1, p_2, \ldots, p_n$ are called *coefficients of the characteristic polynomial*. Hence, the eigenvalues of A are zeros of the characteristic polynomial or *roots* of the *characteristic equation* (2.1).

As the Cayley–Hamilton theorem states, each matrix satisfies its own characteristic equation, i.e.,

$$p(A) = A^n + p_1 A^{n-1} + \cdots + p_{n-1} A + p_n I_n = 0.$$

According to the fundamental theorem of algebra (Horn and Johnson, 2013, Appendix C), each polynomial of nth degree has exactly n zeros, which implies that each nth order matrix has n eigenvalues $\lambda_1,\ \lambda_2, \ldots, \lambda_n$. The set of the eigenvalues of A is called *spectrum* of A and is denoted by

$$\mathrm{spect}(A) := \{\lambda_1,\ \lambda_2, \ldots, \lambda_n\}.$$

The sum of the n eigenvalues of A is equal to the trace of A and the product of these eigenvalues is equal to the determinant of A.

The set of eigenvalues and corresponding vectors of A sometimes is called *eigensystem* of A.

Example 2.2. Given is the matrix

$$A = \begin{bmatrix} -5 & -4 & 6 \\ -2 & -5 & 4 \\ -3 & -4 & 4 \end{bmatrix},$$

whose characteristic equation is

$$\lambda^3 + 6\lambda^2 + 11\lambda + 6 = 0.$$

The roots of this equation, i.e., the eigenvalues of A, are $\lambda_1 = -1$, $\lambda_2 = -2$ and $\lambda_3 = -3$. The right and left eigenvectors, corresponding to these eigenvalues, are

$$\lambda_1 = -1, \quad x_1 = \begin{bmatrix} -2 \\ -1 \\ -2 \end{bmatrix}, \quad y_1^H = [1 \ \ 1 \ -2],$$

$$\lambda_2 = -2, \quad x_2 = \begin{bmatrix} -2 \\ 0 \\ -1 \end{bmatrix}, \quad y_2^H = [-1 \ \ 0 \ 1],$$

$$\lambda_3 = -3, \quad x_3 = \begin{bmatrix} 1 \\ 1 \\ 1 \end{bmatrix}, \quad y_3^H = [1 \ \ 2 \ -2].$$

The following comments can be made in connection with the example considered.

It is not necessary for the eigenvalues to be distinct. It is possible to have one or several eigenvalues which are multiple roots of the characteristic equation, and in such a case, they are counted conformingly with their multiplicity. In the general case, the characteristic polynomial is represented as

$$p(\lambda) = (\lambda - \lambda_1)^{\nu_1}(\lambda - \lambda_2)^{\nu_2} \dots (\lambda - \lambda_p)^{\nu_p},$$

where the numbers $\lambda_1,\ \lambda_2, \dots \lambda_p$ are distinct and

$$\nu_1 + \nu_2 + \cdots \nu_p = n.$$

We say that the eigenvalue λ_i has *algebraic multiplicity* ν_i. If the algebraic multiplicity is equal to one, then such eigenvalue is called *simple.*

If a real matrix has complex eigenvalues, they participate in the matrix spectrum in conjugate pairs. In such a case, the complex eigenvalue can be represented as $\lambda = \alpha + j_o\beta$ and the eigenvector as $x = r + j_o w$. We have that

$$A(r + j_o w) = (\alpha + j_o\beta)(r + j_o w)$$

or

$$Ar = \alpha r - \beta w,\ Aw = \alpha w + \beta r.$$

This result can be represented as

$$A[r, w] = [r, w]\begin{bmatrix} \alpha & \beta \\ -\beta & \alpha \end{bmatrix}.$$

Note that the real and imaginary parts of the complex eigenvector are linearly independent.

2.1.2 *Diagonal Form of a Matrix*

Definition 2.3. Two matrices $A \in \mathbb{C}^{n\times n}$ and $B \in \mathbb{C}^{n\times n}$ are called *similar* if there exists a nonsingular matrix $S \in \mathbb{C}^{n\times n}$ such that

$$B = S^{-1}AS. \tag{2.2}$$

The transformation (2.2) is called *similarity transformation.*

The following theorem characterizes the eigenvalues and eigenvectors of similar matrices.

Theorem 2.4. *Let $A \in \mathbb{C}^{n\times n}$ and $B = S^{-1}AS$ is similar to A. Then A and B have the same characteristic polynomials and hence the same eigenvalues with equal multiplicities. If x is an eigenvector of A, then $S^{-1}x$ is an eigenvector of B for the same eigenvalue λ.*

Let the matrix A have n distinct eigenvalues $\lambda_1, \lambda_2, \ldots, \lambda_n$. In this case, A has exactly n linearly independent eigenvectors $x_1, x_2, \ldots, x_n$ such that

$$\begin{aligned} Ax_1 &= \lambda_1 x_1, \\ Ax_2 &= \lambda_2 x_2, \\ \vdots\ &\quad \vdots \\ Ax_n &= \lambda_n x_n. \end{aligned} \tag{2.3}$$

Equations (2.3) can be represented in the form

$$A[x_1,\ x_2, \ldots, x_n] = [x_1,\ x_2, \ldots, x_n] \begin{bmatrix} \lambda_1 & 0 & \ldots & 0 \\ 0 & \lambda_2 & \ldots & 0 \\ \vdots & \vdots & \ddots & \vdots \\ 0 & 0 & \ldots & \lambda_n \end{bmatrix}. \tag{2.4}$$

Defining the eigenvector matrix as

$$X = [x_1,\ x_2, \ldots, x_n],$$

equation (2.4) is written as

$$AX = X\Lambda$$

or

$$X^{-1}AX = \Lambda, \tag{2.5}$$

where

$$\Lambda = \begin{bmatrix} \lambda_1 & 0 & \ldots & 0 \\ 0 & \lambda_2 & \ldots & 0 \\ \vdots & \vdots & \ddots & \vdots \\ 0 & 0 & \ldots & \lambda_n \end{bmatrix}. \tag{2.6}$$

The matrix Λ is referred to as the *diagonal form* of A. Equation (2.5) shows that in case of distinct eigenvalues, the square matrix is reduced into diagonal form, using a nonsingular similarity transformation with the eigenvector matrix. It is necessary to point out that under some conditions it is also possible to reduce into diagonal form matrices with multiple eigenvalues. More specifically, if the matrix A has n linearly independent eigenvectors, then independently on the eigenvalue multiplicity, A can be reduced into diagonal form. In such case, we say that A has *simple structure* and the corresponding multiple eigenvalues are called *semisimple*. Hence the matrices with simple structure are diagonalizable matrices.

Let

$$Y^H = \begin{bmatrix} y_1^H \\ y_2^H \\ \vdots \\ y_n^H \end{bmatrix}$$

be a matrix composed by the left eigenvectors of A. If the eigenvalues are distinct, then the left eigenvectors are linearly independent and the matrix Y is nonsingular. It may be proved that in this case it is fulfilled that

$$Y^H A (Y^H)^{-1} = \Lambda. \tag{2.7}$$

The comparison of equations (2.5) and (2.7) shows that for a matrix with distinct eigenvalues, one can take the matrix $Y^H = X^{-1}$ as a matrix of left eigenvectors. (Note that this choice is not unique.)

Example 2.5. For the matrix A of Example 2.2, we have that

$$X = \begin{bmatrix} -2 & -2 & 1 \\ -1 & 0 & 1 \\ -2 & -1 & 1 \end{bmatrix}$$

and

$$\Lambda = X^{-1}AX = \begin{bmatrix} -1 & 0 & 0 \\ 0 & -2 & 0 \\ 0 & 0 & -3 \end{bmatrix}.$$

In the program system MATLAB®, the eigenvalues and eigenvectors are evaluated by the function `eig`. The use of the function in the form `[V,D] = eig(A)` produces a diagonal matrix D which contains the eigenvalues and a matrix V whose columns are the corresponding eigenvectors.

According to the considerations in the previous section, if the matrix A is real and has three real and two complex conjugate eigenvalues, it can be reduced into *quasi-diagonal form*

$$\Lambda = \begin{bmatrix} \lambda_1 & 0 & 0 & 0 & 0 \\ 0 & \lambda_2 & 0 & 0 & 0 \\ 0 & 0 & \lambda_3 & 0 & 0 \\ 0 & 0 & 0 & \alpha & \beta \\ 0 & 0 & 0 & -\beta & \alpha \end{bmatrix}.$$

In this case, the last two columns of the matrix X are the real and the imaginary part of the complex eigenvector corresponding to the eigenvalue $\lambda = \alpha + j_o\beta$.

2.1.3 *Eigenvector Normalization*

If (λ, x) is an eigenpair and $s \neq 0$, then (λ, sx) is also an eigenpair, i.e., the eigenvector is unique up to a scalar factor. To remove the trivial nonuniqueness of the eigenvectors, we use *normalization* of x and y. In the following, we consider two of the most frequently used methods for normalization of right and left eigenvectors.

In the first normalization method, the eigenvector x_i is multiplied by the nonzero scalar $s_i = 1/\|x_i\|_2$ and a similar operation is done on the left eigenvectors. In this way, we achieve

$$\|x_i\|_2 = 1, \quad \|y_i\|_2 = 1. \tag{2.8}$$

This way of normalization is used in the program package LAPACK (Anderson *et al.*, 1999) as well as in MATLAB®. In this normalization, the requirement $Y^H = X^{-1}$ is not satisfied.

In the second normalization method, one assumes that the matrix of the left eigenvectors is determined from $Y^H = X^{-1}$ such that $Y^H X = XY^H = I_n$. This case corresponds to a specific normalization of the eigenvectors in which it is fulfilled that

$$y_i^H x_i = 1.$$

This normalization simplifies several expressions involving the left and right eigenvectors, but in this case, it is not possible to guarantee that $\|y_i^H\|_2 = 1$ even when $\|x\|_2 = 1$.

2.1.4 *Eigenspaces*

The equation $Ax = \lambda x$ shows that the subspace of dimension one spanned by the eigenvector x remains invariant under multiplication by A, i.e., the vector Ax has the same direction in $\mathbb{C}^n$ as x. This concept can be generalized with respect to subspaces of dimension greater than one.

Definition 2.6. Let $A \in \mathbb{C}^{n\times n}$ and let $\mathcal{X}$ be a subspace of $\mathbb{C}^n$. Then $\mathcal{X}$ is *invariant subspace* of A if (Figure 2.2)

$$Ax := \{Ax : x \in \mathcal{X}\} \subset \mathcal{X}.$$

Apart from the notion of invariant subspace, we will also use the term *eigenspace* which is also suitable for the case of the generalized eigenvalue problem (Section 2.3).

Clearly, each set of right eigenvectors span an eigenspace. Eigenvectors which are not well defined from numerically point of view may be collected in a well-defined eigenspace. This concept is important in determining the canonical forms of matrices.

Let the columns of $X \in \mathbb{C}^{n\times k}$ form a basis of the eigenspace $\mathcal{X}$ of A. If X has linearly independent columns, then there is a unique

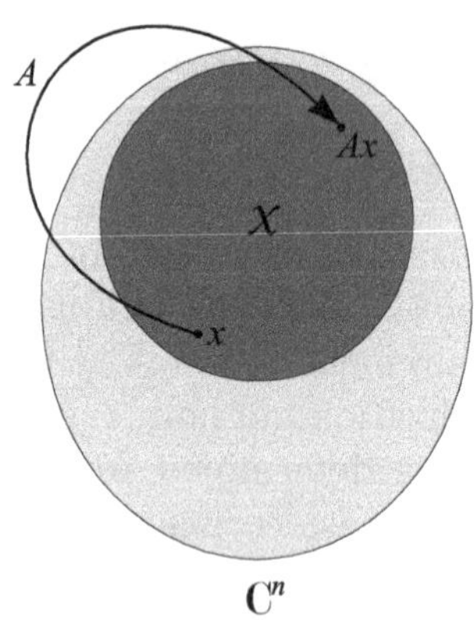

Fig. 2.2. Invariant subspace.

matrix L, such that

$$AX = XL.$$

If (λ, u) is an eigenpair of L, i.e.,

$$Lu = \lambda u,$$

then (λ, Xu) is an eigenpair of A since

$$AXu = XLu = \lambda Xu \subset \mathcal{X}.$$

The matrix L is called *the restriction of A to the eigenspace $\mathcal{X}$ with respect to the basis* X and its eigenvalues are eigenvalues of A.

If $X \in \mathbb{C}^{n\times k}$ has linearly independent columns and $AX = XL$, then we say that (L, X) is a *right eigenpair* of A. If X has orthonormal columns, we say that the eigenpair (L, X) is *orthonormal*.

If (L, X) is a right eigenpair, then $\mathcal{X} = \text{span}(X)$ is a *right eigenspace* of A.

If $Y \in \mathbb{C}^{n\times k}$ has linearly independent columns and $Y^H A = LY^H$, then we say that (L, Y) is a *left eigenpair* of A.

If (L, Y) is a left eigenpair, then $\mathcal{Y} = \text{span}(Y)$ is a *left invariant subspace* or *left eigenspace* in the sense that

$$A^H \mathcal{Y} \subset \mathcal{Y}.$$

In Figure 2.3, we show one-dimensional eigenspaces in $\mathbb{R}^3$ spanned by the right and left eigenvectors of the matrix A given in Example 2.2.

2.1.5 *Schur Decomposition*

The transformations with unitary and orthogonal matrices play an important role in matrix computations. This is due to two reasons. From one hand, the unitary matrices are inverted easily (the inversion is equivalent to Hermitian conjugation in the complex case and transposition in the real case). On the other hand, the spectral norm of the unitary matrices is equal to one and they do not increase the magnitude of the rounding errors during the transformations. For these reasons, the following is interesting: How far can we simplify a square matrix using unitary similarity transformations? The answer of this question is given by the Schur theorem presented in this section.

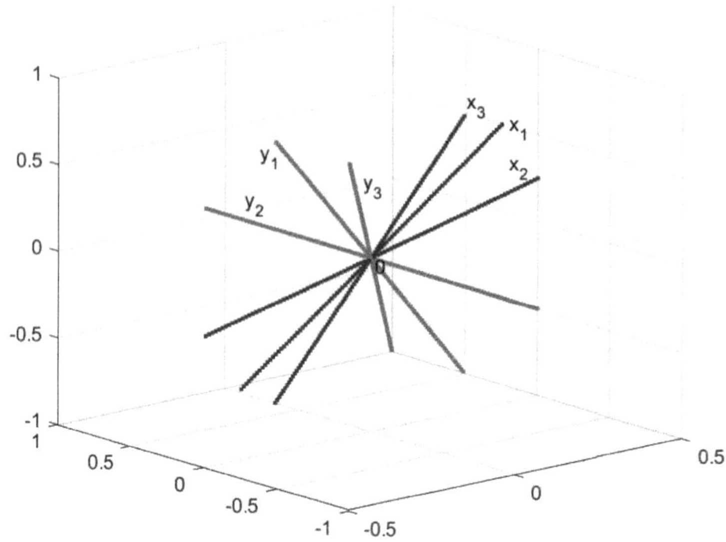

Fig. 2.3. Right and left one-dimensional eigenspaces of a 3 × 3 matrix.

Let the matrix $A \in \mathbb{C}^{n\times n}$ have eigenvalues $\lambda_1, \lambda_2, \ldots, \lambda_n$ and let $x_1 \in \mathbb{C}^n$ be an eigenvector corresponding to λ_1. Compute a Householder transformation $H_1 \in \mathbb{C}^{n\times n}$ (Section 1.5.3, Algorithm 1.6) such that

$$H_1 x_1 = g_1 e_1, \quad g_1 \neq 0.$$

Remember that H_1 is unitary and Hermitian matrix ($H_1^H H_1 = I_n, H_1^H = H_1$). From

$$Ax_1 = \lambda_1 x_1,$$

it follows that

$$U_1^H A U_1 U_1^H x_1 = \lambda_1 U_1^H x_1, \quad U_1 = H_1^H = H_1$$

or

$$U_1^H A U_1 \begin{bmatrix} g_1 \\ 0 \\ \vdots \\ 0 \end{bmatrix} = \lambda_1 \begin{bmatrix} g_1 \\ 0 \\ \vdots \\ 0 \end{bmatrix}.$$

This equation shows that the transformed matrix has the form

$$U_1^H A U_1 = \begin{bmatrix} \lambda_1 & \times & \cdots & \times \\ 0 & \times & \cdots & \times \\ \vdots & \vdots & \ddots & \vdots \\ 0 & \times & \times & \times \end{bmatrix} = \begin{bmatrix} \lambda_1 & t_{12}^H \\ 0 & A_2 \end{bmatrix},$$

where the matrix $A_2 \in \mathbb{C}^{(n-1)\times(n-1)}$ has eigenvalues $\lambda_2, \lambda_3, \ldots, \lambda_n$. Let A_2 have eigenvector x_2 corresponding to λ_2. Then we may compute a Householder transformation H_2 such that

$$H_2 x_2 = \begin{bmatrix} g_2 \\ 0 \\ \vdots \\ 0 \end{bmatrix}, \quad g_2 \neq 0.$$

We have that

$$H_2 A_2 H_2^H H_2 x_2 = \lambda_2 H_2 x_2$$

and

$$H_2 A_2 H_2^H = \begin{bmatrix} \lambda_2 & \times & \cdots & \times \\ 0 & \times & \cdots & \times \\ \vdots & \vdots & \ddots & \vdots \\ 0 & \times & \times & \times \end{bmatrix}.$$

If we set

$$U_2 = \begin{bmatrix} 1 & 0 \\ 0 & H_2 \end{bmatrix},$$

then we obtain

$$U_2^H U_1^H A U_1 U_2 = \begin{bmatrix} \lambda_1 & t_{12}^H H_2^H \\ 0 & H_2 A_2 H_2^H \end{bmatrix} = \begin{bmatrix} \lambda_1 & \times & \times & \cdots & \times \\ 0 & \lambda_2 & \times & \cdots & \times \\ 0 & 0 & \times & \cdots & \times \\ \vdots & \vdots & \vdots & \ddots & \vdots \\ 0 & 0 & \times & \cdots & \times \end{bmatrix}.$$

Continuing this procedure by induction (see Figure 2.4) and denoting $U = U_1 U_2 \ldots U_n$, we obtain the following result.

$$U_4{}^H U_3{}^H U_2{}^H U_1{}^H A U_1 U_2 U_3 U_4 =$$

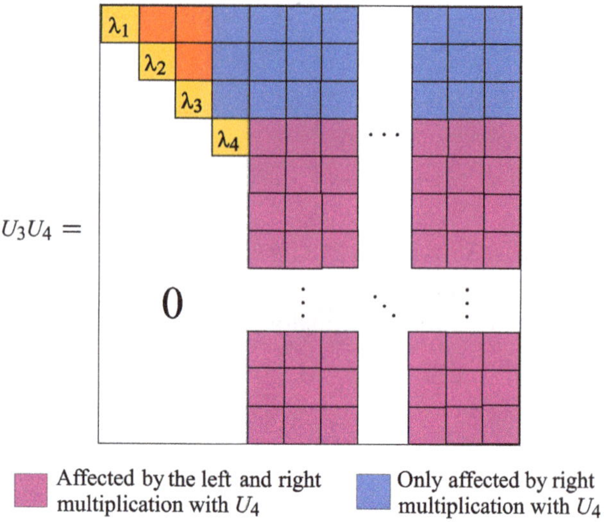

Fig. 2.4. Reduction into Schur form after the 4th step.

Theorem 2.7 (Schur decomposition). *For a given* $A \in \mathbb{C}^{n\times n}$, *there exists a unitary matrix* $U \in \mathbb{C}^{n\times n}$ *such that*

$$U^H AU = T := D + N, \tag{2.9}$$

where $T \in \mathbb{C}^{n\times n}$ *is upper triangular,* N *is strictly upper triangular,*

$$D = \text{diag}(\lambda_1,\ \lambda_2, \ldots, \lambda_n)$$

and $\lambda_i = t_{ii}$, $i = 1,\ 2, \ldots, n$, *are the eigenvalues of* A. *By appropriate choice of the matrix* U, *the eigenvalues can be ordered in an arbitrary way in* D.

In this way, in the Schur decomposition, the eigenvalues are located on the diagonal of T. (Note that some of these eigenvalues can be equal.) The matrix T is called *Schur form* of A and the columns of the matrix

$$U = [u_1,\ u_2, \ldots, u_n]$$

are called *Schur vectors*.

Denoting

$$T_k = \begin{bmatrix} t_{11} & t_{12} & \dots & t_{1k} \\ & t_{22} & \dots & t_{2k} \\ & & \ddots & \vdots \\ & & & t_{kk} \end{bmatrix}, \quad k = 1, 2, \dots, n,$$

we have that

$$\begin{aligned} Au_1 &= T_1 u_1, \\ A[u_1, u_2] &= T_2 [u_1, u_2], \\ \vdots \quad & \quad \vdots \\ A[u_1, u_2, \dots, u_n] &= T_n [u_1, u_2, \dots, u_n]. \end{aligned} \tag{2.10}$$

We see that the nested sequence of subspaces

$$\mathcal{X}_k = \text{span}(u_1, u_2, \dots, u_k), \quad k = 1, 2, \dots, n,$$

consists of invariant subspaces, i.e., $z \in \mathcal{X}_k$ implies that $Az \in \mathcal{X}_k$. Hence, the first k Schur vectors constitute an orthonormal basis of the invariant subspace $\mathcal{X}_k$. Of the Schur vectors, only the first one u_1 is an eigenvector. Nevertheless, since the Schur basis is orthonormal, it is preferred against the eigenvector basis in the applications. Clearly, the matrix T_k is the restriction of A to the eigenspace $\mathcal{X}_k$ with respect to the basis $[u_1, u_2, \dots, u_k]$.

The Schur form is not unique. In fact, if $U^H AU$ is upper triangular matrix, then $V^H AV$, $V = US$ is also upper triangular for each

$$S = \text{diag}(e^{j_o \phi_1}, e^{j_o \phi_2}, \dots, e^{j_o \phi_n}), \quad |e^{j_o \phi_i}| = 1.$$

Moreover, the different ordering of the eigenvalues on the diagonal of the Schur form leads to a different matrix U. The multiple eigenvalues are also a reason for nonuniqueness of the decomposition.

If the triangular matrix T is diagonal, then A is called *normal matrix*.

According to Theorem 2.7, we have the following result.

Theorem 2.8. *Let $A \in \mathbb{C}^{n\times n}$ and $U = [U_1,\ U_2] \in \mathbb{C}^{n\times n}$ be unitary where the columns of U_1 span a right eigenspace $\mathcal{X}$ of A with dimension $r < n$. Then*

$$\begin{bmatrix} U_1^H \\ U_2^H \end{bmatrix} A\,[U_1, U_2] = \begin{bmatrix} T_{11} & T_{12} \\ 0 & T_{22} \end{bmatrix}, \qquad (2.11)$$

From (2.11), we have $AU_1 = U_1T_{11}$, which shows that $\mathcal{X} = \mathcal{R}(U_1)$ is a right eigenspace of A corresponding to the eigenvalues of T_{11}. Since

$$AU_2 = U_1T_{11} + U_2T_{22},$$

the columns of U_2 do not span a right eigenspace. However, from $U_2^H A = T_{22}U_2^H$, it follows that $\mathcal{Y} = \mathcal{R}(U_2)$ is a left eigenspace of A and the rest eigenvalues of A are equal to those of $T_{22} = U_2^H AU_2$.

If the spectra of T_{11} and T_{22} are disjoint, then the eigenspace $\mathcal{X}$ is called *simple.* If $\mathcal{X}$ is a simple eigenspace of A and $\mathcal{Y}$ is the orthogonal complement of $\mathcal{X}$, then $\mathcal{Y}$ is a left eigenspace of A corresponding to the eigenvalues that are not associated with $\mathcal{X}$.

As already mentioned, if the matrix A is real, its complex eigenvalues (if any) appear in conjugate pairs. In such a case, it is possible to reduce A into real *quasi-triangular form* which has 1×1 and 2×2 diagonal blocks. This decomposition is frequently called *real Schur decomposition.*

Theorem 2.9 (Real Schur decomposition). *For a given matrix $A \in \mathbb{R}^{n\times n}$, there exists a real orthogonal matrix $Q \in \mathbb{R}^{n\times n}$ such that*

$$Q^T AQ = T = D + N, \qquad (2.12)$$

where T is a real block upper triangular matrix, D is a block-diagonal with 1×1 and 2×2 diagonal blocks and where all 2×2 blocks have complex conjugate eigenvalues.

Note that in case of real eigenvalues, the real Schur form of a real matrix is upper triangular.

In MATLAB®, the complex Schur decomposition $A = UTU^H$ of the matrix A, where $U^HU = I_n$, is computed by the function `[U,T] = schur(A,'complex')`. If the matrix A is real, the function `[U,T] = schur(A)` computes the real Schur form. The command

`[V,S] = rsfcsf(U,T)` transforms the real Schur form into complex form.

Example 2.10. Let the matrix

$$A = \begin{bmatrix} -5 & -6 & 4 & 8 \\ 1 & -1 & -2 & -2 \\ 3 & 4 & -5 & -6 \\ -4 & -6 & 4 & 7 \end{bmatrix}.$$

By using the function **schur**, we obtain that $A = QTQ^T$, where (up to four decimal digits)

$$Q = \begin{bmatrix} -0.6667 & 0.1169 & 0.2046 & 0.7071 \\ -0.0000 & 0.8682 & -0.4961 & 0.0000 \\ 0.3333 & 0.4678 & 0.8186 & -0.0000 \\ -0.6667 & 0.1169 & 0.2046 & -0.7071 \end{bmatrix}$$

and

$$T = \left[\begin{array}{c|cc|c} 1.0000 & 4.2444 & -11.3840 & 13.4350 \\ \hline 0 & -2.0000 & -2.6928 & 2.8341 \\ 0 & 0.3714 & -2.0000 & 0.6840 \\ \hline 0 & 0 & 0 & -1.0000 \end{array}\right].$$

From the matrix T, we determine that the eigenvalues of A are

$$\lambda_1 = 1, \quad \lambda_{2,3} = -2 \pm 1, \quad \lambda_4 = -1.$$

The eigenspace which corresponds to the eigenvalues λ_1, λ_2 and λ_3 is spanned by the first three columns of the matrix Q.

The matrix A is *nilpotent* if $A^k = 0$ for some $k \geq 1$. In the general case, if T is an upper triangular matrix of nth order with zeros on the diagonal, then the diagonal and the first $k-1$ superdiagonals of T^k are zero. Hence, $T^n = 0$.

Using the Schur form of a nilpotent matrix, it is easy to prove the following result.

Theorem 2.11. *The matrix A of nth order is nilpotent if and only if all its eigenvalues are zero. In such case,*

$$A^n = 0.$$

The numerical reduction of a square matrix into Schur form is discussed latter on in Section 4.1.

2.1.6 *Block-Diagonal Form*

Let (T_{11}, U_1) be a simple orthonormal eigenpair of A and let $[U_1, U_2]$ be a unitary matrix so that

$$\begin{bmatrix} U_1^H \\ U_2^H \end{bmatrix} A\,[U_1, U_2] = \begin{bmatrix} T_{11} & T_{12} \\ 0 & T_{22} \end{bmatrix}, \tag{2.13}$$

where $T_{11} \in \mathbb{C}^{r\times r}, T_{22} \in \mathbb{C}^{(n-r)\times(n-r)}$. Construct a similarity transformation matrix

$$M = \begin{bmatrix} I_r & R \\ 0 & I_{n-r} \end{bmatrix}$$

whose inverse is

$$M^{-1} = \begin{bmatrix} I_r & -R \\ 0 & I_{n-r} \end{bmatrix}.$$

Then

$$M^{-1}TM = \begin{bmatrix} T_{11} & T_{11}R - RT_{22} + T_{12} \\ 0 & T_{22} \end{bmatrix}.$$

We see that if the matrix equation

$$T_{11}R - RT_{22} + T_{12} = 0 \tag{2.14}$$

has a solution with respect to the matrix $R \in \mathbb{C}^{r\times(n-r)}$, then the Schur form of A can be reduced into the *block-diagonal form*

$$\begin{bmatrix} Y_1^H \\ Y_2^H \end{bmatrix} A\,[X_1, X_2] = \begin{bmatrix} T_{11} & 0 \\ 0 & T_{22} \end{bmatrix}. \tag{2.15}$$

Since

$$[X_1, X_2] = [U_1, U_2] \begin{bmatrix} I_r & R \\ 0 & I_{n-r} \end{bmatrix}$$

and

$$\begin{bmatrix} Y_1^H \\ Y_2^H \end{bmatrix} = \begin{bmatrix} I_r & -R \\ 0 & I_{n-r} \end{bmatrix} \begin{bmatrix} U_1^H \\ U_2^H \end{bmatrix},$$

we have that

$$X_1 = U_1, \quad X_2 = U_2 + U_1 R$$

and

$$Y_1 = U_1 - U_2 R^H, \quad Y_2 = U_2.$$

Hence $\mathcal{R}(X_1) = \mathcal{R}(U_1)$ is a simple right eigenspace corresponding to the eigenvalues of T_{11}. If we set

$$X = [X_1, X_2] \quad \text{and} \quad Y = [Y_1, Y_2],$$

then

$$Y^H X - I_n.$$

Equation (2.15) shows that a matrix can be reduced into a block-diagonal form provided that we know the bases X and Y of the right and left eigenspaces. Note that, generally, X and Y are nonunitary matrices. Since $Y^H = X^{-1}$ and $X = (Y^H)^{-1}$, from (2.15) we have that

$$A\,[X_1 X_2] = [X_1 X_2] \begin{bmatrix} T_{11} & 0 \\ 0 & T_{22} \end{bmatrix}$$

and

$$\begin{bmatrix} Y_1^H \\ Y_2^H \end{bmatrix} A = \begin{bmatrix} T_{11} & 0 \\ 0 & T_{22} \end{bmatrix} \begin{bmatrix} Y_1^H \\ Y_2^H \end{bmatrix}.$$

Hence, $AX_2 = X_2 T_{22}$ and it follows that (T_{22}, X_2) is a right eigenpair of A corresponding to the eigenvalues of T_{22}. Also, $Y_1^H A = T_{11} Y_1^H$ so that (T_{11}, Y_1) is a left eigenpair of A and $\mathcal{R}(Y_1)$ is the left eigenspace corresponding to $\mathcal{R}(X_1)$. We have that

$$\mathcal{R}(X_1) \oplus \mathcal{R}(X_2) = \mathbb{C}^n, \tag{2.16}$$

i.e., the space $\mathbb{C}^n$ is decomposed into a direct sum of invariant subspaces. By analogy, the matrix $T = \text{diag}(T_{11}, T_{22})$ is called *direct sum* of the matrices T_{11} and T_{22}. According to equation (2.16), the eigenspace $\mathcal{R}(X_2)$ is complementary to the eigenspace $\mathcal{R}(X_1)$ in the space $\mathbb{C}^n$ and it is called *complementary right eigenspace*. Similarly,

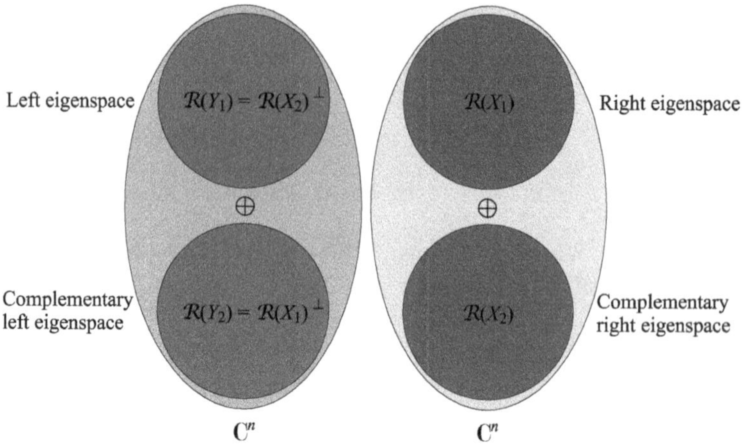

Fig. 2.5. Left and right eigenspaces.

the eigenspace $\mathcal{R}(Y_2)$ is complementary to the left eigenspace $\mathcal{R}(Y_1)$ and it is called *complementary left eigenspace.* Note that both complementary eigenspaces correspond to the eigenvalues of T_{22}. Since

$$Y_1^H X_2 = 0 \quad \text{and} \quad Y_2^H X_1 = 0,$$

it follows that the left eigenspace $\mathcal{R}(Y_1)$ is the orthogonal complement of $\mathcal{R}(X_2)$ and the complementary left eigenspace $\mathcal{R}(Y_2)$ is the orthogonal complement of $\mathcal{R}(X_1)$.

The relations between left and right eigenspaces of a matrix are shown schematically in Figure 2.5.

Equation (2.14) is called *Sylvester matrix equation* and is involved in the analysis and solution of several eigenvalue problems. Expanding the matrices R and T_{12} columnwise as the vectors $\text{vec}(R) \in \mathbb{C}^{r\cdot(n-r)}$ and $\text{vec}(T_{12}) \in \mathbb{C}^{r\cdot(n-r)}$, respectively, (2.14) is represented as the system of linear equations (see Section A.3.2 of the Appendix)

$$(I_{n-r} \otimes T_{11} - T_{22}^T \otimes I_r)\text{vec}(R) = -\text{vec}(T_{12}). \tag{2.17}$$

Using the definition of the Kronecker product (Section A.1 of the Appendix), it is possible to prove that the eigenvalues of the matrix

$$\Omega = I_{n_2} \otimes T_{11} - T_{22}^T \otimes I_{n_1}$$

are of the form

$$\lambda_i - \mu_j, \quad i = 1,\ 2, \ldots, r, \quad j = 1,\ 2, \ldots, n - r,$$

where λ_i belongs to the spectrum of T_{11} and μ_j belongs to the spectrum of T_{22}. Hence the matrix Ω will have all eigenvalues different from zero if and only if T_{11} and T_{22} have no common eigenvalues, i.e., their spectra are disjoint. In such case, Ω is a nonsingular matrix and (2.17) (and consequently (2.14)) has an unique solution. In this way, we come to the conclusion that the Schur form of A can be reduced into block-diagonal form if and only if the corresponding diagonal blocks have no eigenvalues in common. Note that

$$\mathrm{cond}_F(M) = \|M\|_F \|M^{-1}\|_F = n + \|X\|_F^2.$$

The numerical solution of the Sylvester equation (2.14) is described in Section 4.2.

Consider now the general case when the matrix A has multiple eigenvalues and its characteristic polynomial is represented in the form

$$p(\lambda) = (\lambda - \lambda_1)^{\nu_1} (\lambda - \lambda_2)^{\nu_2} \ldots (\lambda - \lambda_p)^{\nu_p},$$

where ν_i is the algebraic multiplicity of λ_i, $i = 1,\ 2, \ldots, p$. Using a sequence of nonsingular similarity transformations involving the solution of the Sylvester equation as shown earlier, it is possible to prove the following theorem.

Theorem 2.12 (Block-diagonal decomposition). *Assume that $A \in \mathbb{C}^{n\times n}$ has p distinct eigenvalues $\lambda_1,\ \lambda_2, \ldots,\ \lambda_p$ and let*

$$U^H A U = T = \begin{bmatrix} T_{11} & T_{12} & \ldots & T_{1p} \\ 0 & T_{22} & \ldots & T_{2p} \\ \vdots & \vdots & \ddots & \vdots \\ 0 & 0 & \ldots & T_{pp} \end{bmatrix}$$

be the Schur decomposition of A, where the diagonal block $T_{ii} \in \mathbb{C}^{\nu_i \times \nu_i}$ has eigenvalue λ_i. Then there exists a nonsingular transformation with the matrix

$$M = [M_1, M_2, \ldots, M_p],\ M_i \in \mathbb{C}^{n\times \nu_i},$$

such that

$$(UM)^{-1}AUM = M^{-1}TM = B,$$

where

$$B = \begin{bmatrix} T_{11} & 0 & \dots & 0 \\ 0 & T_{22} & \dots & 0 \\ \vdots & \vdots & \ddots & \vdots \\ 0 & 0 & \dots & T_{pp} \end{bmatrix}$$

is block-diagonal.

Since $AM_i = M_iT_{ii}$, the columns of M_i span an invariant subspace associated with λ_i. The matrix T_{ii} is the restriction of A which acts on the subspace $\mathcal{R}(M_i)$. The matrix B represents a *direct sum of the transformations* T_{ii}, $i = 1, 2, \dots, p$, acting on the corresponding invariant subspaces $\mathcal{R}(M_i)$.

Note that when A has multiple eigenvalues, it is necessary to identify in practice the clusters of numerically close eigenvalues which can be accepted as multiple ones. This problem is difficult in the presence of rounding errors and its solution is considered latter on in Chapter 6.

2.1.7 *Notes and References*

2.1.7.1 *Eigenvalues and Eigenvectors*

A systematical treatment of the eigenvalue problem is presented in the books of Gantmacher (1959a), Chapters III, IV, VI, VII, Horn and Johnson (2013), Chapters 1–4, Lancaster and Tismenetsky (1985), Chapters 4–6, Stewart (2001), Wilkinson (1965), Björck (2015), Chapter 3, and Chatelin (2012).

The book Gohberg *et al.* (2006) is entirely devoted to the invariant subspaces of matrices and their applications.

The roots of the characteristic equation are used for the first time by the French mathematician Lagrange in 1762 in the solution of a system of differential equations with constant coefficients, arising in the description of perturbed motion of six planets (Lagrange, 1762–1765; Kline, 1972, Chapter 33). The notion of "characteristic equation" itself is introduced by Cauchy in (1840) (Hawkins, 1975).

Also, Cauchy proves for the first time that the eigenvalues of a symmetric matrix can be only real (Cauchy, 1829). The earliest practical method for finding the characteristic polynomial is proposed by the French astronomer Le Verrier (1840). In 1883, the English mathematician James Joseph Sylvester introduced the notion of a *latent root* (Sylvester, 1883). The standard term in English *proper value* is now generally replaced by the word *eigenvalue* which is a (partial) translation of the German term *eigenwert*, introduced by David Hilbert in (1904).

2.1.7.2 *Schur Form*

The Schur form is proposed in 1909 by the German mathematician Issai Schur (1909). In contrast to the Jordan and Weyr forms considered in the following section, the Schur form is not strictly canonical, see Shapiro (1991). The real Schur form of a matrix is proposed in Murnaghan and Wintner (1931).

2.1.7.3 *Sylvester Equation*

The Sylvester equation is named after J.J. Sylvester who considered the homogeneous matrix equation $AX = XB$ in his work (Sylvester, 1884). The solution of the Sylvester equation is discussed by many authors, see Wedderburn (1903), Rutherford (1932), Gantmacher (1959a), Chapter VIII, Bellman (1970), Chapter 10, Bickley and McNamee (1960), Ma (1966), Lancaster (1970), Higham (2002), Chapter 16, and Bhatia and Rosenthall (1997) to mention a few. The Kronecker product to solve this equation is applied by Bellman (1959), Horn and Johnson (1991), Chapter 4, and Lankaster and Tismenetsky (1985), Chapter 12.

A particular case of the Sylvester equation is the Lyapunov matrix equation $A^H X + XA = C$ (Gantmacher, 1959b, Chapter XV; Lancaster and Tismenetsky, 1985, Chapter 13) which plays a prominent role in the stability and optimization of linear control systems.

2.2 Jordan and Weyr Canonical Forms

In this section, we consider the case when the matrix $A \in \mathbb{C}^{n\times n}$ has multiple eigenvalues. In such case, the matrix can or cannot be reduced into diagonal form depending on the number of linearly independent eigenvectors. If the matrix has n linearly independent eigenvectors, then it can be reduced into the diagonal form (2.5), independently of the multiplicities of the eigenvalues. Such matrix is called *diagonalizable* and its multiple eigenvalues are semisimple. Clearly, the matrices with distinct eigenvalues are diagonalizable. If, however, the matrix has less than n eigenvectors, it cannot be reduced into diagonal form by similarity transformations.

2.2.1 *Jordan Structure: A Unique Eigenvector*

Assume first for simplicity that the matrix A has a unique eigenvalue λ with algebraic multiplicity equal to n. In such a case to λ may correspond one or several eigenvectors. (So, each matrix has at least one eigenvector.) The number of the linearly independent eigenvectors corresponding to a given eigenvalue is called *geometric multiplicity* of this eigenvalue in contrast to the algebraic multiplicity. The geometric multiplicity is always less than or equal to the algebraic multiplicity. If the geometric multiplicity of an eigenvalue is less than its algebraic multiplicity, this eigenvalue is called *defective* eigenvalue. If a given matrix has defective eigenvalue, it is also called *defective*. Note that the nondefective matrices are diagonalizable.

Consider first the extreme case when λ corresponds only one eigenvector, i.e., λ has a geometric multiplicity equal to 1 and is a defective eigenvalue. In such a case, the unique eigenvector can be complemented by $n-1$ linearly independent vectors such that the obtained nonsingular transformation matrix has maximum simplified form. One of the possible ways to achieve this is to choose the system of linearly independent vectors according to the equations

$$\begin{aligned} Az_1 &= \lambda z_1, \\ Az_2 &= \lambda z_2 + z_1, \\ Az_3 &= \lambda z_3 + z_2, \\ \vdots\ &\quad \vdots \\ Az_n &= \lambda z_n + z_{n-1}, \end{aligned} \tag{2.18}$$

where z_1 is the eigenvector corresponding to λ and $z_i,\ i > 1$, is a *principal vector of grade* i. The eigenvector z_1 can be considered as a principal vector of grade 1. It can be shown that if λ has only one eigenvector, then the system of vectors $z_1, z_2, \ldots, z_n$ is linearly independent.

We note that some authors instead of the notion *principal vector* use the term *generalized eigenvector*. This term, however, is more appropriate for using in the generalized eigenvalue problem $Ax = \lambda Bx$ (Section 2.3), so we will avoid using it in the consideration of the ordinary eigenvalue problem.

The system of linearly independent vectors (2.18) is called *Jordan chain* of length n, associated with the eigenvalue λ. If from these vectors we compose the matrix

$$Z = [z_1, z_2, z_3, \ldots, z_n], \tag{2.19}$$

then equations (2.18) can be written as

$$AZ = ZJ_n(\lambda), \tag{2.20}$$

where

$$J_n(\lambda) = \begin{bmatrix} \lambda & 1 & 0 & \ldots & 0 & 0 \\ 0 & \lambda & 1 & \ldots & 0 & 0 \\ 0 & 0 & \lambda & \ldots & 0 & 0 \\ \vdots & \vdots & \vdots & \ddots & \vdots & \vdots \\ 0 & 0 & 0 & \ldots & \lambda & 1 \\ 0 & 0 & 0 & \ldots & 0 & \lambda \end{bmatrix} \tag{2.21}$$

is called *Jordan block* or *Jordan ceil* of size n, corresponding to the eigenvalue λ. The superdiagonal elements, equal to 1, are the *coupling elements* of the Jordan block.

In this way, we obtain that

$$Z^{-1}AZ = J_n(\lambda), \tag{2.22}$$

i.e., the matrix A is reduced into the Jordan block $J_n(\lambda)$. The conclusion is that if the matrix has a unique multiple eigenvalue to which corresponds only one eigenvector, then it can be reduced into a form which consists of only one Jordan block.

The eigenvector and the principal vectors are called *root vectors* corresponding to the eigenvalue λ. The collection of root vectors span an invariant subspace whose dimension is equal to the algebraic multiplicity of λ. This subspace is called the *root subspace* associated with λ.

It follows from equations (2.18) that

$$\begin{aligned}
(A-\lambda I)z_1 &= 0, & (A-\lambda I)z_1 &= 0,\\
(A-\lambda I)z_2 &= z_1, & (A-\lambda I)^2 z_2 &= 0,\\
(A-\lambda I)z_3 &= z_2, & (A-\lambda I)^3 z_3 &= 0,\\
\vdots\quad & \quad\vdots & \vdots\quad & \quad\vdots\\
(A-\lambda I)z_n &= z_{n-1}, & (A-\lambda I)^n z_n &= 0.
\end{aligned} \tag{2.23}$$

It is seen that the principal vector z_i has a grade i, where i is the smallest number for which it is fulfilled that

$$(A-\lambda I)^i z_i = 0, \tag{2.24}$$

but

$$(A-\lambda I)^{i-1} z_i \neq 0.$$

The matrix $A-\lambda I$ is a nilpotent matrix since the Jordan ceil $J_n(0)$ satisfies $J_n(0)^n = 0$.

Equation (2.24) can be written as

$$z_i \in \mathcal{N}_i = \mathcal{N}((A-\lambda I)^i), \quad i = 1, 2, \ldots, n, \tag{2.25}$$

where $\mathcal{N}_i$ is the null space (the kernel) of the matrix $(A-\lambda I)^i$. In this way, the eigenvector belongs to $\mathcal{N}_1 = \mathcal{N}(A-\lambda I)$, the principal vector of grade 2 belongs to $\mathcal{N}_2 = \mathcal{N}((A-\lambda I)^2)$, but not to $\mathcal{N}_1$ (to be linearly independent on the eigenvector), the principal vector of grade 3 belongs to $\mathcal{N}_3 = \mathcal{N}((A-\lambda I)^3)$, but not to $\mathcal{N}_2$ (to be linearly independent on the principal vector of grade 2) and so on. These relationships can be written as

$$\begin{aligned}
&z_1 \in \mathcal{N}_1,\\
&z_2 \in \mathcal{N}_2 \setminus \mathcal{N}_1,\\
&\vdots\\
&z_n \in \mathcal{N}_n \setminus \mathcal{N}_{n-1},
\end{aligned}$$

where the subspaces

$$\mathcal{N}_2 \setminus \mathcal{N}_1, \quad \mathcal{N}_3 \setminus \mathcal{N}_2, \ldots, \quad \mathcal{N}_n \setminus \mathcal{N}_{n-1}$$

are called *restricted null subspaces.* The dimensions of these subspaces are given by

$$\begin{aligned} \dim(\mathcal{N}_i \setminus \mathcal{N}_{i-1}) &= \dim(\mathcal{N}_i) - \dim(\mathcal{N}_{i-1}), \\ &= \mathrm{null}((A - \lambda I)^i) - \mathrm{null}((A - \lambda I)^{i-1}), \\ i &= 2, 3, \ldots, n. \end{aligned}$$

In the general case, the restricted subspaces have dimensions greater or equal to one and $(A - \lambda I_n)^h = 0$ for some $h \leq n$. The null spaces satisfy the inclusion (Figure 2.6)

$$\mathcal{N}_1 \subset \mathcal{N}_2 \subset \mathcal{N}_3 \subset \cdots \subset \mathcal{N}_h.$$

In the particular case under consideration, we have that

$$\dim(\mathcal{N}_1) = 1, \quad \dim(\mathcal{N}_2) = 2, \ldots, \dim(\mathcal{N}_n) = n,$$

so that the restricted null spaces have dimensions

$$\dim(\mathcal{N}_2 \setminus \mathcal{N}_1) = \dim(\mathcal{N}_3 \setminus \mathcal{N}_2) = \cdots = \dim(\mathcal{N}_n \setminus \mathcal{N}_{n-1}) = 1.$$

The columns of the matrix Z in (2.19) span an invariant subspace, corresponding to the multiple eigenvalue λ. In fact,

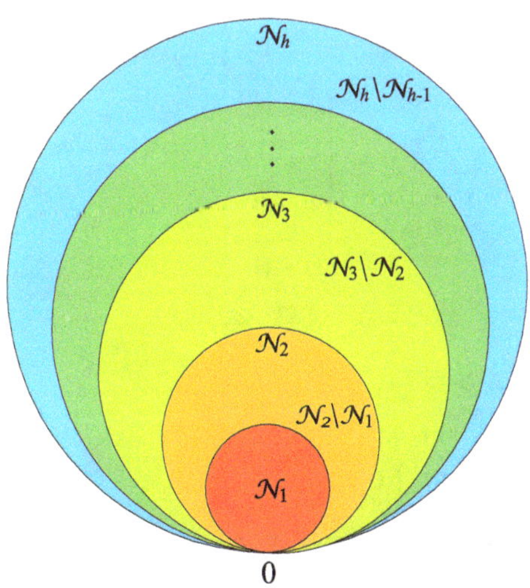

Fig. 2.6. Symbolical representation of the restricted null spaces.

according to (2.24), we have that $(A - \lambda I)^i z_i = 0$, which implies that $(A - \lambda I)^i A z_i = A(A - \lambda I)^i z_i = 0$.

It follows from (2.23) that the root vectors satisfy the equations

$$\begin{aligned} z_1 &= (A - \lambda I) z_2, \\ z_2 &= (A - \lambda I) z_3, \\ &\vdots \quad \vdots \\ z_{n-1} &= (A - \lambda I) z_n. \end{aligned} \tag{2.26}$$

Denote by $x = z_n$ the principal vector of grade n. Writing the root vectors in the reverse order, we obtain that

$$\begin{aligned} z_n &= x, \\ z_{n-1} &= (A - \lambda I) x, \\ &\vdots \quad \vdots \\ z_1 &= (A - \lambda I)^{n-1} x. \end{aligned} \tag{2.27}$$

Subspace, whose basis vectors have the form $x,\ Ax,\ A^2x, \ldots, A^{k-1}x$ for some $1 \leq k \leq n$, is called *A-cyclic subspace generated by* x or *Krylov subspace*. If the dimension of this subspace is equal to n, then the matrix A is similar to the *companion matrix* of its characteristic polynomial,

$$C = S^{-1} A S = \begin{bmatrix} 0 & 0 & \ldots & 0 & -p_n \\ 1 & 0 & \ldots & 0 & -p_{n-1} \\ 0 & 1 & \ldots & 0 & -p_{n-2} \\ \vdots & \vdots & \ddots & \vdots & \vdots \\ 0 & 0 & \ldots & 1 & -p_1 \end{bmatrix},$$

where

$$S = [x,\ Ax,\ A^2x, \ldots, A^{n-1}x]$$

and the elements in the last column of C are the coefficients of the characteristic polynomial of A,

$$p(\lambda) = \det(\lambda I_n - A) = \lambda^n + p_1 \lambda^{n-1} + \cdots + p_{n-1}\lambda + p_n.$$

In case of a unique eigenvalue λ, the transformation of a matrix into companion form is possible only if to this eigenvalue corresponds a unique eigenvector, i.e., if there is only one Jordan block corresponding to λ. In case of distinct eigenvalues, the whole space $\mathbb{C}^n$ is cyclic.

2.2.2 *Jordan Structure: q Eigenvectors*

Consider now the more general case when to the same eigenvalue with multiplicity equal to n corresponds less than n eigenvectors, i.e., we consider the case when the geometric multiplicity of λ is equal to q, $1 < q < n$. Denote these eigenvectors by $z_1^{(1)}$, $z_1^{(2)}, \ldots,$ $z_1^{(q)}$. We complement the first eigenvector $z_1^{(1)}$ with principal vectors until we obtain a system of s_1 linearly independent vectors, then we complement the second eigenvector $z_1^{(2)}$ by principal vectors until we obtain a system of s_2 linearly independent vectors and so on, while the last eigenvector $z_1^{(q)}$ is complemented by principal vectors until obtaining the system of s_q linearly independent vectors. The obtained set of q systems looks as follows:

$$\begin{aligned}
&Az_1^{(1)} = \lambda z_1^{(1)}, && Az_1^{(2)} = \lambda z_1^{(2)}, && \ldots\\
&Az_2^{(1)} = \lambda z_2^{(1)} + z_1^{(1)}, && Az_2^{(2)} = \lambda z_2^{(2)} + z_1^{(2)}, && \ldots\\
&Az_3^{(1)} = \lambda z_3^{(1)} + z_2^{(1)}, && Az_3^{(2)} = \lambda z_3^{(2)} + z_2^{(2)}, && \ldots\\
&\quad\vdots \quad \vdots && \quad\vdots \quad \vdots && \\
&Az_{s_1}^{(1)} = \lambda z_{s_1}^{(1)} + z_{s_1-1}^{(1)}, && Az_{s_2}^{(2)} = \lambda z_{s_2}^{(2)} + z_{s_2-1}^{(2)}, && \ldots
\end{aligned}$$

$$\begin{aligned}
Az_1^{(q)} &= \lambda z_1^{(q)},\\
Az_2^{(q)} &= \lambda z_2^{(q)} + z_1^{(q)},\\
Az_3^{(q)} &= \lambda z_3^{(q)} + z_2^{(q)},\\
&\vdots \quad \vdots\\
Az_{s_q}^{(q)} &= \lambda z_{s_q}^{(q)} + z_{s_q-1}^{(q)}.
\end{aligned} \tag{2.28}$$

Each system of s_j, $j = 1,\ 2,\ ldots,\ q$ equations is called *chain of principal vectors* or shortly *chain relation*. The number s_j is referred to as *chain length*.

Without loss of generality, we may assume that the eigenvectors are ordered in a such way that it is fulfilled

$$s_1 \geq s_2 \geq \cdots \geq s_q.$$

The numbers s_j, $j = 1, 2, \ldots, q$ (the chain lengths), are subject to the relationship

$$1 \leq s_j \leq n, \; s_1 + s_2 + \cdots + s_q = n. \tag{2.29}$$

The list of nonincreasing numbers

$$\mathcal{S}(A, \lambda) = \{s_1, \; s_2, \ldots, s_q\}$$

is called *Segre characteristic* or *Jordan structure* of A associated with λ.

Each of the group of s_j vectors obtained in this way spans a cyclic subspace of dimension s_j, $j = 1, \ldots, q$. The obtained system of n linearly independent vectors is used to construct the transformation matrix

$$Z = \left[\underbrace{z_1^{(1)} \; z_2^{(1)} \; \ldots \; z_{s_1}^{(1)}}_{s_1} \mid \underbrace{z_1^{(2)} \; z_2^{(2)} \ldots \; z_{s_2}^{(2)}}_{s_2} \mid \ldots \mid \underbrace{z_1^{(q)} \; z_2^{(q)} \; \ldots \; z_{s_q}^{(q)}}_{s_q} \right]. \tag{2.30}$$

Equation (2.28) can be written as (for brevity, $\mathcal{S}(A, \lambda)$ is denoted by $\mathcal{S}$)

$$AZ = ZJ(\mathcal{S}, \lambda) \tag{2.31}$$

or

$$Z^{-1}AZ = J(\mathcal{S}, \lambda), \tag{2.32}$$

where

$$J(\mathcal{S}, \lambda) = \begin{bmatrix} J_{s_1}(\lambda) & & & \\ & J_{s_2}(\lambda) & & \\ & & \ddots & \\ & & & J_{s_q}(\lambda) \end{bmatrix} \tag{2.33}$$

and each Jordan block $J_{s_i}(\lambda) \in \mathbb{C}^{s_i \times s_i}$, $i = 1, 2, \ldots, q$, has the form (2.21).

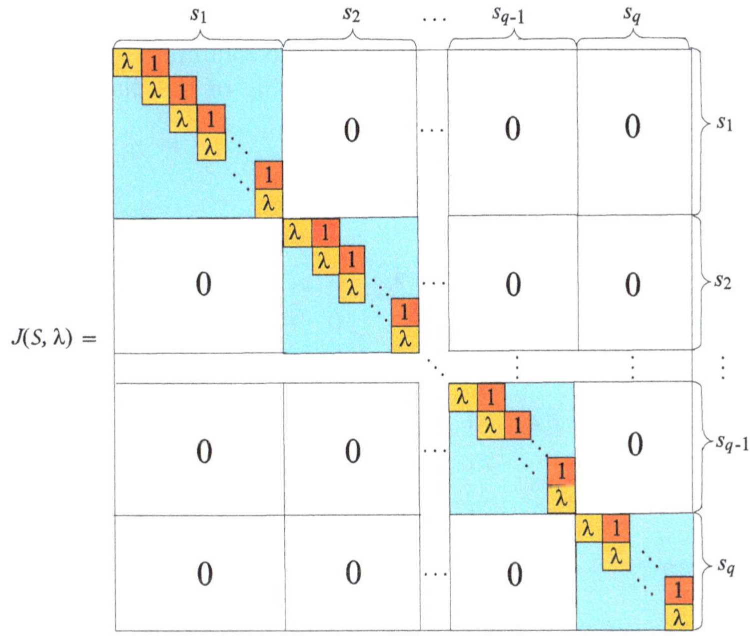

Fig. 2.7. The Jordan normal form in case of n-tuple eigenvalue λ with q eigenvectors.

Equations (2.32) and (2.33) show that in the case of $1 < q < n$ linearly independent eigenvectors, the $n \times n$ matrix A with a unique eigenvalue λ of multiplicity n is reduced into a special form $J(\mathcal{S}, \lambda)$, which contains on its diagonal q Jordan blocks whose sizes are the elements of the Segre characteristic $\mathcal{S} = \{s_1,\ s_2, \ldots, s_q\}$ (Figure 2.7). In case when some of the numbers s_i is equal to 1, the corresponding Jordan block reduces to the scalar value λ. The Jordan blocks are the restrictions of A that act on the corresponding invariant subspaces. To each Jordan block corresponds a cyclic subspace with dimension equal to the size of the block. In this manner, the space $\mathbb{C}^n$ is decomposed in a direct sum of cyclic subspaces.

The matrix $J(\mathcal{S}, \lambda)$ represents the *Jordan normal form* or simply the *Jordan form* of A for the particular case of a single n-tuple eigenvalue. The form is called *normal* in the sense that the superdiagonal elements of the Jordan blocks are equal to 1. Latter on, we shall also make use of the un-normalized Jordan form that may be more convenient from numerical point of view.

The Jordan form is unique up to the ordering of the diagonal blocks. With an appropriate ordering of the columns of the matrix Z, it is possible to achieve the desired ordering of the diagonal blocks of this form.

Example 2.13. The matrix

$$A = \begin{bmatrix} 1 & -4 & 4 & -2 & 0 & -2 & -4 \\ 1 & 0 & 0 & 0 & -1 & -1 & -2 \\ 1 & -2 & 3 & -1 & -1 & -1 & -4 \\ 1 & -1 & 1 & 1 & -1 & 0 & -3 \\ 0 & -1 & 0 & 0 & 1 & -1 & 0 \\ -1 & 1 & 0 & 0 & 1 & 2 & 2 \\ 0 & -1 & 2 & -1 & 0 & 0 & -1 \end{bmatrix}$$

has a unique eigenvalue $\lambda = 1$ with multiplicity 7. To this eigenvalue correspond the eigenvectors

$$z_1^{(1)} = \begin{bmatrix} -2 \\ -1 \\ -2 \\ -1 \\ 0 \\ 1 \\ -1 \end{bmatrix}, \quad z_1^{(2)} = \begin{bmatrix} -2 \\ 0 \\ -1 \\ 0 \\ 0 \\ 0 \\ -1 \end{bmatrix}, \quad z_1^{(3)} = \begin{bmatrix} 1 \\ 0 \\ 1 \\ 0 \\ -1 \\ 0 \\ 1 \end{bmatrix}.$$

This implies that the Jordan form of A has three blocks, hence A is a defective matrix since at least of one of these blocks has a dimension greater than one. Also, A is a *derogatory matrix* since it has more than one Jordan block corresponding to the same eigenvalue. Thus, A is simultaneously defective and derogatory matrix.

The first eigenvector participates in a Jordan chain of three linearly dependent vectors which satisfy equations (2.18). These vectors constitute a basis of an invariant subspace of dimension 3:

$$Z_1 = [z_1^{(1)},\ z_2^{(1)},\ z_3^{(1)}] = \begin{bmatrix} -2 & 0 & -1 \\ -1 & -1 & 1 \\ -2 & 0 & 0 \\ -1 & 0 & 0 \\ 0 & -1 & 1 \\ 1 & 1 & 0 \\ -1 & 1 & -1 \end{bmatrix}.$$

To this subspace corresponds a Jordan ceil with the same dimension,

$$J_3(1) = \begin{bmatrix} 1 & 1 & 0 \\ 0 & 1 & 1 \\ 0 & 0 & 1 \end{bmatrix},$$

i.e., $s_1 = 3$. We have that $AZ_1 = Z_1 J_3(1)$.

The second eigenvector also participates in a chain of three vectors:

$$Z_2 = [z_1^{(2)},\ z_2^{(2)},\ z_3^{(2)}] = \begin{bmatrix} -2 & 2 & 0 \\ 0 & 0 & -1 \\ -1 & 1 & 0 \\ 0 & 1 & 0 \\ 0 & 0 & 0 \\ 0 & 0 & 1 \\ -1 & 1 & 0 \end{bmatrix},$$

which corresponds to a Jordan block of third order, i.e., $s_2 = 3$. For this chain, it is fulfilled that $AZ_2 = Z_2 J_3(1)$.

The third eigenvector has no principal vectors, i.e., $s_3 = 1$ and $Z_3 = z_1^{(3)}$. In this case, $AZ_3 = Z_3 \cdot 1$.

As a result, the transformation matrix takes the form

$$Z = [Z_1,\ Z_2,\ Z_3] = \left[z_1^{(1)},\ z_2^{(1)},\ z_3^{(1)} \middle|,\ z_1^{(2)},\ z_2^{(2)},\ z_3^{(2)} \middle|, z_1^{(3)} \right]$$

$$
= \left[\begin{array}{ccc|ccc|c}
-2 & 0 & -1 & -2 & 2 & 0 & 1 \\
-1 & -1 & 1 & 0 & 0 & -1 & 0 \\
-2 & 0 & 0 & -1 & 1 & 0 & 1 \\
-1 & 0 & 0 & 0 & 1 & 0 & 0 \\
0 & -1 & 1 & 0 & 0 & 0 & -1 \\
1 & 1 & 0 & 0 & 0 & 1 & 0 \\
-1 & 1 & -1 & -1 & 1 & 0 & 1
\end{array}\right].
$$

As a result of the nonsingular transformation with the matrix Z, we obtain the Jordan form

$$
J = Z^{-1}AZ = \left[\begin{array}{ccc|ccc|c}
1 & 1 & 0 & 0 & 0 & 0 & 0 \\
0 & 1 & 1 & 0 & 0 & 0 & 0 \\
0 & 0 & 1 & 0 & 0 & 0 & 0 \\
\hline
0 & 0 & 0 & 1 & 1 & 0 & 0 \\
0 & 0 & 0 & 0 & 1 & 1 & 0 \\
0 & 0 & 0 & 0 & 0 & 1 & 0 \\
\hline
0 & 0 & 0 & 0 & 0 & 0 & 1
\end{array}\right].
$$

2.2.3 *Segre and Weyr Characteristics*

Consider again a matrix $A \in \mathbb{C}^{n\times n}$ with a unique multiple eigenvalue λ and denote by $\mathcal{N}_j$ the null space $\mathcal{N}((A-\lambda I)^j)$ of the matrix $(A-\lambda I)^j$. Note that $\mathcal{N}_0 = \mathcal{N}(I_n) = 0$ and $\mathcal{N}_1$ contains the eigenvectors of A that satisfy $(A-\lambda I)z = 0$. The *index of nilpotence* of A is the smallest positive integer h for which

$$
\dim(\mathcal{N}((A-\lambda I)^h)) = \text{null}((A-\lambda I)^h) = n,
$$

where

$$
\text{null}(A) = n - \text{rank}(A)
$$

is the nullity of the matrix A.

The integers

$$
w_j = \dim(\mathcal{N}_j) \setminus \mathcal{N}_{j-1}) = \dim(\mathcal{N}_j) - \dim(\mathcal{N}_{j-1})
$$

$$= \text{null}((A - \lambda I)^j) - \text{null}((A - \lambda I)^{j-1})$$
$$= \text{rank}((A - \lambda I)^{j-1}) - \text{rank}((A - \lambda I)^j), \quad j = 1,\ 2, \ldots,\ h$$

satisfy the inequalities $w_1 \geq w_2 \geq \cdots \geq w_h$ and their sum is equal to n. The list of numbers

$$\mathcal{W}(A, \lambda) = \{w_1, w_2, \ldots, w_h\}$$

is called *the Weyr characteristic* of A associated with λ or *Weyr structure*. As we see latter on, the Weyr characteristic of A, like the Segre characteristic defines uniquely the Jordan normal form of A.

Example 2.14. Given is the matrix

$$A = \begin{bmatrix} 3 & -4 & 2 & -2 & -4 & -10 & 17 \\ 1 & -5 & 1 & -4 & 2 & 0 & 8 \\ 2 & 0 & -1 & 2 & -4 & -6 & 5 \\ 1 & -2 & 0 & -2 & 0 & -1 & 4 \\ 1 & -3 & 1 & -3 & 0 & -1 & 6 \\ 1 & -1 & 0 & 0 & -1 & -3 & 4 \\ 0 & -2 & 0 & -2 & 2 & 2 & 1 \end{bmatrix},$$

which has 7-tuple eigenvalue $\lambda = -1$. The matrix

$$N = A - \lambda I_7 = \begin{bmatrix} 4 & -4 & 2 & -2 & -4 & -10 & 17 \\ 1 & -4 & 1 & -4 & 2 & 0 & 8 \\ 2 & 0 & 0 & 2 & -4 & -6 & 5 \\ 1 & -2 & 0 & -1 & 0 & -1 & 4 \\ 1 & -3 & 1 & -3 & 1 & -1 & 6 \\ 1 & -1 & 0 & 0 & -1 & -2 & 4 \\ 0 & -2 & 0 & -2 & 2 & 2 & 2 \end{bmatrix}$$

has 7-tuple zero eigenvalue, i.e., it is a nilpotent matrix. The powers of this matrix have the following ranks and nullities:

$$\text{rank}(N^0) = 7, \quad \text{null}(N^0) = 0,$$
$$\text{rank}(N^1) = 4, \quad \text{null}(N^1) = 3,$$

$$\text{rank}(N^2) = 1, \quad \text{null}(N^2) = 6,$$
$$\text{rank}(N^3) = 0, \quad \text{null}(N^3) = 7,$$

$h = 3$ and the Weyr characteristic of A has elements

$$w_1 = \text{null}(N^1) - \text{null}(N^0) = 3,$$
$$w_2 = \text{null}(N^2) - \text{null}(N^1) = 3,$$
$$w_3 = \text{null}(N^3) - \text{null}(N^2) = 1.$$

As shown earlier, the Jordan structure of A is determined by the Segre characteristic

$$\mathcal{S}(A, \lambda) = \{s_1, s_2, \ldots, s_q\}, \quad s_1 \geq s_2 \geq \cdots \geq s_q.$$

The elements of this characteristic are equal to the sizes of the Jordan blocks in the Jordan form of A that have λ as an eigenvalue. The quantity q, $1 \leq q \leq n$, is equal to the number of the linearly independent eigenvectors of A associated with λ (the geometric multiplicity of λ).

Further on we will need the following two definitions.

Definition 2.15. An *integer partition* $\mathcal{P} = \{p_1, p_2, \ldots\}$ of an integer P is a monotonically decreasing sequence of integers $\{p_1 \geq p_2 \geq \cdots \geq 0\}$, where $p_1 + p_2 + \cdots = p$.

Definition 2.16. Let $p_1 \geq p_2 \geq \cdots \geq 0$ be a partition of the positive integer p. Let q_j denote the number of the integers p_i, which are greater or equal to j. Then the integers q_j form a partition of P, known as a *conjugate partition* of P.

The conjugate partition is calculated by Algorithm 2.1.

The relationship between Weyr and Segre characteristics is revealed by the following theorem.

Theorem 2.17. *The Weyr characteristic and the Segre characteristic of A relative to the eigenvalue λ with multiplicity n are conjugated partitions of n.*

Thus the Weyr characteristic determines uniquely the Segre characteristic of A associated with λ and vice versa. This is important in

Algorithm 2.1. Conjugate partition of an integer
For a given partition $\{p_1, p_2, \dots, p_h\}$ *of a positive integer* $p = p_1 + p_2 + \cdots + p_h$*, computes the conjugate partition* $q = \{q_1, q_2, \dots, q_{\max}(p)\}$*, such that* q_1 *is the number of the elements of* p *which are greater or equal to 1,* q_2 *is the number of the elements of* p *which are greater or equal to 2 and so on.*

function $q = \text{conj\_partition}(p)$
Input: $p = \{p_1, p_2, \dots, p_h\}$
Output: $q = \{q_1, q_2, \dots, q_{\max}(p)\}$

```
p_max = max(p)
for i = 1, 2, ..., p_max
    q(i) = 0
    for j = 1, 2, ..., h
        if p(j) >= i
            q(i) = q(i) + 1
        end
    end
end
```

finding the Jordan form (Chapters 6 and 7). Note that the largest element s_1 (the size of the largest Jordan block) of the Segre characteristic is equal to the number h of the elements of Weyr characteristic. In turn, the first element w_1 of the Weyr characteristic (the geometric multiplicity of the eigenvalue λ) is equal to the number q of the elements of Segre characteristic (the number of Jordan blocks) associated with λ.

For the matrix A from Example 2.14 with Weyr characteristic $\mathcal{W}(A, -1) = \{3,\ 3,\ 1\}$, the corresponding Segre characteristic is $\mathcal{S}(A, -1) = \{3,\ 2,\ 2\}$ so that this matrix has $q = w_1 - 3$ Jordan blocks of orders $s_1 = h = 3$, $s_2 = 2$ and $s_3 = 2$ associated with the eigenvalue $\lambda = -1$.

The determination of conjugate partition is illustrated by the *Ferrers diagram* (Figure 2.8), in which s_j is the number of points in the column j and w_k is the number of points in row k.

Let, for example, the Segre characteristic associated with 14-tuple eigenvalue be $\{4,\ 4,\ 3,\ 2,\ 1\}$. Then from the Ferrers diagram, we find that the Weyr characteristic is $\{5,\ 4,\ 3,\ 2\}$.

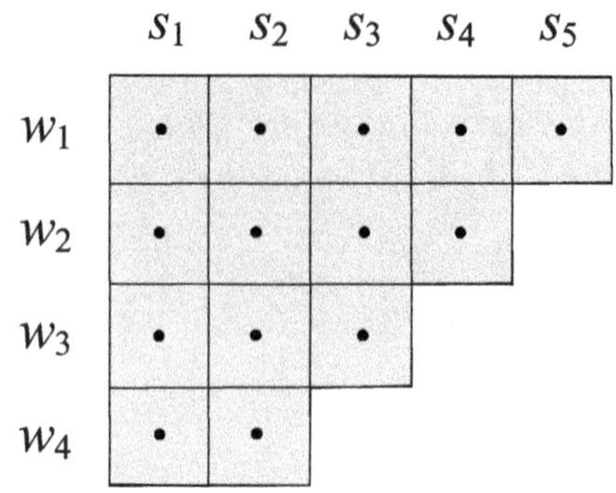

Fig. 2.8. Ferrers diagram.

Consider now the general case when the matrix $A \in \mathbb{C}^{n\times n}$ has $p \leq n$ distinct eigenvalues

$$\lambda_1,\ \lambda_2, \dots, \lambda_p$$

with algebraic multiplicities, respectively,

$$\nu_1,\ \nu_2, \dots, \nu_p,\ \sum_{i=1}^{p} \nu_i = n.$$

Denote by

$$\mathcal{S}_i := \mathcal{S}(A, \lambda_i) = \{s_{1i}, s_{2i}, \dots, s_{q_i,i}\},\ s_{1i} \geq s_{2i} \geq \cdots \geq s_{q_i,i},$$
$$\times \sum_{j=1}^{q_i} s_{ji} = \nu_i$$

and

$$\mathcal{W}_i := \mathcal{W}(A, \lambda_i) = \{w_{i1}, w_{i2}, \dots, w_{i,h_i}\},\ w_{i1} \geq w_{i2} \geq \cdots \geq w_{i,h_i},$$
$$\times \sum_{j=1}^{h_i} w_{ij} = \nu_i$$

the Segre and Weyr characteristics of A, respectively, associated with λ_i, $i = 1,\ 2, \dots, p$.

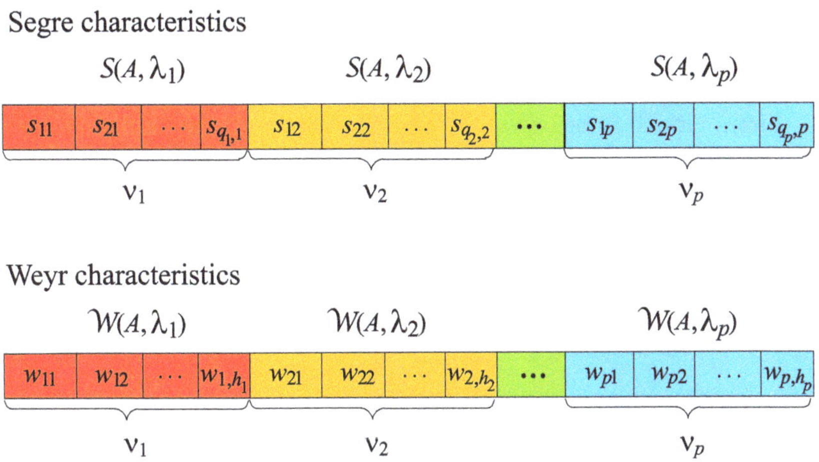

Fig. 2.9. Segre and Weyr characteristics associated with multiple eigenvalues.

The set of Segre characteristics associated with all distinct eigenvalues of A is represented as (see Figure 2.9)

$$\begin{aligned}\mathcal{S} &= [\mathcal{S}_1, \mathcal{S}_2, \ldots, \mathcal{S}_p]\\ &= [\{s_{11}, s_{21}, \ldots, s_{q_1,1}\}, \{s_{12}, s_{22}, \ldots, s_{q_2,2}\}, \ldots, \{s_{1p}, s_{2p}, \ldots, s_{q_p,p}\}]\end{aligned}$$

and the set of Weyr characteristics associated with the same eigenvalues is

$$\begin{aligned}\mathcal{W} &= [\mathcal{W}_1, \mathcal{W}_2, \ldots, \mathcal{W}_p]\\ &= [\{w_{11}, w_{12}, \ldots, w_{1,h_1}\}, \{w_{21}, w_{22}, \ldots, w_{2,h_2}\}, \ldots,\\ &\quad \times \{w_{p1}, w_{p2}, \ldots, w_{p,h_p}\}].\end{aligned}$$

It is possible to show that the index of nilpotence of A satisfies

$$h = \max\{s_{11},\ s_{12}, \ldots, s_{1p}\},$$

i.e., the index of nilpotence is equal to the size of the largest Jordan block of A.

The set of Weyr characteristics of the matrix A, like the set of Segre characteristics, constitutes a complete set of *structural invariants*, i.e., quantities which remain unchanged under nonsingular transformations of A (MacDuffee, 1946, p. 74). Otherwise speaking, two matrices are similar, if they have the same eigenvalues and the same Weyr (or Segre) characteristics.

Further on by the *structure of the matrix* A, we shall mean the Segre or Weyr characteristics of A.

Using the Segre characteristics, the characteristic polynomial of A,

$$p(\lambda) := \det(\lambda I_n - A) = \det(\lambda I_n - J_A),$$

can be represented as

$$p(\lambda) = (\lambda - \lambda_1)^{s_{11}} \dots (\lambda - \lambda_1)^{s_{q_1,1}} \dots (\lambda - \lambda_p)^{s_{1p}} \dots (\lambda - \lambda_p)^{s_{q_p,p}}.$$

The polynomials of the form $(\lambda - \lambda_i)^{s_{ji}} = \det(\lambda I - J_{s_{ji}}(\lambda_i))$ are called *elementary divisors* of A. The degree of an elementary divisor is equal to the size of the corresponding Jordan block. If $s_{ji} = 1$, the elementary divisor is *linear*, and if $s_{ji} > 1$, it is *nonlinear*. An $n \times n$ matrix with distinct eigenvalues always has n linear elementary divisors. If λ_i participates in one or more nonlinear elementary divisors, it is defective and so is the matrix A. If the index of nilpotence h is equal to 1, then all eigenvalues of A participate in linear elementary divisors and the matrix is nondefective even if it has some multiple eigenvalues. The number $q_i = w_{i1}$ of eigenvectors corresponding to the eigenvalue λ_i is the geometric multiplicity of this eigenvalue. Clearly, if the geometric multiplicity is smaller than the algebraic one, i.e., $w_{i1} < \nu_i$, then the eigenvalue is defective since in this case $s_{1i} > 1$. We say that the matrix is *derogatory* if this matrix has one or more algebraically multiple eigenvalues that participate in more than one Jordan block (i.e., in more than one elementary divisor). In such a case, there is at least one i for which $w_{i1} > 1$ and $s_{i1} < \nu_i$. If each eigenvalue λ_i participates in only one nonlinear elementary divisor of degree $s_{1i} = \nu_i > 1$, then the matrix is defective but nonderogatory. An example of nonderogatory matrix is the companion matrix, each distinct eigenvalue of which participates in a single Jordan block (i.e., the geometric multiplicity of each eigenvalue is equal to one). If A has $p > 1$ distinct eigenvalues, it can be reduced to companion form only if it is nonderogatory. If the matrix A is derogatory, i.e., it has more than one Jordan block associated with some eigenvalue, then the reduction into a single companion matrix is not possible. Obviously,

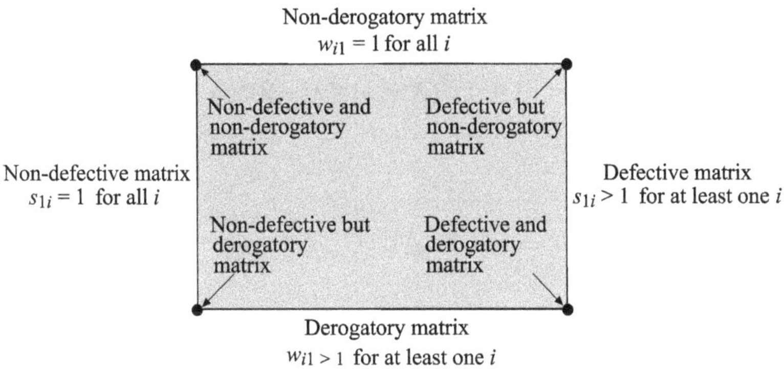

Fig. 2.10. Matrices with different Weyr and Segre characteristics.

a matrix can be both nondefective and nonderogatory (if it has n distinct eigenvalues) or both defective and derogatory (if there is at least one i for which $\nu_i > w_{i1} > 1$ and $1 < s_{1i} < \nu_i$).

The earlier considerations are illustrated in Figure 2.10.

The product

$$p_{\min}(\lambda) = (\lambda - \lambda_1)^{s_{11}}(\lambda - \lambda_1)^{s_{12}} \dots (\lambda - \lambda_p)^{s_{1p}},$$

of the elementary divisors $(\lambda - \lambda_i)^{s_{1i}}$ corresponding to the largest Jordan blocks $J_{s_{1i}}$ for each distinct λ_i, is the polynomial of minimum degree which annihilates A, i.e., $p_{\min}(A) = 0$. This polynomial is called the *minimal polynomial* of A and its degree is obviously equal to $s_{11} + s_{12} + \cdots + s_{1p} \leq n$. The minimal polynomial divides exactly the characteristic polynomial and the similar matrices have the same minimal polynomials. An interesting case is the companion matrix, for which the minimal polynomial coincides with the characteristic polynomial.

Example 2.18. A matrix A with Jordan form

$$J = \left[\begin{array}{ccc|cc|cc|c}
-1 & 1 & 0 & 0 & 0 & 0 & 0 & 0 \\
0 & -1 & 1 & 0 & 0 & 0 & 0 & 0 \\
0 & 0 & -1 & 0 & 0 & 0 & 0 & 0 \\
\hline
0 & 0 & 0 & -1 & 1 & 0 & 0 & 0 \\
0 & 0 & 0 & 0 & -1 & 0 & 0 & 0 \\
\hline
0 & 0 & 0 & 0 & 0 & 2 & 1 & 0 \\
0 & 0 & 0 & 0 & 0 & 0 & 2 & 0 \\
\hline
0 & 0 & 0 & 0 & 0 & 0 & 0 & 2
\end{array}\right]$$

has characteristic polynomial equal to

$$p(\lambda) = (\lambda+1)^3(\lambda+1)^2(\lambda-2)^2(\lambda-2)$$

and minimal polynomial

$$p_{\min}(\lambda) = (\lambda+1)^3(\lambda-2)^2.$$

2.2.4 *Jordan Normal Form*

The determination of the Jordan form in the general case when the matrix has $p < n$ distinct eigenvalues can be done similarly to the case of a unique eigenvalue.

Let the matrix A have p distinct eigenvalues $\lambda_1, \lambda_2, \ldots, \lambda_p$ and the set of Segre characteristics associated with the corresponding eigenvalues. For each distinct eigenvalue λ_i, one finds separately the Jordan chains of eigenvectors and principal vectors whose lengths determine the sizes of the corresponding Jordan blocks (the Segre characteristics). The principal vectors, associated with different Jordan blocks, are linearly independent. The columns of the $n \times \nu_i$ matrix

$$Z_i = \left[z_1^{(1i)}\ z_2^{(1i)} \ldots z_{s_{1i}}^{(1i)} | z_1^{(2i)}\ z_2^{(2i)} \ldots z_{s_{2i}}^{(2i)} | \ldots \right.$$
$$\left. | z_1^{(q_i,i)}\ z_2^{(q_i,i)} \ldots z_{s_{q_i,i}}^{(q_i,i)}\right]$$

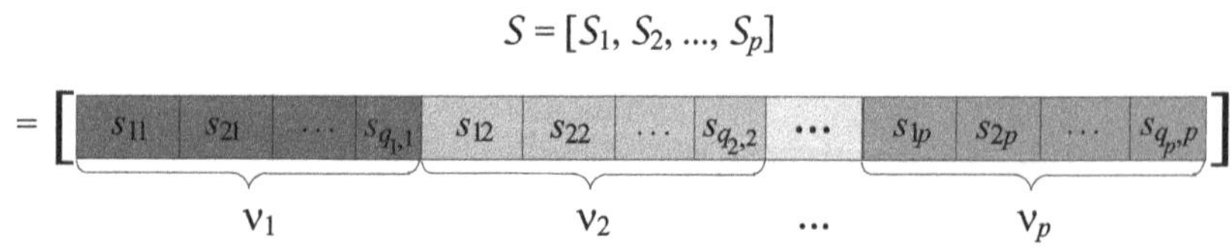

form a basis of the root subspace associated with the eigenvalue λ_i. (Note that the subscript $1 \leq k \leq s_{ji}$, $j = 1, 2, \ldots, q_i$, $i = 1, 2, \ldots, p$, of the principal vector $z_k^{(ji)}$ is equal to its grade.) The bases of the p root subspaces obtained in this way are used to construct the nonsingular matrix Z of the similarity transformation into Jordan form (Figure 2.11).

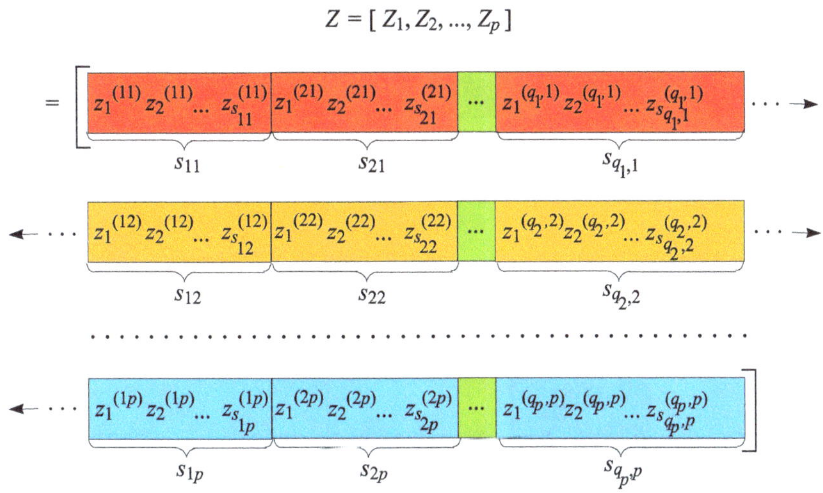

Fig. 2.11. The similarity transformation matrix into Jordan form.

Then the matrix A is reduced into the *Jordan normal form*

$$J = Z^{-1}AZ, \tag{2.34}$$

where the matrix J is shown in Figure 2.12. Each block $J(\mathcal{S}_i, \lambda_i) \in \mathbb{C}^{\nu_i \times \nu_i}$, $i = 1, 2, \ldots, p$, has the form (2.33) and each Jordan block $J_{s_{ji}}(\lambda_i) \in \mathbb{C}^{s_{ji} \times s_{ji}}$, $j = 1, 2, \ldots, q_i$, $i = 1, 2, \ldots, p$, has the form (2.21). The block $J_{s_{ji}}(\lambda_i)$ is the restriction of A to the eigenspace

$$\mathcal{X}^{ji} = \mathcal{R}([z_1^{(ji)}, z_2^{(ji)}, \ldots, z_{s_{ji}}^{(ji)}])$$

with respect to the basis $z_1^{(ji)}, z_2^{(ji)}, \ldots, z_{s_{ji}}^{(ji)}$.

The Jordan normal form gives full information about the eigenvalues, eigenvectors and eigenspaces. This form is also called *Jordan canonical form* since, loosely speaking, it contains the minimum possible number of nonzero elements (the so-called *invariants*) that characterize the class of similarity transformations. Examples of such invariants are the eigenvalues and the associated with them Segre or Weyr characteristics, or the elementary divisors. Obviously, the knowledge of one set of these invariants means the knowledge of the others. Note that the similar matrices have the same invariants.

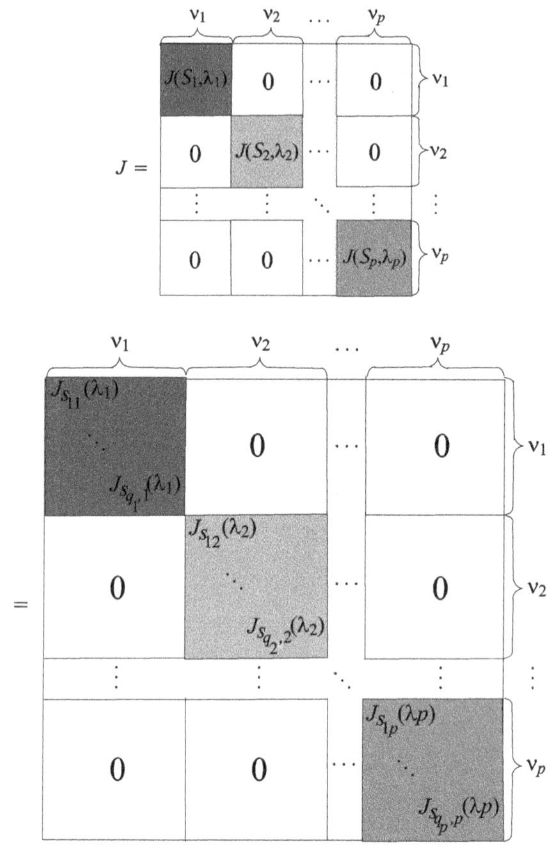

Fig. 2.12. Jordan normal form.

Example 2.19. Given is a tenth-order matrix

$$A = \begin{bmatrix} 1 & -3 & 5 & -7 & 1 & -4 & 3 & 12 & 3 & -18 \\ 10 & 15 & -7 & 18 & -10 & 7 & 12 & -17 & 2 & 59 \\ 4 & 3 & 4 & 2 & -4 & 0 & 8 & 4 & 3 & 10 \\ 7 & 10 & -6 & 17 & -7 & 6 & 7 & -16 & 0 & 47 \\ -12 & -13 & 2 & -14 & 15 & -4 & -21 & 5 & -7 & -54 \\ -3 & -3 & 0 & -3 & 4 & 3 & -9 & -1 & -5 & -14 \\ 8 & 10 & -1 & 10 & -10 & 1 & 22 & -3 & 9 & 43 \\ -12 & -13 & -1 & -12 & 14 & -1 & -28 & 1 & -12 & -53 \\ -3 & -4 & 0 & -3 & 4 & 1 & -9 & 0 & -3 & -16 \\ -11 & -13 & 3 & -15 & 12 & -4 & -20 & 8 & -7 & -54 \end{bmatrix}$$

that has three distinct eigenvalues

$$\lambda_1 = -1; \quad \lambda_2 = 2; \quad \lambda_3 = 3$$

with multiplicities $\nu_1 = 1,\ \nu_2 = 5$ and $\nu_3 = 4$, respectively. The Segre characteristics associated with these eigenvalues are

$$\begin{aligned} \mathcal{S}_1 &= \{s_{11}\} = \{1\}, \\ \mathcal{S}_2 &= \{s_{12}, s_{22}\} = \{3, 2\}, \\ \mathcal{S}_3 &= \{s_{13}, s_{23}\} = \{2, 2\}. \end{aligned}$$

Using a nonsingular transformation with the matrix

$$Z = \left[\underbrace{z_1^{(11)}}_{s_{11}} \,\Big|\, \underbrace{z_1^{(12)} \; z_2^{(12)} \; z_3^{(12)}}_{s_{12}} \,\Big|\, \underbrace{z_1^{(22)} \; z_2^{(22)}}_{s_{22}} \,\Big|\, \underbrace{z_1^{(13)} \; z_2^{(13)}}_{s_{13}} \,\Big|\, \underbrace{z_1^{(23)} \; z_2^{(23)}}_{s_{23}}\right]$$

$$= \left[\begin{array}{r|rrr|rr|rr|rr} 1 & 3 & -1 & 2 & 1 & 0 & -1 & -1 & 0 & 1 \\ -4 & -6 & 1 & -4 & 0 & 0 & 1 & 2 & 0 & -1 \\ -1 & 0 & 0 & 0 & 1 & 0 & 0 & 0 & 1 & 0 \\ -3 & -5 & 1 & -2 & 0 & 0 & 1 & 2 & 0 & -1 \\ 4 & 4 & 0 & 2 & -1 & 1 & 0 & -1 & -1 & 1 \\ 1 & 0 & 0 & 0 & 0 & 0 & 1 & 0 & 1 & 0 \\ -3 & -2 & -1 & -1 & 1 & 1 & -1 & 0 & -1 & -1 \\ 4 & 2 & 1 & 1 & -2 & 0 & 1 & 0 & 0 & 1 \\ 1 & 0 & 1 & 0 & 0 & -1 & 1 & 0 & 1 & 0 \\ 4 & 4 & 0 & 2 & -1 & 0 & 0 & -1 & 0 & 1 \end{array}\right],$$

the matrix A is reduced into the Jordan normal form

$$J = \left[\begin{array}{c|cc|cc} J_{s_{11}}(\lambda_1) & 0 & 0 & 0 & 0 \\ \hline 0 & J_{s_{12}}(\lambda_2) & 0 & 0 & 0 \\ 0 & 0 & J_{s_{22}}(\lambda_2) & 0 & 0 \\ \hline 0 & 0 & 0 & J_{s_{13}}(\lambda_3) & 0 \\ 0 & 0 & 0 & 0 & J_{s_{23}}(\lambda_3) \end{array}\right]$$

$$= \left[\begin{array}{c|ccc|cc|cc|cc}
-1 & 0 & 0 & 0 & 0 & 0 & 0 & 0 & 0 & 0 \\
\hline
0 & 2 & 1 & 0 & 0 & 0 & 0 & 0 & 0 & 0 \\
0 & 0 & 2 & 1 & 0 & 0 & 0 & 0 & 0 & 0 \\
0 & 0 & 0 & 2 & 0 & 0 & 0 & 0 & 0 & 0 \\
\hline
0 & 0 & 0 & 0 & 2 & 1 & 0 & 0 & 0 & 0 \\
0 & 0 & 0 & 0 & 0 & 2 & 0 & 0 & 0 & 0 \\
\hline
0 & 0 & 0 & 0 & 0 & 0 & 3 & 1 & 0 & 0 \\
0 & 0 & 0 & 0 & 0 & 0 & 0 & 3 & 0 & 0 \\
\hline
0 & 0 & 0 & 0 & 0 & 0 & 0 & 0 & 3 & 1 \\
0 & 0 & 0 & 0 & 0 & 0 & 0 & 0 & 0 & 3
\end{array}\right].$$

The characteristic polynomial of A is

$$p(\lambda) = (\lambda + 1)(\lambda - 2)^5(\lambda - 3)^4$$

and its minimal polynomial is

$$p(\lambda) = (\lambda + 1)(\lambda - 2)^3(\lambda - 3)^2.$$

2.2.5 *Weyr Canonical Form*

Apart of the Jordan form, it is possible to define other canonical forms characterized by a simple structure. One such form is the *Weyr canonical form*, which may be obtained from the Jordan form by an appropriate permutation of the columns of the nonsingular similarity transformation. The Weyr form is introduced by the following example.

Example 2.20. Consider again the seventh-order matrix A with 7-tuple eigenvalue $\lambda = 1$, given in Example 2.13. Let us collect in the matrix X_1 the three eigenvectors of A such that

$$X_1 = [z_1^{(1)}, z_1^{(2)}, z_1^{(3)}] = \begin{bmatrix} -2 & -2 & 1 \\ -1 & 0 & 0 \\ -2 & -1 & 1 \\ -1 & 0 & 0 \\ 0 & 0 & -1 \\ 1 & 0 & 0 \\ -1 & -1 & 1 \end{bmatrix}.$$

The columns of this matrix belong to the null space of $A - \lambda I$,

$$\mathcal{N}_1 = \mathcal{N}(A - \lambda I), \quad \dim(\mathcal{N}_1) = \mathrm{null}(A - \lambda I) = 3.$$

We have that

$$AX_1 = X_1 \begin{bmatrix} \lambda & 0 & 0 \\ 0 & \lambda & 0 \\ 0 & 0 & \lambda \end{bmatrix}. \tag{2.35}$$

In the next step, we collect in the matrix X_2 the principal vectors of grade 2, i.e.,

$$X_2 = [z_2^{(1)}, z_2^{(2)}] = \begin{bmatrix} 0 & 2 \\ -1 & 0 \\ 0 & 1 \\ 0 & 1 \\ -1 & 0 \\ 1 & 0 \\ 1 & 1 \end{bmatrix}.$$

The columns of this matrix belong to the restricted null space $\mathcal{N}_2 \backslash \mathcal{N}_1$, where

$$\mathcal{N}_2 = \mathcal{N}((A - \lambda I)^2), \quad \dim(\mathcal{N}_2) = \mathrm{null}((A - \lambda I)^2) = 5,$$
$$\dim(\mathcal{N}_2 \setminus \mathcal{N}_1) = 2.$$

It follows from the properties of the principal vectors of grade 2 that

$$AX_2 = X_2 \begin{bmatrix} \lambda & 0 \\ 0 & \lambda \end{bmatrix} + S_1 \begin{bmatrix} 1 & 0 \\ 0 & 1 \\ 0 & 0 \end{bmatrix}. \tag{2.36}$$

In the third step, we collect in the matrix X_3 all principal vectors of grade 3, i.e.,

$$X_3 = [z_3^{(1)}, z_3^{(2)}] = \begin{bmatrix} -1 & 0 \\ 1 & -1 \\ 0 & 0 \\ 0 & 0 \\ 1 & 0 \\ 0 & 1 \\ -1 & 0 \end{bmatrix}.$$

The columns of the matrix X_3 belong to the restricted null space $\mathcal{N}_3 \setminus \mathcal{N}_2$, where

$$\mathcal{N}_3 = \mathcal{N}((A - \lambda I)^3), \quad \dim(\mathcal{N}_3) = \text{null}((A - \lambda I)^3) = 7,$$
$$\dim(\mathcal{N}_3 \setminus \mathcal{N}_2) = 2.$$

From the properties of the principal vectors of grade 3, it follows that

$$AX_3 = X_3 \begin{bmatrix} \lambda & 0 \\ 0 & \lambda \end{bmatrix} + S_2 \begin{bmatrix} 1 & 0 \\ 0 & 1 \end{bmatrix}. \tag{2.37}$$

Combining the relationships (2.35), (2.36) and (2.37), we obtain

$$A[X_1,\ X_2,\ X_3] = [X_1,\ X_2,\ X_3]W,$$

where

$$W = \left[\begin{array}{ccc|cc|cc} \lambda & 0 & 0 & 1 & 0 & 0 & 0 \\ 0 & \lambda & 0 & 0 & 1 & 0 & 0 \\ 0 & 0 & \lambda & 0 & 0 & 0 & 0 \\ \hline 0 & 0 & 0 & \lambda & 0 & 1 & 0 \\ 0 & 0 & 0 & 0 & \lambda & 0 & 1 \\ \hline 0 & 0 & 0 & 0 & 0 & \lambda & 0 \\ 0 & 0 & 0 & 0 & 0 & 0 & \lambda \end{array}\right].$$

Setting

$$X_W = [X_1,\ X_2,\ X_3],$$

we obtain that

$$X_W^{-1} A X_W = W.$$

The matrix W is the Weyr form of the matrix A. The diagonal blocks of this form have sizes 3, 2, 2 which are equal to the number of the eigenvectors, the number of principal vectors of grade 2 and the number of the principal vectors of grade 3, respectively. These sizes are equal to the dimensions of the restricted subspaces

$$\mathcal{N}_1, \quad \mathcal{N}_2 \setminus \mathcal{N}_1, \quad \mathcal{N}_3 \setminus \mathcal{N}_2,$$

which in turn are equal to the elements of the Weyr characteristic $\mathcal{W} = \{3,\ 3,\ 2\}$ associated with λ. Thus, the sizes of the diagonal blocks in the Weyr form are equal to the elements of the Weyr characteristic.

The earlier example shows that the Weyr form is permutation similar to the Jordan form, since the reduction into this form is done by using a matrix whose columns are the permuted columns of the matrix used in the reduction to Jordan form. Regardless of this fact, the Weyr form represents an independent interest since its determination can be done by a numerically reliable procedure. That is why in practice the Jordan form is obtained reducing the matrix into Weyr form after that over it, and over the transformation matrix, are performed the corresponding permutations.

Let the matrix A with n-tuple eigenvalue λ be given. Reordering the eigenvectors and the principal vectors of A presented in (2.30), so that the eigenvectors are in the first place, after them are the principal vectors of grade 2, after them the principal vectors of grade 3 and so on, we obtain (see Figure 2.13)

$$X_W = \Big[\underbrace{z_1^{(1)}\ z_1^{(2)}\ \dots\ z_1^{(w_1)}}_{w_1} \,|\, \underbrace{z_2^{(1)}\ z_2^{(2)}\ \dots\ z_2^{(w_2)}}_{w_2} \,|\, \dots$$

$$|\, \underbrace{z_{w_h}^{(1)}\ z_{w_h}^{(2)}\ \dots\ z_{w_h}^{(w_h)}}_{w_h} \Big], \tag{2.38}$$

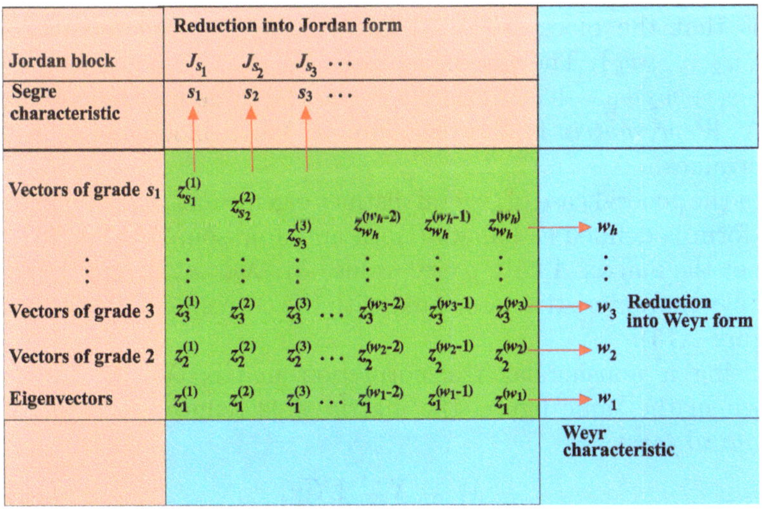

Fig. 2.13. Ordering of the eigenvectors and principal vectors in Jordan and Weyr decomposition for n-tuple eigenvalue.

where the numbers $w_1 \geq w_2 \geq \cdots \geq w_h$ are the elements of the Weyr characteristic $\mathcal{W} = \{w_1, w_2, \ldots, w_h\}$ of A associated with λ.

Using a nonsingular transformation with the matrix X_W, A is reduced into the form

$$W(\mathcal{W}, \lambda) = X_W^{-1} A X_W, \tag{2.39}$$

where the upper block bidiagonal matrix

$$W(\mathcal{W}, \lambda) = \begin{bmatrix} \lambda I_{w_1} & G_{w_1,w_2} & & & \\ & \lambda I_{w_2} & G_{w_2,w_3} & & \\ & & \ddots & \ddots & \\ & & & \ddots & G_{w_{h-1},w_h} \\ & & & & \lambda I_{w_h} \end{bmatrix} \tag{2.40}$$

is called *Weyr block* corresponding to the eigenvalue λ and

$$G_{w_i,w_j} = \begin{bmatrix} I_{w_j} \\ 0_{(w_i - w_j) \times w_j} \end{bmatrix} \in \mathbb{C}^{w_i \times w_j}, \quad 1 \leq i < j$$

(see Figure 2.14). Note that $\text{rank}(G_{w_i,w_j}) = w_j$ and if $w_i = w_{i+1}$, then $G_{w_i,w_{i+1}} = I_{w_i}$. The diagonal blocks $\lambda_i I_{w_i}$ are called *scalar matrices* and the superdiagonal blocks G_{w_i,w_j} are *full-column-rank matrices.*

The direct evaluation of the powers of $W(\mathcal{W}, \lambda) - \lambda I$ confirms that the block $W(\mathcal{W}, \lambda)$ has the Weyr characteristic $\mathcal{W} = \{w_1, w_2, \ldots, w_h\}$. The size of the block $W(\mathcal{W}, \lambda)$ is equal to the sum $\nu = w_1 + w_2 + \cdots + w_h$, where ν is the algebraic multiplicity of λ.

A *Weyr matrix* is a direct sum of Weyr blocks with distinct eigenvalues.

In the general case of several distinct eigenvalues, the Weyr canonical form is defined as follows (Horn and Johnson, 2013, p. 203).

Let the matrix A have p eigenvalues $\lambda_1, \lambda_2, \ldots, \lambda_p$ and the set $\mathcal{W}$ of Weyr characteristics associated with the corresponding eigenvalues (Figure 2.15).

Using a nonsingular transformation matrix X_W of the form, shown in the same figure, the matrix A is reduced into the *Weyr canonical form*

$$W = X_W^{-1} A X_W, \tag{2.41}$$

where the matrix W is shown in Figure 2.16 and each block $W(\mathcal{W}_i, \lambda_i)$ is of the form (2.40). The columns of the matrix X_W

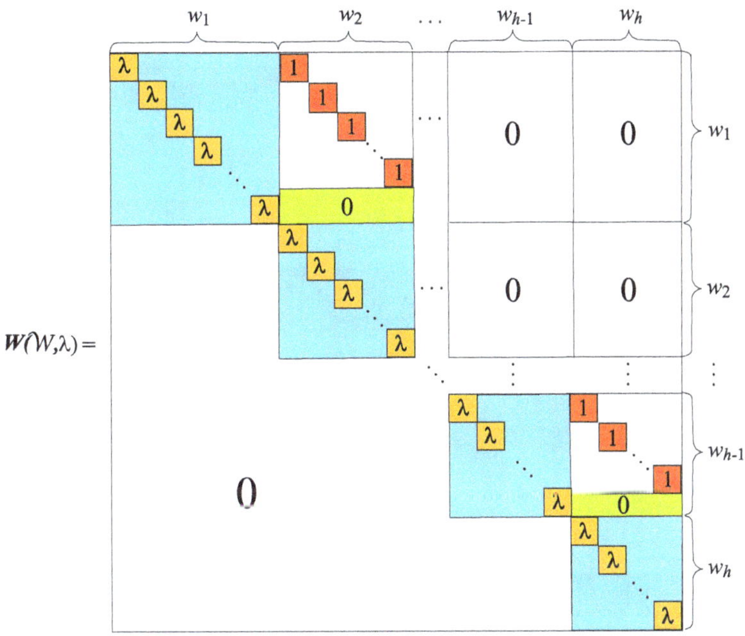

Fig. 2.14. Weyr block.

are permutations of the columns of the matrix Z of the similarity transformation into Jordan normal form. In case of a nonderogatory matrix (matrix which has only one Jordan block corresponding to a given eigenvalue), the two canonical forms are identical.

Example 2.21. Consider again the tenth-order matrix given in Example 2.19. The Weyr characteristics, associated with the eigenvalues

$$\lambda_1 = -1; \quad \lambda_2 = 2; \quad \lambda_3 = 3,$$

are, respectively,

$$\mathcal{W}_1 = \{w_{11}\} = \{1\},$$
$$\mathcal{W}_2 = \{w_{21}, w_{22}, w_{23}\} = \{2, 2, 1\},$$
$$\mathcal{W}_3 = \{w_{31}, w_{32}\} = \{2, 2\}.$$

This means that the Weyr canonical form of A has three blocks of order $\nu_1 = 1$, $\nu_2 = 5$, $\nu_3 = 4$.

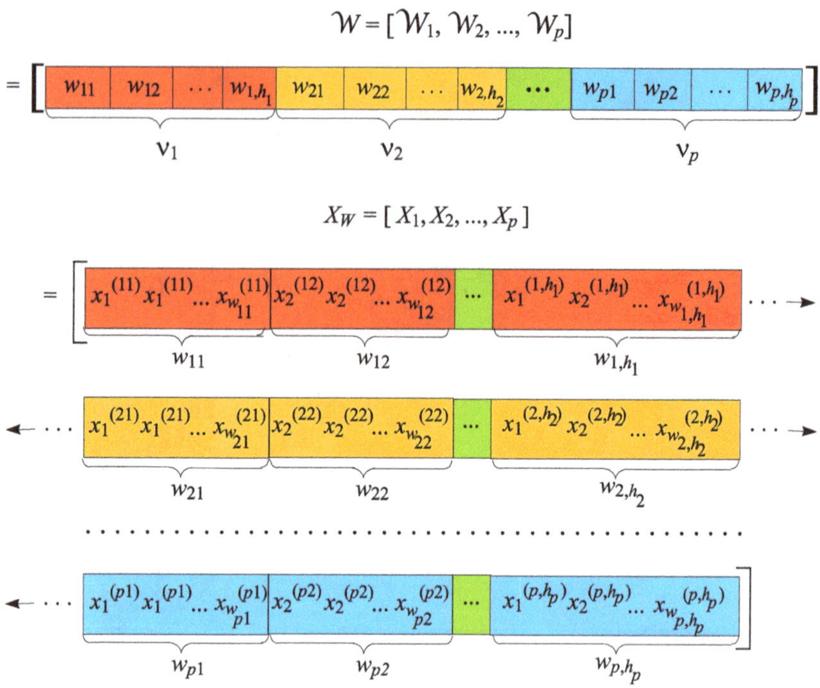

Fig. 2.15. The Weyr characteristics and the similarity transformation matrix into Weyr form.

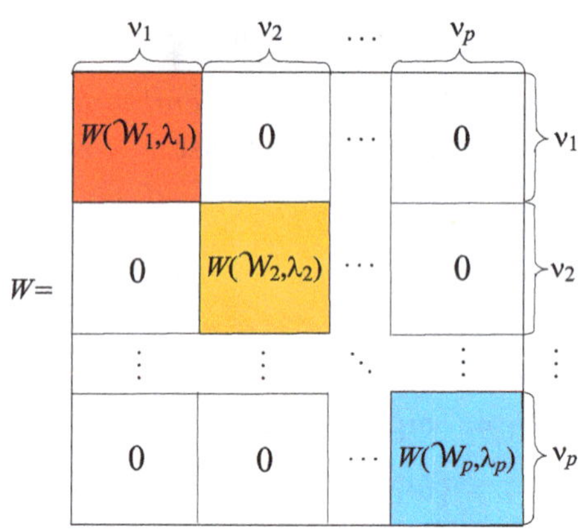

Fig. 2.16. Weyr canonical form.

Permuting appropriately the columns of the matrix Z of the similarity transformation into Jordan form, we obtain the similarity transformation matrix into Weyr form:

$$X_W = \left[\underbrace{z_1^{(11)}}_{\nu_1} \,\middle|\, \underbrace{z_1^{(12)}\; z_1^{(22)}\; z_2^{(12)}\; z_1^{(22)}\; z_2^{(22)}}_{\nu_2} \,\middle|\, \underbrace{z_1^{(13)}\; z_1^{(23)}\; z_2^{(13)}\; z_2^{(23)}}_{\nu_3} \right]$$

$$= \left[\begin{array}{r|rrrrr|rrrr} 1 & 3 & 1 & -1 & 0 & 2 & -1 & 0 & -1 & 1 \\ -4 & -6 & 0 & 1 & 0 & -4 & 1 & 0 & 2 & -1 \\ -1 & 0 & 1 & 0 & 0 & 0 & 0 & 1 & 0 & 0 \\ -3 & -5 & 0 & 1 & 0 & -2 & 1 & 0 & 2 & -1 \\ 4 & 4 & -1 & 0 & 1 & 2 & 0 & -1 & -1 & 1 \\ 1 & 0 & 0 & 0 & 0 & 0 & 1 & 1 & 0 & 0 \\ -3 & -2 & 1 & -1 & 1 & -1 & -1 & -1 & 0 & -1 \\ 4 & 2 & -2 & 1 & 0 & 1 & 1 & 0 & 0 & 1 \\ 1 & 0 & 0 & 1 & -1 & 0 & 1 & 1 & 0 & 0 \\ 4 & 4 & -1 & 0 & 0 & 2 & 0 & 0 & -1 & 1 \end{array}\right].$$

The Weyr form of the matrix A is obtained as

$$W = X_W^{-1} A X_W = \left[\begin{array}{r|r|r} W_{\mathcal{W}_1}(\lambda_1) & 0 & 0 \\ \hline 0 & W_{\mathcal{W}_2}(\lambda_2) & 0 \\ \hline 0 & 0 & W_{\mathcal{W}_3}(\lambda_2) \end{array}\right]$$

$$= \left[\begin{array}{r|rrrrr|rrrr} -1 & 0 & 0 & 0 & 0 & 0 & 0 & 0 & 0 & 0 \\ \hline 0 & 2 & 0 & 1 & 0 & 0 & 0 & 0 & 0 & 0 \\ 0 & 0 & 2 & 0 & 1 & 0 & 0 & 0 & 0 & 0 \\ 0 & 0 & 0 & 2 & 0 & 1 & 0 & 0 & 0 & 0 \\ 0 & 0 & 0 & 0 & 2 & 0 & 0 & 0 & 0 & 0 \\ 0 & 0 & 0 & 0 & 0 & 2 & 0 & 0 & 0 & 0 \\ \hline 0 & 0 & 0 & 0 & 0 & 0 & 3 & 0 & 1 & 0 \\ 0 & 0 & 0 & 0 & 0 & 0 & 0 & 3 & 0 & 1 \\ 0 & 0 & 0 & 0 & 0 & 0 & 0 & 0 & 3 & 0 \\ 0 & 0 & 0 & 0 & 0 & 0 & 0 & 0 & 0 & 3 \end{array}\right].$$

Consider the relationships between the number and sizes of the blocks of Jordan and Weyr forms. Assume that the set of Segre characteristics associate with the distinct eigenvalues $\lambda_1,\ \lambda_2,\dots,\lambda_p$ is

$$\mathcal{S} = [\{s_{11}, s_{21}, \dots, s_{q_1,1}\}, \{s_{12}, s_{22}, \dots, s_{q_2,2}\}, \dots, \{s_{1p}, s_{2p}, \dots, s_{q_p,p}\}]$$

and the set of Weyr characteristics associated with the same eigenvalues is

$$\begin{aligned}\mathcal{W} = [&\{w_{11}, w_{12}, \dots, w_{1,h_1}\}, \{w_{21}, w_{22}, \dots, w_{2,h_2}\}, \dots,\\ &\{w_{p1}, w_{p2}, \dots, w_{p,h_p}\}].\end{aligned}$$

We have that $q_1 = w_{11},\ q_2 = w_{21},\dots,q_p = w_{p1}$ and $h_1 = s_{11},\ h_2 = s_{12},\dots,h_p = s_{1p}$. Thus, the first element w_{i1} of the Weyr characteristic associated with λ_i (the size of the maximum order Weyr block corresponding to λ_i) is equal to the number q_i of the elements of the Segre characteristic associated with the same eigenvalue, i.e., to the number of the Jordan blocks in the canonical form of A. In turn, the number h_i of the elements of the Weyr characteristic associated with λ_i is exactly equal to the first element s_{1i} of the Segre characteristic associated with this eigenvalue, i.e., to the size of the maximum order Jordan block corresponding to λ_i.

The relationship between the number of elements of Segre and Weyr characteristics associated with λ_i is shown in Figure 2.17.

2.2.6 *Frobenius Canonical Form*

Another canonical form that may be considered as an alternative to the Jordan normal form, is the so called Frobenius form of a matrix. For simplicity, we shall presented this form first for the particular case of a matrix A with an n-tuple eigenvalue. Assume that this eigenvalue has geometric multiplicity equal to q and associate Segre characteristic $\mathcal{S} = \{s_1, s_2, \dots, s_q\}$. Then the matrix of the similarity transformation of A into Jordan form is

$$Z = \left[\underbrace{z_1^{(1)}\ z_2^{(1)}\ \dots\ z_{s_1}^{(1)}}_{s_1} \mid \underbrace{z_1^{(2)}\ z_2^{(2)}\ \dots\ z_{s_2}^{(2)}}_{s_2} \mid \dots \mid \underbrace{z_1^{(q)}\ z_2^{(q)}\ \dots\ z_{s_q}^{(q)}}_{s_q}\right].$$

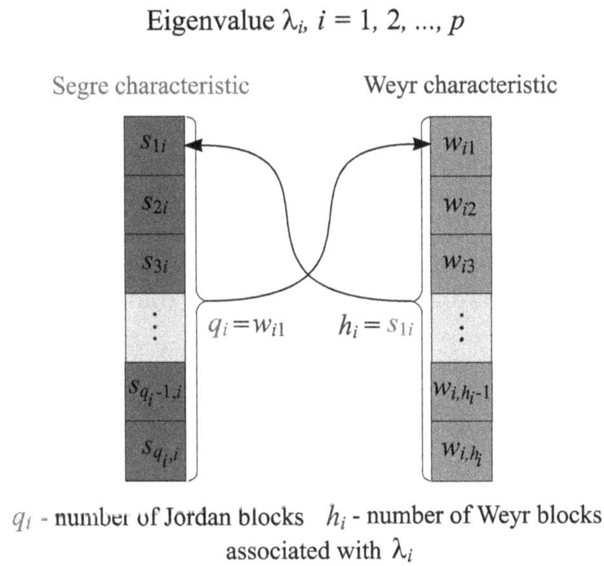

Fig. 2.17. Relation between the number of elements of Segre and Weyr characteristics.

Similar to the case of reducing a matrix into companion form considered in Section 2.2.1, we construct a similarity transformation matrix consisting of the bases of the corresponding cyclic subspaces in the form

$$X_F = \Big[\underbrace{z_{s_1}^{(1)}\ Az_{s_1}^{(1)}\ \dots\ A^{s_1-1}z_{s_1}^{(1)}}_{s_1} \mid \underbrace{z_{s_2}^{(2)}\ Az_{s_2}^{(2)}\ \dots\ A^{s_2-1}z_{s_2}^{(2)}}_{s_2} \mid \dots$$

$$\mid \underbrace{z_{s_q}^{(q)}\ Az_{s_q}^{(q)}\ \dots\ A^{s_q-1}z_{s_q}^{(q)}}_{s_q}\Big], \tag{2.42}$$

where $z_{s_i}^{(i)}$ is the last principal vector of each Jordan chain. Now it may be shown that using a similarity transformation with the matrix X_F, the matrix A is reduced into the *Frobenius normal form* or *rational canonical form*

$$F = X_F^{-1}AX_F = \operatorname{diag}(C_1, C_2, \dots, C_q),$$

representing a direct sum of q companion matrices C_j of order s_j, $s_1 \geq s_2 \geq \cdots \geq s_q$, associated with the corresponding cyclic subspaces. As far as the matrices A and F have the same characteristic polynomials, the characteristic polynomial of A is equal to the product of the characteristic polynomials of the diagonal blocks and the minimal polynomial of A is equal to the characteristic polynomial of the largest block C_1. Since the order of each block $C_2, C_3, \ldots, C_q$ is less than or equal to the order of the previous block and since the eigenvalues of these blocks are equal, the characteristic polynomial of each block divides without remainder the characteristic polynomial of the previous block. The maximum dimension of the cyclic subspace

$$\text{span}(x, Ax, A^2x, \ldots, A^nx)$$

for an arbitrary vector x is equal to the order of the block C_1, i.e., to the degree of the minimal polynomial of A.

Example 2.22. Consider again the seventh-order matrix A with 7-tuple eigenvalue equal to 1, whose eigenvectors and principal vectors are given in Example 2.13. Let us construct the nonsingular matrix

$$X_F = \left[z_3^{(1)}, Az_3^{(1)}, A^2z_3^{(1)} |, z_3^{(2)}, Az_3^{(2)}, A^2z_3^{(2)} |, z_1^{(3)}\right]$$

$$= \left[\begin{array}{ccc|ccc|c} -1 & -1 & -3 & 0 & 2 & 2 & 1 \\ 1 & 0 & -2 & -1 & -1 & -1 & 0 \\ 0 & 0 & -2 & 0 & 1 & 1 & 1 \\ 0 & 0 & -1 & 0 & 1 & 2 & 0 \\ 1 & 0 & -1 & 0 & 0 & 0 & -1 \\ 0 & 1 & 3 & 1 & 1 & 1 & 0 \\ -1 & 0 & 0 & 0 & 1 & 1 & 1 \end{array}\right]$$

consisting of the bases of the cyclic subspaces associated with the principal vectors $z_3^{(1)}, z_3^{(2)}$ and $z_1^{(3)}$. Performing a similarity transformation of A with this matrix, we obtain the Frobenius normal form

$$F = X_F^{-1} A X_F$$

$$= \left[\begin{array}{ccc|ccc|c} 0 & 0 & 1 & 0 & 0 & 0 & 0 \\ 1 & 0 & -3 & 0 & 0 & 0 & 0 \\ 0 & 1 & 3 & 0 & 0 & 0 & 0 \\ \hline 0 & 0 & 0 & 0 & 0 & 1 & 0 \\ 0 & 0 & 0 & 1 & 0 & -3 & 0 \\ 0 & 0 & 0 & 0 & 1 & 3 & 0 \\ \hline 0 & 0 & 0 & 0 & 0 & 0 & 1 \end{array}\right],$$

which in the given case is a direct sum of three companion matrices of order $s_1 = 3$, $s_2 = 3$ and $s_3 = 1$. The characteristic polynomial of A is

$$p(\lambda) = (\lambda - 1)^3(\lambda - 1)^3(\lambda - 1) = \lambda^7 - 7\lambda^6 + 21\lambda^5 - 35\lambda^4 + 35\lambda^3 - 21\lambda^2 + 7\lambda - 1$$

and its minimal polynomial is

$$p_{\min}(\lambda) = (\lambda - 1)^3 = \lambda^3 - 3\lambda^2 + 3\lambda - 1.$$

Consider now the general case of a matrix A with p distinct eigenvalues $\lambda_1, \lambda_2, \ldots, \lambda_p$ and a set of the associated Segre characteristics

$$\mathcal{S} = [\{s_{11}, s_{21}, \ldots, s_{q_1,1}\}, \{s_{12}, s_{22}, \ldots, s_{q_2,2}\}, \ldots, \{s_{1p}, s_{2p}, \ldots, s_{q_p,p}\}].$$

To each s_{ji} corresponds a Jordan block $J_{s_{ji}}(\lambda_i)$, $j = 1, 2, \ldots, q_i$, $i = 1, 2, \ldots, p$. Let us order these blocks as

$$\begin{array}{cccc} J_{s_{11}}(\lambda_1), & J_{s_{21}}(\lambda_1), & \ldots & J_{s_{q_1,1}}(\lambda_1), \\ J_{s_{12}}(\lambda_2), & J_{s_{22}}(\lambda_2), & \ldots & J_{s_{q_2,2}}(\lambda_2), \\ \cdots & \cdots & \cdots & \cdots \\ J_{s_{1m}}(\lambda_m), & J_{s_{2m}}(\lambda_m), & \ldots & J_{s_{q_{\max},m}}(\lambda_m), \\ \cdots & \cdots & \cdots & \cdots \\ J_{s_{1p}}(\lambda_p), & J_{s_{2p}}(\lambda_p), & \ldots & J_{s_{q_p,p}}(\lambda_p), \end{array}$$

where the integer $q_{\max} = \max\{q_1, q_2, \ldots, q_p\}$ is equal to the maximum number of blocks, associated with an eigenvalue.

The Jordan canonical form of A can be written as

$$
\left[\begin{array}{cccc|ccc|c|cc}
J_{s_{11}}(\lambda_1) & & & & & & & & & \\
& J_{s_{12}}(\lambda_2) & & & & & & & & \\
& & \ddots & & & & & & & \\
& & & J_{s_{1p}}(\lambda_p) & & & & & & \\
\hline
& & & & J_{s_{21}}(\lambda_1) & & & & & \\
& & & & & J_{s_{22}}(\lambda_2) & & & & \\
& & & & & & \ddots & & & \\
\hline
& & & & & & & \ddots & & \\
\hline
& & & & & & & & J_{s_{q_{\max},m}}(\lambda_m) & \\
& & & & & & & & & \ddots
\end{array}\right]
$$

$$
= \left[\begin{array}{c|c|c|c}
G_1 & & & \\
\hline
& G_2 & & \\
\hline
& & \ddots & \\
\hline
& & & G_{q_{\max}}
\end{array}\right],
$$

where we have grouped together the Jordan blocks of highest order in each eigenvalue to produce the matrix G_1, then those of highest order to produce G_2 and so on. Each of the matrices G_k has only one Jordan block associated with each eigenvalue λ_i and, consequently, it can be transformed to a companion matrix. In this way, each matrix A can be reduced into the Frobenius canonical form

$$
F = \left[\begin{array}{c|c|c|c}
C_1 & & & \\
\hline
& C_2 & & \\
\hline
& & \ddots & \\
\hline
& & & C_{q_{\max}}
\end{array}\right],
$$

representing a direct sum of $q_{\max}$ companion matrices C_k, $k = 1, 2, \ldots, q_{\max}$. If the matrix A is nonderogatory, then $q_{\max} = 1$ and

$C_1 \in \mathbb{C}^{n \times n}$. Different from Jordan and Weyr form, the Frobenius form is rational in the field of A, which means that if A is real, then the blocks C_k are also real even if A has complex conjugate eigenvalues.

Example 2.23. Let the 11×11 matrix A have three distinct eigenvalues λ_1, λ_2 and λ_3 with corresponding Segre characteristics $S_1 = \{2,\ 1\}, S_2 = \{3,\ 2,\ 1\}$ and $S_3 = \{2\}$. Ordering the Jordan blocks as

$$\begin{aligned}
&J_2(\lambda_1),\ J_1(\lambda_1),\\
&J_3(\lambda_2),\ J_2(\lambda_2),\ J_1(\lambda_2),\\
&J_2(\lambda_3),
\end{aligned}$$

we obtain that the matrix

$$\left[\begin{array}{c|c|c} G_1 & & \\ \hline & G_2 & \\ \hline & & G_3 \end{array}\right] = \left[\begin{array}{ccc|cc|c} J_2(\lambda_1) & & & & & \\ & J_3(\lambda_2) & & & & \\ & & J_2(\lambda_3) & & & \\ \hline & & & J_1(\lambda_1) & & \\ & & & & J_2(\lambda_2) & \\ \hline & & & & & J_1(\lambda_2) \end{array}\right]$$

and hence the matrix A is similar to the Frobenius matrix

$$F = \left[\begin{array}{c|c|c} C_1 & & \\ \hline & C_2 & \\ \hline & & C_3 \end{array}\right] = \left[\begin{array}{ccccccc|ccc|c}
 & & & & & & \times & & & & \\
1 & & & & & & \times & & & & \\
 & 1 & & & & & \times & & & & \\
 & & 1 & & & & \times & & & & \\
 & & & 1 & & & \times & & & & \\
 & & & & 1 & & \times & & & & \\
 & & & & & 1 & \times & & & & \\ \hline
 & & & & & & & & & \times & \\
 & & & & & & & 1 & & \times & \\
 & & & & & & & & 1 & \times & \\ \hline
 & & & & & & & & & & \times
\end{array}\right].$$

Note that if A is real (complex), F is also real (complex) independent of the types of the eigenvalues.

The characteristic polynomials of the blocks C_1, C_2, C_3 are

$$
\begin{aligned}
p(C_1) &= (\lambda - \lambda_1)^2(\lambda - \lambda_2)^3(\lambda - \lambda_3)^2,\\
p(C_2) &= (\lambda - \lambda_1)^1(\lambda - \lambda_2)^2,\\
p(C_3) &= (\lambda - \lambda_2)^1.
\end{aligned}
$$

The polynomial $p(C_2)$ is a factor of the polynomial $p(C_1)$ and $p(C_3)$ is a factor of $p(C_2)$, i.e., each of these polynomials divides without reminder the characteristic polynomial of the preceding block.

The Frobenius form is not appropriate for numerical computations, since the reduction to this form requires similarity transformations with a matrix X_F of the type (2.42) which may be ill conditioned (Wilkinson, 1965).

2.2.7 *Spectral Projections*

Let the matrix A be diagonalizable, i.e.,

$$A = X\Lambda Y^H,$$

where $Y^H = X^{-1}$. Then it may be represented in the form

$$A = \sum_{i=1}^{n} \lambda_i P_i, \tag{2.43}$$

where

$$P_i = x_i y_i^H. \tag{2.44}$$

It follows from $y_i^H x_i = 1$ that

$$P_i^2 = x_i(y_i^H x_i)y_i^H = P_i, \quad \sum_{i=1}^{n} P_i = I_n. \tag{2.45}$$

Hence, $P_i = x_i y_i^H$ is an elementary projection (see Section A.9 of the Appendix). It is called *eigenprojection* or *spectral projection* on the eigenspace, spanned by the eigenvector x_i, parallel to the rest eigenspaces. It is possible to show that this projection satisfies the conditions

$$AP_i = P_i A = P_i A P_i. \tag{2.46}$$

It is necessary to point out that the expression (2.45) for the spectral projection is valid only if the left eigenvectors are chosen so that $y_i^H x_i = 1$. If the eigenvectors are normalized so that $\|x_i\|_2 = 1$, $\|y_i\|_2 = 1$, then this condition is generally not fulfilled and the spectral projection should be computed from

$$P_i = \frac{x_i y_i^H}{y_i^H x_i}. \tag{2.47}$$

The representation (2.43) is called *spectral decomposition* of the matrix A and is valid for a matrix with a simple structure.

The spectral projections play an important role in studying the sensitivity of eigenvalues and eigenspaces (see Chapter 3).

Example 2.24. For the matrix

$$A = \begin{bmatrix} -5 & -4 & 6 \\ -2 & -5 & 4 \\ -3 & -4 & 4 \end{bmatrix}$$

of Example 2.2, the projections on the three eigenspaces spanned by the eigenvectors are

$$P_1 = x_1 y_1^T = \begin{bmatrix} -2 & -2 & 4 \\ -1 & -1 & 2 \\ -2 & -2 & 4 \end{bmatrix}, \quad P_2 = x_2 y_2^T = \begin{bmatrix} 2 & 0 & -2 \\ 0 & 0 & 0 \\ 1 & 0 & -1 \end{bmatrix},$$

$$P_3 = x_3 y_3^T = \begin{bmatrix} 1 & 2 & -2 \\ 1 & 2 & -2 \\ 1 & 2 & -2 \end{bmatrix}.$$

The projections P_i satisfy the conditions (2.45) and (2.46).

The spectral projections can also be determined for eigenspace with dimension larger than 1.

Let X and Y^H be the respective unitary bases of the right $\mathcal{X}$ and left $\mathcal{Y}$ eigenspaces of A with dimension k. Suppose that L is the restriction of A to $\mathcal{X}$. Providing that $Y^H X = I_k$, the matrix $P = XY^H$ is a spectral projection on $\mathcal{X}$ and

$$PAP = XY^H AXY^H = XLY^H.$$

In case of defective eigenvalues, the spectral decomposition of a matrix can be done using the Jordan canonical form.

Let the matrix A have $p < n$ distinct eigenvalues $\lambda_1, \lambda_2, \ldots, \lambda_p$ with multiplicities $\nu_1, \nu_2, \ldots, \nu_p$ and associated Segre characteristics $\mathcal{S}_1, \mathcal{S}_2, \ldots, \mathcal{S}_p$, respectively. Then A is reduced into the Jordan canonical form

$$J = Z^{-1}AZ,$$

where

$$J = \operatorname{diag}(J(\mathcal{S}_1, \lambda_1), J(\mathcal{S}_2, \lambda_2), \ldots, J(\mathcal{S}_p, \lambda_p)).$$

We have that $\mathcal{S}_i = \{s_{1i}, s_{2i}, \ldots, s_{q_i,i}\}$, $i = 1, 2, \ldots, p$. The block $J(\mathcal{S}_i, \lambda_i) \in \mathbb{C}^{\nu_i \times \nu_i}$ contains all Jordan blocks, associated with λ_i and can be represented as

$$J(\mathcal{S}_i, \lambda_i) = \lambda_i I_{\nu_i} + N_i,$$

where the matrix N_i has the form

$$N_i = \operatorname{diag}\left(\underbrace{\begin{bmatrix} 0 & 1 & 0 & \ldots & 0 \\ 0 & 0 & 1 & \ldots & 0 \\ \vdots & \vdots & \vdots & \ddots & \vdots \\ 0 & 0 & 0 & \ldots & 1 \\ 0 & 0 & 0 & \ldots & 0 \end{bmatrix}}_{s_{1i}}, \underbrace{\begin{bmatrix} 0 & 1 & 0 & \ldots & 0 \\ 0 & 0 & 1 & \ldots & 0 \\ \vdots & \vdots & \vdots & \ddots & \vdots \\ 0 & 0 & 0 & \ldots & 1 \\ 0 & 0 & 0 & \ldots & 0 \end{bmatrix}}_{s_{2i}}, \ldots, \underbrace{\begin{bmatrix} 0 & 1 & 0 & \ldots & 0 \\ 0 & 0 & 1 & \ldots & 0 \\ \vdots & \vdots & \vdots & \ddots & \vdots \\ 0 & 0 & 0 & \ldots & 1 \\ 0 & 0 & 0 & \ldots & 0 \end{bmatrix}}_{s_{q_i,i}}\right).$$

The diagonal blocks of the matrix N_i have sizes which are given by the Segre characteristic associated with the eigenvalue λ_i. Clearly, N_i is a nilpotent matrix with index of nilpotence equal to s_{1i}.

Let Z_i be the matrix, formed by the ν_i columns of Z, corresponding to λ_i, and Y_i^H be the matrix, formed by the ν_i rows of Z^{-1}, corresponding to λ_i. The columns of Z_i constitute a basis of the eigenspace $\mathcal{Z}_i$ corresponding to λ_i. We have that

$$Y_i^H Z_i = I_{\nu_i}.$$

Then the matrix

$$P_i = Z_i Y_i^H \tag{2.48}$$

represents a spectral projection on the subspace $\mathcal{Z}_i$ in accordance with the following scheme:

$$A = \left[\; \Bigg| \; Z_i \; \Bigg| \; \right] \left[\begin{array}{c|c|c} \times & 0 & 0 \\ \hline 0 & J(\mathcal{S}_i, \lambda_i) & 0 \\ \hline 0 & 0 & \times \end{array} \right] \left[\begin{array}{c} \hline Y_i^H \\ \hline \end{array} \right].$$
$$= \quad Z \qquad\qquad J \qquad\qquad Z^{-1}$$

We have that

$$A = \sum_{i=1}^{p} [Z_i(\lambda_i I_{\nu_i} + N_i)Y_i^H] = \sum_{i=1}^{p} (\lambda_i P_i + D_i),$$
$$D_i = Z_i N_i Y_i^H. \tag{2.49}$$

The matrix N_i has at most $\ell_i - 1$ consecutive units on its superdiagonal, where ℓ_i is equal to the maximum element s_{1i} of the Segre characteristic associated with the eigenvalues λ_i. That is why

$$N_i^{\ell_i} = 0 \quad \text{and} \quad D_i^{\ell_i} = Z_i N_i^{\ell_i} Y_i^H = 0.$$

It is possible to show that

$$P_i^2 = P_i, \; AP_i = P_i A = P_i A P_i = \lambda_i P_i + D_i, \; D_i = (A - \lambda_i I_n) P_i.$$

Equation (2.49) is analog of equation (2.43) for the case of a matrix with defective eigenvalues.

Example 2.25. The matrix

$$A = \begin{bmatrix} -2 & -1 & 2 & 1 & 0 \\ 2 & -1 & -2 & 1 & -4 \\ 2 & 1 & -4 & 0 & -3 \\ 1 & -1 & 1 & 0 & -2 \\ -1 & -1 & 2 & 1 & -1 \end{bmatrix}$$

has a double eigenvalue $\lambda_1 = -1$ and a triple eigenvalue $\lambda_2 = -2$. To each of these eigenvalues corresponds only one eigenvector so that the Jordan form

$$J = \left[\begin{array}{cc|ccc} -1 & 1 & 0 & 0 & 0 \\ 0 & -1 & 0 & 0 & 0 \\ \hline 0 & 0 & -2 & 1 & 0 \\ 0 & 0 & 0 & -2 & 1 \\ 0 & 0 & 0 & 0 & -2 \end{array}\right]$$

has two blocks. The basis vectors of the right eigenspaces are

$$Z_1 = \begin{bmatrix} -1 & -1 \\ -1 & -1 \\ 0 & -1 \\ -2 & -1 \\ -1 & 0 \end{bmatrix}, \quad Z_2 = \begin{bmatrix} 1 & -1 & 1 \\ 1 & 1 & 0 \\ 0 & 1 & -1 \\ 1 & 0 & 1 \\ 1 & -1 & 1 \end{bmatrix},$$

respectively, and the basis vectors of the left eigenspaces are

$$Y_1^H = \begin{bmatrix} 1 & 1 & -1 & -1 & -1 \\ -1 & 0 & 0 & 0 & 1 \end{bmatrix}, \quad Y_2^H = \begin{bmatrix} 0 & 1 & 0 & -1 & 1 \\ 0 & 1 & -1 & 0 & -1 \\ 1 & 1 & -2 & 0 & -2 \end{bmatrix},$$

respectively. For the projections on the right subspaces, we obtain

$$P_1 = \begin{bmatrix} 0 & -1 & 1 & 1 & 0 \\ 0 & -1 & 1 & 1 & 0 \\ 1 & 0 & 0 & 0 & -1 \\ -1 & -2 & 2 & 2 & 1 \\ -1 & -1 & 1 & 1 & 1 \end{bmatrix}, \quad P_2 = \begin{bmatrix} 1 & 1 & -1 & -1 & 0 \\ 0 & 2 & -1 & -1 & 0 \\ -1 & 0 & 1 & 0 & 1 \\ 1 & 2 & -2 & -1 & -1 \\ 1 & 1 & -1 & -1 & 0 \end{bmatrix}.$$

The nilpotent matrices N_i, $i = 1,\ 2$, have the form

$$N_1 = \begin{bmatrix} 0 & 1 \\ 0 & 0 \end{bmatrix}, \quad N_2 = \begin{bmatrix} 0 & 1 & 0 \\ 0 & 0 & 1 \\ 0 & 0 & 0 \end{bmatrix}$$

and $\ell_1 = 2,\ \ell_2 = 3$.

For the matrices D_i, $i = 1, 2$, we obtain

$$D_1 = \begin{bmatrix} 1 & 0 & 0 & 0 & -1 \\ 1 & 0 & 0 & 0 & -1 \\ 0 & 0 & 0 & 0 & 0 \\ 2 & 0 & 0 & 0 & -2 \\ 1 & 0 & 0 & 0 & -1 \end{bmatrix}, \quad D_2 = \begin{bmatrix} -1 & 0 & 1 & 0 & 1 \\ 1 & 2 & -3 & 0 & -3 \\ 1 & 1 & -2 & 0 & -2 \\ 0 & 1 & -1 & 0 & -1 \\ -1 & 0 & 1 & 0 & 1 \end{bmatrix}.$$

An efficient numerical method for computing the spectral projection utilizing the Schur form is presented in Section 4.3.1.

2.2.8 *Angles between Eigenspaces*

Consider now the application of the spectral projections to the computation of the angles between eigenspaces. The acute angle between two vectors is defined as (see Section A.9.3 of the Appendix)

$$\Theta = \arccos\left(\frac{|y^H x|}{\|x\|_2 \|y\|_2}\right). \tag{2.50}$$

Let (λ, x) be a simple eigenpair of the matrix A. According to (2.50), the angle between the left and right eigenvectors of A, corresponding to λ, is given by

$$\Theta = \arccos\left(1/\|P\|_2\right), \tag{2.51}$$

where P is the spectral projection, corresponding to λ. In this way, the cosine of the angle between the left and right eigenspaces is inversely proportional of the norm of the eigenprojection on the right eigenspace. That is why the angle between the left y and right x eigenvectors is large if the norm of the spectral projection P is large. If the eigenvectors are normalized so that $y^H x = 1$, then (2.51) is reduced to

$$\Theta = \arccos(\|x\|_2 \|y\|_2).$$

In particular, from this expression it follows that if x is a simple eigenvector, then its left eigenvector y cannot be orthogonal to x since $y^H x \neq 0$.

Example 2.26. For the matrix

$$A = \begin{bmatrix} -5 & -4 & 6 \\ -2 & -5 & 4 \\ -3 & -4 & 4 \end{bmatrix},$$

whose spectral projections are determined in Example 2.24, we have that

i	$1/\|P_i\|_2$	Θ_i deg
1	0.1361	82.1878
2	0.3162	71.5651
3	0.1925	78.9042

where

$$\Theta_i = \arccos(\|x_i\|_2 \|y_i\|_2).$$

The mutual disposition of the vectors x_i and y_i, $i = 1, 2, 3$, is shown in Figure 2.3. Note that Θ_i are the acute angles between the corresponding left and right eigenvectors.

Consider now the case of eigenspaces whose dimension is greater than one. The canonical angles between two subspaces $\mathcal{X}$ and $\mathcal{Y}$ of $\mathbb{C}^n$ with dimension p are obtained from equation (A.56) in the Appendix as

$$\Theta_i = \arccos(\sigma_i(Y^H X)), \tag{2.52}$$

where σ_i are the singular values of the matrix $Y^H X$ and X, Y are orthonormal bases of these subspaces.

The expression (2.52) has the disadvantage that if Θ_i is small, then $\cos(\Theta_i) \approx 1$ and the angle Θ_i is not well conditioned. To overcome this difficulty, instead of using $\cos(\Theta_i(\mathcal{X}, \mathcal{Y}))$, it is preferable to use $\sin(\Theta_i(\mathcal{X}, \mathcal{Y}))$. It is possible to show that

$$\sin(\Theta_i(\mathcal{X}, \mathcal{Y})) = \sigma_i({Y^{\perp}}^H X), \tag{2.53}$$

where $Y^{\perp}$ is the orthogonal complementary subspace of Y, ${Y^{\perp}}^H Y = 0$.

2.2.9 *Notes and References*

2.2.9.1 *Multiple Eigenvalues*

There is an interesting episode in the history of eigenvalue problem related to the way in which the multiple roots of the characteristic equation are treated by the great French mathematicians Lagrange and Laplace. In solving the problem of the infinitely small oscillations of a system of bodies about their positions of equilibrium, they suppose that the characteristic roots are real and distinct. As pointed out in Kelvin and Tait (1912), p. 381, it is remarkable that both Lagrange (1811, p. 360; 1853, p. 333) and Laplace (1799, p. 304; 1829, p. 303) fell into the error of affirming the incompatibility of multiple roots and stable solutions to the corresponding system of differential equations. This erroneous claim formed the starting point of the investigations that culminated in the theory of elementary divisors developed by the German mathematician Karl Weierstrass (1815–1897) (Weierstrass, 1868). The method used by Weierstrass involved the key ideas subsequently developed further in his theory of elementary divisors that was the basis of the Jordan form (Hawkins, 1977).

2.2.9.2 *Jordan Canonical Form*

The French mathematician Camille Jordan (1838–1922) proposed the canonical form which is later named after him in 1870 (Jordan, 1870, Livre II, Chapitre II). A detailed history of the Jordan form is presented in Brechenmacher (2006) and interesting facts about the controversy between Jordan and the German mathematician Kronecker as a result of their different views on the generality in Mathematics are given in Brechenmacher (2016).

Systematic derivation of the Jordan normal form can be found in Gantmacher (1959a), Chapter 7, Horn and Johnson (2013), Chapter 3, Lancaster and Tismenetsky (1985), Chapter 6, and Ortega (1987), Chapter 3. An elementary derivation of the Jordan form is due to Filippov (1971) (also presented in Strang (2006)),

Appendix B. A combinatorial approach to the determination of the Jordan form is proposed in Brualdi (1987) and an algorithmic derivation of this form is given in Fletcher and Sorensen (1983). The books Weintraub (2008, 2009) are the only books devoted entirely to the Jordan form of a matrix and its applications.

The Segre characteristics are named after the Italian mathematician Corrado Segre who introduced them in Segre (1883–84). The properties of the Segre characteristics are discussed in MacDuffee (1946), p. 48, and Horn and Johnson (2013), p. 170, and the Weyr characteristics are described in Turnbull and Aitken (1932), p. 80, MacDuffee (1946), pp. 73–74, Horn and Johnson (2013), p. 170, and Shapiro (1999). Theorem 2.11 is formulated in Turnbull and Aitken (1932), p. 80. The Ferrers diagram is explained in MacDuffee (1946), p. 74, Shapiro (1999), and Horn and Johnson (2013). A survey of the properties of defective and derogatory matrices is given in Gregory (1960).

The canonical forms of matrices under similarity and in particular unitary transformations are surveyed in Shapiro (1991).

2.2.9.3 *Weyr Canonical Form*

The Weyr canonical form is proposed by the Czech mathematician Edward Weyr (1852–1903) in a short publication in French in 1885 (Weyr, 1885) and with more details in a much longer publication in German (Weyr, 1890). Latter on, these publications were forgotten and "rediscovered" in 1999 by the American mathematician Helene Shapiro (1999). It is interesting to note the using of Weyr and Segre characteristics before 1999 in linear control theory by Nicos Karkanias and his collaborators, see, for instance, Karkanias (1987) and Karkanias and Kalogeropoulos (1986). The Weyr canonical form is discussed in detail in the books Horn and Johnson (2013), Chapter 3, and O'Meara *et al.* (2011).

2.2.9.4 *Frobenius Canonical Form*

The Frobenius canonical form in the complex case is proposed by the German mathematician Ferdinand Georg Frobenius in (1879). A clear presentation of the reduction into Frobenius form is given in Wilkinson (1965), Chapter 1. An elaborate theory of this form can be found in Gantmacher (1959a), Chapter VII, and Lancaster and Tismenetsky (1985), Chapter 6.

2.2.9.5 *Spectral Decomposition and Angles between Eigenspaces*

The spectral decomposition of a matrix and the spectral projections are described in full in Chatelin (2012). The computation of the angles between eigenspaces is discussed in Stewart (2001).

2.3 The Generalized Eigenvalue Problem

2.3.1 *Generalized Eigenvalues and Eigenvectors*

Let A and B be $m \times n$ matrices. The set of all matrices of the form $A - \lambda B$, where λ is an arbitrary complex scalar, is referred to as a *matrix pencil* or just a *pencil.* The *generalized eigenvalue problem* consists of determining those values of λ for which the equation

$$Ax = \lambda Bx \tag{2.54}$$

has a nontrivial solution for x. Such values of λ are called eigenvalues of the pencil $A - \lambda B$ or *generalized eigenvalues.* The solution x of (2.54) is a right eigenvector of $A - \lambda B$ or *right generalized eigenvector.* The eigenvalues of $A - \lambda B$ are the roots of the characteristic equation

$$p(\lambda) \equiv \det(A - \lambda B) = 0. \tag{2.55}$$

The set of all eigenvalues is called *spectrum* of the generalized eigenvalue problem.

Consider the case when $m = n$. If $B = I_n$, then the generalized eigenvalue problem is reduced to the ordinary eigenvalue problem for A. If B is nonsingular, then (2.54) is equivalent to

$$B^{-1}Ax = \lambda x$$

so that the eigenvalues of $A - \lambda B$ are the eigenvalues of $B^{-1}A$. They are also eigenvalues of $AB^{-1} = B(B^{-1}A)B^{-1}$.

The generalized eigenvalue problem has two special features. First, both sides of (2.54) may be equal to zero. This happens when there is a nonzero vector z such that $Az = \lambda Bz = 0$. In this case, every λ is an eigenvalue and it is natural to regard $0/0$ as an eigenvalue of $A - \lambda B$. Second, if $m = n$ and B is singular, then the degree of the characteristic polynomial $\det(A - \lambda B)$ is less than n, i.e., the pencil $A - \lambda B$ has less than n eigenvalues. In such a case, if A is nonsingular, equation (2.54) is equivalent to

$$A^{-1}Bx = \lambda^{-1}x = \mu x$$

and the missing eigenvalues correspond to zero eigenvalues of the reciprocal problem $B - \mu A$. That is why the missing eigenvalues of $A - \lambda B$ may be considered as *infinite eigenvalues.*

Based on the earlier considerations, we say that the algebraic multiplicity of a finite eigenvalue of $A - \lambda B$ is its multiplicity as a zero of the characteristic equation (2.55). If m is the degree of the characteristic polynomial and $m < n$, then we say that $A - \lambda B$ has an infinite eigenvalue of algebraic multiplicity $n - m$. An eigenvalue of algebraic multiplicity one is called a *simple eigenvalue.*

Example 2.27. If

$$A = \begin{bmatrix} 1 & 2 \\ 0 & 0 \end{bmatrix} \quad \text{and} \quad B = \begin{bmatrix} 1 & 1 \\ 0 & 0 \end{bmatrix},$$

then $\det(A - \lambda B) \equiv 0$ and every λ is an eigenvalue of $A - \lambda B$.

If

$$A = \begin{bmatrix} 1 & -1 \\ 0 & 2 \end{bmatrix}, \quad B = \begin{bmatrix} 0 & 1 \\ 0 & 0 \end{bmatrix},$$

then $\det(A - \lambda B) = 2$ and the pencil $A - \lambda B$ has two infinite eigenvalues.

If the matrices A and B are square and $\det(A - \lambda B)$ is not identically equal to zero, then the pencil $A - \lambda B$ is called *regular.* Otherwise it is called *singular.* Note that the regular pencils are always square but the singular pencils can be square (in such case $\det(A - \lambda B) \equiv 0$) or rectangular. When the pencil $A - \lambda B$ is regular, its eigenvalues are defined to be

(1) the roots of $\det(A - \lambda B) = 0$,
(2) ∞ (with multiplicity $n - \deg(p)$) if $\deg(p) < n$.

Thus, a regular matrix pencil can have only a finite number of eigenvalues. If the pencil is singular, its columns or rows or both of them are linearly dependent.

Example 2.28. Let

$$A = \begin{bmatrix} 2 & & \\ & 0 & \\ & & 1 \end{bmatrix} \quad \text{and} \quad B = \begin{bmatrix} 3 & & \\ & 1 & \\ & & 0 \end{bmatrix}.$$

Then $p(\lambda) = \det(A - \lambda B) = (2 - 3\lambda)(0 - \lambda)(1 - 0\lambda)$ so that the generalized eigenvalues are $\lambda = 2/3$, 0 and ∞.

For the case of square pencils, the earlier considerations are summarized in Table 2.1.

Let P and Q be nonsingular matrices. Then pencils $A - \lambda B$ and $P^{-1}(A - \lambda B)Q = P^{-1}AQ - \lambda P^{-1}BQ$ are called *equivalent*. Alternatively, we say that $P^{-1}AQ - \lambda P^{-1}BQ$ is obtained from $A - \lambda B$ by an *equivalence transformation*. The equivalent regular pencils have the same eigenvalues with the same multiplicities. The vector x is a right eigenvector of $A - \lambda B$ if and only if $Q^{-1}x$ is a right eigenvector of $P^{-1}AQ - \lambda P^{-1}BQ$. The vector y is a *left eigenvector* of $A - \lambda B$,

Table 2.1. Generalized eigenvalues of square pencils.

Given A and B	$\det(A - \lambda B)$	Pencil type	Generalized eigenvalues
A is regular or singular B is regular	$\neq 0$	Regular pencil	Finite eigenvalues, the same as the eigenvalues of AB^{-1} or $B^{-1}A$
A is singular B is regular	$\neq 0$	Regular pencil	Zero eigenvalue with multiplicity $n - \text{rank}(A)$
A is regular or singular B is singular	$\neq 0$	Regular pencil	Infinite eigenvalue with multiplicity $n - \text{rank}(B)$
A is regular B is singular	$\neq 0$	Regular pencil	The eigenvalues are the same as the reciprocals of the eigenvalues of BA^{-1} or $A^{-1}B$
A is singular B is singular	0	Singular pencil	In general, there are some finite, some infinite, and some arbitrary eigenvalues

i.e., $y^H(A-\lambda B)=0$, if and only if $P^H y$ is a left eigenvector of $P^{-1}AQ-\lambda P^{-1}BQ$.

2.3.2 *Generalized Schur Form*

Similar to the Schur form of a matrix, it is possible to reduce a regular pencil to upper triangular form using unitary equivalence transformations.

Let $A-\lambda B$ be a regular pencil. Then there exist unitary matrices U and V such that $S=U^HAV$ and $T=U^HBV$ are both upper triangular. The eigenvalues of $A-\lambda B$ are then s_{ii}/t_{ii}, the ratios of the diagonal entries of S and T.

If A and B are real, then there exist orthogonal matrices U and V such that U^TAV is quasi-upper triangular and U^TBV is upper triangular. The pencils corresponding to the blocks of U^TAV of order two contain a pair of complex conjugate eigenvalues of $A-\lambda B$.

An example of reduction into generalized Schur form is given later on in Chapter 4 (Example 4.14).

Using unitary equivalence transformations, the eigenvalues of $A-\lambda B$ can be reordered in the desired way on the diagonal of $S-\lambda T$. In such a case, it is possible to find orthonormal bases of the subspaces associated with a given group of eigenvalues. Let

$$U^HAV=\begin{bmatrix} S_{11} & S_{12} \\ 0 & S_{22}\end{bmatrix}, \quad U^HBV=\begin{bmatrix} T_{11} & T_{12} \\ 0 & T_{22}\end{bmatrix}, \tag{2.56}$$

where $S_1, T_1 \in \mathbb{C}^{r\times r}$, and let Σ_1 be the spectrum of $S_{11}-\lambda T_{11}$. Partition the columns of U and V as $U=[U_1,\ U_2]$ and $V=[V_1,\ V_2]$, where $U_1,\ V_1 \in \mathbb{C}^{n\times r}$. Then the subspaces $\mathcal{Y}$ and $\mathcal{X}$, spanned by the columns of U_1 and V_1, respectively, are called *left and right deflating subspaces* of $A-\lambda B$ corresponding to Σ_1. The deflating subspaces are generalizations of the invariant subspaces of a matrix.

2.3.3 *Generalized Block-Diagonal Form*

Let the regular pencil $A-\lambda B$ be reduced by using unitary equivalence transformations into the generalized Schur form

$$\begin{bmatrix} U_1^H \\ U_2^H \end{bmatrix}(A-\lambda B)\begin{bmatrix} V_1 & V_2\end{bmatrix}=\begin{bmatrix} S_{11}-\lambda T_{11} & S_{12}-\lambda T_{12} \\ 0 & S_{22}-\lambda T_{22}\end{bmatrix}$$

and assume that the pencils $S_{11} - \lambda T_{11}$ and $S_{22} - \lambda T_{22}$ have disjoint spectra. Then we say that $\mathcal{Y} = \mathcal{R}(U_2)$ and $\mathcal{X} = \mathcal{R}(V_1)$ are *simple left and right eigenspaces*, respectively. In such a case, it is possible to find matrices $R, W \in \mathbb{C}^{r\times(n-r)}$ such that

$$\begin{bmatrix} I_r & W \\ 0 & I_{n-r} \end{bmatrix} \begin{bmatrix} S_{11} & S_{12} \\ 0 & S_{22} \end{bmatrix} \begin{bmatrix} I_r & R \\ 0 & I_{n-r} \end{bmatrix} = \begin{bmatrix} S_{11} & 0 \\ 0 & S_{22} \end{bmatrix} \tag{2.57}$$

and

$$\begin{bmatrix} I_r & W \\ 0 & I_{n-r} \end{bmatrix} \begin{bmatrix} T_{11} & T_{12} \\ 0 & T_{22} \end{bmatrix} \begin{bmatrix} I_r & R \\ 0 & I_{n-r} \end{bmatrix} = \begin{bmatrix} T_{11} & 0 \\ 0 & T_{22} \end{bmatrix}. \tag{2.58}$$

If we set

$$X_1 = U_1, \quad X_2 = U_2 + U_1 R$$

and

$$Y_1 = V_1 + V_2 W^H, \quad Y_2 = V_2,$$

then we will have that

$$\begin{bmatrix} Y_1^H \\ Y_2^H \end{bmatrix} (A - \lambda B) \left[X_1 \; X_2\right] = \begin{bmatrix} S_{11} - \lambda T_{11} & 0 \\ 0 & S_{22} - \lambda T_{22} \end{bmatrix}.$$

Note that the matrices

$$\left[Y_1 \; Y_2\right] \quad \text{and} \quad \left[X_1 \; X_2\right]$$

are nonsingular. The eigenspaces $\mathcal{R}(X_2)$ and $\mathcal{R}(Y_2)$ are called *complementary* to the eigenspaces $\mathcal{R}(X_1)$ and $\mathcal{R}(Y_1)$, respectively.

From (2.57) and (2.58), it follows that the matrices R and W are solutions of the equations

$$S_{11}R + WS_{22} + S_{12} = 0, \quad T_{11}R + WT_{22} + T_{12} = 0.$$

The matrix equation

$$S_{11}R + WS_{22} + S_{12} = 0, \tag{2.59}$$

$$T_{11}R + WT_{22} + T_{12} = 0 \tag{2.60}$$

is called *generalized Sylvester equation*. It has unique solutions for R and W, if and only if the pencils $S_{11} - \lambda T_{11}$ and $S_{22} - \lambda T_{22}$ have no eigenvalues in common.

In the general case, by using nonsingular equivalence transformations, the regular pencil $A - \lambda B$ can be represented in the block-diagonal form

$$P^{-1}(A - \lambda B)Q = S - \lambda T,$$

where $S = \mathrm{diag}(S_1, S_2, \ldots, S_p)$ and $T = \mathrm{diag}(T_1, T_2, \ldots, T_p)$. Denote by Σ_i the spectrum of $S_i - \lambda_i T_i$ and partition the columns of R and W according to the dimensions of the blocks of $S - \lambda T$ as $R = [R_1, R_2, \ldots, R_p], W = [W_1, W2, \ldots, W_p]$. Then the subspaces $\mathcal{R}_i$ and $\mathcal{W}_i$, spanned by the columns of R_i and W_i, respectively, are left and right deflating subspaces of $A - \lambda B$ corresponding to Σ_i.

2.3.4 *Weierstrass Canonical Form*

Let $A - \lambda B$ be a regular pencil. Then there exists an equivalence transformation with matrices P and Q, such that the pencil is reduced into the *Weierstrass canonical form*

$$P^{-1}(A - \lambda B)Q = \begin{bmatrix} J_{n_1}(\lambda_1) - \lambda I_{n_1} & & & & & \\ & \ddots & & & & \\ & & J_{n_k}(\lambda_k) - \lambda I_{n_k} & & & \\ & & & N_{m_1} & & \\ & & & & \ddots & \\ & & & & & N_{m_r} \end{bmatrix},$$

where

$$J_{n_i}(\lambda_i) = \begin{bmatrix} \lambda_i & 1 & & \\ & \lambda_i & \ddots & \\ & & \ddots & 1 \\ & & & \lambda_i \end{bmatrix}, \quad i = 1, 2, \ldots, k,$$

is a Jordan block with a finite eigenvalue λ_i and

$$N_{m_i} = I_{m_i} - \lambda J_{m_i}(0) = \begin{bmatrix} 1 & -\lambda & & \\ & 1 & \ddots & \\ & & \ddots & -\lambda \\ & & & 1 \end{bmatrix}, \quad i = 1, 2, \ldots, r,$$

can be considered as a Jordan block for the infinite eigenvalue with multiplicity m_i. Like the ordinary eigenvalue problem, the characteristic polynomial of the pencil $A - \lambda B$ can be presented as a product of elementary divisors. In the given case, the blocks $J_{n_i}(\lambda_i) - \lambda I_{n_i}$ contain the finite elementary divisors of the pencil and the blocks $N_{m_i} = I_{m_i} - \lambda J_{m_i}(0)$ contain the infinite elementary divisors of the pencil.

The Weierstrass canonical form can be regarded as a generalization of the Jordan canonical form to regular pencils.

Example 2.29. Given is a regular matrix pencil $A - \lambda B$ with matrices

$$A = \begin{bmatrix} 3 & 2 & 2 & 2 & -3 & 5 \\ -3 & -2 & 1 & 0 & 4 & -3 \\ 4 & 2 & -1 & 2 & -4 & 6 \\ 0 & 0 & -3 & -2 & -2 & -1 \\ 0 & 2 & 6 & 1 & 1 & 1 \\ 5 & 2 & -4 & 0 & -4 & 4 \end{bmatrix},$$

$$B = \begin{bmatrix} -1 & 1 & 1 & 0 & 0 & -2 \\ 0 & -1 & 0 & 1 & 0 & 2 \\ 1 & 1 & 0 & 0 & 0 & 0 \\ 0 & 0 & -1 & -1 & 0 & -1 \\ -4 & 1 & 2 & 0 & 0 & -5 \\ 3 & 1 & -1 & -1 & 0 & 1 \end{bmatrix}.$$

Using the nonsingular matrices

$$P = \begin{bmatrix} 1 & 1 & 0 & 1 & -3 & -1 \\ -1 & 0 & 1 & 0 & 2 & 2 \\ 1 & 0 & 0 & 0 & -1 & -2 \\ 0 & -1 & -1 & 0 & 0 & -1 \\ 1 & 2 & 0 & 3 & -5 & 1 \\ 1 & -1 & -1 & -2 & 0 & -3 \end{bmatrix},$$

$$Q = \begin{bmatrix} 0 & 1 & -1 & -2 & 0 & -1 \\ 1 & -1 & 1 & 2 & 0 & 1 \\ 0 & 1 & 0 & 0 & -1 & 0 \\ 0 & 1 & 0 & -2 & 2 & -1 \\ 0 & -1 & 1 & 1 & 0 & 1 \\ 0 & -1 & 1 & 2 & -1 & 1 \end{bmatrix},$$

the pencil $A - \lambda B$ is reduced into the Weierstrass canonical form

$$\begin{aligned} P^{-1}(A - \lambda B)Q &= \left[\begin{array}{ccc|cc|c} 2 & 1 & 0 & & & \\ 0 & 2 & 1 & & & \\ 0 & 0 & 2 & & & \\ \hline & & & 1 & 0 & \\ & & & 0 & 1 & \\ \hline & & & & & 1 \end{array}\right] - \lambda \left[\begin{array}{ccc|cc|c} 1 & 0 & 0 & & & \\ 0 & 1 & 0 & & & \\ 0 & 0 & 1 & & & \\ \hline & & & 0 & 1 & \\ & & & 0 & 0 & \\ \hline & & & & & 0 \end{array}\right] \\ &= \left[\begin{array}{ccc|cc|c} 2-\lambda & 1 & 0 & & & \\ 0 & 2-\lambda & 1 & & & \\ 0 & 0 & 2-\lambda & & & \\ \hline & & & 1 & -\lambda & \\ & & & 0 & 1 & \\ \hline & & & & & 1 \end{array}\right]. \end{aligned}$$

It follows from the Weierstrass form that $n_1 = 3, m_1 = 2$ and $m_2 = 1$. The pencil has three finite generalized eigenvalues at 2 and three infinite eigenvalues, one with multiplicity 2 and one scalar.

2.3.5 *Kronecker Canonical Form*

Let $A - \lambda B$ be a singular pencil. Then there exists an equivalent transformation with matrices P and Q, such that the pencil is reduced to the *Kronecker canonical form*

$$\begin{aligned} &P^{-1}(A-\lambda B)Q \\ &\quad = \operatorname{diag}\Big\{L_{\varepsilon_1},\ldots,L_{\varepsilon_p},L_{\eta_1}^T,\ldots,L_{\eta_q}^T,J_{n_1}(\lambda_1)-\lambda I_{n_1},\ldots,J_{n_k}(\lambda_k) \\ &\qquad\qquad -\lambda I_{n_k},\ldots,N_{m_1},\ldots,N_{m_r}\Big\}, \end{aligned}$$

where

$$L_{\varepsilon_i} = \begin{bmatrix} -\lambda & 1 & & & \\ & -\lambda & \ddots & & \\ & & \ddots & 1 & \\ & & & -\lambda & 1 \end{bmatrix}$$

is a $\varepsilon_i \times (\varepsilon_i + 1)$ bidiagonal block,

$$L_{\eta_j}^T = \begin{bmatrix} -\lambda & & & \\ 1 & -\lambda & & \\ & 1 & \ddots & \\ & & \ddots & -\lambda \\ & & & 1 \end{bmatrix}$$

is a $(\eta_j + 1) \times \eta_j$ bidiagonal block,

$$J_{n_i} - \lambda I_{n_i} = \begin{bmatrix} \lambda_i - \lambda & 1 & & \\ & \lambda_i - \lambda & \ddots & \\ & & \ddots & 1 \\ & & & \lambda_i - \lambda \end{bmatrix},$$

$$N_{m_i} = I_{m_i} - \lambda J_{m_i}(0) = \begin{bmatrix} 1 & -\lambda & & \\ & 1 & \ddots & \\ & & \ddots & -\lambda \\ & & & 1 \end{bmatrix}.$$

Similar to the Weierstrass canonical form, $J_{n_i} - \lambda I_{n_i}$ is an $n_i \times n_i$ Jordan block where λ_i is a finite generalized eigenvalue and N_{m_i} is an $m_i \times m_i$ block where λ_i is an infinite generalized eigenvalue of multiplicity m_i. The $J_{n_i} - \lambda I_{n_i}$ and N_{m_i} blocks together constitute the *regular structure* of the pencil.

The notation ε_i is used to denote the number of the rows in the blocks L_{ε_i} and the notation η_j is used to denote the number of columns in the blocks $L_{\eta_j}^T$. Note that the blocks L_{ε_i} and $L_{\eta_j}^T$ are rectangular and when A and B are square, there must be the same number of such blocks. However, in general, the dimensions of the two sets of blocks are not related. The block L_{ε_i} is called a *singular block of right (column) minimal index* ε_i. Since

$$\begin{bmatrix} -\lambda & 1 & & & \\ & -\lambda & \ddots & & \\ & & \ddots & 1 & \\ & & & -\lambda & 1 \end{bmatrix} \begin{bmatrix} 1 \\ \lambda \\ \lambda^2 \\ \vdots \\ \lambda^{\varepsilon_i} \end{bmatrix} = \begin{bmatrix} 0 \\ 0 \\ \vdots \\ 0 \end{bmatrix},$$

this block has a one-dimensional right null space, $[1, \lambda, \lambda^2, \ldots, \lambda^{\varepsilon_i}]^T$ for any λ. Hence, this block reveals the linear dependencies between columns of the pencil.

The block $L_{\eta_j}^T$ is called a *singular block of left (row) minimal index* η_j. Since

$$[1, \lambda, \lambda^2, \ldots, \lambda^{\eta^j}] \begin{bmatrix} -\lambda & & & \\ 1 & -\lambda & & \\ & 1 & \ddots & \\ & & \ddots & -\lambda \\ & & & 1 \end{bmatrix} = [0, 0, \ldots, 0],$$

this block has a one-dimensional left null space for any λ and thus reveals the linear dependencies between rows of the pencil.

The right L_{ε_i} and left $L_{\eta_i}^T$ singular blocks together constitute the *singular structure* of the pencil. The minimal indices of a singular pencil are also called *Kronecker indices* of the pencil. There are no elementary divisors corresponding to L_{ε_i} and L_{η_j}. It is important that the equivalent pencils have the same minimal indices.

The Kronecker canonical form can be regarded as a generalization of the Jordan and Weierstrass canonical form to singular pencils.

Example 2.30. Given is a singular matrix pencil $A - \lambda B$ with matrices

$$A = \begin{bmatrix} -1 & 1 & -1 & -1 & -1 & 1 & -2 & -2 & -1 & -4 \\ -2 & 0 & 0 & -1 & -1 & 2 & -3 & -5 & 0 & -8 \\ 0 & 0 & 1 & 0 & 1 & -1 & 0 & 0 & 1 & 1 \\ 1 & 0 & -1 & 1 & 0 & -1 & 2 & 3 & -1 & 5 \\ -1 & 0 & 0 & -1 & 1 & 1 & -2 & -3 & 1 & -5 \\ -1 & 0 & 0 & -1 & 0 & 1 & -2 & -3 & 0 & -5 \\ 2 & -1 & 0 & 2 & 0 & 1 & 0 & -1 & 0 & 3 \\ -1 & 3 & -1 & -1 & -1 & 1 & 1 & 0 & -1 & -2 \\ 3 & -4 & 3 & 2 & 3 & -4 & 2 & 4 & 3 & 10 \\ 1 & 1 & 0 & 0 & 0 & -3 & 5 & 9 & -1 & 9 \end{bmatrix}$$

and

$$B = \begin{bmatrix} 0 & 0 & 0 & -1 & 0 & 1 & -1 & -2 & 0 & -3 \\ -1 & 1 & 0 & -1 & 1 & 1 & -2 & -2 & 1 & -4 \\ 1 & 0 & 0 & 2 & 0 & -1 & 3 & 4 & 0 & 7 \\ 0 & -1 & 0 & 0 & 0 & 0 & 0 & -1 & 0 & -1 \\ 0 & 0 & 0 & 0 & 0 & 1 & 0 & 0 & 0 & -1 \\ 0 & 0 & 0 & 1 & 0 & 0 & 1 & 1 & 0 & 2 \\ -1 & 0 & 0 & -1 & 0 & 0 & -1 & -1 & 0 & -3 \\ -1 & -1 & 0 & -1 & 0 & 1 & -3 & -4 & 0 & -6 \\ 2 & 1 & 0 & 3 & -1 & -3 & 6 & 9 & -1 & 14 \\ 2 & -1 & 0 & 1 & -1 & -2 & 2 & 2 & -1 & 6 \end{bmatrix}.$$

Note that $\operatorname{rank}(A) = 9$ and $\operatorname{rank}(B) = 8$.
The Kronecker canonical form of this pencil is

$$P^{-1}(A - \lambda B)Q$$
$$= \left[\begin{array}{ccc|ccc|cc|cc} 0 & 1 & 0 & & & & & & & \\ 0 & 0 & 1 & & & & & & & \\ \hline & & & 0 & 0 & 0 & & & & \\ & & & 1 & 0 & 0 & & & & \\ & & & 0 & 1 & 0 & & & & \\ & & & 0 & 0 & 1 & & & & \\ \hline & & & & & & -3 & 1 & & \\ & & & & & & 0 & -3 & & \\ \hline & & & & & & & & 1 & 0 \\ & & & & & & & & 0 & 1 \end{array}\right] - \lambda \left[\begin{array}{ccc|ccc|cc|cc} 1 & 0 & 0 & & & & & & & \\ 0 & 1 & 0 & & & & & & & \\ \hline & & & 1 & 0 & 0 & & & & \\ & & & 0 & 1 & 0 & & & & \\ & & & 0 & 0 & 1 & & & & \\ & & & 0 & 0 & 0 & & & & \\ \hline & & & & & & 1 & 0 & & \\ & & & & & & 0 & 1 & & \\ \hline & & & & & & & & 0 & 1 \\ & & & & & & & & 0 & 0 \end{array}\right]$$

$$
= \left[\begin{array}{ccc|ccc|cc|cc}
-\lambda & 1 & 0 & & & & & & & \\
0 & -\lambda & 1 & & & & & & & \\
\hline
 & & & -\lambda & 0 & 0 & & & & \\
 & & & 1 & -\lambda & 0 & & & & \\
 & & & 0 & 1 & -\lambda & & & & \\
 & & & 0 & 0 & 1 & & & & \\
\hline
 & & & & & & -3-\lambda & 1 & & \\
 & & & & & & 0 & -3-\lambda & & \\
\hline
 & & & & & & & & 1 & -\lambda \\
 & & & & & & & & 0 & 1
\end{array}\right].
$$

For this example, we have that $\varepsilon_1 = 2$, $\eta_1 = 3$. The pencil has a regular part of dimension 4 and a singular part of dimension 6. More specifically, it has two finite generalized eigenvalues at -3, two infinite eigenvalues and six generalized eigenvalues of the type $0/0$ corresponding to the singular part of the pencil. Depending on the software used, the computed values of the $0/0$ eigenvalues can be arbitrary.

In case of a singular pencil, instead of the deflating subspaces, one can use the so-called *reducing subspaces* (Van Dooren, 1983).

2.3.6 *Notes and References*

A systematic exposition of the theory of the generalized eigenvalue problem is given in Stewart (2001), Chapter 2, and a detailed presentation of this theory and the related numerical methods can be found in Ikramov (1993).

The canonical form of a regular pencil is proposed by Weierstrass in the classical paper Weierstrass (1868). The existence of singular pencils was mentioned by him but excluded from consideration.

The canonical form of a singular pencil is proposed by the German mathematician Leopold Kronecker in (1890). A detailed derivation of the Kronecker canonical form can be found in Turnbull and Aitken (1932), Chapter 9, and Gantmacher (1959b), Chapter XII. Different aspects of this form and its applications are discussed in Aitken (1933), Wilkinson (1978), Van Dooren (1981, 1983) and Demmel (1997), Chapter 4.

The generalized Schur form is due to Stewart (1972). Other types of important eigenvalue problems are considered in Tisseur and Meerbergen (2001) and Mehrmann and Voss (2005).

Chapter 3

The Eigenvalue Sensitivity Problem

In this chapter, we study the sensitivity of the eigenvalues of a matrix to small changes (perturbations) of its elements. Since the multiple eigenvalues associated with nonlinear elementary divisors are usually very sensitive to perturbations of the matrix, the eigenvalue sensitivity is closely related to the problem of determining the Jordan structure of the matrix. The investigation of the matrix eigenvalue sensitivity is a subject to the *eigenvalue perturbation theory* which is an important part of the matrix analysis. There exist two reasons to study this problem in connection with determining the Jordan canonical form. At first, the mathematical models involving matrices are not known exactly due to measurement errors and approximations in their derivation. Also, due to different factors, these models may undergo changes in the time of the entries of the corresponding matrices. On second, the numerical algorithms implemented for solving the eigenvalue problem produce eigenvalues $\widetilde{\lambda}$ and eigenvectors $\widetilde{x}$, which satisfy the equation

$$(A + \Delta A)\widetilde{x} = \widetilde{\lambda}\widetilde{x},$$

where $\|\Delta A\|/\|A\|$ is at best of order of the unit roundoff error of the computer arithmetic. In both cases, instead of the exact eigenvalue λ of the matrix A, one obtains the perturbed eigenvalue $\widetilde{\lambda} = \lambda + \Delta\lambda$ of the matrix $A + \Delta A$. For some matrices, a small perturbation ΔA may cause large changes in the eigenvalues and eigenvectors. These changes may provoke qualitative changes in the behavior of the investigated mathematical model. For instance, the changes of the eigenvalues may convert a stable linear dynamical system into unstable one which leads to fundamental changes in the model behavior. Hence, the knowledge of the eigenvalue sensitivity allows to predict the changes in the behavior of the model under investigation.

Also, in a separate section, we consider the sensitivity of the generalized eigenvalue problem, which arises in connection with the determining of Weierstrass and Kronecker canonical form.

The literature on the eigenvalue perturbation analysis is extremely huge, and in this chapter, we present in brief only some of the most important results. Detailed and systematic expositions of this subject may be found in the references given at the end of this chapter.

3.1 Measuring Eigenvalue Sensitivity

The following notion is useful in describing how the eigenvalues of the perturbed and unperturbed matrix are disposed to each other.

Let A have eigenvalues $\lambda_1, \lambda_2, \ldots, \lambda_n$ and B have eigenvalues $\mu_1, \mu_2, \ldots, \mu_n$. Then the *spectral deviation* of B with respect to A is defined as

$$\mathrm{sv}_A(B) := \max_i\{\min_j|\mu_i - \lambda_j|\}. \tag{3.1}$$

The function $\mathrm{sv}_A(B)$ can be zero, although the eigenvalues of A and B are different, which means that it does not have the properties of a distance. Geometrically, the eigenvalues of B lie in the union of disks with radii $\mathrm{sv}_A(B)$, whose centers are in the eigenvalue of A.

The *Hausdorff distance* between the spectra of A and B is the quantity

$$\mathrm{hd}(A, B) = \max\{\mathrm{sv}_A(B), \mathrm{sv}_B(A)\}. \tag{3.2}$$

The *eigenvalue variation* or *matching distance* between the eigenvalues of A and B is the quantity

$$\mathrm{v}(A, B) = \min_\pi\{\max_i|\mu_{\pi(i)} - \lambda_i|\}, \tag{3.3}$$

Algorithm 3.1. Computing the spectral variation of the matrix B with respect to matrix A

Computes the spectral variation of the matrix B with eigenvalues stored in the vector w, with respect to matrix A with eigenvalues, stored in the vector v

function $s = \text{spectral\_var}(w, v)$
Input: Vector $w \in \mathbb{C}^{nw}$, containing the eigenvalues of B,
vector $v \in \mathbb{C}^{nv}$, containing the eigenvalues of A.
Output: s contains the spectral variation of B with respect to A.

```
n = max{nw, nv}
for i = 1, 2, ..., n
    d_i = |w_i - v_1|
    for j = 1, 2, ..., n
        d_i = min{d_i, |w_i - v_j|}
    end
end
s = max_i{d_i}
```

where π is taken over all permutations of $1, 2, \ldots, n$. One has that $\text{v}(A, B) = \text{v}(B, A)$ and

$$\text{sv}_A(B) \leq \text{v}(A, B)$$

for all matrices A and B.

The Hausdorff distance and the eigenvalue variation have the properties of metrics, i.e., they are zeros only when the spectra of both matrices coincide.

Most of the results in eigenvalue perturbation theory are derived with respect to the spectral variation between two matrices. Algorithm 3.1 computes the spectral variation of B with respect to A provided that the eigenvalues of the two matrices are known. Computing consecutively $\text{sv}_A(B)$ and $\text{sv}_B(A)$ with this algorithm, one can determine also the Hausdorff distance $\text{hd}(A, B)$.

Example 3.1. Let the eigenvalues of A be $-1, -2, -3$ and those of B be - $-1, -1, -3$. Then

$$\text{sv}_B(A) = 1, \quad \text{sv}_A(B) = 0$$

and

$$\text{hd}(A, B) = 1.$$

3.2 Gershgorin Theorem

The Gershgorin theorem is one of the most popular methods to localize the eigenvalues of a matrix. In the strict sense, this theorem is not related to the perturbation analysis of eigenvalues, but it can be used to obtain very accurate perturbation bounds.

Theorem 3.2 (Gershgorin Theorem). *All eigenvalues of the complex matrix* $A \in \mathbb{C}^{n\times n}$ *lie in the union of the* n Gershgorin disks

$$\mathcal{D}_i = \{z \in \mathbb{C} : |z - a_{ii}| \leq r_i\}, \quad r_i = \sum_{j\neq i}^{n} |a_{ij}|, \quad i = 1, 2, \ldots, n,$$

in the complex plane.

It is also possible to show that if the union of k Gershgorin disks is isolated from the other disks, then this union contains exactly k eigenvalues of the matrix A.

Example 3.3. Given is the matrix

$$A = \begin{bmatrix} -1 + \jmath_o & -0.2 & \jmath_o 0.4 & 0.3 \\ -0.3 & 0.5 & -0.1 & -\jmath_o 0.2 \\ -\jmath_o 0.8 & -\jmath_o 0.2 & 2.0 & 0.5 \\ -\jmath_o 0.1 & -0.3 & 0.3 & -\jmath_o 1.5 \end{bmatrix}$$

whose eigenvalues are

$$\begin{aligned} \lambda_1 &= -1.1222 + \jmath_o 0.9605, \\ \lambda_2 &= 2.1391 - \jmath_o 0.0032, \\ \lambda_3 &= 0.5669 + \jmath_o 0.0194, \\ \lambda_4 &= -0.0838 - \jmath_o 1.4767. \end{aligned}$$

The Gershgorin disks for this matrix are (see Figure 3.1)

$$\begin{aligned} \mathcal{D}_1 &= \{z : |z + 1 - \jmath_o| \leq 0.9\}, \\ \mathcal{D}_2 &= \{z : |z - 0.5| \leq 0.6\}, \\ \mathcal{D}_3 &= \{z : |z - 2| \leq 1.5\}, \\ \mathcal{D}_4 &= \{z : |z + \jmath_o 1.5| \leq 0.7\}. \end{aligned}$$

Since the disks $\mathcal{D}_1$ and $\mathcal{D}_4$ are isolated from the rest disks, they must contain one eigenvalue each (these are λ_1 and λ_4, respectively).

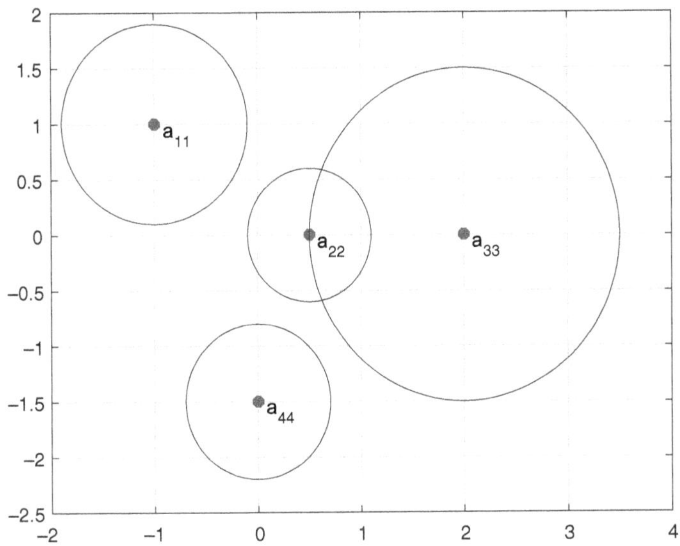

Fig. 3.1. Localization of eigenvalues by the Gershgorin theorem.

The rest two eigenvalues (λ_2 and λ_3) must lie in the union of the disks $\mathcal{D}_2$ and $\mathcal{D}_3$.

The Gershgorin theorem is a convenient tool for obtaining bounds on the groups of eigenvalues.

Example 3.4. Consider the matrix

$$A = \begin{bmatrix} 1 & 2.10^{-4} & -3.10^{-4} & 2.10^{-3} \\ -2.10^{-4} & 2 & 10^{-3} & 3.10^{-4} \\ 10^{-3} & 2.10^{-3} & 2 & -5.10^{-4} \\ 10^{-4} & -4.10^{-4} & -5.10^{-4} & 2 \end{bmatrix}.$$

This matrix can be considered as derived from the diagonal matrix $\text{diag}(1, 2, 2, 2)$, to which small off-diagonal elements are added. The radii of the Gershgorin disks for A are

$$\begin{aligned} r_1 &= 2.5 \cdot 10^{-3}, \\ r_2 &= 1.5 \cdot 10^{-3}, \\ r_3 &= 3.5 \cdot 10^{-3}, \\ r_4 &= 1.0 \cdot 10^{-3}. \end{aligned}$$

It is seen that the first disk is isolated by the rest, from where it follows that A has eigenvalue λ_1 which satisfies

$$|\lambda_1 - 1| \leq 2.5 \cdot 10^{-3}.$$

The rest three disks have centers at 2 and intersect. Hence A has three eigenvalues λ_2, λ_3 and λ_4 that satisfy

$$|\lambda_i - 2| \leq 3.5 \cdot 10^{-3}, \quad i = 2,\ 3,\ 4. \tag{3.4}$$

Unfortunately, the estimates obtained are not very tight since the actual deviations of the eigenvalues compared to the diagonal elements (up to three decimal digits) are

$$\begin{aligned} |\lambda_1 - 1| &= 1.40 \cdot 10^{-7}, \\ |\lambda_2 - 2| &= 1.48 \cdot 10^{-3}, \\ |\lambda_3 - 2| &= 1.43 \cdot 10^{-3}, \\ |\lambda_4 - 2| &= 4.74 \cdot 10^{-5}. \end{aligned}$$

It is seen that the Gershgorin disks localize the eigenvalues which are near to 2 sufficiently well but largely overestimate the change in the eigenvalue which is near to 1.

The estimates obtained by the Gershgorin theorem can be greatly improved if one uses diagonal similarity transformations.

Example 3.5. Let A is the same matrix as in Example 3.4 and let

$$D = \operatorname{diag}(10^{-3}, 1, 1, 1).$$

Then

$$DAD^{-1} = \begin{bmatrix} 1 & 2.10^{-7} & -3.10^{-7} & 2.10^{-6} \\ -2.10^{-1} & 2 & 10^{-3} & 3.10^{-4} \\ 1 & 2.10^{-3} & 2 & -5.10^{-4} \\ 10^{-1} & -4.10^{-4} & -5.10^{-4} & 2 \end{bmatrix}$$

and the radii of the Gershgorin disks are

$$\begin{aligned} r_1 &= 2.5 \cdot 10^{-6}, \\ r_2 &= 2.013 \cdot 10^{-1}, \\ r_3 &= 1.0025 \cdot 10^{0}, \\ r_4 &= 1.009 \cdot 10^{-1}. \end{aligned}$$

The similarity transformation DAD^{-1} decreases the radius of the first disk but increases the radii of the rest disks. As a result, we obtain that the matrix A, which has the same eigenvalues as DAD^{-1}, has an eigenvalue which is in a distance less or equal to $2.5 \cdot 10^{-6}$ from 1. On the other hand, it is known from (3.4) that the rest eigenvalues are at a distance not greater than $3.5 \cdot 10^{-3}$ from 2.

The Gershgorin theorem in conjunction with diagonal scaling transformations can be used for clustering of multiple eigenvalues (Section 6.4).

3.3 Sensitivity of Simple Eigenvalues

In the eigenvalue sensitivity analysis, one uses the fact that the perturbations of the eigenvalues are continuous functions of the elements of the perturbations of the matrices under interest. Also, it is necessary to have in mind that the simple eigenvalues remain simple for sufficiently small real perturbations of the matrix.

If λ is a simple eigenvalue of A and ΔA is small, then it is possible to find a perturbed eigenvalue $\widetilde{\lambda} = \lambda + \Delta\lambda$ of $A + \Delta A$, which is closest to λ. In such a case, the perturbation $\Delta\lambda$ satisfies the following theorem.

Theorem 3.6. *Let λ be a simple eigenvalue of A with right eigenvector x and left eigenvector y. Let $\lambda + \Delta\lambda$ be the corresponding eigenvalue of $A + \Delta A$. Then*

$$\Delta\lambda = \frac{y^H \Delta A x}{y^H x} + O(\|\Delta A\|^2). \tag{3.5}$$

In (3.5), the quantity $O(\|\Delta A\|^2)$ reflects the influence of the second-order terms with respect to ΔA on the size of the change of the simple eigenvalue. If $\|\Delta A\|$ is sufficiently small, its square is very small and can be neglected in the perturbation analysis. Estimates of the form (3.5) are called *asymptotic*, since they are valid for $\|\Delta A\| \to 0$.

It follows from equation (3.5) that in a first-order approximation it is fulfilled

$$|\widetilde{\lambda} - \lambda| \leq c_\lambda \|\Delta A\|_2, \tag{3.6}$$

where

$$c_\lambda = \frac{\|x\|_2 \|y\|_2}{|y^H x|}. \tag{3.7}$$

Expression (3.6) is an example of a *linear* or *first-order perturbation analysis* and the right-hand side of the expression is called *eigenvalue sensitivity bound.* This bound is also referred to as *eigenvalue perturbation bound.*

The number c_λ is called *condition number of the eigenvalue* λ since it determines the size of the change of λ, due to the perturbation ΔA. If the condition number c_λ is large, then the eigenvalue λ is called *ill conditioned.* If this number is small, the eigenvalue is *well conditioned.* In the general case, the spectrum of a matrix may consists of well-conditioned as well as ill-conditioned eigenvalues.

The estimate (3.6) shows that for small $\|\Delta A\|_2$, the perturbed eigenvalues $\widetilde{\lambda}_i$, $i = 1, 2, \ldots, n$, of the matrix $A + \Delta A$ lie in circles with centers λ_i and radii $c_{\lambda_i} \|\Delta A\|_2$.

It follows from (3.6) that the size of the change in a simple eigenvalue has the order of the size of the perturbation.

It should be pointed out that the expression (3.7) for the condition number does not depend on the way in which the right and left eigenvectors are normalized.

In accordance with (3.7), the condition number of the eigenvalue λ can be expressed as

$$c_\lambda = \|P\|_2, \tag{3.8}$$

where

$$P = \frac{xy^H}{y^H x}$$

is the eigenprojection, corresponding to λ.

Since (see Section 2.2.7)

$$\|P\|_2 = 1/\cos\angle(x, y),$$

where $\angle(x, y)$ is the acute angle between the left y and right x eigenvectors corresponding to λ, it follows that

$$c_\lambda = 1/\cos\angle(x, y),$$

i.e., the condition number is the secant of the angle between the left and right eigenvectors. With the increase in the angle between

y and x, the condition number increases, and for $\angle(x, y) \to \pi/2$, one has that $c_\lambda \to \infty$. This bound corresponds to the moment when the simple eigenvalues coalesce and the presented perturbation analysis is not valid further on.

The following example demonstrates that in the computation of eigenvalues and eigenvectors, it is preferable to compute the invariant subspace belonging to a cluster of eigenvalues which is better conditioned than the individual eigenvectors that span it.

Example 3.7. Consider the matrix

$$\widetilde{A} = \begin{bmatrix} -1-\varepsilon & 1 \\ 0 & -1+\varepsilon \end{bmatrix}$$

for some small $\varepsilon > 0$, which is obtained by perturbing a 2×2 Jordan block A having eigenvalue equal to -1. The eigenvalues of $\widetilde{A}$, $\lambda_1 = -1 - \varepsilon$ and $\lambda_2 = -1 + \varepsilon$ are distinct and the corresponding two linearly independent eigenvectors are

$$x_1 = \begin{bmatrix} 1 \\ 0 \end{bmatrix} \quad \text{and} \quad x_2 = \begin{bmatrix} 1/2\varepsilon \\ 1 \end{bmatrix}.$$

The angle between the two eigenvectors can be determined from

$$\cos(\Theta) = \frac{|x_2^H x_1|}{\|x_1\|\|x_2\|} = \frac{1}{\sqrt{1+4\varepsilon^2}}$$

or, equivalently, from

$$\sin(\Theta) = \frac{2\varepsilon}{\sqrt{1+4\varepsilon^2}}.$$

We see that with the decrease in ε, the eigenvalues become closer and the angle between the eigenvectors tends to zero, i.e., the eigenvectors become almost parallel (linearly dependent). At the same time, the eigenvalue condition numbers satisfy

$$c_{\lambda_1} = c_{\lambda_2} = \frac{\sqrt{1+4\varepsilon^2}}{2\varepsilon}$$

so that with $\varepsilon \to 0$ it holds c_{λ_1}, $c_{\lambda_2} \to \infty$. To say it another way, the eigenvalue sensitivity tends to infinity when the distance to a double eigenvalue tends to zero. (At the same time, the sensitivity of the multiple eigenvalue $\lambda = -1$ for $\varepsilon = 0$ is finite.) This is a common situation in many numerical problems when the condition number of the problem is proportional to the reciprocal of the shortest distance from that problem to an ill-posed one (Demmel, 1987a).

On the other hand, in the ε-vicinity of A lies the Jordan block

$$J = \begin{bmatrix} -1 & 1 \\ 0 & -1 \end{bmatrix},$$

which fulfills $\|\widetilde{A} - J\|_2 = \varepsilon$. The eigenvector of J corresponding to $\lambda = -1$ is $z_1 = x_1$ and the principal vector of grade 2 is

$$z_2 = \begin{bmatrix} 0 \\ 1 \end{bmatrix}.$$

Thus the invariant subspace corresponding to λ (which in the given case is the mean value of λ_1 and λ_2) has as basis vectors z_1 and z_2 which are orthogonal ($\Theta = \pi/2$) and hence perfectly conditioned. This confirms that in the case of ill-conditioned eigenproblems, the eigenvectors are not suitable as basis vectors and it is better to look for a near matrix with a defective eigenvalue.

The eigenvalue condition numbers of a matrix are invariant to unitary similarity transformations since these transformations preserve the angles between the corresponding left and right eigenvectors. In fact, let

$$A' = UAU^H,$$

where the matrix U is unitary. The eigenvalue λ remains unchanged under the transformation and its condition number (with respect to perturbations in A') is given by

$$c'_\lambda = \frac{\|x'\|_2 \|y'\|_2}{|y'^H x'|},$$

where

$$x' = Ux, \quad y' = Uy$$

are the right and left eigenvectors of A', respectively, corresponding to λ. Hence

$$c'_\lambda = \frac{\|Ux\|_2 \|Uy\|_2}{|y^H U^H Ux|} = \frac{\|x\|_2 \|y\|_2}{|y^H x|} = c_\lambda.$$

The eigenvalue condition numbers can be represented in different ways, depending on the normalization of the eigenvectors.

If the eigenvectors are normalized such that

$$\|x\|_2 = 1, \quad \|y\|_2 = 1,$$

then it follows from (3.7) that

$$c_\lambda = 1/|y^H x|.$$

The number

$$s = 1/\|P\|_2 = \cos \angle(x, y)$$

is introduced by Wilkinson (1965), pp. 68–69, as a reciprocal of the eigenvalue condition number.

For the other possible way of normalizing the eigenvectors when $y^H x = 1$, one has that

$$c_\lambda = \|x\|_2 \|y\|_2.$$

Both ways of representing the condition number are used in the literature.

In most cases, the condition number produces a realistic estimation of the perturbation in the eigenvalue. However, it may give also a pessimistic estimate if the specific perturbation does not influence significantly the eigenvalue.

In LAPACK, the eigenvalue condition numbers are computed by the routines `xGEESX` and in MATLAB® by the function `condeig`.

Example 3.8. Given is the matrix

$$A = \begin{bmatrix} 4.5 & 0.9 & -1.9 & 4.4 & -9.7 \\ 4.3 & 1.9 & -1.9 & 5.2 & -11.3 \\ -3.1 & 0.0 & 2.0 & -3.6 & 6.7 \\ 1.7 & 0.9 & -1.9 & 4.6 & -7.1 \\ 3.0 & 0.9 & -1.9 & 4.4 & -8.2 \end{bmatrix},$$

with simple eigenvalues

$$\lambda_1 = 2, \quad \lambda_2 = 1.5, \quad \lambda_3 = 0.2, \quad \lambda_4 = 0.1, \quad \lambda_5 = 1$$

Table 3.1. Estimates of the sensitivity of simple eigenvalues, obtained by using the condition numbers.

i	λ_i	c_{λ_i}	$c_{\lambda_i}\|\Delta A\|_2$	$\|\Delta\lambda_i\|$
1	2.0	$8.66 \cdot 10^0$	$1.77 \cdot 10^{-6}$	$1.40 \cdot 10^{-7}$
2	1.5	$4.47 \cdot 10^0$	$9.15 \cdot 10^{-7}$	$4.10 \cdot 10^{-7}$
3	0.2	$1.59 \cdot 10^1$	$3.25 \cdot 10^{-6}$	$8.20 \cdot 10^{-7}$
4	0.1	$4.00 \cdot 10^0$	$8.19 \cdot 10^{-7}$	$1.20 \cdot 10^{-7}$
5	1.0	$6.48 \cdot 10^0$	$1.33 \cdot 10^{-6}$	$8.00 \cdot 10^{-8}$

and let on this matrix act the perturbation

$$\Delta A = 10^{-8} \cdot \begin{bmatrix} -3 & 1 & 7 & -4 & 1 \\ 6 & 0 & 4 & 2 & 9 \\ -3 & -2 & 7 & 1 & -5 \\ 8 & 6 & -9 & -3 & 4 \\ 7 & 4 & -3 & 2 & 6 \end{bmatrix}.$$

One has that

$$\|\Delta A\|_2 = 2.05 \cdot 10^{-7}.$$

The eigenvalue condition numbers, computed according to (3.7), the estimates of the changes of these eigenvalues obtained by (3.6), as well as the actual changes $|\Delta\lambda_i|$, are shown in Table 3.1. The third eigenvalue 0.2 is the most sensitive one to the perturbations of matrix elements. Clearly, the estimates based on the condition numbers give a right impression for the actual changes in the eigenvalues.

As pointed out earlier, the bounds on the changes of the eigenvalues obtained by using the condition numbers are asymptotic and are valid for sufficiently small perturbations of the matrix. Such bounds are also called *local*, since they are valid for perturbations whose size is in the vicinity of the zero. In the perturbation theory for eigenvalues, there are results which allow obtaining also *nonlocal* or *global bounds* on the eigenvalue sensitivity. These bounds are also valid for large perturbations of the matrix elements.

Global bounds on the eigenvalue sensitivity can by obtained by the Bauer–Fike theorem (Bauer and Fike, 1960).

Theorem 3.9 (Bauer–Fike Theorem). *Let the matrix $A \in \mathbb{C}^{n\times n}$ be diagonalizable,*

$$X^{-1}AX = \Lambda = \operatorname{diag}(\lambda_1, \lambda_2, \ldots, \lambda_n),$$

and let μ be an eigenvalue of the perturbed matrix $A + \Delta A$. Then there exists an eigenvalue λ_i of A with corresponding spectral projector P_i, such that

$$\min_i |\mu - \lambda_i| \le n\|P_i\|_2\|\Delta A\|_2. \tag{3.9}$$

Equation (3.9) shows that the eigenvalues of the perturbed matrix $\widetilde{A} = A + \Delta A$ lie in the union of the circles $\mathcal{D}_i$, where $\mathcal{D}_i$ has a center λ_i and a radius $n\|P_i\|_2\|\Delta A\|_2$. These circles are exactly n times larger than the bound (3.5).

Despite the fact that the Bauer–Fike does not impose restrictions on the size of the perturbation ΔA, this theorem is useful only when all radii $n\|P_i\|_2\|\Delta A\|_2$ are sufficiently small. If one or several circles are very large and cross the rest circles, then the information for the position of the individual eigenvalues is lost since they lie in the union of all circles.

If the matrix A is Hermitian ($A^H = A$) or normal ($AA^H = A^H A$), then its eigenvectors are orthonormal, i.e.,

$$Q^H AQ = \Lambda, \quad \Lambda = \operatorname{diag}(\lambda_1, \lambda_2, \ldots, \lambda_n)$$

with $Q^H Q = I_n$. That is why the right eigenvectors x (the columns of Q) and the left eigenvectors y (the conjugate-transpose rows of Q^H) are identical and $1/|y^H x| = 1$. Thus, if A is Hermitian (or symmetric) matrix, the eigenvalues' condition numbers are equal to one and are valid the asymptotic estimates

$$|\Delta\lambda_i| \lesssim \|\Delta A\|_2. \tag{3.10}$$

In this way, *the eigenvalues of Hermitian (symmetric) matrix are perfectly conditioned.*

The following global estimate is valid for symmetric matrices (Wilkinson, 1965, Chapter 2; Stewart and Sun, 1990, Chapter 4).

Theorem 3.10 (Theorem of Wielandt–Hoffman). *If A and $\widetilde{A} = A + \Delta A$ are $n \times n$ symmetric matrices, then*

$$\sum_{i=1}^{n} (\lambda_i(\widetilde{A}) - \lambda_i(A))^2 \le \|\Delta A\|_F^2.$$

3.4 Significance of the Jordan Structure

As pointed out previously, there exist matrices for which small perturbation of the elements lead to large changes in the eigenvalues. Such kind of matrices are those whose eigenvalues participate in nonlinear elementary divisors. The following example illustrates the significance of the Jordan structure associated with a given eigenvalue on its sensitivity.

Example 3.11 (This example is due to G.E. Forsythe). Consider a Jordan block $J_n(\beta)$ of nth order with eigenvalue β. If the nonzero element in position $(n, 1)$ is replaced by the small number $\tau > 0$, then the characteristic equation of the perturbed block becomes

$$(\beta - \lambda)^n - (-1)^n \tau = 0$$

and the multiple eigenvalue β is split into n distinct eigenvalues

$$\lambda_k = \beta + \tau^{1/n}(\cos(2k\pi/n) + j_o \sin(2k\pi/n)), \quad k = 0, 1, \ldots, n-1.$$

The perturbed eigenvalues satisfy the equation

$$|\lambda_k - \beta| = |\tau|^{1/n},$$

i.e., they lie on a circle with center β and radius $\tau^{1/n}$. If, for instance,

$$J_6(-1) = \begin{bmatrix} -1 & 1 & 0 & 0 & 0 & 0 \\ 0 & -1 & 1 & 0 & 0 & 0 \\ 0 & 0 & -1 & 1 & 0 & 0 \\ 0 & 0 & 0 & -1 & 1 & 0 \\ 0 & 0 & 0 & 0 & -1 & 1 \\ 0 & 0 & 0 & 0 & 0 & -1 \end{bmatrix},$$

with $n = 6$ and $\beta = -1$, for $\tau = 10^{-6}$, one obtains the eigenvalues

$$\begin{aligned}
&-0.900000000000000,\\
&-0.950000000000000 + j_o 0.086602540378444,\\
&-0.950000000000000 - j_o 0.086602540378444,\\
&-1.050000000000000 + j_o 0.086602540378444,\\
&-1.050000000000000 - j_o 0.086602540378444,\\
&-1.100000000000000.
\end{aligned}$$

The spectral deviation of the perturbed Jordan block with respect to the unperturbed one is exactly equal to $\tau^{1/6} = 0.1$. The same value has the Hausdorf distance between the spectra of both blocks. The changes in the eigenvalues have modules equal to 0.1, which is 10^5 larger than the perturbation $\tau = 10^{-6}$.

In Figure 3.2, we show the eigenvalue perturbations of the block $J_6(-1)$ for 20 equally spaced values of τ between 10^{-8} (the innermost circle) and 10^{-6} (the outermost circle). Note that the number of the eigenvalue loci is equal to the size of the Jordan block.

In this way, a perturbation of size τ can introduce a perturbation of order $\tau^{1/n}$ in the matrix eigenvalues, which even for small n leads to large changes in the eigenvalues. Since the trace of the matrix remains unchanged, the mean value of the eigenvalues $\sum_1^n \lambda_i/n$ remains the same and is equal to β.

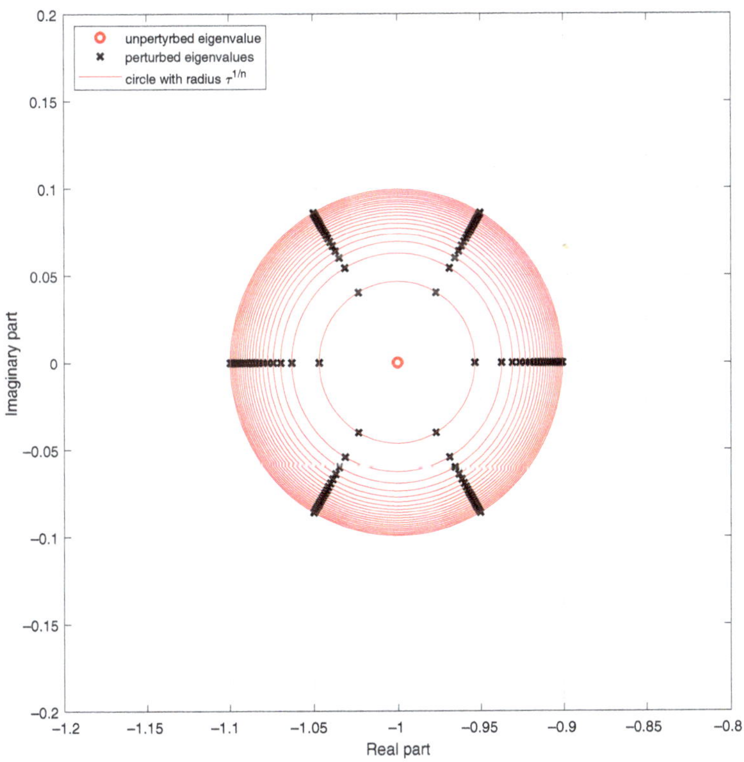

Fig. 3.2. Sensitivity of the eigenvalues of a Jordan block for values of τ between 10^{-8} and 10^{-6}.

In this example, the eigenvalues are very sensitive to perturbations since the matrix is entirely defective. If the eigenvalues have linear elementary divisors, their sensitivity is generally small. This shows that the analysis of the eigenvalue sensitivity should take into account explicitly or implicitly the Jordan structure of the matrix.

3.5 Sensitivity of Multiple Eigenvalues

A global bound on the spectral variation of the perturbed matrix with respect to the unperturbed one that is valid independently on the eigenvalue multiplicity can be obtained by the Henrichi theorem (Henrici, 1962). This theorem is formulated in terms of the *deviation of the matrix A from normality.* Remember that if a matrix is normal, its Schur form is diagonal. That is why the size of the off-diagonal elements of the Schur form can be used as a measure of the deviation from normality. Specifically, if the Schur form of the matrix A is represented as

$$T = \Lambda + R,$$

where $\Lambda = \mathrm{diag}(\lambda_1, \lambda_2, \ldots, \lambda_n)$ and R is strictly upper triangular matrix, then in case of using the 2-norm, the deviation from normality is determined as

$$\mathrm{dev}_2(A) = \|R\|_2.$$

Since the Schur form of a matrix is not unique, the matrix R is also not unique and it is possible to find different values of $\mathrm{dev}_2(A)$ depending on the numerical algorithm for finding the Schur form.

In case of using the Frobenius norm, the deviation from normality is obtained uniquely as

$$\mathrm{dev}_F(A) = \|R\|_F = \sqrt{\|A\|_F - \sum |\lambda_i|^2}.$$

It is convenient for the real variable $x \geq 0$ to define the function $\omega = \omega(x)$ as the (unique) nonnegative root of the equation

$$\omega^n - x\omega^{n-1} - x\omega^{n-2} - \cdots - x\omega - x = 0.$$

Now it is possible to formulate the following theorem.

Theorem 3.12 (Henrichi Theorem). *Let A be a matrix, for which* $\mathrm{dev}(A) > 0$ *is the deviation from normality. If* $\widetilde{A} = A + \Delta A$ *and*

$$x = \|\Delta A\|/\mathrm{dev}(A),$$

then the spectral deviation of $\widetilde{A}$ *with respect to* A *is given by*

$$\mathrm{sv}_A(\widetilde{A}) \le \omega(x)\mathrm{dev}(A). \tag{3.11}$$

In (3.11), *we may use the 2-norm or the Frobenius norm in computing* $\|\Delta A\|$ *and* $\mathrm{dev}(A)$.

Equation (3.11) gives a nonlocal bound that may be used in estimating the perturbations of multiple eigenvalues.

Algorithm 3.2 estimates the eigenvalue sensitivity by the Henrichi theorem.

Consider two special cases covered by the Henrichi theorem (Stewart and Sun, 1990, Chapter 4). From

$$x = \omega^n/(1 + \omega + \cdots + \omega^{n-1}),$$

it follows that for small ω it is fulfilled that $x \approx \omega^n$, i.e.,

$$\frac{\mathrm{sv}_A(\widetilde{A})}{\mathrm{dev}(A)} \lesssim \left(\frac{\|\Delta A\|}{\mathrm{dev}(A)}\right)^{\frac{1}{n}}.$$

This case corresponds to a matrix A, which is similar to a single Jordan block with size n. The estimate obtained shows that if a multiple eigenvalue participates in a single Jordan block of size n, then the appearance of the term $(\|\Delta A\|)^{1/n}$ in the sensitivity estimate is unavoidable. Note, however, that even in this extreme case, the eigenvalue perturbation is finite.

When ω is large, $x \approx \omega$ and the bound (3.11) takes the asymptotic form

$$\mathrm{sv}_A(\widetilde{A}) \lesssim \|\Delta A\|.$$

This is the case when the multiple eigenvalue has linear elementary divisors (i.e., the Jordan form has n scalar blocks and the matrix A is derogatory).

The Henrichi theorem may give pessimistic result if the Jordan block associated with the multiple eigenvalue is of order smaller than n. If the Jordan normal form of A is known, then $\mathrm{dev}_2(A) = 1$ and we have the following theorem.

Algorithm 3.2. Determine the eigenvalue sensitivity by the Heinrichi theorem

For a given matrix A and the norm $\|\Delta A\|$ of the perturbation ΔA, this algorithm computes an upper bound on the spectral variation of the matrix $A + \Delta A$ with respect to A.

function sv = henrici(A, $\|\Delta A\|$)

Input: Matrix $A \in \mathbb{C}^{n \times n}$, whose eigenvalue sensitivity is analyzed; the norm $\|\Delta A\|$ of the perturbation matrix ΔA.

Output: Spectral deviation sv of the perturbed matrix with respect to the unperturbed one

Compute the Schur form of A, $T = \Lambda + R$, where
$\Lambda = \mathrm{diag}(\lambda_1, \lambda_2, \ldots, \lambda_n)$
contains the diagonal elements of T (the eigenvalues of A).
Compute the deviation from normality
$\mathrm{dev}(A) = \|R\|_2$ or $\mathrm{dev}(A) = \sqrt{\|A\|_F^2 - \sum_{i=1}^{n} |\lambda_i|^2}$, depending on the norm used.
$x = \delta/\mathrm{dev}(A)$
Set the coefficients of the polynomial
$p_1\omega^n + p_2\omega^{n-1} + \cdots + p_n\omega + p_{n+1}$
$p_1 = 1$
for $i = 2 : (n+1)$
  $p_i = -x$
end
Compute the zeros $r_1, r_2, \ldots, r_n$ of the polynomial
Determine the positive real root
for $i = 1, 2, \ldots, n$
  if $\mathrm{Re}(r_i) > 0$ & $\mathrm{Im}(r_i) = 0$
    $\omega = r_i$
  end
end
$\mathrm{sv} = \mathrm{dev}(A) \cdot \omega$

Theorem 3.13 (Stewart and Sun (1990)). *Let $Z^{-1}AZ = J$ be the Jordan normal form of A and let m be the order of the largest Jordan block of J. Then it is fulfilled that*

$$\mathrm{sv}_A(\widetilde{A}) \leq \omega(x), \tag{3.12}$$

where $\omega(x)$ is the unique positive root of

$$\psi^m - x\psi^{m-1} - x\psi^{m-2} - \cdots - x\psi - x = 0$$

and $x = \|Z^{-1}\Delta AZ\|_2$.

In (3.12), we can bound x as

$$x = \text{cond}_2(Z)\|\Delta A\|_2.$$

This relationship shows that the eigenvalue sensitivity increases with the increase in the conditioning of the similarity transformation matrix Z. In case of large cond(Z), we shall say that the matrix eigensystem is *ill conditioned.*

Example 3.14. Given is again the Jordan block of sixth order with eigenvalue -1, considered in Example 3.11. For this block, a perturbation of 10^{-6} in the element $(6, 1)$ leads to a spectral deviation, equal to 0.1. If we apply the Henrichi theorem to the block, using Theorem 3.13, the deviation from normality is

$$\text{dev}(A)(J_6(-1)) = \|J_6(-1) - (-1)I_6\|_2 = 1,$$

and the bound on the spectral deviation of the perturbed block with respect to the unperturbed one is equal to

$$\text{sv}_A(\widetilde{A}) = 0.101805566813350.$$

The obtained bound on the sensitivity of multiple eigenvalues is shown in Figure 3.3. Note that the bound produced by the Henrichi theorem reflects the influence of all entries of the perturbation matrix ΔA on the spectral variation. The fact that this bound is close to the quantity $\tau^{1/6} = 0.1$ shows that the single perturbation in the $(6, 1)$ entry dominates the sensitivity of the multiple eigenvalue.

The following example illustrates the application of Theorem 3.13 to estimate the eigenvalue sensitivity when the Jordan form of the matrix is known.

Example 3.15. Consider the matrix

$$A = \begin{bmatrix} -2 & -4 & 1 & -2 & 2 & 2 \\ -1 & -3 & 0 & -2 & 2 & 1 \\ 0 & -1 & -2 & -2 & 1 & 2 \\ -1 & -2 & 1 & -5 & 3 & 2 \\ -1 & -4 & 1 & -4 & 2 & 3 \\ 0 & 0 & 0 & -2 & 1 & 0 \end{bmatrix},$$

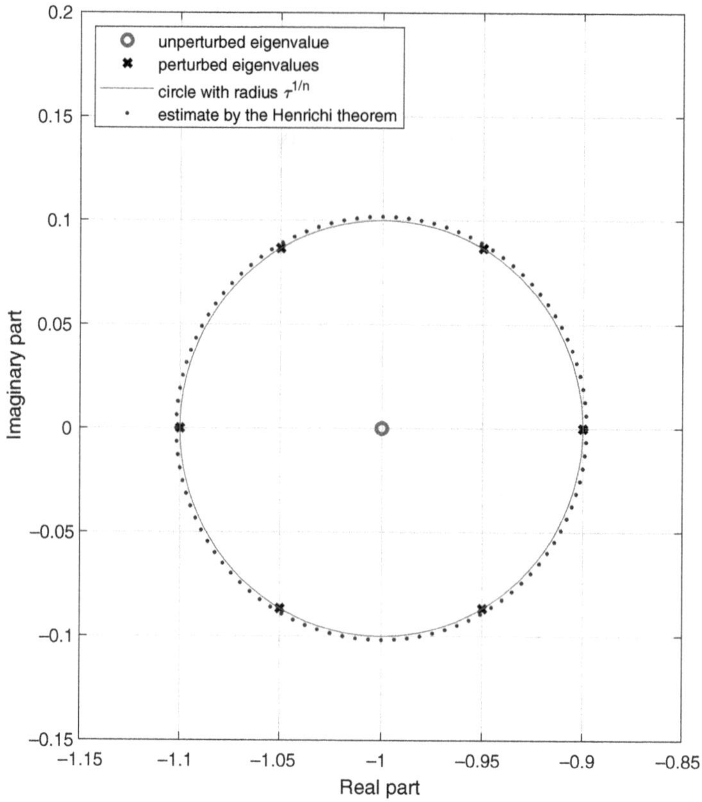

Fig. 3.3. Estimating the multiple eigenvalue sensitivity by the Henrichi theorem.

which has the Jordan decomposition

$$A = QJQ^{-1}$$

with a Jordan form

$$J = \left(\begin{bmatrix} -1 & 1 \\ & -1 \end{bmatrix}, \begin{bmatrix} -2 & 1 & & \\ & -2 & 1 & \\ & & -2 & 1 \\ & & & -2 \end{bmatrix} \right),$$

and the matrix of the nonsingular similarity transformation

$$Q = \begin{bmatrix} -2 & -1 & 1 & 0 & 1 & -1 \\ 0 & 0 & 0 & -1 & 0 & 0 \\ 0 & -1 & 0 & -1 & 0 & 1 \\ -1 & -1 & -1 & -1 & 1 & 1 \\ -2 & -1 & 0 & -1 & 1 & 0 \\ 0 & -1 & -1 & -1 & 0 & 1 \end{bmatrix}.$$

The family of perturbation matrices is given as

$$\Delta A = 10^s \cdot \Delta A_0,$$

where s varies from -8 to -6 and

$$\Delta A_0 = \begin{bmatrix} 0.117473319440409 & -0.075642022914664 & 0.033881566254010 \\ -0.311158913908956 & 0.104297549314854 & 0.029764138089774 \\ -0.013057114818279 & 0.029764138089774 & 0.064770238938190 \\ 0.029764138089774 & 0.029764138089774 & 0.029764138089774 \\ 0.715727670251472 & 0.029764138089774 & 0.029764138089774 \\ -0.391860505927979 & 0.029764138089774 & 0.029764138089774 \end{matrix} \cdots \rightarrow$$

$$\leftarrow \cdots \begin{matrix} 0.451388782107528 & 0.019882310495608 & 0.438213011981974 \\ 0.065997505935050 & 0.082467218591994 & 0.398685701605309 \\ -0.391860505927979 & 0.029764138089774 & -0.395154448459368 \\ -0.106514772694023 & 0.029764138089774 & 0.135170299094213 \\ -0.088817793040219 & 0.104297549314854 & 0.029764138089774 \\ 0.293279540600871 & 0.029764138089774 & 0.091121779189299 \end{bmatrix}.$$

The matrix A_0 is scaled so that $\|\Delta A_0\| = 1$.

For $s = -6$, the absolute changes in the multiple eigenvalue at -1 are

$$\text{err}_1 = \begin{bmatrix} 0.001120732584260, \\ 0.001120732584260 \end{bmatrix}$$

and the absolute changes in the multiple eigenvalue at -2 are

$$\mathrm{err}_2 = \begin{bmatrix} 0.044327569679282, \\ 0.044363837993670, \\ 0.044363837993670, \\ 0.044392661000395, \end{bmatrix}$$

so that the corresponding spectral variations are

$$\mathrm{sv}_1 = 0.001120732584260$$

and

$$\mathrm{sv}_2 = 0.044392661000395.$$

For this perturbation, one has that $\|\Delta A\|_2 = 10^{-6}$, $\mathrm{cond}_2(Q) = 18.816$ and the estimates of the spectral deviations for $\lambda = -1$ and $\lambda = -2$, obtained by Theorem 3.12, are respectively

$$\mathrm{est}_1 = 0.004347145889714$$

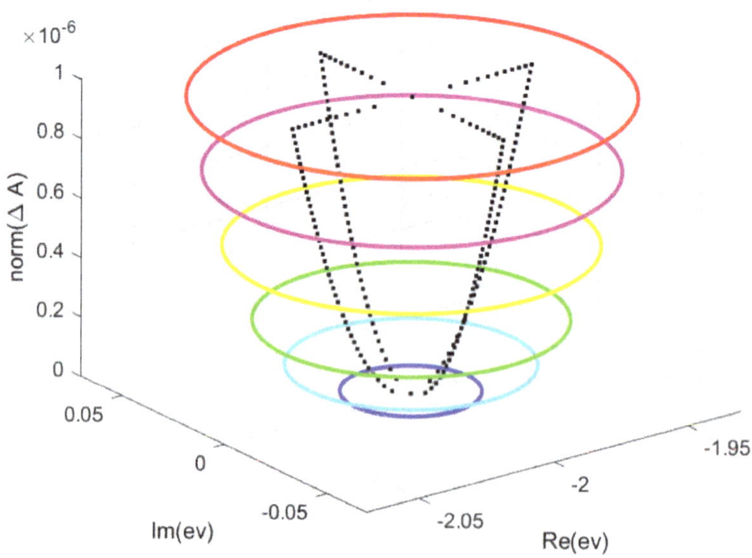

Fig. 3.4. Perturbations of a defective eigenvalue at -2 pertaining to a Jordan block of order four for different perturbation sizes. The perturbed eigenvalues are denoted by dots. The circles at different levels represent the Henrichi bound for the corresponding perturbation sizes.

and

$$\text{est}_2 = 0.067013156632194.$$

The estimates found are close to the exact spectral variations.

In Figure 3.4, we show the perturbations of the defective eigenvalue at -2 of the Jordan block of size 4 as functions of the perturbation size $\|\Delta A\|_2$. There are four eigenvalue loci corresponding to the perturbed eigenvalue. The circles at different levels have radii equal to the corresponding Henrichi estimate est_2.

3.6 Sensitivity of Invariant Subspaces

Consider now the problem of computing sensitivity estimates of eigenvectors and invariant subspaces under perturbations of the matrix elements. This is done by implementing the following perturbation analysis:

Let

$$U^H AU = \begin{bmatrix} T_{11} & T_{12} \\ 0 & T_{22} \end{bmatrix} \tag{3.13}$$

be the Schur decomposition of $A \in \mathbb{C}^{n\times n}$, where $T_{11} \in \mathbb{C}^{r\times r}$ contains the desired eigenvalues of A. The matrix of the unitary similarity transformation can be partitioned as

$$U = [U_1, U_2], \quad U_1 \in \mathbb{C}^{n\times r},$$

where the columns of U_1 are the basis vectors of the invariant subspace $\mathcal{X}$, corresponding to the block T_{11}. Note that $U_2^H AU_1 = 0$.

If the matrix A is a subject of a perturbation ΔA, then instead of the decomposition (3.13), one will have the decomposition

$$\widetilde{U}^H \widetilde{A}\widetilde{U} = \begin{bmatrix} \widetilde{T}_{11} & \widetilde{T}_{12} \\ 0 & \widetilde{T}_{22} \end{bmatrix}, \quad \widetilde{T}_{11} \in \mathbb{C}^{r\times r}, \quad \widetilde{A} = A + \Delta A \tag{3.14}$$

with a perturbed matrix of the unitary transformation

$$\widetilde{U} = [\widetilde{U}_1, \widetilde{U}_2].$$

The columns of the matrix $\widetilde{U}_1$ are basis vectors of the perturbed invariant subspace $\widetilde{\mathcal{X}}$.

If $\|\Delta A\|$ is small, we may expect that the spaces $\widetilde{U}_1$ and $\widetilde{U}_2$ will be close to the spaces U_1 and U_2, respectively. Taking into account the representation of the right eigenspace and its orthogonal complement present in Section 2.1.6, we look for a matrix $R \in \mathbb{C}^{(n-r)\times r}$ so that

$$\widetilde{U}_1 = (U_1 + U_2 R)(I_r + R^H R)^{-1/2},$$
$$\widetilde{U}_2 = (U_2 - U_1 R^H)(I_{n-r} + RR^H)^{-1/2}$$

and $\mathcal{R}(\widetilde{U}_1)$ is an invariant subspace of $\widetilde{A}$. It may be shown that due to the use of the scaling factors $(I_r + R^H R)^{-1/2}$ and $(I_{n-r} + RR^H)^{-1/2}$, the matrix $[\widetilde{U}_1, \widetilde{U}_2]$ is unitary. One has that $\mathcal{R}(\widetilde{U}_1)$ is an invariant subspace of $\widetilde{A}$ if and only if

$$\widetilde{U}_2^H A \widetilde{U}_1 = 0. \tag{3.15}$$

Define

$$E = U^H \Delta A U = \begin{bmatrix} E_{11} & E_{12} \\ E_{21} & E_{22} \end{bmatrix}, \quad E_{11} \in \mathbb{C}^{r\times r}, \quad E_{22} \in \mathbb{C}^{(n-r)\times(n-r)}.$$

Equation (3.15) can be written as

$$(T_{22} + E_{22})R - R(T_{11} + E_{11}) = -E_{21} + R(T_{12} + E_{12})R$$

or

$$T_{22}R - RT_{11} = -E_{21} - E_{22}R + RE_{11} + R(T_{12} + E_{12})R. \tag{3.16}$$

For small R, this equation may be approximated by a Sylvester matrix equation in respect to R. By using the Kronecker product, (3.16) can be represented as

$$\begin{aligned}(I_r \otimes (T_{22} - (T_{11})^T \otimes I_{n-r})\mathrm{vec}(R) \\ = -\mathrm{vec}(E_{21}) - (I_r \otimes E_{22} - E_{11}^T \otimes I_{n-r})\mathrm{vec}(R) \\ + \mathrm{vec}(R(T_{12} + E_{12})R).\end{aligned}$$

Equation (3.16) can also be written as the operator equation

$$\mathbf{\Omega} R = g - \varphi(R), \tag{3.17}$$

where $g = -E_{21}$ and the mappings $\mathbf{\Omega}$, $\varphi : \mathbb{C}^{(n-r)\times r} \to \mathbb{C}^{(n-r)\times r}$ are defined by

$$\mathbf{\Omega} R = T_{22}R - RT_{11}$$

and

$$\varphi(R) = E_{22}R - RE_{11} - R(T_{12} + E_{12})R,$$

respectively. For sufficiently small $\|\varphi(R)\|$, the size of R can be estimated using the contraction mapping theorem (see Ortega and Rheinboldt, 2000, Chapter 5 or Stewart and Sun, 1990, Chapter 5), provided that the operator $\mathbf{\Omega}^{-1}$ exists. According to Example A.2 in the Appendix, the Sylvester operator $\mathbf{\Omega} R = T_{22}R - RT_{11}$ can be represented as the linear transformation

$$\mathrm{vec}(R) \to (I_r \otimes T_{22} - T_{11}^T \otimes I_{n-r})\mathrm{vec}(R)$$

from $\mathbb{C}^{(n-r)\cdot r}$ to $\mathbb{C}^{(n-r)\cdot r}$ so that $\mathbf{\Omega} R$ is nonsingular if and only if the matrices T_{11} and T_{22} have no eigenvalues in common (see Section 2.1.6). Thus, the size of the matrix R depends on the norm of the matrix Ω^{-1}, where

$$\Omega = I_r \otimes T_{22} - T_{11}^T \otimes I_{n-r}.$$

The norm of Ω^{-1} is closely related to the quantity *separation between two matrices.* The separation between given matrices $A \in \mathbb{C}^{n\times n}$ and $B \in \mathbb{C}^{m\times m}$ characterizes the distance between the spectra $\lambda(A)$ and $\lambda(B)$ and is defined by

$$\mathrm{sep}(A, B) = \min_{X\neq 0} \frac{\|AX - XB\|_F}{\|X\|_F}. \tag{3.18}$$

Note that $\mathrm{sep}(A, B) = 0$ if and only if A and B have common eigenvalues. It is possible to prove that

$$\mathrm{sep}(A, B) - \delta \leq \mathrm{sep}(A + E, B + F) \leq \mathrm{sep}(A, B) + \delta,$$
$$\delta = \|E\|_2 + \|F\|_2.$$

In the given case, the separation

$$\mathrm{sep}(T_{11}, T_{22}) = \min_{X\neq 0} \frac{\|T_{22}X - XT_{11}\|_F}{\|X\|_F}, \tag{3.19}$$

between the two blocks T_{11} and T_{22}, can be determined from

$$\text{sep}(T_{11}, T_{22}) = \sigma_{\min}(\Omega) \tag{3.20}$$

which is equivalent to

$$\|\Omega^{-1}\|_2 = 1/\text{sep}(T_{11}, T_{22}).$$

Assume that the spectra of T_{11} and T_{22} are disjoint so that $\text{sep}(T_{11}, T_{22}) > 0$. The following theorem gives an estimate of the size of R in (3.16) thus providing an estimate of the sensitivity of an invariant subspace of A.

Theorem 3.16 (Stewart (1973a)). *Let the matrix A be decomposed as in* (3.13). *Given the perturbation E, set*

$$\begin{aligned} \gamma &= \|E_{21}\|_2, \\ \eta &= \|T_{12} + E_{12}\|_2, \\ \delta &= \text{sep}(T_{11}, T_{22}) - \|E_{11}\|_2 - \|E_{22}\|_2. \end{aligned}$$

If $\delta > 0$ and

$$\frac{\eta\gamma}{\delta^2} < \frac{1}{4}, \tag{3.21}$$

there is a unique matrix R satisfying

$$\|R\|_2 \le \frac{2\gamma}{\delta + \sqrt{\delta^2 - 4\gamma\eta}} < 2\frac{\gamma}{\delta}, \tag{3.22}$$

such that the columns of $\widetilde{U}_1 = (U_1 + U_2 R)(I_r + R^H R)^{-1/2}$ span a right invariant subspace of $A + \Delta A$.

According to (2.53), the singular values of the matrix

$$U_2^H \widetilde{U}_1 = R(I_r + R^H R)^{-1/2}$$

are the sines of the canonical angles $\Theta_1, \Theta_2, \ldots, \Theta_r$ between the invariant subspaces $\mathcal{R}(U_1)$ and $\mathcal{R}\widetilde{U}_1$. If R has singular values $\sigma_1, \sigma_2, \ldots \sigma_r$, then the singular values of $I_r + R^H R$ are

$$1 + \sigma_1^2, 1 + \sigma_2^2, \ldots, 1 + \sigma_r^2$$

and

$$\frac{\sigma_i}{\sqrt{1+\sigma_i^2}} = \sin(\Theta_i).$$

Hence

$$\sigma_i = \tan(\Theta_i), \quad i = 1,\ 2, \ldots, r.$$

Thus, Theorem 3.16 bounds the tangents of the canonical angles between the unperturbed $\mathcal{R}(U_1)$ and perturbed $\mathcal{R}(\widetilde{U}_1)$ invariant subspace of A.

The practical implementation of Theorem 3.16 is considered latter on in Section 4.3.

Example 3.17. Let

$$A = T = \begin{bmatrix} T_{11} & T_{12} \\ 0 & T_{22} \end{bmatrix},$$

where

$$T_{11} = \begin{bmatrix} -3 & -6 \\ 0 & 2 \end{bmatrix}, \quad T_{12} = \begin{bmatrix} 5 & -4 & 8 \\ 7 & 1 & 4 \end{bmatrix}, \quad T_{22} = \begin{bmatrix} 5 & 3 & -9 \\ 0 & 2.1 & -7 \\ 0 & 0 & 1 \end{bmatrix}.$$

Since the matrix A is already in Schur form, we can take as an orthonormal basis of the invariant subspace, associated with the eigenvalues 3 and 2, the columns of the matrix

$$U_1 = \begin{bmatrix} 1 & 0 \\ 0 & 1 \\ 0 & 0 \\ 0 & 0 \\ 0 & 0 \end{bmatrix}.$$

We have that (up to seven decimal digits)

$$\text{sep}(T_{11}, T_{22}) = 6.532697 \cdot 10^3.$$

Suppose that the matrix A is perturbed so that

$$E_{21} = 10^{-7} \begin{bmatrix} 1 & 1 \\ 1 & 1 \\ 1 & 1 \end{bmatrix}$$

and $E_{11} = E_{21} = E_{22} = 0$. Since $\|E_{21}\|_2 = 2.449490 \cdot 10^{-7}$, the condition (3.22) is satisfied. After reduction into Schur form of the matrix $\widetilde{A} = A + \Delta A$ using the MATLAB®function `schur` and reordering this form so that the eigenvalues -3 and 2 appear in leading positions, we obtain that

$$\widetilde{U}_1 = \begin{bmatrix} 1.000000 \cdot 10^{0} & -5.935053 \cdot 10^{-8} \\ 5.935048 \cdot 10^{-8} & 1.000000 \cdot 10^{0} \\ -2.040441 \cdot 10^{-8} & 1.122235 \cdot 10^{-6} \\ -5.392156 \cdot 10^{-8} & -1.264755 \cdot 10^{-6} \\ -2.500000 \cdot 10^{-8} & -4.999940 \cdot 10^{-8} \end{bmatrix}.$$

Thus, the exact perturbation of the matrix U_1 is

$$\|\widetilde{U}_1 - U_1\|_2 = 1.692866 \cdot 10^{-6},$$

and according to (3.22), we have the estimate

$$\|R\|_F < 2\frac{\gamma}{\delta} = 7.499168 \cdot 10^{-5}.$$

The exact angles between $\mathcal{R}(U_1)$ and $\mathcal{R}(\widetilde{U}_1)$ are

$$\Theta_1 = 9.693446 \cdot 10^{-5} \text{ deg}, \quad \Theta_2 = 3.236479 \cdot 10^{-6} \text{ deg},$$

while the estimate of the maximum angle between these subspaces is

$$\widehat{\Theta}_{\max} = \arctan(\|R\|_2) = 4.296707 \cdot 10^{-3} \text{ deg}.$$

3.7 Componentwise Sensitivity of the Schur Decomposition

In this section, we present a perturbation analysis of the elements of the Schur decomposition of a matrix.

Suppose that a matrix $A \in \mathbb{C}^{n\times n}$ is a subject to a perturbation ΔA so that $\widetilde{A} = A + \Delta A$. Then instead of the Schur decomposition $T = U^H AU$, we obtain the perturbed decomposition $\widetilde{T} = \widetilde{U}^H \widetilde{A}\widetilde{U}$. Apart from the sensitivity of a specific invariant subspace, one can be interested in the sensitivity of the whole unitary transformation matrix U measured by $\Delta U = \widetilde{U} - U$ and/or by the sensitivity of the Schur form T measured by $\Delta T = \widetilde{T} - T$, both considered in respect to the perturbation ΔA. This problem has a solution in a linear approximation only for a matrix A with distinct eigenvalues since in case of multiple eigenvalues the elements of ΔU and ΔT depend nonlinearly on $\|\Delta A\|$. Hence, the presented analysis is valid only for sufficiently small perturbations, i.e., it is an asymptotic (linear) analysis.

In a first-order approximation, we have that

$$\|\Delta U\|_F \leq c_U \|\Delta A\|_F, \tag{3.23}$$

where the quantity c_U is the normwise condition number of the unitary transformation matrix U with respect to the perturbation of A. It is possible to show that (Konstantinov *et al.*, 1994)

$$c_U = \sqrt{2}\|M^{-1}\|_2, \tag{3.24}$$

where

$$M = \Omega(I_n \otimes T - T^T \otimes I_n)\Omega^T \in \mathbb{C}^{\nu\times\nu}, \quad \nu = n(n-1)/2,$$

and

$$\begin{aligned}
\Omega &:= \left[\mathrm{diag}(\Omega_1, \Omega_2, \ldots, \Omega_{n-1}), 0_{\nu\times n}\right] \in \mathbb{R}^{\nu\times n^2}, \\
\Omega_k &:= \left[0_{(n-k)\times k}, I_{n-k}\right] \in \mathbb{R}^{(n-k)\times n}, \quad k = 1, 2, \ldots, n-1, \\
&\Omega\Omega^T = I_\nu, \quad \|\Omega\|_2 = 1.
\end{aligned}$$

Since in a linear approximation it is fulfilled that

$$\Delta T = U^H A\Delta U + \Delta U^H AU + U^H \Delta AU,$$

taking into account (3.23), one obtains

$$\|\Delta T\|_F \leq c_T \|\Delta A\|_F, \tag{3.25}$$

where

$$c_T = 1 + 2\sqrt{2}\|M^{-1}\|_2\|A\|_F \tag{3.26}$$

is the normwise condition number of the Schur form T with respect to the perturbation ΔA.

The numerical experiments with the normwise perturbation bounds of the Schur decomposition show that while in most cases the estimate (3.23) yields a realistic upper bound of ΔU, the estimate (3.25) may produce very pessimistic estimates for ΔT.

A disadvantage of the normwise perturbation bound is that it produces a single number to estimate the size of the all n^2 elements of the corresponding matrix which can be of different magnitudes. That is why it is preferable to use *componentwise perturbation bounds* of the form

$$|\Delta U| \preceq \Delta U^{\mathrm{lin}}, \quad |\Delta T| \preceq \Delta T^{\mathrm{lin}},$$

where the elements of the $n \times n$ matrices ΔU^{lin} and ΔT^{lin} bound the absolute values of the corresponding elements of ΔU and ΔT, respectively.

In the following, we represent the determination of asymptotic perturbation bounds of $|\Delta U|$ and $|\Delta T|$.

3.7.1 *Sensitivity of the Unitary Transformation Matrix*

Let the unperturbed and perturbed unitary transformation matrices be presented as

$$U = [u_1, u_2, \ldots, u_n], \quad u_j \in \mathbb{C}^n,$$
$$\widetilde{U} = [\widetilde{u}_1, \widetilde{u}_2, \ldots, \widetilde{u}_n], \quad \widetilde{u}_j = u_j + \Delta u_j, \quad j = 1, 2, \ldots n,$$

respectively, and

$$\Delta U = \widetilde{U} - U = [\Delta u_1, \Delta u_2, \ldots, \Delta u_n], \quad \Delta u_j \in \mathbb{C}^n.$$

From the Schur decomposition of the unperturbed and perturbed matrices

$$A := UTU^H \quad \text{and} \quad \widetilde{A} = \widetilde{U}\widetilde{T}\widetilde{U}^H,$$

it follows that

$$u_i^H A \Delta u_j + \Delta u_i^H A u_j + \Delta u_i^H A \Delta u_j = -\widetilde{u}_i^H \Delta A \widetilde{u}_j. \tag{3.27}$$

Since $U^H A = TU^H$ and $AU = UT$, we have that

$$u_i^H A = \sum_{k=i}^{n} t_{ik} u_k^H, \quad Au_j = \sum_{k=1}^{j} t_{kj} u_k, \quad 1 \le j < i \le n.$$

Hence, equation(3.27) can be written as

$$\sum_{k=i}^{n} t_{ik} u_k^H \Delta u_j + \sum_{k=1}^{j} t_{kj} \Delta u_i^H u_k + \Delta u_i^H A \Delta u_j = -\widetilde{u}_i^H \Delta A \widetilde{u}_j. \quad (3.28)$$

From the unitarity of the matrices U and $\widetilde{U}$, it follows that

$$\Delta u_i^H u_j = -u_i^H \Delta u_j - \Delta u_i^H \Delta u_j, \quad 1 \le i < j \le n. \quad (3.29)$$

Using (3.29), equation (3.28) can be written in the form

$$\sum_{k=i}^{n} t_{ik} u_k^H \Delta u_j - \sum_{k=1}^{j} t_{kj} u_i^H \Delta u_k - \sum_{k=1}^{j} t_{kj} \Delta u_i^H \Delta u_k$$
$$+ \Delta u_i^H A \Delta u_j = -\widetilde{u}_i^H \Delta A \widetilde{u}_j, \quad 1 \le j < i \le n. \quad (3.30)$$

Equation (3.30) represent a system of ν nonlinear algebraic equations for the ν perturbation parameters

$$x_\ell := u_i^H \Delta u_j, \quad \ell = i + (j-1)n - \frac{j(j+1)}{2}, \quad 1 \le j < i \le n.$$

These parameters are related to the matrix $\Delta W = U^H \Delta U$. More precisely, we have that

$$x = \mathrm{vec}(\mathrm{Low}(\Delta W)),$$

or, equivalently,

$$x = \Omega \mathrm{vec}(\Delta W),$$

where $\mathrm{Low}(Z)$ is the strictly lower triangular part of $Z \in \mathbb{C}^{n \times n}$ and Ω is defined in (3.24).

Define the lower triangular matrices

$$L_1 = \Omega(I_n \otimes T)\Omega^T \in \mathbb{C}^{\nu\times\nu}, \quad L_2 = \Omega(T^T \otimes I_n)\Omega^T \in \mathbb{C}^{\nu\times\nu}.$$

It can be shown that

$$\sum_{k=i}^{n} t_{ik} u_k^H \Delta u_j = L_1 x, \quad \sum_{k=1}^{j} t_{kj} u_i^H \Delta u_k = L_2 x.$$

From (3.30), we obtain that

$$(L_1 - L_2)x = f + \Delta^x, \tag{3.31}$$

where

$$f = \mathrm{vec}(\mathrm{Low}(F)) \in \mathbb{C}^{\nu}, \quad F = -\widetilde{U}^H \Delta A \widetilde{U}$$

and Δ^x contains second-order terms in x_ℓ. Since $L_1 - L_2 = M$, (3.31) is represented as

$$Mx = f + \Delta^x.$$

The linear componentwise bounds ΔU^{lin} and ΔT^{lin} can be found by using the matrix

$$\Delta W = U^H \Delta U. \tag{3.32}$$

It is easy to show that a linear approximation of this matrix is given by

$$|\Delta W^{\mathrm{lin}}| = \begin{bmatrix} 0 & |x_1| & |x_2| & \dots & |x_{n-1}| \\ |x_1| & 0 & |x_n| & \dots & |x_{2n-3}| \\ |x_2| & |x_n| & 0 & \dots & |x_{3n-6}| \\ \vdots & \vdots & \vdots & \ddots & \vdots \\ |x_{n-1}| & |x_{2n-3}| & |x_{3n-6}| & \dots & 0 \end{bmatrix} \in \mathbb{R}^{n\times n}.$$

Since it is fulfilled that

$$x_\ell^{\mathrm{lin}} = M_{\ell,1:\nu}^{-1} \|\Delta A\|_F, \quad \ell = 1, 2, \dots \nu,$$

we have the linear bound

$$|x_\ell^{\mathrm{lin}}| \le \|M_{\ell,1:\nu}^{-1}\|_2 \|\Delta A\|_F, \quad \ell = 1, 2, \dots, \nu. \tag{3.33}$$

Now, it follows from (3.32) that a linear approximation of the matrix $|\Delta U|$ can be found as

$$|\Delta U| \preceq \Delta U^{\text{lin}} = |U||U^H \Delta U| = |U||\Delta W^{\text{lin}}|. \tag{3.34}$$

The elements of the matrix ΔU^{lin} give componentwise bounds on the absolute values of the corresponding elements of the perturbation ΔU. Hence, equation (3.34) is an example of a *componentwise perturbation analysis.*

The asymptotic componentwise perturbation bound of the matrix U of the unitary transformation of A into Schur form is determined by Algorithm 3.3.

3.7.2 Sensitivity of Eigenvalues and Invariant Subspaces

The perturbation bounds of the elements of the matrix U derived in the previous section allow finding perturbation bounds of the invariant subspaces of A. These bounds are alternative to the bounds determined in Section 3.6.

Let the unperturbed right invariant subspace corresponding to the first r eigenvalues be denoted as $\mathcal{X}$ and its perturbed counterpart as $\widetilde{\mathcal{X}}$ and let U_X and $\widetilde{U_X}$ be the orthonormal bases for $\mathcal{X}$ and $\widetilde{\mathcal{X}}$, respectively. Then the maximum angle Θmax between $\widetilde{\mathcal{X}}$ and $\mathcal{X}$ is determined from

$$\sin(\Theta\text{max}(\widetilde{\mathcal{X}}, \mathcal{X})) = \|U_{\overline{X}}^{\perp H} \widetilde{U}_X\|_2, \tag{3.35}$$

where $U_{\overline{X}}^{\perp}$ is the unitary complement of U_X, $U_{\overline{X}}^{\perp H} U_X = 0$. Since

$$\widetilde{U}_X = U_X + \Delta U_X,$$

one has that

$$\sin(\Theta\text{max}(\widetilde{\mathcal{X}}, \mathcal{X})) = \|U_{\overline{X}}^{\perp H} \Delta U_X\|_2. \tag{3.36}$$

Equation (3.36) shows that the sensitivity of the invariant subspace $\mathcal{X}$ is related to the values of perturbation parameters

$$x_\ell = u_i^H \Delta u_j, \quad \ell = i + (j-1)n - \frac{j(j+1)}{2}, \quad i > r, \quad j = 1, 2, \ldots, r.$$

Algorithm 3.3. Componentwise perturbation bounds of the matrix of unitary transformation into Schur form

Given the Schur decomposition $A = UTU^H$ of an $n \times n$ complex matrix A with simple eigenvalues and the Frobenius norm of the perturbation ΔA, this algorithm computes asymptotic componentwise perturbation bound of the matrix of unitary transformation of A into Schur form.

function $[x^{\mathrm{lin}}, M, \Delta U^{\mathrm{lin}}] = \mathit{pert\_schur\_unit}(U, T, \|\Delta A\|_F)$

Input: The matrix $U \in \mathbb{C}^{n\times n}$ of the unitary transformation of A into Schur form, the upper triangular matrix T (the Schur form of A), the norm $\|\Delta A\|_F$ of the perturbation ΔA

Output: The quantities x^{lin}, M, obtained from the perturbation analysis of the unitary reduction into Schur form; the matrix ΔU^{lin} containing the asymptotic perturbation bounds of the elements of U.

$\nu = n(n-1)/2$

Construct the perturbation operator matrix M

$\Omega = 0_{\nu\times 2\nu},\ ii = 0,\ jj = 0$

for $i = 1, 2, \ldots, n-1$

  $\Omega_i = [0_{(n-i)\times i},\ I_{n-i}],\ \ \Omega_{ii+1:ii+n-i,jj+1:jj+n} = \Omega_i$

  $ii = ii + n - i,\ jj = jj + n$

end

$\Omega = [\Omega,\ 0_{\nu\times n}]$

$M = \Omega(I_n \otimes T - T^T \otimes I_n)\Omega^T$

Find linear approximations of the perturbation parameters

for $\ell = 1, 2, \ldots, \nu$

  $x_\ell^{\mathrm{lin}} = \|M^{-1}_{\ell,1:\nu}\|_2 \|\Delta A\|_F$

end

Find linear bounds on the elements of the matrix U

$\ell = 0,\ Z = 0_{n\times n}$

for $j = 1, 2, \ldots, n-1$

  for $i = j+1 : n$

    $\ell = \ell + 1$

    $Z_{ij} = x_\ell^{\mathrm{lin}}$

  end

end

$\Delta W^{\mathrm{lin}} = Z + Z^T$

$\Delta U^{\mathrm{lin}} = |U| \Delta W^{\mathrm{lin}}$

In particular, for $r = 1$ the sensitivity of the first column of U (the right eigenvector corresponding to t_{11}) is determined by

$$\sin(\Theta\max(\widetilde{\mathcal{X}}, \mathcal{X})) = \|\Delta W_{2:n,1}\|_2,$$

for $r = 2$ one has

$$\sin(\Theta\max(\widetilde{\mathcal{X}}, \mathcal{X})) = \|\Delta W_{3:n,1:2}\|_2$$

and so on.

In this way, if the perturbation parameters are known, it is possible to find together the sensitivity estimates if all consecutive invariant subspaces have dimension $r = 1, 2, \ldots, n-1$. Note that the set of these subspaces is exactly the nested sequence of subspaces shown in (2.10). Let

$$\Delta W = \begin{bmatrix} \times & \times & \times & \cdots & \times \\ x_1 & \times & \times & \cdots & \times \\ x_2 & x_n & \times & \cdots & \times \\ \vdots & \vdots & \vdots & \ddots & \vdots \\ x_{n-1} & x_{2n-3} & x_{3n-6} & \cdots & \times \end{bmatrix}.$$

Then we have that the maximum angle between the perturbed and unperturbed invariant subspaces of dimension r is

$$\Theta\max(\widetilde{\mathcal{X}}, \mathcal{X}) = \arcsin(\|\Delta W_{r+1:n,1:r}\|_2). \tag{3.37}$$

An asymptotic estimate of the maximum angle can be obtained, if in the expression for the matrix ΔW the elements $x_\ell, \ell = 1, 2, \ldots, \nu$, are replaced by their linear approximations (3.33). Representing the matrix M^{-1} as

$$M^{-1} = \begin{bmatrix} \boxed{M^{-1}_{1,1:\nu}} \\ \boxed{M^{-1}_{2,1:\nu}} \\ \boxed{M^{-1}_{3,1:\nu}} \\ \vdots \\ \boxed{M^{-1}_{\nu,1:\nu}} \end{bmatrix},$$

where the rows are highlighted in boxes, the matrix ΔW can be written as

$$\delta W = \begin{bmatrix} \times & \times & \times & \cdots & \times \\ \boxed{M^{-1}_{1,1:\nu}}\,f & \times & \times & \cdots & \times \\ \boxed{M^{-1}_{2,1:\nu}}\,f & \boxed{M^{-1}_{n,1:\nu}}\,f & \times & \cdots & \times \\ \vdots & \vdots & \vdots & \ddots & \vdots \\ \boxed{M^{-1}_{n-1,1:\nu}}\,f & \boxed{M^{-1}_{2n-3,1:\nu}}\,f & \boxed{M^{-1}_{3n-6,1:\nu}}\,f & \cdots & \times \end{bmatrix}$$

$$= L(I_n \otimes f),$$

where

$$L = \begin{bmatrix} \times & \times & \times & \cdots & \times \\ \boxed{M^{-1}_{1,1:\nu}} & \times & \times & \cdots & \times \\ \boxed{M^{-1}_{2,1:\nu}} & \boxed{M^{-1}_{n,1:\nu}} & \times & \cdots & \times \\ \vdots & \vdots & \vdots & \ddots & \vdots \\ \boxed{M^{-1}_{n-1,1:\nu}} & \boxed{M^{-1}_{2n-3,1:\nu}} & \boxed{M^{-1}_{3n-6,1:\nu}} & \cdots & \times \end{bmatrix} \in \mathbb{C}^{n\times n\nu}$$

and

$$I_n \otimes f = \begin{bmatrix} f & & & \\ & f & & \\ & & \ddots & \\ & & & f \end{bmatrix} \in \mathbb{C}^{n\nu\times n}.$$

Using the fact that

$$\|I_n \otimes f\|_2 = \|f\|_2,$$

we obtain the following asymptotic estimate:

$$\begin{aligned} \Theta\max(\widetilde{\mathcal{X}}, \mathcal{X}) &\le \arcsin(\|L_{r+1:n,1:r\nu}\|_2 \|f\|_2) \\ &\le \arcsin(\|L_{r+1:n,1:r\nu}\|_2 \|\Delta A\|_F). \end{aligned} \tag{3.38}$$

Thus, an asymptotic bound of $\Theta\max(\widetilde{\mathcal{X}}, \mathcal{X})$ is determined as

$$\Theta\max(\widetilde{\mathcal{X}}, \mathcal{X}) \leq \Theta\max^{\text{lin}}(\widetilde{\mathcal{X}}, \mathcal{X}) = \text{cond}(\Theta\max)\|\Delta A\|_F, \tag{3.39}$$

where the number

$$\text{cond}(\Theta\max) = \|L_{r+1:n,1:r\nu}\|_2$$

can be considered as a condition number of the invariant subspace $\mathcal{X}$. The derivation of cond(Θmax) is done to find its possible minimum value.

Consider now how to find componentwise estimates of the perturbations of the diagonal elements (the eigenvalues) of T.

The perturbations of the elements of the Schur form T satisfy

$$\Delta t_{ij} = u_i^H A\Delta u_j + \Delta u_i^H A u_j + \Delta u_i^H A\Delta u_j + \widetilde{u}_i^H \Delta A\widetilde{u}_j. \tag{3.40}$$

Using equations (3.28) and (3.29), one obtains for the perturbations of the diagonal ($i = j$) elements of T the expressions

$$\begin{aligned}\Delta t_{ii} = &\sum_{k=i}^{n} t_{ik} u_k^H \Delta u_i - \sum_{k=1}^{i} t_{ki} u_i^H \Delta u_k - \sum_{k=1}^{i} t_{ki} \Delta u_i^H \Delta u_k \\ &+ \Delta u_i^H A\Delta u_i + \widetilde{u}_i^H \Delta A\widetilde{u}_i, \quad i = 1, 2, \ldots, n.\end{aligned} \tag{3.41}$$

The system of equation (3.41) can be represented as

$$\Delta\lambda = Nx + g + \Delta^d, \tag{3.42}$$

where

$$N = \Pi(I_n \otimes T - T^T \otimes I_n)\Omega^T \in \mathbb{C}^{n\times\nu},$$

$$\Pi = \text{diag}\left(e_1^T, e_2^T, \ldots, e_n^T\right) = \left[e_1e_1^T, e_2e_2^T, \ldots, e_ne_n^T\right] \in \mathbb{R}^{n\times n^2},$$

$$g = \begin{bmatrix} \widetilde{u}_1^H \Delta A\widetilde{u}_1 \\ \widetilde{u}_2^H \Delta A\widetilde{u}_2 \\ \vdots \\ \widetilde{u}_n^H \Delta A\widetilde{u}_n \end{bmatrix} \in \mathbb{C}^n$$

and Δ^d contains second-order terms in the elements of x.

Equation (3.42) can be represented in the compact form

$$\Delta\lambda = [NM^{-1}, I_n]\begin{bmatrix} f \\ g \end{bmatrix}. \tag{3.43}$$

Using (3.43), one can derive condition numbers of the eigenvalues. Let

$$Q = [NM^{-1}, I_n] \in \mathbb{R}^{n\times(\nu+n)}.$$

Since

$$\left\| \begin{bmatrix} f \\ g \end{bmatrix} \right\|_2 \leq \|\Delta A\|_F,$$

it follows from (3.43) that the asymptotic eigenvalue perturbation $\Delta\lambda_i = \Delta t_{ii}$ satisfies

$$|\Delta\lambda_i| \leq \Delta\lambda_i^{\text{lin}} = \text{cond}(\lambda_i)\|\Delta A\|_F, \quad i = 1, 2, \ldots, n, \tag{3.44}$$

where

$$\text{cond}(\lambda_i) = \|Q_{i,1:\nu+n}\|_2 \tag{3.45}$$

is the condition number of λ_i.

Note that the eigenvalue condition numbers (3.45) are exactly equal to the corresponding condition numbers defined by using the eigenvectors in Section 3.3.

The componentwise perturbation bounds of the n eigenvalues and the $n-1$ nested invariant subspaces are determined by Algorithm 3.4.

3.7.3 *Sensitivity of the Superdiagonal Elements*

Based on equation (3.40), the superdiagonal elements of the perturbed and unperturbed Schur form can be determined as (note that the elements are taken row-wise)

$$\widetilde{t}_{ij} = \widetilde{u}_i^H(A + \Delta A)\widetilde{u}_j,$$

$$t_{ij} = u_i^H A u_j, \quad 1 \leq i < j \leq n.$$

Algorithm 3.4. Perturbation bounds of eigenvalues and invariant subspaces

Given the Schur form $T = U^H AU$ of an $n \times n$ complex matrix A with simple eigenvalues and the Frobenius norm of the perturbation ΔA, this algorithm computes asymptotic componentwise perturbation bounds of the n eigenvalues and $n-1$ invariant subspaces.

function $[\mathrm{cond}(\lambda), \Delta\lambda^{\mathrm{lin}}, \mathrm{cond}(\Theta\mathrm{max}), \Delta\Theta\mathrm{max}^{\mathrm{lin}}]$
$= \mathit{pert\_schur\_eig}(T, M, \|\Delta A\|_F)$

Input: The Schur form T of the matrix A; the matrix $M \in \mathbb{C}^{\nu\times\nu}$, $\nu = n(n-1)/2$, obtained by Algorithm 3.3; the norm $\|\Delta A\|_F$ of the perturbation ΔA

Output: The n-vectors $\mathrm{cond}(\lambda)$, $\Delta\lambda^{\mathrm{lin}}$ containing the condition numbers and the asymptotic perturbation bounds of the eigenvalues; the $(n-1)$-vectors $\mathrm{cond}(\Theta\mathrm{max})$, $\Delta\Theta\mathrm{max}^{\mathrm{lin}}$ containing the condition numbers and the asymptotic perturbation bounds of the nested invariant subspaces.

Set $\ell = 0$, $L = 0_{n\times n\cdot\nu}$
for $j = 1, 2, \ldots, n-1$
  for $i = j+1, j+2, \ldots, n$
    $\ell = \ell + 1$
    $L_{i,1+(j-1)\cdot\nu:j\cdot\nu} = (M^{-1})_{\ell,1:\nu}$
  end
end
for $k = 1, 2, \ldots, n-1$
  $\mathrm{cond}(\Theta\mathrm{max}_k) = \|L_{k+1:n,1:k\cdot\nu}\|_2$
  $\Delta\Theta\mathrm{max}_k^{\mathrm{lin}} = \mathrm{cond}(\Theta\mathrm{max}_k)\|\Delta A\|_F$
end
$\Pi = [\ ]$, **for** $i = 1:n$, $\Pi = [\Pi,\ e_i e_i^T]$, **end**
$\Omega = 0_{\nu\times 2\nu}$, $ii = 0$, $jj = 0$
for $i = 1, 2, \ldots, n-1$
  $\Omega_i = [0_{(n-i)\times i},\ I_{n-i}]$, $\Omega_{ii+1:ii+n-i,jj+1:jj+n} = \Omega_i$
  $ii = ii + n - i$, $jj = jj + n$
end
$\Omega = [\Omega,\ 0_{\nu\times n}]$
$N = \Pi(I_n \otimes T - T^T \otimes I_n)\Omega^T$, $Q = [NM^{-1},\ I_n]$
Compute eigenvalue sensitivity estimates
for $k = 1, 2, \ldots, n$
  $\mathrm{cond}(\lambda_k) = \|Q_{k,1:\nu+n}\|_2$
  $\Delta\lambda_k^{\mathrm{lin}} = \mathrm{cond}(\lambda_k)\|\Delta A\|_F$
end

From these equations, the perturbations of the superdiagonal elements of the Schur form can be determined as

$$\Delta t_{ij} = \widetilde{t}_{ij} - t_{ij} = \sum_{k=i}^{n} t_{ik} u_k^H \Delta u_j - \sum_{k=1}^{j} t_{kj} u_i^H \Delta u_k - \sum_{k=1}^{j} t_{kj} \Delta u_i^H \Delta u_k$$
$$+ \Delta u_i^H A \Delta u_j + \widetilde{u}_i^H \Delta A \widetilde{u}_j, \quad 1 \le i < j \le n. \quad (3.46)$$

Denoting by $\mathrm{Up}(Z)$, the strictly upper triangular part of $Z \in \mathbb{C}^{n\times n}$, equations (3.46) may be represented as the system of ν nonlinear algebraic equations

$$\Delta t = M_1 y + M_2 x + M_3 d + h + \Delta^t, \quad 1 \le i < j \le n, \qquad (3.47)$$

where

$$\Delta t = \mathrm{vec}((\mathrm{Up}(\Delta T))^T) \in \mathbb{C}^\nu,$$
$$y = \mathrm{vec}((\mathrm{Up}(U^H \Delta U))^T) \in \mathbb{C}^\nu,$$
$$d = \begin{bmatrix} u_1^H \Delta u_1 \\ u_2^H \Delta u_2 \\ \vdots \\ u_n^H \Delta u_n \end{bmatrix} \in \mathbb{C}^n,$$
$$h = \mathrm{vec}((\mathrm{Up}(\widetilde{U}^H \Delta A \widetilde{U}))^T) \in \mathbb{C}^\nu,$$
$$M_1 = \Omega P_n (I_n \otimes T - T^T \otimes I_n) P_n \Omega^T \in \mathbb{C}^{\nu\times\nu}, \quad M_1 = -M^T,$$
$$M_2 = \Omega P_n (I_n \otimes T - T^T \otimes I_n) \Omega^T, \quad M_2 = -M_2^T,$$
$$M_3 = -N^T = -\Omega(I_n \otimes T^T - T \otimes I_n)\Pi^T \in \mathbb{C}^{\nu\times n}$$

and Δ^t contains second-order terms in x. The matrix P_n in the expressions for M_1 and M_2 is the vec-permutation matrix, determined from (see Section A.1 of the Appendix)

$$\mathrm{vec}(A^T) = P_n \mathrm{vec}(A).$$

In a linear approximation, it is fulfilled that

$$y = -\bar{x} = -\overline{M}^{-1} \bar{f}.$$

Neglecting the second-order terms in equation (3.47), we obtain

$$\Delta t = -M_1 \bar{x} + M_2 x + h = -M_1 \overline{M}^{-1} \bar{f} + M_2 M^{-1} f + h.$$

Thus, we determine the asymptotic estimate

$$|\Delta t| \preceq (|M_1 \overline{M}^{-1}| + |M_2 M^{-1}|)|f| + |h|, \quad 1 \le i < j \le n.$$

Let us denote

$$Q = \left[|M_1 \overline{M}^{-1}| + |M_2 M^{-1}|, I_\nu \right] \in \mathbb{R}^{\nu \times 2\nu}.$$

Since

$$\left\| \begin{bmatrix} |f| \\ |h| \end{bmatrix} \right\|_2 \le \|\Delta A\|_F,$$

one concludes that in a first-order approximation the superdiagonal elements of $|\Delta T|$ fulfill

$$|\Delta t_{ij}| \le \Delta t_{ij}^{\text{lin}} = \text{cond}(t_{ij}) \|\Delta A\|_F, \quad 1 \le i < j \le n, \tag{3.48}$$

where the quantities

$$\text{cond}(t_{ij}) = \|Q_{\ell,1:2\nu}\|_2, \quad \ell = j + (i-1)n - \frac{i(i+1)}{2}, \quad 1 \le i < j \le n, \tag{3.49}$$

can be considered as componentwise condition numbers of the superdiagonal elements.

The determining of the condition numbers and asymptotic perturbation bounds of the superdiagonal elements of the Schur form is done by Algorithm 3.5.

Example 3.18. In Figure 3.5, we give the actual changes of the entries of the unitary transformation matrix U of the Schur decomposition of a 50×50 matrix and their asymptotic bounds evaluated by Algorithm 3.3 for a random perturbation $\|\Delta A\|$ with $\|\Delta A\|_F = 10^{-6}$. As is seen from the figure, the linear bound estimates correctly the actual perturbations of the individual elements which differ very much in magnitude. The changes of the entries of U vary between $1.5 \cdot 10^{-14}$ and $5 \cdot 10^{-8}$, while the componentwise bound varies between

Algorithm 3.5. Perturbation bounds of the superdiagonal elements of the Schur form of a matrix

Given the Schur form $T = U^H AU$ *of an* $n \times n$ *complex matrix* A *with simple eigenvalues and the Frobenius norm of the perturbation* ΔA*, this algorithm computes asymptotic perturbation bounds of the* $n(n-1)/2$ *super diagonal elements of the Schur form.*

function $[\text{cond}(t_{\text{supd}}), \Delta t_{\text{supd}}^{\text{lin}}] = \textit{pert\_schur\_supd}(T, M, \|\Delta A\|_F)$

Input: The Schur form T of the matrix A; the matrix $M \in \mathbb{C}^{\nu \times \nu}$, $\nu = n(n-1)/2$, obtained by Algorithm 3.3; the norm $\|\Delta A\|_F$ of the perturbation ΔA.

Output: The ν-vectors $\text{cond}(t_{\text{supd}})$, $\Delta t_{\text{supd}}^{\text{lin}}$, containing the condition numbers and the asymptotic perturbation bounds of the super diagonal elements, respectively.

Construct the permutation matrix Ω

$\Omega = 0_{\nu \times 2\nu}, \; ii = 0, \; jj = 0$

for $i = 1, 2, \ldots, n-1$

  $\Omega_i = [0_{(n-i)\times i}, \; I_{n-i}], \; \Omega_{ii+1:ii+n-i, jj+1:jj+n} = \Omega_i$

  $ii = ii + n - i, \; jj = jj + n$

end

$\Omega = [\Omega, \; 0_{\nu \times n}]$

Construct the vec-permutation matrix P^{vec}

$ij = 0, \; P^{\text{vec}} = 0_{n^2 \times n^2}$

for $i = 1, 2, \ldots, n$

  $k = i - n$

  for $j = 1, 2, \ldots, n$

    $ij = ij + 1, \; k = k + n$

    $P_{ij,k}^{\text{vec}} = 1$

  end

end

Construct the operator matrices M_1, M_2

$M_1 = \Omega P^{\text{vec}}(I_n \otimes T - T^T \otimes I_n)P^{\text{vec}}\Omega^T$

$M_2 = \Omega P^{\text{vec}}(I_n \otimes T - T^T \otimes I_n)\Omega^T$

Determine the condition numbers and asymptotic estimates of the super diagonal elements

$Q = [|M_1 \bar{M}^{-1}| + |M_2 M^{-1}|, I_\nu]$

for $\ell = 1, 2, \ldots, \nu$

  $\text{cond}(t_{\text{supd}})_\ell = \|Q_{\ell,1:2\nu}\|_2$

  $(\Delta t_{\text{supd}}^{\text{lin}})_\ell = \text{cond}(t_{\text{supd}})_\ell \|\Delta A\|_F$

end

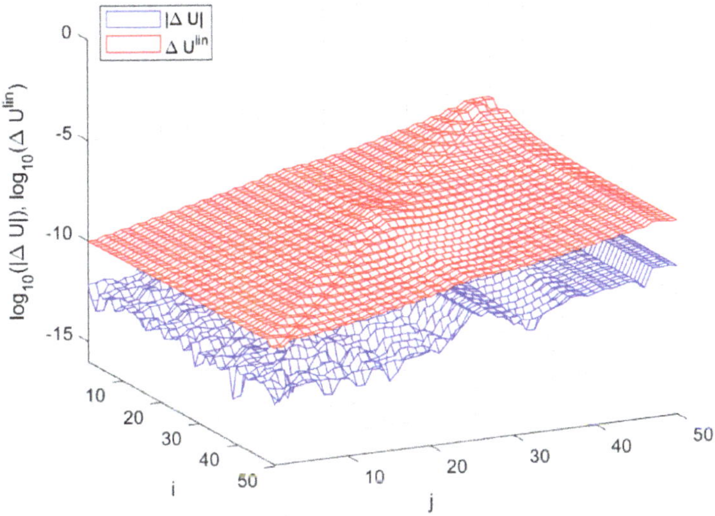

Fig. 3.5. Componentwise perturbation bound of the unitary transformation matrix U (the surface above) and the actual changes in the elements of $|U|$ (the surface beneath).

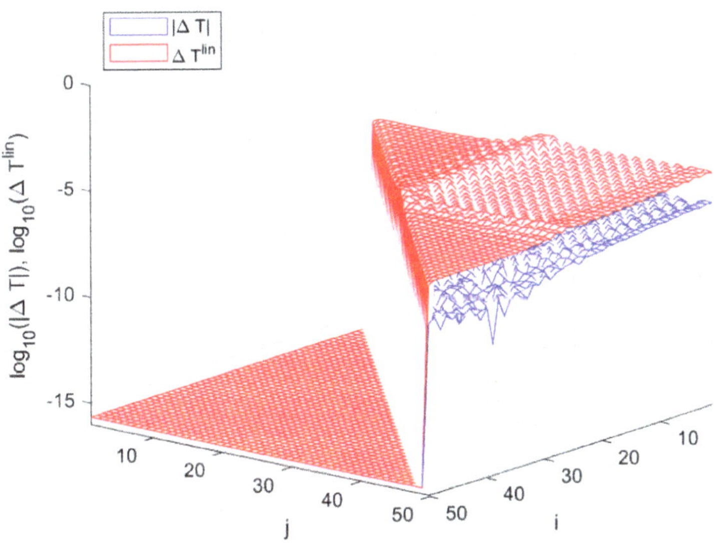

Fig. 3.6. Componentwise perturbation bound of the Schur form T (the surface above) and the actual changes in the elements of $|T|$ (the surface beneath).

$7 \cdot 10^{-11}$ and $4 \cdot 10^{-6}$. At the same time, the normwise bound (3.23) is $6 \cdot 10^{-6}$, which is a very pessimistic estimate of the smallest elements of ΔU.

In Figure 3.6, we show the actual changes of the entries of the Schur form T and their linear bounds for the same perturbation. The maximum actual change of the elements of T is approximately $7 \cdot 10^{-7}$ and the maximum componentwise bound $\Delta t_{ij}^{\text{lin}}$ is $1.5 \cdot 10^{-5}$. If we compute the normwise bound of ΔT for the same problem, we obtain that $c_T = 8 \cdot 10^5$ and $c_T \|\Delta A\|_F = 8 \cdot 10^{-1}$ which is much larger than the actual change $\|\Delta T\|_F = 6 \cdot 10^{-6}$. This large overestimate of the change in T is a result of the large norm of the matrix A, $\|A\|_F \approx 6.5 \cdot 10^4$, whose value participates in the normwise condition number (3.26).

3.8 Sensitivity of the Generalized Eigenvalue Problem

In this section, we consider the sensitivity of the regular generalized eigenvalue problem $Ax = \lambda Bx$.

Let

$$\alpha = y^H Ax, \quad \beta = y^H Bx,$$

where x and y are right and left eigenvectors of the regular pencil $A - \lambda B$. Then it is possible to show that

$$\lambda = \frac{\alpha}{\beta} = \frac{y^H Ax}{y^H Bx}$$

is a simple eigenvalue of $A - \lambda B$. We are interested in the sensitivity of λ under perturbations of the matrices A and B. If the perturbed pencil is denoted by

$$\widetilde{A} + \lambda \widetilde{B} = A + \Delta A + \lambda (B + \Delta B),$$

we wish to know how the difference between $\widetilde{\lambda}$ and λ depends on the size of the perturbation measured by the quantity

$$\varepsilon = \sqrt{\|\Delta A\|_F^2 + \|\Delta B\|_F^2}.$$

The distance between two generalized eigenvalues $\lambda = \alpha/\beta$ and $\mu = \gamma/\delta$ is measured by the so-called *chordal distance* defined as

$$\chi(\lambda, \mu) = \frac{|\alpha\delta - \beta\gamma|}{\sqrt{|\alpha|^2 + |\beta|^2}\sqrt{|\gamma|^2 + |\delta|^2}}.$$

We have that if $\gamma = \alpha$ and $\delta = \beta$, then $\chi = 0$. Also, it is possible to show that

$$\chi(\lambda, \mu) = \frac{|\lambda - \mu|}{\sqrt{1 + |\lambda|^2}\sqrt{1 + |\mu|^2}}.$$

Let $\lambda = \alpha/\beta$ be a simple eigenvalue (possibly infinite) of the regular pencil $A - \lambda B$ with right and left eigenvectors x and y. Then for ε sufficiently small, there is an eigenvalue $\widetilde{\lambda} = \widetilde{\alpha}/\widetilde{\beta}$ of the pencil $\widetilde{A} - \widetilde{\lambda}\widetilde{B}$ satisfying

$$\chi(\lambda, \widetilde{\lambda}) \leq \nu\varepsilon + O(\varepsilon^2),$$

where

$$\nu = \frac{\|x\|_2 \|y\|_2}{\sqrt{|\alpha|^2 + |\beta|^2}}.$$

The number ν reveals the effect of the perturbation ε on the change of the eigenvalue and can be considered as a condition number of the simple generalized eigenvalue. Note that this number is independent of the scaling of x and y. If both α and β are small, they are sensitive to the perturbation ΔA and ΔA and the condition number ν is large. In such case, λ is ill conditioned.

Example 3.19. Consider the regular matrix pencil $A - \lambda B$ with matrices

$$A = \begin{bmatrix} 3 & 2 & 2 & 2 & -3 & 5 \\ -3 & -2 & 1 & 0 & 4 & -3 \\ 4 & 2 & -1 & 2 & -4 & 6 \\ 0 & 0 & -3 & -2 & -2 & -1 \\ 0 & 2 & 6 & 1 & -1 & 1 \\ 5 & 2 & -4 & 0 & -4 & 4 \end{bmatrix},$$

Table 3.2. Condition numbers of the generalized eigenvalues.

k	α	β	λ_k	ν_k
1	2.0	1.0	2.0	2.280350850198276
2	2.0	1.0	2.0	5.585696017507575
3	2.0	1.0	2.0	3.464101615137754
4	1.0	0.0	∞	8.246211251235321
5	1.0	0.0	∞	8.124038404635959
6	1.0	0.0	∞	4.472135954999580

$$B = \begin{bmatrix} -1 & 1 & 1 & 0 & 0 & -2 \\ 0 & -1 & 0 & 1 & 0 & 2 \\ 1 & 1 & 0 & 0 & 0 & 0 \\ 0 & 0 & -1 & -1 & 0 & -1 \\ -4 & 1 & 2 & 0 & 0 & -5 \\ 3 & 1 & -1 & -1 & 0 & 1 \end{bmatrix}$$

given in Example 2.20.

The exact values of α, β, the generalized eigenvalues and the corresponding condition numbers are shown in Table 3.2. Clearly, the finite eigenvalues at 2 are well conditioned.

In the applications of the generalized Schur form, it is often important to know the sensitivity of specific deflating subspaces to perturbations in A and B. Further on, we present a perturbation analysis of the deflating subspaces corresponding to specified generalized eigenvalues.

Suppose that the generalized Schur form of the pencil $A - \lambda B$ is reordered as

$$U^H AV = \begin{bmatrix} A_{11} & A_{12} \\ 0 & A_{22} \end{bmatrix}, \quad U^H BV = \begin{bmatrix} B_{11} & B_{12} \\ 0 & B_{22} \end{bmatrix}, \tag{3.50}$$

where $A_{11} - \lambda B_{11}$ contains r specified eigenvalues. Partitioning the unitary matrices of the equivalence transformation as $U = [U_1, U_2], V = [V_1, V_2]$, we obtain orthonormal bases $U_1 \in \mathbb{C}^{n\times r}$ and $V_1 \in \mathbb{C}^{n\times r}$ of the left $\mathcal{Y}$ and right $\mathcal{X}$ deflating subspaces, respectively. The subspaces $\mathcal{R}(U_2)$ and $\mathcal{R}(V_2)$ are complementary subspaces to

$\mathcal{R}(U_1)$ and $\mathcal{R}(V_1)$, respectively. Note that

$$U_i A V_j = A_{ij}, \quad U_i B V_j = B_{ij},$$
$$U_2 A V_1 = 0, \quad U_2 B V_1 = 0.$$

Suppose that the matrices of the pencil $A - \lambda B$ are perturbed as $\widetilde{A} = A + \Delta A$ and $\widetilde{B} = B + \Delta B$. The upper triangular matrices in the generalized Schur decomposition are represented as

$$\widetilde{U}^H \widetilde{A} \widetilde{V} = \begin{bmatrix} \widetilde{A}_{11} & \widetilde{A}_{12} \\ 0 & \widetilde{A}_{22} \end{bmatrix}, \quad \widetilde{U}^H \widetilde{B} \widetilde{V} = \begin{bmatrix} \widetilde{B}_{11} & \widetilde{B}_{12} \\ 0 & \widetilde{B}_{22} \end{bmatrix},$$

where

$$\widetilde{U} = \left[\widetilde{U}_1, \widetilde{U}_2\right], \quad \widetilde{V} = \left[\widetilde{V}_1, \widetilde{V}_2\right]$$

are the modified equivalence transformation matrices.

If $\|\Delta A\|$ and $\|\Delta B\|$ are small, we may expect that the spaces $\mathcal{R}(\widetilde{U}_2)$ and $\mathcal{R}(\widetilde{V}_1)$ will be close to the spaces $\mathcal{R}(U_2)$ and $\mathcal{R}(V_1)$, respectively. That is why we look for matrices R and W so that

$$\widetilde{V}_1 = V_1 + V_2 R,$$
$$\widetilde{U}_2 = U_2 - U_1 W^H,$$

and

$$\widetilde{U}_2^H \widetilde{A} \widetilde{V}_1 = 0, \qquad (3.51)$$

$$\widetilde{U}_2^H \widetilde{B} \widetilde{V}_1 = 0. \qquad (3.52)$$

The earlier equations show that the matrices $\widetilde{V}_1$ and $\widetilde{U}_2$ provide bases for the right and left complementary deflating subspaces, respectively, of $\widetilde{A} - \lambda \widetilde{B}$.

Define

$$E = U^H \Delta A V = \begin{bmatrix} E_{11} & E_{12} \\ E_{21} & E_{22} \end{bmatrix}, \quad F = U^H \Delta B V = \begin{bmatrix} F_{11} & F_{12} \\ F_{21} & F_{22} \end{bmatrix}.$$

Equations (3.51) and (3.52) can be represented as

$$A_{22} R - W A_{11} = -E_{21} - E_{22} R + W E_{11} + W(A_{12} + E_{12}) R, \quad (3.53)$$

$$B_{22} R - W B_{11} = -F_{21} - F_{22} R + W F_{11} + W(B_{12} + F_{12}) R. \quad (3.54)$$

Equations (3.53) and (3.54) may be considered as a generalized Sylvester equation (see Section 2.3.3) in respect to the matrices

(R, W). Like the ordinary Sylvester equation (Section 2.1.6), by using the Kronecker product, the generalized Sylvester equation may be written as

$$\begin{bmatrix} I_r \otimes A_{22} - A_{11}^T \otimes I_{n-r} \\ I_r \otimes B_{22} - B_{11}^T \otimes I_{n-r} \end{bmatrix} \begin{bmatrix} \text{vec}(R) \\ \text{vec}(W) \end{bmatrix}$$
$$= - \begin{bmatrix} \text{vec}(E_{21}) \\ \text{vec}(F_{21}) \end{bmatrix}$$
$$- \begin{bmatrix} I_r \otimes E_{22} - E_{11}^T \otimes I_{n-r} \\ I_r \otimes F_{22} - F_{11}^T \otimes I_{n-r} \end{bmatrix} \begin{bmatrix} \text{vec}(R) \\ \text{vec}(W) \end{bmatrix}$$
$$+ \begin{bmatrix} \text{vec}(W(A_{12} + E_{12})R) \\ \text{vec}(W(B_{12} + F_{12})R) \end{bmatrix}.$$

The solution of this equation exists if the matrix

$$\Omega = \begin{bmatrix} I_r \otimes A_{22} - A_{11}^T \otimes I_{n-r} \\ I_r \otimes B_{22} - B_{11}^T \otimes I_{n-r} \end{bmatrix}$$

is invertible and this is fulfilled if and only if the pencils $A_{11} - \lambda B_{11}$ and $A_{22} - \lambda B_{22}$ have no eigenvalues in common.

Similar to the perturbation analysis of an invariant subspace (Section 3.6), the size of the matrices R and W which are solutions of (3.53) and (3.54) can be estimated by using the contraction mapping theorem (Stewart, 1972).

Define the *difference between the spectra of* $A_{11} - \lambda B_{11}$ *and* $A_{22} - \lambda B_{22}$ as

$$\text{dif}(A_{11}, B_{11}; A_{22}, B_{22}) = \min_{[R,W]\neq 0} \frac{\left\| \begin{bmatrix} A_{22}R - WA_{11} \\ B_{22}R - WB_{11} \end{bmatrix} \right\|}{\|R, W\|_F}.$$

Note that the function dif satisfies

$$\begin{aligned} &\text{dif}(A_{11} + E_1, B_{11} + F_1; A_{22} + E_2, B_{22} + F_2) \\ &\geq \text{dif}(A_{11}, B_{11}; A_{22}, B_{22}) \\ &\quad - \max\{\|E_1\|_2 + \|E_2\|_2, \|F_1\|_2 + \|F_2\|_2\}. \end{aligned}$$

It is possible to prove the relationship

$$\mathrm{dif}(A_{11}, B_{11}; A_{22}, B_{22}) = \sigma_{\min}(\Omega)$$

so that

$$\|\Omega^{-1}\|_2 = 1/\mathrm{dif}(A_{11}, B_{11}; A_{22}, B_{22}).$$

Assume that the spectra of $A_{11} - \lambda B_{11}$ and $A_{22} - \lambda B_{22}$ are disjoint so that

$$\mathrm{dif}(A_{11}, B_{11}; A_{22}, B_{22}) > 0.$$

The following theorem gives an estimate of the size of matrices (R, W) that are solution of equations (3.53) and (3.54), thus providing estimate of the sensitivity of a deflating subspace of a regular pencil.

Theorem 3.20 (Stewart (1972)). *Let the regular pencil $A - \lambda B$ be represented as in* (3.50). *Given the perturbation* (E, F), *set*

$$\begin{aligned}
\gamma &= \|(E_{21}, F_{21})\|_F, \\
\eta &= \|(A_{12} + E_{12}, B_{12} + F_{12})\|_F, \\
\delta &= \mathrm{dif}(A_{11}, B_{11}; A_{22}, B_{22}) \\
&\quad - \max\{\|E_{11}\|_F + \|E_{22}\|_F, \|F_{11}\|_F + \|F_{22}\|_F\}.
\end{aligned}$$

If $\delta > 0$ and

$$\frac{\eta\gamma}{\delta^2} < \frac{1}{4},$$

there are matrices R and W satisfying

$$\|(R, W)\|_F \le \frac{2\gamma}{\delta + \sqrt{\delta^2 - 4\gamma\eta}} < 2\frac{\gamma}{\delta}, \tag{3.55}$$

such that the columns of $\widetilde{V}_1 = V_1 + V_2 R$ and $\widetilde{U}_2 = U_2 - U_1 W^H$ span left and right complementary deflating subspaces of $A + \Delta A - \lambda(B + \Delta B)$.

It is possible to show that equation (3.55) bounds the tangents of the canonical angles between $\mathcal{R}(V_1)$ and $\mathcal{R}(\widetilde{V}_1)$ or $\mathcal{R}(U_2)$ and $\mathcal{R}(\widetilde{U}_2)$, similar to the ordinary eigenvalue problem (see the discussion after Theorem 3.16).

Theorem 3.20 can be considered as a generalization of Theorem 3.16 for the ordinary eigenvalue problem.

Since it is fulfilled that (Stewart, 1972)

$$\mathrm{dif}(A_{11}, I; A_{22}, I) \leq \mathrm{sep}(A_{11}, A_{22}),$$

the sensitivity of the deflation subspaces for the generalized eigenvalue problem $Ax = \lambda Ix$ is higher or equal to the sensitivity of the corresponding invariant subspaces for the ordinary eigenvalue problem $Ax = \lambda x$.

The solution of the generalized Sylvester equation and the computation of an estimate of the function dif is done in LAPACK by the routine `xTGSYL`.

3.9 Notes and References

Different measures of closeness between the eigenvalues of two matrices are considered in Stewart and Sun (1990), Chapter 4.

3.9.1 *Perturbation Theory for Eigenvalues*

The basics of the perturbation theory for eigenvalues and eigenvectors are set by the physicists J.W. Strutt, Baron Rayleigh and E. Schrödinger. Rayleigh derived a formula for the natural frequencies and modes of an oscillating system which deviates slightly from the simpler system allowing complete determination of the frequencies and modes (Strutt, 1945, Sections 90 and 91). From mathematical point of view, this method is equivalent to the approximate solution of the eigenvalue and eigenvector problems for a matrix which is slightly different from a matrix for which this problem is completely solved. Schrödinger developed a similar method with larger generality and systematism for problems which appear in the quantum mechanics (Schrödinger, 1928, Paper 1). More on the Rayleigh–Schrödinger perturbation theory can be found in McCartin (2009).

The first paper in which perturbation bounds on the eigenvalues of an arbitrary matrix are derived is published by Ostrowski in (1957). Several important results in this area are then published in Henrici (1962), Taussky (1964), Lidskii (1966), Wilkinson (1984),

Wilkinson (1986), Elsner (1982; 1985; 1986), Moro *et al.* (1997), Sun (1998) and Nakatsukasa (2011), to name a few. Theoretical and practical aspects of the eigenvalue conditioning analysis are considered in Smith (1967), Chen *et al.* (1977), Sun (1992), Van Loan (1987), Bai *et al.* (1993), Lin *et al.* (1996) and Karow *et al.* (2006).

There exist several books which are entirely devoted to the eigenvalue perturbation theory in more general setting, among them are the classic monograph of Kato (1995) and the books Baumgärtel (1985), Collatz (1949) and Rellich (1969). A large number of results on the matrix eigenvalues sensitivity can be found in the books Wilkinson (1965), Stewart (2001), Golub and Van Loan (2013) and Bhatia (2007). The book of Stewart and Sun (1990) presents in detail different methods for perturbation analysis of several eigenvalue and eigenvector problems. A comprehensive description of such methods is contained also in the survey (Li, 2014).

3.9.2 *Gershgorin Theorem*

The history and various applications of the Gershgorin theorem are presented in the book of Varga (2004). This theorem is formulated in Geršhgorin (1931) and different aspects of its application to the analysis of eigenvalue sensitivity are discussed in Porsching (1966), Householder *et al.* (1970), Johnston and Smith (1970), Solov'ev (1984) and Parlett (2009). An especially interesting problem is to find the smallest isolated Gershgorin disks containing the eigenvalues (Levinger and Varga, 1966; Medley and Varga, 1968; Varga, 1965), see also Todd (1965). A Gershgorin theory for the generalized eigenvalue problem $Ax = \lambda Bx$ is developed in Stewart (1975).

The Gershgorin theorem is a particular approach to the larger problem for determining the areas in which the matrix eigenvalues are included without computing them (*the eigenvalue localization problem*), see Beattie and Ipsen (2003), Brualdi and Mellendorf (1994), Marcus and Minc (1964), Chapter III, Parodi (1959) and Obrechkoff (2003).

3.9.3 *First-Order Perturbation Analysis*

The perturbation analysis through first-order term expansion (the asymptotic analysis) is used traditionally in the linearization of nonlinear differential equations. This analysis may produce sufficiently accurate linear (they are also called *local* or *asymptotic*) bounds and is widely used in the practice of matrix computations. The nonlocal (*global*) perturbation analysis produces guaranteed sensitivity estimates of the eigenvalues and eigenvectors but is usually valid in narrow bounds on the perturbations. The recent results of the first-order perturbation theory for eigenvalues and eigenvectors can be found in the rich survey Greenbaum *et al.* (2020) and in the paper Moro and Dopico (1990). A sharper version of the Bauer–Fike theorem is presented in Shi and Wei (2012).

In case of random perturbations, more realistic sensitivity estimates can be obtained using the stochastic perturbation theory (Stewart, 1990).

3.9.4 *Sensitivity of Multiple Eigenvalues*

The link between the eigenvalue multiplicity and its ill conditioning is revealed long ago in the literature, see, for instance, Wilkinson (1965) and Ruhe (1970c). The sensitivity of Jordan blocks is studied in Erxiong (1994), Ma and Edelman (1998), Markus and Parilis (1983) and Moro *et al.* (1997). The modified Henrichi theorem (Theorem 3.13) demonstrates that the sensitivity analysis of multiple eigenvalues may produce reliable results only if one knows the Jordan structure of the matrix.

3.9.5 *Sensitivity of Invariant Subspaces*

Theorem 3.16 is derived by Stewart (1973a). In the same paper, he introduced the notion of separation between two matrices and the properties of this quantity are discussed in Varah (1979). Improved versions of Theorem 3.16 are presented in Stewart (2001), Stewart and Sun (1990) and Demmel (1986). Methods for refining the estimates of invariant subspaces are discussed in Chatelin (1984), Dongarra *et al.* (1992) and Demmel (1987b). The practical implementation of Theorem 3.16 is considered in Bai *et al.* (1993). Methods for estimating $\mathrm{sep}_F(A, B)$ and the sensitivity of invariant subspaces for high-order problems are presented in Bai *et al.* (1993) and implemented in the LAPACK package (Anderson *et al.*, 1999; Bai *et al.*, 2000).

3.9.6 *Sensitivity of the Schur Decomposition*

The sensitivity of the Schur decomposition is analyzed in Konstantinov *et al.* (1994), Konstantinov and Petkov (2020), Chapter 5, and Petkov (2021). More on the componentwise perturbation analysis in matrix computations can be found in Higham (1994).

3.9.7 *Sensitivity of the Generalized Eigenvalue Problem*

The sensitivity of the generalized eigenvalue problem is considered for the first time by Stewart, who introduced the chordal metric and the quantity dif characterizing the distance between the spectra of two pencils (Stewart, 1972). Improved versions of Theorem 3.20 are presented in Stewart and Sun (1990), Chapter VI, Demmel and Kågström (1987) and Stewart (2001), Chapter 2, Section 4. The generalized eigenvalue sensitivity is discussed in Sun (1995b), Demmel (1983a), Elsner and Sun (1982), Elsner and Lancaster (1985), Boley (1990) and Zhang *et al.* (2022).

The sensitivity of the generalized Sylvester equation is studied in Kågström (1994) and Lin and Wei (2007) and generalized Schur methods with condition estimators for the numerical solution of this equation are proposed in Kågström and Westin (1989) and Kågström and Poromaa (1996b). The sensitivity of rectangular pencils is considered in Stewart (1994) and componentwise perturbation analysis for the generalized Schur decomposition is presented in Zhang *et al.* (2022).

Chapter 4

Numerical Solution of the Eigenvalue Problem

The transformation into upper triangular (Schur) form and determination of the eigenvalues is the first step in the reduction of a square matrix into Jordan canonical form. In this chapter, we discuss in brief some basic properties of the QR algorithm used in the reduction into Schur form and computing approximations of the eigenvalues of a matrix. We present also the solution of the Sylvester matrix equation and its application to the computation of spectral projections and estimating the sensitivity of invariant subspaces.

4.1 Reduction into Schur Form

The numerical reduction into Schur form is a basic task which is involved in the solution of several important problems arising in matrix computations. Further on, we shall implement a method for reducing a complex matrix into triangular (Schur) form using unitary similarity transformations. In the basis of this method lies a remarkable iterative procedure called the *QR algorithm* which has obtained wide application for computing the eigenvalues and eigenvectors of dense matrices. As it is well known, for matrices of order five and higher, it is not possible to find the eigenvalues with a finite number of arithmetic operations, i.e., the reduction into Schur form cannot be done by using a *direct numerical method.* Instead of this, it is necessary to implement an *iterative method* in which the subdiagonal elements become smaller and smaller in module by a sequence of similarity transformations. After a certain number of such transformations, the subdiagonal elements become so small that they can be neglected and the matrix can be accepted as triangular.

Since the detailed description of the QR method is out of the scope of this book, we shall give only a brief description of this method and its properties, noting that a systematic presentation of the QR algorithm can be found in the references given at the end of this section.

4.1.1 *QR Algorithm with Explicit Shift*

Let $A_0 = A$ be a complex $n \times n$ matrix. The QR algorithm produces a sequence of similar matrices $A_0,\ A_1,\ A_2, \ldots$ in the following way. For a given matrix A_k, from its elements is determined a scalar $\varkappa_k$ called *origin shift* (with the convergence of the iterative process $\varkappa_k$ approximates an eigenvalue of A), after that the matrix $A_k - \varkappa_k I_n$ is decomposed into the form

$$A_k - \varkappa_k I_n = Q_k R_k, \tag{4.1}$$

where Q_k is unitary and R_k is upper triangular. Such decomposition is unique provided that $A_k - \varkappa_k I_n$ is nonsingular (i.e., $\varkappa_k$ is not equal to an eigenvalue of A_k). Next, a matrix A_{k+1} is computed that satisfies

$$A_{k+1} = R_k Q_k + \varkappa_k I_n. \tag{4.2}$$

Equations (4.1) and (4.2) can be written as the following algorithm:

> **for** $k = 0,\ 1,\ 2, \ldots$
>   Choose a shift $\varkappa_k$
>   Decompose $A_k - \varkappa_k I_n = Q_k R_k$, where
> Q is unitary and R is upper triangular
>   Compute $A_{k+1} = R_k Q_k + \varkappa_k I_n$
> **end**

It is seen that to pass from one iteration to another, i.e., to perform one *QR step*, it is necessary to subtract the shift $\varkappa_k$ from the matrix diagonal, to compute the QR decomposition of $A_k - \varkappa_k I_n$, to find the product $Q_k R_k$ of the factors in inverse order and to add back the diagonal shift. This method is called *QR algorithm with explicit shift* since the shift is subtracted explicitly from the diagonal of A and is added back at the end of the step.

From (4.1), we have that

$$R_k = Q_k^H (A_k - \varkappa_k I_n)$$

and consequently from (4.2), it follows that

$$\begin{aligned} A_{k+1} &= Q_k^H (A_k - \varkappa_k I_n) Q_k + \varkappa_k I_n \\ &= Q_k^H A_k Q_k, \end{aligned}$$

which shows that A_{k+1} is unitary similar to A_k.

On the basis of these considerations, it is possible to show that

$$A_{k+1} = (Q_1 Q_2 \cdots Q_k)^H A Q_1 Q_1 \cdots Q_k$$

and

$$(Q_1 Q_2 \cdots Q_k)(R_k \cdots R_2 R_1) = (A_1 - \varkappa_1 I_n)(A_1 - \varkappa_2 I_n) \cdots (A_1 - \varkappa_k I_n).$$

Under certain conditions, it is possible to achieve the off-diagonal elements in the last row of A_k to approach the zero very quickly. Let us partition the matrix $A_{k+1} = Q_k^H A Q$ in the form

$$A_{k+1} = \begin{bmatrix} B_k & h_k \\ g_k^H & \mu_k \end{bmatrix}, \tag{4.3}$$

where $B_k \in \mathbb{C}^{(n-1)\times(n-1)}$ and $g_k \in \mathbb{C}^{n-1}$. It can be proved that once $\|g_k\| < 1$, then by a suitable choice of the shift $\varkappa_k$, it is possible to achieve that the norm of g_{k+1} is bounded by a quantity which is proportional to the square of the norm of g_k. In such a case, it is said that $\|g_k\|_2$ *converges at least quadratically* to the zero. If appeared, the quadratic convergence is very fast. Assuming that $\|g_0\|_2 = 3 \cdot 10^{-2}$, we obtain that

$$\begin{aligned} \|g_1\|_2 &\leq 9.0 \cdot 10^{-4}, \\ \|g_2\|_2 &\lesssim 8.1 \cdot 10^{-7}, \end{aligned}$$

$$\|g_3\|_2 \leq 6.6 \cdot 10^{-13},$$
$$\|g_4\|_2 \leq 4.4 \cdot 10^{-25}.$$

In this way, four iterations are sufficient to reduce the elements in the last row of the transformed A to a size 10^8 times smaller than the unit roundoff of the double precision IEEE arithmetic.

Once g_k is accepted as zero, the approximation μ_k can be accepted as eigenvalue of A. In this way, we *"deflate"* the problem, i.e., separate the already known eigenvalue from the rest and reduce the dimension of the problem with 1.

4.1.2 *Reduction into Hessenberg Form*

When the QR algorithm is applied to a full matrix, it requires a large number of computational operations. If the QR decomposition of $A_k - \varkappa_k I_n$ is obtained by Algorithms 1.9 and 1.10 and the product $R_k Q_k$ is formed from the factorized form of Q_k, one step of the iteration will require a multiple of n^3 operations. In such case, the determination of all eigenvalues will require $O(n^4)$ operations which is undesirable.

This situation can be overcome if the matrix A is preliminary reduced to some form for which the QR algorithm requires a small volume of computations. This form should be close to the Schur form and it is desirable to be obtained by a direct algorithm. A form which satisfies these requirements is the *upper Hessenberg form*, shown in the following for $n = 6$:

$$H = \begin{bmatrix} \times & \times & \times & \times & \times & \times \\ \times & \times & \times & \times & \times & \times \\ & \times & \times & \times & \times & \times \\ & & \times & \times & \times & \times \\ & & & \times & \times & \times \\ & & & & \times & \times \end{bmatrix}.$$

The reduction into Hessenberg form $H = Q^H A Q$ can be done in $n - 2$ steps by using unitary similarity transformations with Householder reflections or plane rotations. The algorithm implementing

Householder transformations requires about $\frac{10}{3}n^3$ flops for the reduction of A and $\frac{4}{3}n^3$ flops for accumulation of Q. It is numerically stable from backward point of view. The computed Hessenberg matrix $\widehat{H}$ satisfies

$$\widehat{H} = \widehat{Q}^H(A+E)\widehat{Q},$$

where $\widehat{Q}$ is a matrix close to an unitary matrix and

$$\frac{\|E\|_2}{\|A\|_2} = \gamma_n \mathbf{u},$$

where γ_n is a slowly increasing function of n and $\mathbf{u}$ is the machine precision.

The reduction into upper Hessenberg form can be done by the Fortran routine `xGEHRD` of LAPACK or by the function `hess` of MATLAB®. In the following example, we use the M-function `hessreduce` implementing Algorithm 2.2 presented in Stewart (2001), p. 85.

Example 4.1. Consider a fourth-order matrix

$$A = \begin{bmatrix} 2-j_o4 & -1+j_o2 & -2+j_o3 & 5-j_o7 \\ -6+j_o8 & 0-j_o3 & 3-j_o4 & 2+j_o6 \\ 1-j_o7 & 4-j_o1 & -1+j_o5 & -6-j_o3 \\ -3+j_o5 & 6+j_o6 & -4+j_o8 & 1-j_o4 \end{bmatrix}.$$

This matrix is reduced into Hessenberg form (up to five decimal digits)

$$H = \left[\begin{matrix} 2.0000-j_o4.0000 & -1.2533+j_o3.6123 \\ 8.1388-j_o10.8517 & 1.7391+j_o3.5435 \\ & 1.9688-j_o5.6655 \\ & \end{matrix}\right. \quad \cdots \rightarrow$$

$$\leftarrow \cdots \quad \left.\begin{matrix} -5.4123+j_o1.6882 & 2.8192-j_o6.1065 \\ 1.8962+j_o4.7282 & -0.7297-j_o3.5707 \\ -2.2373-j_o0.4727 & 10.7246+j_o9.5888 \\ -1.2765-j_o0.6545 & 0.4982-j_o5.0708 \end{matrix}\right].$$

The reduction is performed using a transformation with the unitary matrix

$$Q = \begin{bmatrix} 1.0000 + \jmath_o 0.0000 & 0.0000 + \jmath_o 0.0000 \\ 0.0000 + \jmath_o 0.0000 & -0.7372 + \jmath_o 0.0000 \\ 0.0000 + \jmath_o 0.0000 & 0.4571 - \jmath_o 0.2507 \\ 0.0000 + \jmath_o 0.0000 & -0.4276 + \jmath_o 0.0442 \end{bmatrix} \cdots \rightarrow$$

$$\leftarrow \cdots \begin{matrix} 0.0000 + \jmath_o 0.0000 & 0.0000 + \jmath_o 0.0000 \\ 0.0658 + \jmath_o 0.0707 & -0.6680 - \jmath_o 0.0320 \\ -0.5516 - \jmath_o 0.0091 & -0.5744 + \jmath_o 0.3066 \\ -0.7368 - \jmath_o 0.3788 & 0.3588 + \jmath_o 0.0091 \end{matrix} \Bigg].$$

The backward error in the reduction of A into Hessenberg form is

$$\|QHQ^H - A\|_F = 3.92 \cdot 10^{-16}.$$

4.1.3 *QR Algorithm for Hessenberg Matrices*

Consider the application of the explicit QR algorithm on a Hessenberg matrix. If A_0 is an upper Hessenberg matrix, such are also the next matrices A_k, i.e., the QR algorithm preserves the Hessenberg form. This allows to organize the computations such that each Hessenberg QR step requires a multiple of n^2 operations and the reduction into Schur form is done by $O(n^3)$ operations. The development of a working algorithm requires attention to lot off details involving the detection of negligible subdiagonal elements, deflation of the eigenvalue problem to decrease the volume of computations and determining an appropriate shift to speed up the computations. As a result, the implementation of the explicitly shifted QR algorithm on the Hessenberg form leads to an efficient method for reducing a matrix into Schur form. The reduction of a complex matrix into Hessenberg form and the subsequent reduction into Schur form by the explicitly shifted QR algorithm requires approximately $\frac{20}{3}n^3 + 12kn^3$ flops, where k is the average number of iterations necessary to find an eigenvalue (usually k does not exceed 2) (Stewart, 2001, Section 2.4). The computed Schur form T is unitary similar to a matrix close to A, i.e.,

$$T = Q^H(A + E)Q,$$

where Q is a matrix which is close to an unitary matrix and the backward error is $\|E\|_2 \approx \mathbf{u}\|A\|_2$. This implies that the QR method for reducing a matrix to a Schur form is numerically stable from backward error analysis point of view.

The reduction into Schur form of a nonsymmetric matrix is implemented by using the LAPACK routine **xGEES** and the desired ordering of the eigenvalues on the diagonal of the Schur form is done by the routine **xGEESX**. In MATLAB®, the reduction into Schur form is done by the function `schur` implementing the option `'complex'` and the desired ordering of the eigenvalues can be performed by the function `ordschur`.

Example 4.2. Consider the fourth-order matrix

$$A = \begin{bmatrix} 2 - j_o4 & -1 + j_o2 & -2 + j_o3 & 5 - j_o7 \\ -6 + j_o8 & 0 - j_o3 & 3 - j_o4 & 2 + j_o6 \\ 1 - j_o7 & 4 - j_o1 & -1 + j_o5 & -6 - j_o3 \\ -3 + j_o5 & 6 + j_o6 & -4 + j_o8 & 1 - j_o4 \end{bmatrix}.$$

First, the matrix A is reduced into Hessenberg form as it is shown in Example 4.1. Then we use the function `hqr` that implements the explicitly shifted QR algorithm as realized by Algorithm 2.8 presented in Stewart (2001), p. 99. This function determines the Schur form totally for 9 iterations.

The evolution of the subdiagonal elements of the Hessenberg form during the QR iterations is shown in Table 4.1 (see also Figure 4.1). After the first 5 iterations, the element H_{43} is accepted as negligible and we find the eigenvalue

$$\lambda_4 = 0.257841701385934 - j_o7.811433574563048.$$

Table 4.1. Convergence to zero of the subdiagonal elements of the Hessenberg form.

k	$\lvert H_{43}\rvert$	$\lvert H_{32}\rvert$	$\lvert H_{21}\rvert$
0	$1.4346 \cdot 10^{0}$	$5.9978 \cdot 10^{0}$	$1.3565 \cdot 10^{1}$
1	$3.2988 \cdot 10^{-1}$	$6.0708 \cdot 10^{0}$	$7.9408 \cdot 10^{0}$
2	$5.4477 \cdot 10^{-2}$	$2.1375 \cdot 10^{0}$	$5.6843 \cdot 10^{0}$
3	$4.7895 \cdot 10^{-4}$	$5.7129 \cdot 10^{-1}$	$6.8518 \cdot 10^{0}$
4	$1.3291 \cdot 10^{-8}$	$1.4899 \cdot 10^{-1}$	$8.3724 \cdot 10^{0}$
5	$2.7083 \cdot 10^{-18}$	$4.6973 \cdot 10^{-2}$	$6.7140 \cdot 10^{0}$
6		$1.3654 \cdot 10^{-4}$	$3.5555 \cdot 10^{0}$
7		$4.5982 \cdot 10^{-10}$	$2.3733 \cdot 10^{0}$
8		$2.5441 \cdot 10^{-21}$	$1.6283 \cdot 10^{0}$
9			$2.5289 \cdot 10^{-17}$

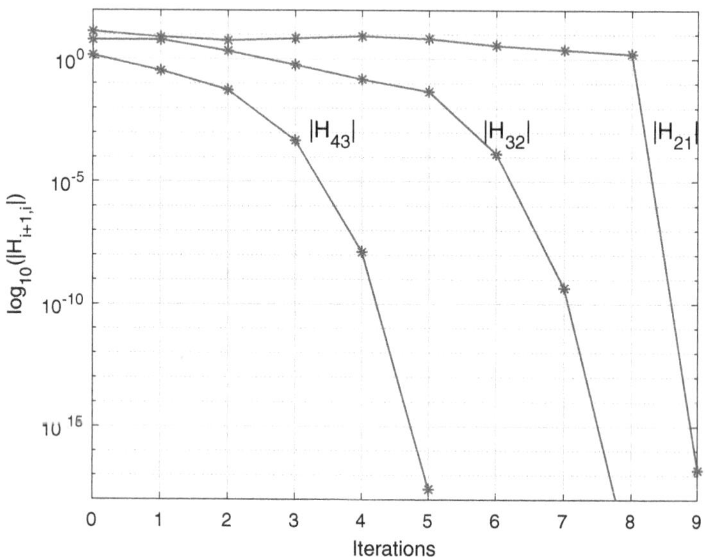

Fig. 4.1. Convergence to zero of the subdiagonal elements of the Hessenberg matrix.

As a result of these iterations, the Hessenberg form takes the form (the entries are shown up to five decimal digits)

$$H = \left[\begin{array}{c|c} 9.1865 + j_o 4.9416 & -4.8326 + j_o 3.4537 \\ \hline -6.6538 + j_o 0.8974 & -5.0337 + j_o 1.6844 \\ \hline & 0.0400 + j_o 0.0246 \\ \hline & \end{array}\right. \cdots \rightarrow$$

$$\leftarrow \cdots \left.\begin{array}{c|c} 2.2394 + j_o 7.3645 & -0.4882 + j_o 2.1334 \\ \hline -4.5357 - j_o 2.0333 & -11.3950 + j_o 1.8518 \\ \hline -2.4106 - j_o 4.8146 & 8.7894 - j_o 4.5054 \\ \hline 0 & 0.2578 - j_o 7.8114 \end{array}\right].$$

After the deflation of the last eigenvalue, the problems of order 3 remain. Performing further three iterations, the element H_{32} also becomes negligible and we find the eigenvalue

$$\lambda_3 = -2.406167746468483 - j_o 4.820825070077387.$$

The Hessenberg form becomes

$$H = \left[\begin{array}{c|c} 11.0494 + j_o 2.6886 & 1.1930 + j_o 4.5467 \\ \hline 1.2481 + j_o 1.0457 & -6.9011 + j_o 3.9437 \\ \hline & 0 \\ \hline & \end{array}\right. \cdots \rightarrow$$

$$\leftarrow \cdots \left.\begin{array}{c|c} 2.6140 + j_o 8.1039 & 4.0703 + j_o 1.8668 \\ \hline -2.8722 + j_o 1.8083 & -9.9044 + j_o 4.4071 \\ \hline -2.4062 - j_o 4.8208 & 8.7893 - j_o 4.5600 \\ \hline 0 & 0 \end{array}\right].$$

After the deflation of λ_3, the problem order becomes equal to 2 and after a single iteration, the element H_{21} also becomes negligible.

In this way, we obtain the Schur form

$$T = \begin{bmatrix} 10.8467 + j_o 3.0687 & 0.0537 + j_o 5.7101 & \cdots \rightarrow \\ & -6.6984 + j_o 3.5636 & \\ & & \\ & & \end{bmatrix}$$

$$\begin{matrix} & 2.5228 + j_o 8.3692 & 3.6631 + j_o 2.7681 \\ \leftarrow \cdots & -2.5330 + j_o 1.0960 & -10.0200 + j_o 4.0114 \\ & -2.4062 - j_o 4.8208 & 8.7893 - j_o 4.5600 \\ & & 0.2578 - j_o 7.8114 \end{matrix} \Bigg] ,$$

and with this the last two eigenvalues

$$\lambda_2 = -6.698360427204300 + j_o 3.563605782963195$$

and

$$\lambda_1 = 10.846686472286859 + j_o 3.068652861677232$$

are determined. The matrix of the unitary transformation from A to T is

$$Q = \begin{bmatrix} -0.3336 - j_o 0.2786 & 0.1224 - j_o 0.2729 & \\ 0.5852 - j_o 0.1138 & 0.1105 - j_o 0.2763 & \cdots \rightarrow \\ -0.3326 + j_o 0.1381 & 0.1824 + j_o 0.6852 & \\ 0.3980 - j_o 0.4094 & 0.1461 + j_o 0.5457 & \end{bmatrix}$$

$$\begin{matrix} & 0.6163 - j_o 0.5597 & 0.0536 + j_o 0.1604 \\ \leftarrow \cdots & -0.1338 - j_o 0.3591 & -0.6368 - j_o 0.0612 \\ & -0.0030 - j_o 0.1494 & -0.5345 + j_o 0.2438 \\ & -0.0997 - j_o 0.3576 & 0.4582 - j_o 0.0840 \end{matrix} \Bigg] .$$

The backward error corresponding to the computed Schur form is

$$\|QTQ^H - A\|_F = 2.461 \cdot 10^{-14},$$

and the error characterizing the deviation from unitarity of the matrix Q is

$$\|Q^H Q - I_4\|_F = 1.485 \cdot 10^{-15}.$$

In the following two sections, we consider the numerical issues related to the QR algorithm.

4.1.4 *Accuracy of the Computed Eigenvalues*

Regardless of the backward stability of the QR algorithm, its accuracy with respect to the eigenvalues can be very poor in the presence of multiple eigenvalues, as seen from the example given in the following.

Example 4.3. Consider the matrix

$$A = \begin{bmatrix} -2 & 0 & 0 & -1 & 2 & -1 \\ -1 & -1 & 0 & -1 & 1 & 1 \\ -1 & -1 & -1 & 0 & 1 & 0 \\ 2 & -1 & 1 & -2 & 0 & -4 \\ 0 & -1 & 0 & -2 & 1 & -2 \\ -1 & 0 & 0 & 0 & 1 & -1 \end{bmatrix},$$

which is similar to a Jordan block of sixth-order with eigenvalue equal to -1.

After reduction into Hessenberg form H by the function `hessreduce`, we use the function `hqr` to reduce H into Schur form. It appears that for number of iterations less than 27, the QR algorithm does not converge. For 27 iterations, we obtain for the diagonal elements of Schur form (the eigenvalues of A) the following values:

$$\begin{aligned}
&-1.003281962111736 - \jmath_o 0.001700002926268,\\
&-1.003111858937610 + \jmath_o 0.001989853757381,\\
&-1.000167382232817 - \jmath_o 0.003694665811722,\\
&-0.999835449018137 + \jmath_o 0.003689900449601,\\
&-0.996882598488896 - \jmath_o 0.001989922978971,\\
&-0.996720749210805 + \jmath_o 0.001704837509981.
\end{aligned}$$

The maximum absolute error of the computed eigenvalues is equal to 0.0036985. This large error is due to the high sensitivity of the eigenvalues participating in a Jordan block of order 6 (see Section 3.4). This sensitivity is characterized by the quantity $(\|A\|_2\mathbf{u})^{1/6} \approx 0.0033075$, which corresponds to the error $(\|A\|_2\mathbf{u})^{1/6}$ of storing the matrix A into the computer memory. We see that the maximum error in the computed eigenvalues is close to this quantity.

In this way, the computation of the Schur form is not sufficient to determine accurately the eigenvalues, since the error in them depends on the Jordan structure of the matrix in which these eigenvalues participate. In this sense, the QR algorithm should be considered only as a first step in the reliable determination of the eigenvalues. We return to this important topic in the following chapters.

4.1.5 *Accuracy of the Computed Invariant Subspaces*

The following example demonstrates that some of the invariant subspaces related to the computed Schur form can be very sensitive to the rounding errors accompanying the reduction done by the QR algorithm.

Example 4.4. Given is a family of matrices of order $n = 5$ constructed as

$$A = QT_0Q^{-1},$$

where T_0 is an upper triangular matrix whose diagonal elements are the eigenvalues of A,

$$Q = H_2 \Sigma H_1,$$

H_1 and H_2 are elementary reflections of the form

$$\begin{aligned}
H_1 &= I_n - \frac{2}{n} ee^T, \quad e = [1,\ 1, \dots, 1]^T, \\
H_2 &= I_n - \frac{2}{n} ff^T, \quad f = [1,\ -1, \dots, (-1)^{n-1}]^T, \\
\Sigma &= \operatorname{diag}(1, \sigma, \sigma^2, \dots, \sigma^{n-1}),
\end{aligned}$$

and $\sigma > 0$ is a given parameter. Since H_1 and H_2 are symmetric and orthogonal, the condition number of the nonsingular transformation matrix Q is equal to the condition number of the matrix Σ given by $\operatorname{cond}_2(\Sigma) = \sigma^{n-1}$. Such matrices are proposed in Bavely and Stewart (1979) and they have the advantage that Q and $Q^{-1} = H_1 \Sigma^{-1} H_2$ are obtained with very small rounding errors. Also, the conditioning of the similarity transformation matrix is easily controlled by changing σ.

In the given case, the matrix T_0 is chosen as

$$T_0 = \begin{bmatrix} -1 & 1 & -4 & 2 & -5 \\ & -1 & 1 & -4 & 3 \\ & & -0.999 & -3 & -2 \\ & & & 2 & 1 \\ & & & & 2 \end{bmatrix}$$

and we are interested by the behavior of the invariant subspace corresponding to the double eigenvalue at -1 when σ varies. To perform the analysis, the matrix A is reduced to upper triangular form

$$T = U^H A U = \begin{bmatrix} T_{11} & T_{12} \\ 0 & T_{22} \end{bmatrix}, \quad T_{11} \in \mathbb{C}^{2\times 2}$$

by using the MATLAB® function `schur`, after that the diagonal of T is reordered by the function `ordschur`, such that in the leading position appear the eigenvalues equal to -1. In this way, the first two columns of the unitary matrix U, which reduces A into ordered Schur form, constitute a basis of the invariant subspace whose sensitivity is under interest. The unperturbed invariant subspace for each σ is obtained as span$(q_1,\ q_2)$, where q_1 and q_2 are the first two columns of Q. The sensitivity of an invariant subspace can be measured by the angles between the perturbed and unperturbed subspace (see Section 2.2.8). In the given case, the maximum angle between the unperturbed and perturbed space is obtained from (2.52) as

$$\Theta_{\max} = \arccos(\sigma_{\min}(Y^H X)),$$

where $\sigma_{\min}$ is the minimum singular value of the matrix $Y^H X$, X is the orthonormal basis of the first two columns $[q_1,\ q_2]$ of the similarity transformation matrix Q and $Y = [u_1, u_2]$ is the orthonormal basis of the perturbed invariant subspace consisting of the first two columns of U. The basis X is obtained by using the singular value decomposition of $[q_1, q_2]$.

In Table 4.2, we give the values of the maximum angle between the perturbed and unperturbed invariant subspaces of dimension $r = 2$ corresponding to the double eigenvalue at -1 along with the condition number $\text{cond}_2(Q)$ of the similarity transformation matrix Q and

Table 4.2. Invariant subspace sensitivity.

i	σ	$\text{cond}_2(Q)$	*berr*	$\text{sep}(T_{11}, T_{22})$	$\theta_{\max}(U, \widetilde{U})$ deg
1	1.0	$1.00 \cdot 10^0$	$2.15 \cdot 10^{-15}$	$6.88 \cdot 10^{-7}$	$1.36 \cdot 10^{-8}$
2	7.0	$2.40 \cdot 10^3$	$5.84 \cdot 10^{-16}$	$1.15 \cdot 10^{-10}$	$2.87 \cdot 10^{-2}$
3	13.0	$2.86 \cdot 10^4$	$3.59 \cdot 10^{-16}$	$2.71 \cdot 10^{-10}$	$2.18 \cdot 10^0$
4	19.0	$1.30 \cdot 10^5$	$9.76 \cdot 10^{-16}$	$1.54 \cdot 10^{-10}$	$6.91 \cdot 10^0$
5	25.0	$3.91 \cdot 10^5$	$1.00 \cdot 10^{-15}$	$2.42 \cdot 10^{-10}$	$2.19 \cdot 10^1$
6	31.0	$9.24 \cdot 10^5$	$3.98 \cdot 10^{-16}$	$3.22 \cdot 10^{-10}$	$4.11 \cdot 10^1$

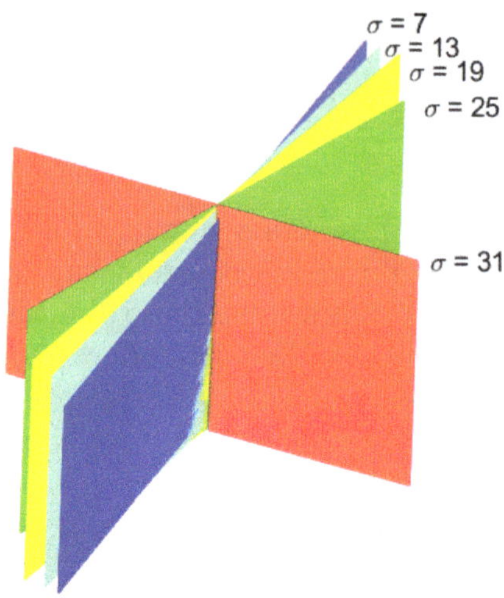

Fig. 4.2. Perturbed invariant subspaces for different values of s.

the relative backward error

$$berr = \frac{\|UTU^H - A\|_2}{\|A\|_2}$$

in computing the Schur form. In Figure 4.2, we show parts of the perturbed two-dimensional invariant subspaces for different values of σ. The subspace image is obtained as the span of the subspace basis vectors u_1 and u_2 by computing 10^4 linear combinations of the form $v = \delta_1 u_1 + \delta_2 u_2$, where δ_1, δ_2 are free parameters. The first three components of v are used as x, y and z coordinates of the points lying on the two-dimensional subspace in the three-dimensional space.

The results indicate that with the increase in $\text{cond}_2(Q)$, the maximum angle between the perturbed and unperturbed invariant subspaces for a given σ also increases and for $\sigma = 31$, it exceeds 40 deg. This gradual increase in the sensitivity of the invariant subspace to rounding errors is due to the increasing conditioning of the similarity transformations matrix Q as well as to the presence of the eigenvalue -0.999 which is close to the double eigenvalue at -1 (A becomes closer to a matrix with a triple eigenvalue at -1). On the other hand, for all values of σ, the backward error in computing the Schur form remains of the order of the unit roundoff error $\mathbf{u}$ as a result of the numerical stability of the QR method.

The increasing sensitivity of the invariant subspace shows that in the given case the conditioning of the Schur form worsens with the increase in the parameter σ and reflects not only on the accuracy of the invariant subspaces but also on the accuracy of the eigenvalues and superdiagonal elements of T. This fact contradicts the usual opinion that it is unconditionally preferable to use the Schur form since it is obtained by a numerically stable method and since it is much more reliable than the Jordan form. We see that the Schur form can also be very sensitive and its use requires some care.

4.1.6 *Notes and References*

4.1.6.1 *Hessenberg Form*

The Hessenberg form of a matrix is introduced by the German mathematician and electrical engineer Karl Hessenberg in the technical report Hessenberg (1940).

4.1.6.2 *QR Algorithm*

The QR algorithm for finding the eigenvalues of a nonsymmetric matrix is proposed by the English mathematician Francis (1961, 1962) and the Russian mathematician Kublanovskaya (1961, 1962). It may be considered as "orthogonalized" variant of the LR algorithm proposed earlier by Rutishauser (1958). For a detailed history of this remarkable algorithm, see Golub and Uhlig (2009). Profound descriptions of the QR algorithm and its explicit and implicit shift versions are given in Stewart (2001), Chapter 2, Golub and Van Loan (2013), Chapter 7, and Wilkinson (1965),

Chapter 8. A systematical presentation of the different types of the QR algorithm can be found in Watkins (2010). More efficient versions of the QR algorithm are proposed in Bai and Demmel (1989) and Braman *et al.* (2002a, 2002b) and methods for improving the accuracy of computed eigenvalues and eigenvectors are presented in Dongarra *et al.* (1983). Algorithms for reordering the real and complex Schur form are developed by several authors, see Ruhe (1970a), Stewart (1976), Bai and Demmel (1993), Bojanczyk and Van Dooren (1993), Brandts (2001), Granat *et al.* (2009) and Kressner (2006). Numerical methods for solving different eigenvalue problems are described in Saad (2011), Kressner (2005), Bai and Pan (2021), Bai and Demmel (1998), Bai *et al.* (1997), Byers *et al.* (1997) and Dieci and Russell (1987). The solution of the symmetric eigenvalue problem is presented systematically in Parlett (1998) and Stewart (2001), Chapter 3.

The development of a software for computing eigenvalues and eigenvectors of different types of matrices has its roots in the Algol procedures presented in Wilkinson and Reinsch (1971) and the Fortran programs contained in the EISPACK package (Garbow *et al.*, 1977; Smith *et al.*, 1976). High quality routines for this purpose are included in the LAPACK package (Anderson *et al.*, 1999) and its scalable version ScaLAPACK (Blackford *et al.*, 1997) containing block-partitioned algorithms intended to achieve higher efficiency on distributed-memory computer architectures. Useful templates for the solution of eigenvalue problems for different kinds of matrices are given in Bai *et al.* (2000).

It is interesting to note that the largest matrix involved in computations which is of order several billions is the Google matrix used to rank the pages and to present to the user the more important (and typically most relevant and helpful) pages on the web first (Bryan and Leise, 2006; Gleich and Constantine, 2015; Moler, 2002). The Google matrix is a *stochastic matrix*, i.e., its elements are all strictly between zero and one and its column sums are equal to one. The largest eigenvalue of this matrix is equal to one and the components of the corresponding left eigenvector quantitatively rate the importance of each page on the web. Google's PageRank algorithm (Page *et al.*, 1999) finds this eigenvector by the simple *power method* (Golub and Van Loan, 2013, Chapter 7; Stewart, 2001, Chapter 2) which is one of the predecessors of the QR algorithm. More on the eigenvalues and Jordan form of the Google matrix can be found in Serra-Capizzano (2005) and Horn and Serra-Capizzano (2006).

4.2 Numerical Solution of the Sylvester Equation

The Sylvester matrix equation plays an important role is solving several matrix problems like the reduction into block-diagonal form (Section 2.1.6) and estimating the sensitivity of invariant subspaces (Section 3.6). In this section, we consider the conditioning of this equation and its numerical solution.

As shown in Section 2.1.6, the Sylvester matrix equation

$$AX + XB = C, \tag{4.4}$$

where $A \in \mathbb{C}^{n\times n}$, $B \in \mathbb{C}^{m\times m}$ and $C \in \mathbb{C}^{n\times m}$ has a unique solution for X if and only if the matrices A and $-B$ have no eigenvalues in common, i.e.,

$$\lambda_i + \mu_j \neq 0, \tag{4.5}$$

where λ_i, $i = 1, 2, \ldots, n$, and μ_j, $j = 1, 2, \ldots, m$, are the eigenvalues of A and B, respectively.

4.2.1 *Sensitivity and Conditioning*

An asymptotic perturbation analysis of the Sylvester equation can be done as follows:

Let

$$\Omega = I_m \otimes A + B^T \otimes I_n \in \mathbb{C}^{nm\times nm}.$$

Then it may be shown that the relative change $\|\Delta X\|_F/\|X\|_F$ in the solution of the Sylvester equation due to small relative perturbations $\Delta A/\|A\|_F, \Delta B/\|B\|_F$ and $\Delta C/\|C\|_F$ is given by

$$\|\Delta X\|_F/\|X\|_F \leq \sqrt{3}\text{cond}_S \eta, \tag{4.6}$$

where

$$\text{cond}_S = \frac{\|\Omega^{-1}\|_2 \| \left[\|A\|_F (X^T \otimes I_n), \|B\|_F (I_m \otimes X), -\|C\|_F I_{nm} \right] \|_2}{\|X\|_F}$$

is the *condition number of the Sylvester equation* and

$$\eta = \max\{\|\Delta A\|_F / \|A\|_F, \|\Delta B\|_F / \|B\|_F, \|\Delta C\|_F / \|C\|_F\}$$

is the maximum relative perturbation in the data.

The condition number of the Sylvester equation is closely related to the separation between matrices A and B (Section 3.6). In the case of the Sylvester equation (4.4), we have that

$$\|\Omega^{-1}\|_2 = 1/\text{sep}(A, -B),$$

where

$$\text{sep}(A, -B) = \min_{X \neq 0} \frac{\|AX + XB\|_F}{\|X\|_F}.$$

Thus, we get the expression

$$\text{cond}_S = \frac{\left\| \left[\|A\|_F (X^T \otimes I_n), \|B\|_F (I_m \otimes X), -\|C\|_F I_{nm} \right] \right\|_2}{\text{sep}(A, -B) \|X\|_F}, \quad (4.7)$$

which shows that if the separation between A and $-B$ is small, then the condition number of the Sylvester equation is large.

The exact value of $\text{sep}(A, -B)$ can be evaluated according to the expression

$$\text{sep}(A, -B) = \sigma_{\min}(I_m \otimes A + B^T \otimes I_n), \quad (4.8)$$

where $\sigma_{\min}(\Omega)$ is the minimum singular value of Ω. (Note that even in the case of complex matrices, one should take in the earlier expression the transposed of the matrix B.) The computation of $\text{sep}(A, -B)$ using (4.8) involves manipulation of matrices of order $mn \times mn$ which makes it suitable only for low- and medium-sized order problems.

Example 4.5. Consider the conditioning of a family of Sylvester equations $AX + XB = C$ with matrix A of twentieth order and matrix B of tenth order. The 20×10 solution matrix X is chosen with unit entries and the matrix C is determined from $C = AX+XB$. The matrices A and B are constructed as

$$A = Q_A J_A Q_A^{-1}, \quad B = Q_B J_B Q_B^{-1},$$

where

$$J_A = \begin{bmatrix} \lambda & 1 & & & \\ & \lambda & 1 & & \\ & & \ddots & \ddots & \\ & & & & 1 \\ & & & & \lambda \end{bmatrix}, \quad J_B = \begin{bmatrix} \mu & & & & \\ 1 & \mu & & & \\ & 1 & \ddots & & \\ & & \ddots & \mu & \\ & & & 1 & \mu \end{bmatrix},$$

and $\lambda = 1$, $\mu = 2$. Similar to Example 4.4, the transformation matrices Q_A and Q_B are constructed as

$$\begin{aligned} &Q_A = H_2 \Sigma_A H_1, && Q_B = M_2 \Sigma_B M_1, \\ &H_1 = I_n - 2uu^T/n, && M_1 = I_m - 2ww^T/m, \\ &H_2 = I_n - 2vv^T/n, && M_2 = I_m - 2zz^T/m, \\ &u = [1, 1, 1, \ldots, 1]^T, && w = [1, 1, 1, \ldots, 1]^T, \\ &v = [1, -1, 1, \ldots, (-1)^{n-1}]^T, && z = [1, -1, 1, \ldots, (-1)^{m-1}]^T, \\ &\Sigma_A = \text{diag}(1, \sigma_A, \sigma_A^2, \ldots, \sigma_A^{n-1}), && \Sigma_B = \text{diag}(1, \sigma_B, \sigma_B^2, \ldots, \sigma_B^{m-1}), \end{aligned}$$

where H_1, H_2, M_1, M_2 are orthogonal elementary reflections. The condition numbers of Q_A and Q_B in respect to the inversion and hence the condition number of the Sylvester equation are controlled by the parameters $\sigma_A > 0$ and $\sigma_B > 0$, respectively.

In Figure 4.3, we show the condition number of the Sylvester equation for σ_A varying between 1.0 and 2.0 and σ_B varying between 1.0 and 4.0. With the increase in the parameters σ_A and σ_B, the condition number cond_S of the equation increases its value varying between 16.3 for $\sigma_A = 1.0$, $\sigma_B = 1.0$ and $1.19 \cdot 10^{13}$ for $\sigma_A = 2.0$, $\sigma_B = 4.0$.

4.2.2 *Solution Methods*

The Sylvester equation is a linear matrix equation with respect to the matrix X and can be represented as a system of nm scalar linear equations. Let the nm vectors

$$c = \text{vec}(C) = [c_{11}, c_{21}, \ldots, c_{n1}, c_{12}, c_{22}, \ldots, c_{nm}]^T,$$

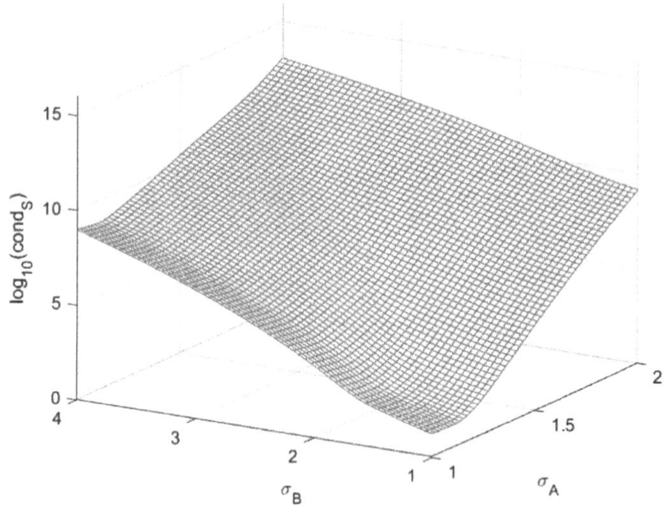

Fig. 4.3. Condition number of the Sylvester equation for different values of the parameters σ_A and σ_B.

$$x = \text{vex}(X) = [x_{11}, x_{21}, \ldots, x_{n1}, x_{12}, x_{22}, \ldots, x_{nm}]^T$$

be constructed from the entries of the matrices C and X. Then according to the presentation in Section 2.1.6, equation (4.4) can be written as a system of nm equations in the nm unknown elements of X:

$$\Omega x = c, \tag{4.9}$$

where $\Omega = I_m \otimes A + B^T \otimes I_n$ and the $nm \times nm$ matrices

$$I_m \otimes A = \begin{bmatrix} A & & 0 \\ & A & \\ & & \vdots & \\ 0 & & A \end{bmatrix},$$

$$B \otimes I_n = \begin{bmatrix} b_{11}I_n & b_{21}I_n & \ldots & b_{m1}I_n \\ b_{12}I_n & b_{22}I_n & \ldots & b_{m2}I_n \\ \vdots & \vdots & & \vdots \\ b_{1m}I_n & b_{2m}I_n & \ldots & b_{mm}I_n \end{bmatrix}$$

are the Kronecker products of A and B^T, respectively, with unit matrices. The matrix Ω is nonsingular if and only if A and $-B$ have no eigenvalues in common, since its eigenvalues are given by the sums $\lambda_i + \mu_j$.

The solution of (4.4) through (4.9) allows in solving accurately the Sylvester equation but requires computational work proportional to n^3m^3 and storage proportional to n^2m^2, which is not acceptable for large n and m.

Example 4.6. Consider the solution of the Sylvester equation presented in Example 4.5, using the Kronecker product. In the given case, the system of linear equations (4.9) is of order $n \times m = 200$.

In Figure 4.4, we show the relative error

$$\text{rerr} = \|\widehat{X} - X\|_F / \|X\|_F$$

of the computed solution $\widehat{X}$ as a function of the parameters σ_A and σ_B. As one can expect, the value of the error for specific values of σ_A and σ_B is of order approximately equal to $-16 + \log_{10}(\text{cond}_S)$ (compare with Figure 4.3).

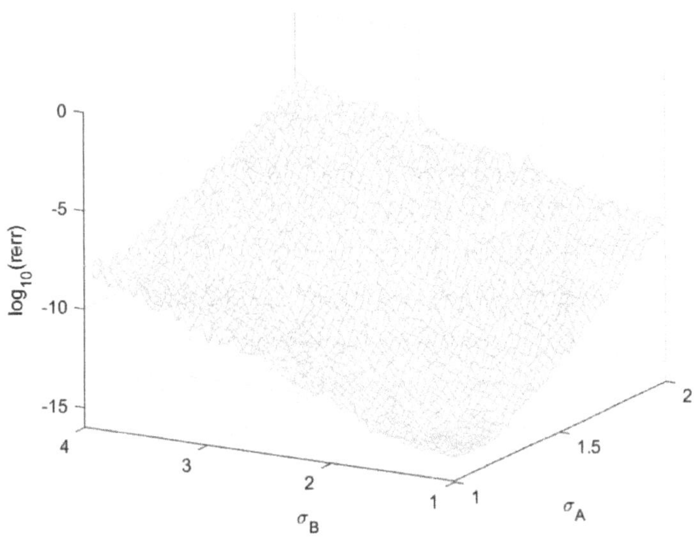

Fig. 4.4. Relative error in solving Sylvester equation by using the Kronecker product for different values of the parameters σ_A and σ_B.

The Sylvester equation can be solved efficiently, if it is transformed into the equivalent equation

$$\widetilde{A}\widetilde{X} + \widetilde{X}\widetilde{B} = \widetilde{C}, \tag{4.10}$$

where

$$\widetilde{A} = U^{-1}AU, \quad \widetilde{B} = V^{-1}BV,$$
$$\widetilde{C} = U^{-1}CV, \quad \widetilde{X} = U^{-1}XV,$$

U and V are certain nonsingular matrices and the matrices $\widetilde{A}$ and $\widetilde{B}$ have some simple form. If, for instance, the matrices A and B are reduced into diagonal form by using the eigenvector matrices U and V, respectively, then the elements of the matrix $\widetilde{X}$ in equation (4.10) are found simply from

$$\widetilde{x}_{ij} = \frac{\widetilde{c}_{ij}}{\widetilde{a}_{ii} + \widetilde{b}_{jj}}, \quad i = 1, 2, \ldots, n, \quad j = 1, 2, \ldots, m.$$

Since the matrices U and V may be ill conditioned, this method is not recommended for numerical solution of (4.10). In this way, we come to the idea to use unitary U and V, which leads to an efficient and numerically reliable solution of the Sylvester equation.

4.2.3 *Bartels–Stewart Algorithm*

Let A and B be reduced into the Schur form

$$S = U^H AU, \quad T = V^H BV, \tag{4.11}$$

where S and T are upper triangular and U, V are unitary matrices. Then (4.11) is transformed to

$$SY + YT = D, \tag{4.12}$$

where $Y = U^H XV$ and $D = U^H CV$. Partitioning the matrices Y and D by columns, equation (4.12) is represented as

$$S[y_1, y_2, \ldots, y_m] + [y_1, y_2, \ldots, y_m] \begin{bmatrix} t_{11} & t_{12} & \cdots & t_{1m} \\ & t_{22} & \cdots & t_{2m} \\ & & \vdots & \ddots \\ & & & t_{mm} \end{bmatrix}$$
$$= [d_1, d_2, \ldots, d_m]. \tag{4.13}$$

The first column of this system is

$$Sy_1 + t_{11}y_1 = (S + t_{11}I_n)y_1 = d_1,$$

where

$$S + t_{11}I_n = \begin{bmatrix} s_{11}+t_{11} & s_{12} & \cdots & s_{1n} \\ & s_{22}+t_{11} & \cdots & s_{2n} \\ & & \ddots & \vdots \\ & & & s_{nn}+t_{11} \end{bmatrix}$$

and $s_{11}, s_{22}, \ldots, s_{nn}$ are the eigenvalues of S. Since t_{11} is an eigenvalue of T, it follows from (4.5) that $s_{ii} + t_{11} \neq 0$, $i = 1, 2, \ldots, n$, so that the matrix $S + t_{11}I_n$ is nonsingular. That is why y_1 can be found as the solution of the triangular linear system $(S+t_{11}I_n)y_1 = d_1$ using Algorithm 1.2.

For the second column of equation (4.13), we obtain

$$Sy_2 + t_{22}y_2 + t_{12}y_1 = d_2$$

so that

$$(S + t_{22}I_n)y_2 = d_2 - t_{12}y_1.$$

The matrix $S + t_{22}I_n$ is nonsingular which allows finding y_2 as a solution of another triangular system of linear equations.

In general, suppose that we have found $y_1, y_2, \ldots, y_{k-1}$. For the kth column of (4.13), we get

$$Sy_k + t_{kk}y_k + \sum_{i=1}^{k-1} t_{ik}y_i = d_k.$$

Now the kth column of Y can be found by solving the upper triangular linear system

$$(S + t_{kk}I_n)y_k = d_k - \sum_{i=1}^{k-1} t_{ik}y_i.$$

After evaluating all columns of Y, the solution of (4.4) is found from $X = UYV^H$. Note that each triangular system of equations with a

Algorithm 4.1. Bartels–Stewart algorithm for solving the Sylvester equation

Given matrices A, B *and* C*, this algorithm solves the Sylvester equation* $AX + XB = C$.

function $X = \text{sylv\_solv}(A, B, C)$
Input: Matrices $A \in \mathbb{C}^{n\times n}, B \in \mathbb{C}^{m\times m}, C \in \mathbb{C}^{n\times m}$
Output:$X \in \mathbb{C}^{n\times m}$ contains the solution matrix
Compute the Schur decomposition $A = USU^H$
Compute the Schur decomposition $B = VTV^H$
$D = UCV^H$
for $k = 1, 2, \ldots, n$
  Solve the triangular system $(S + t_{kk}I_n)y_k = d_k - \sum_{i=1}^{k-1} t_{ik}y_i$
  using Algorithm 1.2
end
$X = UYV^H$

matrix $S + t_{kk}I_n$ can be solved in $n^2 - n$ computational operations by Algorithm 1.2.

Thus, we have the famous Bartels–Stewart algorithm for solving the Sylvester equation (Algorithm 4.1).

The solution of the Sylvester equation (4.12) in which the matrices S and T are in Schur form requires

$$n(n-1)m + m(m-1)n$$

flops.

The analysis done in Higham (1993, 2002, Chapter 16) shows that the relative residual

$$\frac{\|C - (A\widehat{X} + \widehat{X}B)\|_F}{(\|A\|_F + \|B\|_F)\|\widehat{X}\|_F}$$

corresponding to the solution $\widehat{X}$ computed by Algorithm 4.1 is always small and is of order $\mathbf{u}$. However, the backward error corresponding to the computed solution $\widehat{X}$ can be large in some cases leading to a large forward error in the solution. This means that the Bartels–Stewart algorithm is only conditionally backward stable which requires performing a backward error analysis.

The Bartels–Stewart algorithm for the transformed Sylvester equation (4.12) is implemented in LAPACK by the routine `xTRSYL`.

The solution of Sylvester and Lyapunov equations in MATLAB® is done by the function `lyap`.

4.2.4 *Backward Error Estimate*

According to the definition given in Section 1.3.2, Algorithm 4.1 is numerically stable if the relative backward error in solving the Sylvester equation (4.4) is sufficiently small. Since the Sylvester equation can be represented as a system of linear equations (4.9), the backward error in its solution can be determined as in the case of solving linear equations. However, this approach may produce pessimistic results since the Sylvester equation is a structured system of linear equations. This leads to the necessity to define the backward error in a way which takes into account the special structure of this equation.

Assume that Algorithm 4.1 produced an approximate solution $\widehat{X}$. The normwise relative backward error of this solution is defined by

$$\begin{aligned}\eta(\widehat{X}) = \min\{\varepsilon : (A+\Delta A)\widehat{X} + \widehat{X}(B+\Delta B) = C+\Delta C\},\\ \|\Delta A\|_F \leq \varepsilon\|A\|_F,\ \|\Delta B\|_F \leq \varepsilon\|B\|_F,\\ \|\Delta C\|_F \leq \varepsilon\|C\|_F\}.\end{aligned} \tag{4.14}$$

The equation $(A+\Delta A)\widehat{X}+\widehat{X}(B+\Delta B) = C+\Delta C$ can be written as

$$\Delta A\widehat{X} + \widehat{X}\Delta B - \Delta C = R, \tag{4.15}$$

where R is the residual corresponding to the computed solution $\widehat{X}$. To estimate the sizes of ΔA, ΔB and ΔC, equation (4.15) is transformed using the SVD of $\widehat{X}$, $\widehat{X} = U\Sigma V^H$, where $U \in \mathbb{C}^{n\times n}$ and $V \in \mathcal{C}^{m\times m}$ are unitary matrices and $\Sigma = \operatorname{diag}(\sigma_i) \in \mathbb{R}^{n\times m}$. It is convenient, in addition to the singular values $\sigma_1 \geq \sigma_2 \geq \cdots \geq \sigma_{\min(n,m)}$, to define the numbers $\sigma_{\min(n,m)+1} = \cdots = \sigma_{\max(n,m)} = 0$. As a result, we obtain

$$\widetilde{\Delta A}\widehat{X} + \widehat{X}\widetilde{\Delta B} - \widetilde{\Delta C} = \widetilde{R}, \tag{4.16}$$

where

$$\widetilde{\Delta A} = U^H\Delta AU,\ \ \widetilde{\Delta B} = V^H\Delta BV,\ \ \widetilde{\Delta C} = U^H\Delta CV,\ \ \widetilde{R} = U^HRV.$$

Equation (4.16) represents an undetermined system of $n \cdot m$ linear equations with respect to the $n^2 + m^2 + n \cdot m$ unknown elements of $\widetilde{\Delta A}$, $\widetilde{\Delta B}$, $\widetilde{\Delta C}$. Each scalar equation of this system can be written in the form

$$\frac{\widetilde{\Delta a}_{ij}}{\|A\|_F} \cdot \|A\|_F \sigma_j + \|B\|_F \sigma_i \cdot \frac{\widetilde{\Delta b}_{ij}}{\|B\|_F} - \|C\|_F \cdot \frac{\widetilde{\Delta c}_{ij}}{\|C\|_F} = \widetilde{r}_{ij},$$

$$i = 1, 2, \ldots, n, \quad j = 1, 2, \ldots, m. \tag{4.17}$$

It is possible to prove that the minimum of the quadratic functional

$$(\widetilde{\Delta a}_{ij}/\|A\|_F)^2 + (\widetilde{\Delta b}_{ij}/\|B\|_F)^2 + (\widetilde{\Delta c}_{ij}/\|C\|_F)^2$$

subject to (4.17) is attained for

$$\frac{\widetilde{\Delta a}_{ij}}{\|A\|_F} = \frac{\|A\|_F \sigma_j}{\|A\|_F^2 \sigma_j^2 + \|B\|_F^2 \sigma_i^2 + \|C\|_F^2} \widetilde{r}_{ij},$$

$$\frac{\widetilde{\Delta b}_{ij}}{\|B\|_F} = \frac{\|B\|_F \sigma_i}{\|A\|_F^2 \sigma_j^2 + \|B\|_F^2 \sigma_i^2 + \|C\|_F^2} \widetilde{r}_{ij},$$

$$\frac{\widetilde{\Delta c}_{ij}}{\|C\|_F} = - \frac{\|C\|_F}{\|A\|_F^2 \sigma_j^2 + \|B\|_F^2 \sigma_i^2 + \|C\|_F^2} \widetilde{r}_{ij}.$$

Note that if $n > m$, then the matrix $\widetilde{\Delta B}$ is extended to dimension $n \times m$ and if $n < m$, then the matrix $\widetilde{\Delta A}$ is extended to the same dimension with fictitious elements which do not affect the minimization. According to (4.14), the backward error corresponding to $\widehat{X}$ is determined from

$$\eta(\widehat{X}) = \max\{\|\widetilde{\Delta A}\|_F/\|A\|_F, \|\widetilde{\Delta B}\|_F/\|B\|_F, \|\widetilde{\Delta C}\|_F/\|C\|_F\}. \tag{4.18}$$

Example 4.7. Consider again the Sylvester equation presented in Example 4.5.

In Figure 4.5, we show the backward error bound for σ_A varying between 1.0 and 2.0 and σ_B varying between 1.0 and 4.0. In the given case, the Bartels–Stewart algorithm produces small backward error for all parameter values.

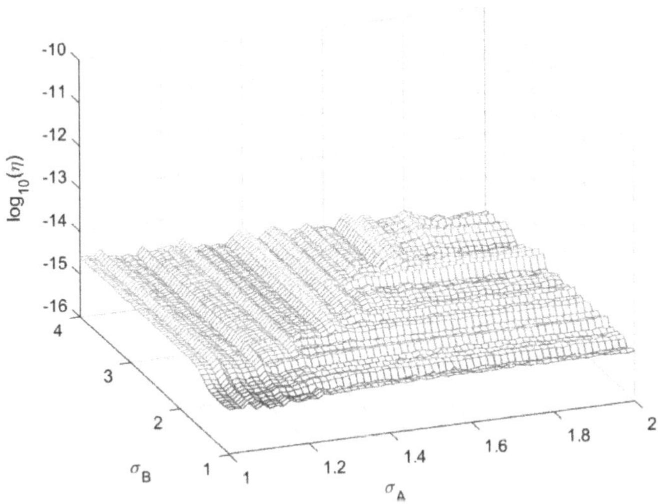

Fig. 4.5. Backward error in solving Sylvester equation for different values of the parameters σ_A and σ_B.

4.2.5 *Forward Error Estimate*

As a results of the backward error analysis accompanied by a sensitivity analysis, it is possible to obtain a forward error estimate of the computed solution $\widehat{X}$ of the Sylvester equation. The normwise relative error in the solution of this equation can be estimated by

$$\|\widehat{X} - X\|_F / \|X\|_F \leq \mathit{ferr},$$

where

$$\mathit{ferr} = \mathrm{cond}_S \eta(\widehat{X})$$

is the forward error bound.

Algorithm 4.2 solves the Sylvester equation providing in addition the equation condition number and estimates of the forward and backward errors.

An estimate close to the forward error of the solution can also be obtained by the following practical error bound. Let the computed residual of the computed solution $\widehat{X}$ be represented as

$$\widehat{R} = \mathrm{fl}(C - (\mathrm{fl}(A\widehat{X} + \mathrm{fl}(\widehat{X}B))) = R + \Delta R,$$

$$|\Delta R| \leq \mathbf{u}|C| + \gamma_A|A||\widehat{X}| + \gamma_B|\widehat{X}||B| =: R_u,$$

Algorithm 4.2. Solution of Sylvester equation with condition and accuracy estimates

Given matrices A, B and C, this algorithm solves the Sylvester equation $AX + XB = C$ with condition and accuracy estimates

function $[\widehat{X}, \text{cond}_S, ferr, \eta] = \text{sylv\_est}(A, B, C)$

Input: Matrices $A \in \mathbb{C}^{n\times n}, B \in \mathbb{C}^{m\times m}, C \in \mathbb{C}^{n\times m}$

Output: $\widehat{X}$ contains the equation solution; cond_S contains the condition number of the Sylvester equation; $ferr$ contains an estimate of the relative forward error of the solution $\widehat{X}$; η contains the backward error corresponding to $\widehat{X}$

Compute the solution $\widehat{X}$ of $AX + XB = C$ using Algorithm 4.1

Compute the condition number of the Sylvester equation

$$\Omega = I_m \otimes A + B^T \otimes I_n$$

$$\text{sep}(A, -B) = \sigma_{\min}(\Omega)$$

$$\text{cond}_S = \left\| \left[\|A\|_F(\widehat{X}^T \otimes I_n), \|B\|_F(I_m \otimes \widehat{X}), -\|C\|_F I_{nm} \right] \right\|_2 \Big/ (\text{sep}(A, -B)\|\widehat{X}\|_F)$$

Decompose $\widehat{X} = U\Sigma V^H,\ U^H U = I_n,\ V^H V = I_m,$

$$\Sigma = \text{diag}(\sigma_i),\ \sigma_1 \geq \sigma_2 \geq \cdots \geq \sigma_{\min(n,m)}$$

Set $\sigma_{\min(n,m)+1} = \cdots = \sigma_{\max(n,m)} = 0$

Compute and transform the residual

$$R = A\widehat{X} + \widehat{X}B - C,\ R \leftarrow U^H R V$$

Compute the backward error $\eta(\widehat{X})$

for $i = 1, 2, \ldots, n$

  for $j = 1, 2, \ldots, m$

$$\widetilde{\Delta a}_{ij}/\|A\|_F = \|A\|_F \sigma_j r_{ij}/(\|A\|_F^2 \sigma_j^2 + \|B\|_F^2 \sigma_i^2 + \|C\|_F^2)$$
$$\widetilde{\Delta b}_{ij}/\|B\|_F = \|B\|_F \sigma_i r_{ij}/(\|A\|_F^2 \sigma_j^2 + \|B\|_F^2 \sigma_i^2 + \|C\|_F^2)$$
$$\widetilde{\Delta c}_{ij}/\|C\|_F = -\|C\|_F r_{ij}/(\|A\|_F^2 \sigma_j^2 + \|B\|_F^2 \sigma_i^2 + \|C\|_F^2)$$

  end

end

$$\eta(\widehat{X}) = \max\left\{ \|\widetilde{\Delta A}\|_F/\|A\|_F, \|\widetilde{\Delta B}\|_F/\|B\|_F, \|\widetilde{\Delta C}\|_F/\|C\|_F \right\}$$

Compute the forward error estimate

$ferr = \text{cond}_S \eta(\widehat{X})$

where

$$\gamma_A = (n+2)\mathbf{u}/(1-(n+2)\mathbf{u}), \quad \gamma_B = (m+2)\mathbf{u}/(1-(m+2)\mathbf{u}).$$

Then the relative error in the computed solution satisfies (Higham, 2002, Equation (16.29))

$$\frac{\|\widehat{X}-X\|}{\|\widehat{X}\|} \leq \frac{\||\Omega^{-1}|(|\mathrm{vec}(\widehat{R})|+\mathrm{vec}(R_u))\|}{\|\widehat{X}\|}, \tag{4.19}$$

where $\|X\| := \max_{i,j}|x_{ij}|$.

Example 4.8. Given is the Sylvester equation considered in Examples 4.5 and 4.7.

The forward error bound $ferr = \mathrm{cond}_S \cdot \eta$ along with the exact relative error

$$\mathrm{rerr} = \frac{\|\widehat{X}-X\|_F}{\|X\|_F}$$

is shown in Figure 4.6. The actual error is close to the forward error estimate.

For this example, the practical error bound (4.19) produces worse results.

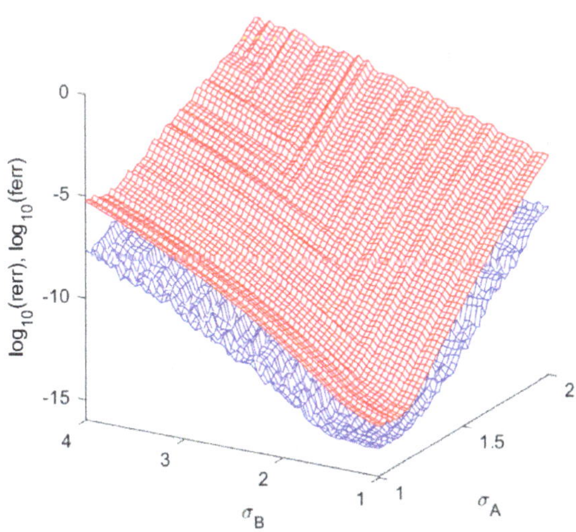

Fig. 4.6. Forward error bound (the surface above) and the actual relative error (the surface beneath) in the solution of Sylvester equation for different values of the parameters σ_A and σ_B.

4.2.6 *Notes and References*

Bartels–Stewart algorithm for solving the Sylvester equation is proposed in Bartels and Stewart (1972) and is one of the most reliable algorithms in the area of matrix computations. The idea to use the triangular form of the matrix A in solving the Lyapunov matrix equation $A^H X + BX = C$, $C^H = C$ goes back to Hahn (1955) and Lefschetz (1965), p. 134, who assumed, however, that this form is obtained by general nonsingular, not by unitary transformations. The backward error analysis of the Sylvester equation is done in Ghavimi and Laub (1995) and the sensitivity of the solution of this equation to perturbations in the data is analyzed in Byers (1984), Golub *et al.* (1979) and Higham (1993), see also Higham (2002), Chapter 16. To avoid the use of large matrices in computing the condition number cond_S and the bound (4.6), one can use the technique described in Higham (2002), Section 16.4, see also Kågström and Poromaa (1992). The practical error bound in the solution of the Sylvester equation is presented in Higham (2002), Chapter 16. A Schur–Hessenberg algorithm for solving the Sylvester equation which is slightly more efficient than the Bartels–Stewart algorithm is proposed in Golub *et al.* (1979). A modification of the Bartels–Stewart algorithm to solve the Lyapunov matrix equation is described in the original publication Bartels and Stewart (1972). Other numerical methods for solving the Sylvester and Lyapunov equations are considered in Bini *et al.* (2012) and Simoncini (1996, 2007), see also the rich survey Simoncini (2016).

4.3 Computing Perturbation Bounds of Eigenspaces

In this section, we consider the computation of practical local and nonlocal perturbation bounds of desired invariant subspaces. For this aim, we use the notion of spectral projection introduced in Section 2.2.7.

4.3.1 *Computing Spectral Projections*

The reduction of a matrix into Schur form makes it possible to compute efficiently spectral projections. Such computation involves the solution of the Sylvester matrix equation.

Assume that the eigenvalues under interest participate in a group of r numbers and let

$$U^H AU = \begin{bmatrix} T_{11} & T_{12} \\ 0 & T_{22} \end{bmatrix}, \quad T_{11} \in \mathbb{C}^{r\times r} \tag{4.20}$$

be the Schur decomposition of the matrix $A \in \mathbb{C}^{n\times n}$, where T_{11} contains the desired eigenvalues. As shown latter on in Section 6.2, such ordering of the diagonal elements of the Schur form can be always achieved by using unitary similarity transformations. The matrix of the unitary transformation can be written as

$$U = [U_1, U_2], \quad U_1 \in \mathbb{C}^{n\times r},$$

where the columns of U_1 are the basis vectors of the invariant subspace $\mathcal{X}$, corresponding to the block T_{11}.

The spectral projection on the invariant subspace corresponding to the block T_{11} can be determined in the following way.

Consider the $n \times n$ matrix

$$X = \begin{bmatrix} I_r & R \\ 0 & I_{n-r} \end{bmatrix}.$$

If the matrix $R \in \mathbb{C}^{r\times(n-r)}$ is a solution of the Sylvester matrix equation

$$T_{11}R - RT_{22} + T_{12} = 0, \tag{4.21}$$

then one has that

$$\begin{aligned} X^{-1}TX &= \begin{bmatrix} I_r & -R \\ 0 & I_{n-r} \end{bmatrix} \begin{bmatrix} T_{11} & T_{12} \\ 0 & T_{22} \end{bmatrix} \begin{bmatrix} I_r & R \\ 0 & I_{n-r} \end{bmatrix} \\ &= \begin{bmatrix} T_{11} & T_{11}R - RT_{22} + T_{12} \\ 0 & T_{22} \end{bmatrix} \\ &= \begin{bmatrix} T_{11} & 0 \\ 0 & T_{22} \end{bmatrix}. \end{aligned}$$

Thus, using a nonsingular transformation with the matrix X, the Schur form T is reduced to a block-diagonal form.

Since

$$\begin{bmatrix} T_{11} & T_{12} \\ 0 & T_{22} \end{bmatrix} \begin{bmatrix} I_r \\ 0 \end{bmatrix} = \begin{bmatrix} I_r \\ 0 \end{bmatrix} T_{11}$$

and

$$[I_r \quad -R] \begin{bmatrix} T_{11} & T_{12} \\ 0 & T_{22} \end{bmatrix} = T_{11}[I_r \quad -R],$$

it follows that the matrices

$$X_1 = \begin{bmatrix} I_r \\ 0 \end{bmatrix}$$

and

$$Y_1^H = [I_r - R]$$

span the right and left eigenspaces corresponding to T_{11}, respectively, and satisfy the normalization condition $Y_1^H X_1 = I_r$. That is why these matrices can be used to determine the spectral projection P on the invariant subspace corresponding to the block T_{11}. More specifically, one has that

$$P = X_1 Y_1^H = \begin{bmatrix} I_r & -R \\ 0 & 0 \end{bmatrix}. \tag{4.22}$$

It is easy to check that the spectral projection has the properties of a projection,

$$P = P^2, \quad TP = PT, \quad PTP = PT.$$

The norm of the spectral projection can be found from

$$\|P\|_2 = \sqrt{1 + \|R\|_2^2}.$$

For a given ordered Schur decomposition and dimension of the invariant subspace, Algorithm 4.3 obtains the spectral projection on this subspace. The projection P is determined according to (4.22),

Algorithm 4.3. Computing spectral projection
Given the Schur decomposition $A = UTU^H$, $U^HU = I_n$ *of a complex* $n \times n$ *matrix* A, *where*

$$T = \begin{bmatrix} T_{11} & T_{12} \\ 0 & T_{22} \end{bmatrix}, \quad T_{11} \in \mathbb{C}^{r \times r}, \quad T_{22} \in \mathbb{C}^{(n-r)\times(n-r)},$$

$$U = [U_1, \ U_2], \quad U_1 \in \mathbb{C}^{n \times r}, U_2 \in \mathbb{C}^{n \times (n-r)},$$

this algorithm computes the spectral projection on the invariant subspace with dimension r, *spanned by the columns of the matrix* U_1 *and its 2-norm. The Schur form should be reordered so that the block* T_{11} *contains the desired eigenvalues.*
function $[P, p_{\mathrm{norm}}] = \mathrm{projector}(T, U, r)$
Input: T contains the ordered Schur form $T = U^H AU$;
U contains the unitary matrix of the similarity transformation which reduces A into Schur form;
r contains the dimension of the invariant subspace.
Output:P contains the projection on the invariant subspace;
p_{norm} contains the 2-norm of P.
Solve the Sylvester equation

$$T_{11}R - RT_{22} + T_{12} = 0$$

Compute the spectral projection

$$P = \begin{bmatrix} I_r & -R \\ 0_{(n-r)\times r} & 0_{(n-r)\times(n-r)} \end{bmatrix}$$

Compute the 2-norm of the projection P

$$p_{\mathrm{norm}} = \sqrt{1 + \|R\|_2^2}$$

while the solution of the Sylvester equation (4.21) can be done by Algorithm 4.1.

Example 4.9. Given is the fifth-order matrix

$$A = \begin{bmatrix} -14 & -21 & 13 & 18 & -26 \\ -2 & -5 & 2 & 4 & -5 \\ -6 & -12 & 6 & 10 & -12 \\ 2 & 3 & -2 & -4 & 4 \\ 5 & 8 & -5 & -7 & 9 \end{bmatrix}$$

whose Jordan form is

$$J = \left(\begin{bmatrix} -1 & 1 \\ 0 & -1 \end{bmatrix}, \begin{bmatrix} -2 & 1 & 0 \\ 0 & -2 & 1 \\ 0 & 0 & -2 \end{bmatrix} \right).$$

We want to determine the spectral projection on the invariant subspace associated with the triple eigenvalue at -2. To perform the analysis, the matrix A is reduced into upper triangular form T by the MATLAB® function `schur` and then the diagonal of T is reordered by the function `ordschur`, such that in the leading position appear the eigenvalues equal to -2. Applying Algorithm 4.3, we find

$$P = \begin{bmatrix} 1 & 0 & 0 & -9.626173661996573 & -20.269843507558189 \\ 0 & 1 & 0 & -2.622657398195327 & -2.091585956011424 \\ 0 & 0 & 1 & 1.415586243610690 & 4.267701577597039 \\ 0 & 0 & 0 & 0 & 0 \\ 0 & 0 & 0 & 0 & 0 \end{bmatrix}$$

and $p_{\text{norm}} = \|P\|_2 = 23.09841$.

4.3.2 *Computing Bounds of Eigenspaces*

According to the presentation in Sections 2.2.8 and 3.6, the sensitivity of an invariant subspace is measured by the angles between the perturbed and unperturbed subspaces. The canonical angles between two eigenspaces $\mathcal{X}$ and $\mathcal{Y}$ of $\mathbb{C}^n$ with dimension r are obtained as (see (2.53))

$$\Theta_i = \arcsin(\sigma_i({Y^{\perp}}^H X)), \tag{4.23}$$

where σ_i, $i = 1, 2, \ldots, p$, are the singular values of the matrix ${Y^{\perp}}^H X$, X is an orthonormal basis of $\mathcal{X}$ and $Y^{\perp}$ is an orthogonal basis of the complementary subspace of Y, ${Y^{\perp}}^H Y = 0$.

Let us apply this result to the problem for perturbation analysis of invariant subspaces. In this case, $\mathcal{Y} = \widetilde{\mathcal{X}}$ and the most important is the maximum angle

$$\Theta_{\max}(\mathcal{X}, \widetilde{\mathcal{X}}).$$

This angle is determined by the relationship

$$\sin(\Theta_{\max}(\mathcal{X}, \widetilde{\mathcal{X}})) = \sigma_{\max}((\widetilde{U}_1^{\perp})^H U_1).$$

Since

$$\widetilde{U}_1^{\perp} = \widetilde{U}_2,$$

it follows that

$$\Theta_{\max}(\mathcal{X}, \widetilde{\mathcal{X}}) = \arcsin(\sigma_{\max}(\widetilde{U_2}^H U_1)). \qquad (4.24)$$

Using the notion of spectral projection and separation between two square matrices, it is possible to prove the following theorem which is a slight modification of Theorem 3.16.

Theorem 4.10 (Bai *et al.* (1993)). *Let the perturbation ΔA satisfy the condition*

$$\|\Delta A\|_F \leq \frac{\text{sep}(T_{11}, T_{22})}{4\|P\|_2}. \qquad (4.25)$$

Then

$$\Theta_{\max}(\mathcal{X}, \widetilde{\mathcal{X}}) \leq \arctan(2\|\Delta A\|_F/(\text{sep}(T_{11}, T_{22}) - 4\|P\|_2\|\Delta A\|_F)). \qquad (4.26)$$

The estimate (4.26) gives a nonlocal bound on the sensitivity of the invariant subspace which is valid only for perturbations which satisfy the condition (4.25).

For small perturbations, instead of (4.26), one may use the asymptotic estimate

$$\Theta_{\max}(\mathcal{X}, \widetilde{\mathcal{X}})^{\text{lin}} \leq \|\Delta A\|_F/\text{sep}(T_{11}, T_{22}). \qquad (4.27)$$

The bound (4.27) shows that the condition number of the invariant subspace $\mathcal{X}$ is inversely proportional to the separation $\text{sep}(T_{11}, T_{22})$ between the matrices T_{11} and T_{22}.

Algorithm 4.4. Global perturbation bound of an invariant subspace

Given the Schur decomposition $A = UTU^H$ of an $n \times n$ complex matrix A and the norm $\|\Delta A\|_F$ of the perturbation ΔA, this algorithms compute a global perturbation bound of the invariant subspace of A spanned by the first $k < n$ columns of the unitary transformation matrix U. The Schur form should be reordered so that the k eigenvalues associated with the desired invariant subspace are in leading positions.

function $[\Theta\text{max}, sep] = pert\_invar\_subspace(T, U, \|\Delta A\|_F)$

Input: The Schur form T of the matrix $A \in \mathbb{C}^{n\times n}$; the unitary matrix U of the similarity transformation to Schur form; the norm $\|\Delta A\|_F$ of the perturbation ΔA.

Output: The quantity Θmax containing an estimate of the maximum angle between the perturbed and unperturbed subspace of dimension k; the quantity *sep* containing the value of the separation between the diagonal blocks $T_{11} \in \mathbb{C}^{k\times k}$ and $T_{22} \in \mathbb{C}^{(n-k)\times(n-k)}$ of the Schur form T.

$T_{11} = T_{1:k,1:k}$
$T_{12} = T_{1:k,k+1:n}$
$T_{22} = T_{k+1:n,k+1:n}$
Determine the separation between T_{11} and T_{22}
$\text{sep} = \sigma_{\min}(I_{n-k} \otimes T_{11} - T_{22}^T \otimes I_k)$
Compute the spectral projection P onto the subspace spanned by the first k columns of U using Algorithm 4.3
$[P, p_{\text{norm}}] = projector(T, U, k)$
if $\|\Delta A\|_F \leq \text{sep}_k/(4p_{\text{norm}})$
    $\Theta\text{max}(k) = 2\|\Delta A\|_F/(\text{sep}_F(k) - 4p_{\text{norm}}\|\Delta A\|_F)$
else
    The global perturbation bound doesn't exist
end

It is necessary to emphasis that the invariant subspace corresponding to a set of sensitive eigenvalues can be unsensitive to perturbations provided that the corresponding group of eigenvalues is isolated.

Computing global estimates of the sensitivity of an invariant subspaces is done by Algorithm 4.4.

In the following, we present two examples of invariant subspaces with different conditioning.

Example 4.11. Consider first the matrix

$$A = \begin{bmatrix} -14 & -21 & 13 & 18 & -26 \\ -2 & -5 & 2 & 4 & -5 \\ -6 & -12 & 6 & 10 & -12 \\ 2 & 3 & -2 & -4 & 4 \\ 5 & 8 & -5 & -7 & 9 \end{bmatrix},$$

given in Example 4.9. In the given case, we are interested in the sensitivity of the invariant subspace corresponding to the group of three eigenvalues at -2 to perturbations of the matrix A in the form

$$\Delta A = 10^{-11+k} \cdot \begin{bmatrix} 8 & 4 & -9 & 1 & -1 \\ -6 & -3 & -5 & -4 & -6 \\ -7 & 5 & 6 & -3 & 8 \\ 3 & -8 & -3 & -8 & -2 \\ 6 & -4 & 8 & -6 & 7 \end{bmatrix},$$

where $k = 1, 2, \ldots, 5$.

We begin with reordering of the Schur form T so that the eigenvalue -2 appears in the upper left corner of T. In this way, the first three columns of the unitary matrix U, which reduces A into ordered Schur form, constitute a basis of the invariant subspace, whose sensitivity is under interest.

The computation of the estimate $\widehat{\Theta}_{\max}$ of the invariant subspace sensitivity is done by Algorithm 4.4 according to (4.26). In the given case, we have that

$$\|P\|_2 = 23.098405$$

and

$$\text{sep}(T_{11}, T_{22}) = 0.024796.$$

In Table 4.3, we give the global estimate $\widehat{\Theta}_{\max}$ of the maximum angle $\Theta_{\max}$ between the subspaces $\mathcal{X}$ and $\widetilde{\mathcal{X}}$, the linear estimate $\widehat{\Theta}^{\text{lin}}_{\max,k}$, as well as the exact value of the angle $\Theta_{\max,k}$ for different values of k.

Table 4.3. Invariant subspace sensitivity (Example 4.11).

k	$\|\Delta A\|_F$	$\widehat{\Theta}_{\max,k}$	$\widehat{\Theta}^{\text{lin}}_{\max,k}$	$\Theta_{\max,k}$
1	$2.86 \cdot 10^{-9}$	$2.31 \cdot 10^{-7}$	$1.15 \cdot 10^{-7}$	$2.26 \cdot 10^{-8}$
2	$2.86 \cdot 10^{-8}$	$2.31 \cdot 10^{-6}$	$1.15 \cdot 10^{-6}$	$2.26 \cdot 10^{-7}$
3	$2.86 \cdot 10^{-7}$	$2.31 \cdot 10^{-5}$	$1.15 \cdot 10^{-5}$	$2.26 \cdot 10^{-6}$
4	$2.86 \cdot 10^{-6}$	$2.33 \cdot 10^{-4}$	$1.15 \cdot 10^{-4}$	$2.26 \cdot 10^{-5}$
5	$2.86 \cdot 10^{-5}$	$2.58 \cdot 10^{-3}$	$1.15 \cdot 10^{-3}$	$2.26 \cdot 10^{-4}$

Consider now an example of ill-conditioned invariant subspace.

Example 4.12. Given is the matrix

$$A = \begin{bmatrix} 315 & 212 & 258 & -1520 & 3988 \\ -402 & -273 & -336 & 1964 & -5144 \\ 334 & 234 & 304 & -1736 & 4506 \\ -28 & -12 & 0 & 41 & -144 \\ -36 & -22 & -22 & 143 & -387 \end{bmatrix}$$

with a Jordan form

$$J = \left(\begin{bmatrix} -1 & 1 \\ 0 & -1 \end{bmatrix}, 0, \begin{bmatrix} 1 & 1 \\ 0 & 1 \end{bmatrix} \right).$$

In this case, we are interested in analyzing the sensitivity of the invariant subspace corresponding to the double eigenvalue -1 to the perturbation

$$\Delta A = 10^{-7+k} \cdot \begin{bmatrix} -5 & 9 & -7 & -4 & 2 \\ 8 & -7 & -6 & 3 & -8 \\ -4 & -2 & 8 & -6 & 5 \\ -7 & 4 & -9 & 5 & -4 \\ 9 & -5 & -3 & 6 & -8 \end{bmatrix},$$

where $k = 1, 2, 3$.

In Table 4.4, we give the linear (asymptotic) estimate $\widehat{\Theta}^{\text{lin}}_{\max,k}$ of the maximum angle between the unperturbed and perturbed invariant subspace and the value of the true angle $\Theta_{\max,k}$ as functions of the

Table 4.4. Invariant subspace sensitivity (Example 4.12).

k	$\|\Delta A\|_F$	$\widehat{\Theta}^{\text{lin}}_{\max,k}$ deg	$\Theta_{\max,k}$ deg
1	$3.07 \cdot 10^{-5}$	$9.84 \cdot 10^{0}$	$2.71 \cdot 10^{0}$
2	$3.07 \cdot 10^{-4}$	$6.00 \cdot 10^{1}$	$9.43 \cdot 10^{0}$
3	$3.07 \cdot 10^{-3}$	$8.67 \cdot 10^{1}$	$1.57 \cdot 10^{1}$

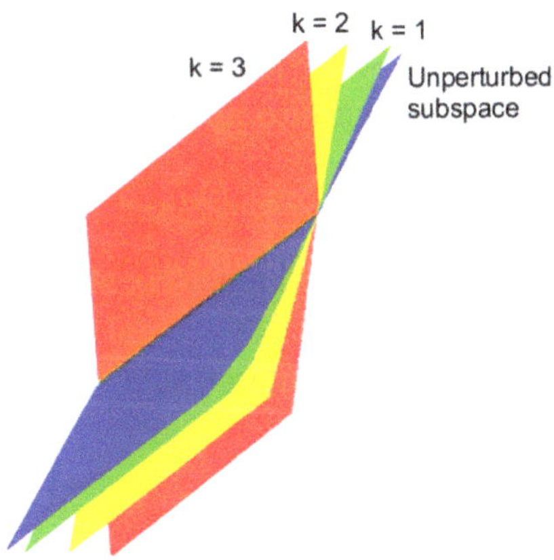

Fig. 4.7. Unperturbed and perturbed invariant subspaces for different values of k.

parameter k. In the given case, the deviations of the perturbed subspace from the unperturbed one are much larger than in the previous example due to the ill-conditioned eigensystem of the matrix A. Note that the linear estimate of the maximum angle between the two subspaces exceeds significantly the value of the true angle which is usual in estimating ill-conditioned problems. For this example, the global sensitivity estimate (4.26) is not valid since the perturbations do not satisfy the condition (4.25).

Parts of the perturbed invariant subspaces for different k are shown in Figure 4.7. The images of the subspace are obtained as described in Example 4.4.

The problem of the sensitivity analysis of an eigenvector represents a particular case for investigating the sensitivity of an invariant subspace with dimension $r = 1$, in which the desired eigenvalue λ is

Table 4.5. Estimates of the eigenvector sensitivity.

k	λ_i	$\text{sep}(\lambda_k, T_{22})$	$\widehat{\Theta}_{\max,k}$	$\Theta_{\max,k}$
1	2.0	$2.38 \cdot 10^{-1}$	$1.72 \cdot 10^{-6}$	$4.31 \cdot 10^{-7}$
2	1.5	$8.87 \cdot 10^{-2}$	$4.62 \cdot 10^{-6}$	$4.29 \cdot 10^{-7}$
3	0.2	$6.15 \cdot 10^{-2}$	$6.66 \cdot 10^{-6}$	$2.39 \cdot 10^{-6}$
4	0.1	$1.78 \cdot 10^{-2}$	$2.30 \cdot 10^{-5}$	$2.39 \cdot 10^{-6}$
5	1.0	$6.82 \cdot 10^{-2}$	$6.00 \cdot 10^{-6}$	$1.78 \cdot 10^{-7}$

in the position $(1, 1)$ of the Schur form T. In this case, one can again use the estimate (4.26), taking into account that

$$\text{sep}(T_{11}, T_{22}) = \sigma_{\min}(\lambda I_{n-r} - T_{22}).$$

Example 4.13. Consider again the matrix A and perturbation ΔA, given in Example 4.12. In Table 4.5, we give the values of $\text{sep}(\lambda_k, T_{22})$ for different values of λ_k, the estimate $\widehat{\Theta}_{\max,k}$ of the eigenvector sensitivity, as well as the actual change $\Theta_{\max,k}$.

4.3.3 *Notes and References*

The computation of the canonical angles between eigenspaces is considered in Stewart (2001), Chapter 4. Algorithm 4.2 for computing spectral projections by using the Schur form is proposed in Bai *et al.* (1993). The computation of sensitivity estimates of desired right invariant subspaces by Algorithm 4.4 is done in LAPACK by the routine `xGEESX`.

4.4 Reduction into Generalized Schur Form

In this section, we consider some numerical aspects of the reduction of a matrix pencil into generalized Schur form. We present briefly the QZ algorithm for reduction of a regular pencil to generalized Schur form. We also give a short description of the GUPTRI form of a singular pencil.

4.4.1 *QZ Algorithm*

In case of nonsymmetric matrices A and B, the generalized Schur form of a regular pencil $A - \lambda B$ is computed by the so-called QZ algorithm, which is a generalization of the QR algorithm for the ordinary eigenvalue problem. The first step of this algorithm is a unitary reduction of the pencil into a form in which A is upper Hessenberg matrix and B is triangular. This form of the pencil is called *Hessenberg-triangular form.* It is easy to show that if A is upper Hessenberg and B is upper triangular matrix, then AB^{-1} is upper Hessenberg. Thus the reduction into Hessenberg-triangular form corresponds to the reduction into Hessenberg form of the matrix AB^{-1} but it avoids the inversion of B. First, the matrix B is reduced to triangular matrix using Householder transformations, and after that A is reduced to Hessenberg form using plane rotations. This reduction and the accumulation of the transformations require approximately $8n^3$ flops.

Like the QR algorithm, the QZ algorithm is an iterative method that reduces a Hessenberg-triangular pencil into generalized Schur form. Let the pencil $A - \lambda B$ be in Hessenberg-triangular form and assume that B is nonsingular. Then the matrix $C = AB^{-1}$ is in Hessenberg form and its is possible to apply on it the QR algorithm. The idea of the QZ algorithm is to imitate a QR step on the matrix C by equivalence transformations applied separately on the matrices A and B. The QZ step requires about $13n^3$ flops, i.e., it is approximately 2.5 times more expensive than the QR step. After several iterations, the algorithms find unitary matrices Q and Z such that

$$Q^H(A - \lambda B)Z = S - \lambda T,$$

where S and T are upper triangular. The generalized eigenvalues are then found as

$$\lambda_i = s_{ii}/t_{ii}.$$

The QZ algorithm is numerically stable, the computed S and T satisfying

$$\widehat{S} - \lambda\widehat{T} = \widehat{Q}(A + E - \lambda(B + F))\widehat{Z},$$

where $\widehat{Q}$, $\widehat{Z}$ are almost unitary and the backward errors $\|E\|$ and $\|F\|$ are functions of the problem size n and the machine precision $\mathbf{u}$.

If the matrices A and B are real, then it is possible to use orthogonal instead of unitary transformations. In such a case, the matrix S may have 2×2 blocks on the diagonal corresponding to complex eigenvalues.

The reduction into generalized Schur form and the determination of the generalized eigenvalues in LAPACK is done by the routine `xGGES`. The desired ordering of the generalized eigenvalues on the diagonal of the pencil and the computing of condition numbers of eigenvalues and desired deflating subspaces are done by the routine `xGGESX`. In MATLAB®, the reduction into generalized Schur form and the computing of the generalized eigenvalues are done by the function `eig` and the desired ordering of the eigenvalues on the diagonal of the generalized Schur form is performed by the function `ordqz`.

The experiments with the QZ algorithm show that generally its convergence is slower than that of the QR algorithm. Taking into account the fact that one QZ step requires more than two times computations than one QR step, it is reasonable to conclude that the QZ algorithm is not efficient in solving the ordinary eigenvalue problem $Ax = \lambda I x$.

Example 4.14. Consider the regular pencil whose matrices are given in Example 3.19. Using appropriate unitary matrices U and V, the pencil $A - \lambda B$ is reduced into the generalized Schur form $S - \lambda T$, where

$$
\begin{aligned}
S &= U^H A V \\
&= \left[\begin{array}{ccc}
0.9075 + \jmath_o 0.0000 & -0.7065 + \jmath_o 0.1335 & 3.2611 + \jmath_o 5.6920 \\
 & 4.9658 + \jmath_o 0.0000 & -0.7364 - \jmath_o 2.1252 \\
 & & 1.8960 + \jmath_o 0.0000 \\
 & & \\
 & &
\end{array}\right. \cdots \rightarrow
\end{aligned}
$$

$$
\leftarrow\cdots\begin{array}{ccc}
2.3438-j_o2.7432 & -0.3381+j_o0.0000 & -2.2819+j_o0.0000\\
1.6795-j_o1.3500 & -7.0227-j_o1.3269 & -10.9992-j_o2.0782\\
-1.2106-j_o3.3845 & 0.6930-j_o1.2097 & -0.4914+j_o0.8577\\
1.5350-j_o0.0000 & 1.0486+j_o1.2273 & 1.6791+j_o1.9654\\
 & 0.7953+j_o0.0000 & 1.8012+j_o0.0000\\
 & & 0.7670+j_o0.0000
\end{array}\Bigg],
$$

$$
T=U^HBV
$$

$$
=\left[\begin{array}{ccc}
0.0000+j_o0.0000 & -0.6006+j_o0.1135 & 1.0594+j_o1.8491\\
 & 2.4829+j_o0.0000 & -0.0478-j_o0.1380\\
 & & 0.9480+j_o0.0000\\
 & & \\
 & & \\
 & &
\end{array}\right.\cdots\rightarrow
$$

$$
\leftarrow\cdots\begin{array}{ccc}
0.6511-j_o0.7620 & -6.1274+j_o0.0000 & -3.8032+j_o0.0000\\
1.1795-j_o0.9481 & 0.3067+j_o0.0579 & 1.9422+j_o0.3670\\
-0.4282-j_o1.1971 & 0.2140-j_o0.3736 & 0.2967-j_o0.5178\\
0.7675+j_o0.0000 & 0.7883+j_o0.9226 & 0.7883+j_o0.9226\\
 & 0.0000+j_o0.0000 & 0.0000+j_o0.0000\\
 & & 0.0000+j_o0.0000
\end{array}\Bigg].
$$

The eigenvalues of $A-\lambda B$ are determined as

$$
\begin{aligned}
\lambda_1 &= s_{11}/t_{11} = 0.9075+j_o0.0000/0.0000+j_o0.0000=\infty,\\
\lambda_2 &= s_{22}/t_{22} = 4.9658+j_o0.0000/2.4829+j_o0.0000\\
&= 2.0000+j_o0.0000,\\
\lambda_3 &= s_{33}/t_{33} = 1.8960+j_o0.0000/0.9480+j_o0.0000\\
&= 2.0000+j_o0.0000,\\
\lambda_4 &= s_{44}/t_{44} = 1.5350-j_o0.0000/0.7675+j_o0.0000\\
&= 2.0000+j_o0.0000,\\
\lambda_5 &= s_{55}/t_{55} = 0.7953+j_o0.0000/0.0000+j_o0.0000=\infty,\\
\lambda_6 &= s_{66}/t_{66} = 0.7670+j_o0.0000/0.0000+j_o0.0000=\infty.
\end{aligned}
$$

4.4.2 *Reduction into GUPTRI Form*

In case of a singular pencil $A - \lambda B$, the full information about its structure is provided by the Kronecker canonical form (Section 2.3.5). Unfortunately, the computation of the Kronecker form, like the determination of the Jordan form, is an ill-posed problem since infinitely small perturbations of the matrices A and B may lead to changes (even disappearing) of the Kronecker indices, changes of the number of finite and infinite eigenvalues and changes of the Jordan structure of the finite eigenvalues.

Example 4.15. Consider a singular pencil with matrices

$$A = \begin{bmatrix} 0 & 0 \\ 0 & 0 \end{bmatrix} \quad \text{and} \quad B = \begin{bmatrix} 1 & 0 \\ 0 & 0 \end{bmatrix}$$

so that $A - \lambda B$ has one finite eigenvalue at 0. Introducing small perturbations to get

$$\widetilde{A} = \begin{bmatrix} 0 & \varepsilon_1 \\ \varepsilon_2 & 0 \end{bmatrix} \quad \text{and} \quad \widetilde{B} = \begin{bmatrix} 1 & \varepsilon_3 \\ \varepsilon_4 & 0 \end{bmatrix},$$

the eigenvalues become $\varepsilon_1/\varepsilon_3$ and $\varepsilon_2/\varepsilon_4$ which can be arbitrary numbers.

Instead of using general equivalence transformations into Kronecker form that are numerically unstable, it is possible to use unitary equivalence transformations to transform further the generalized Schur form into a special form that reveals the Kronecker structure of the singular pencil. This form, which is called GUPTRI (Generalized UPper TRIangular), can be used to solve several problems that make use of the Kronecker structure of a pencil. In this section, we give a brief description of this form since its determination is closely related to the obtaining of the Jordan form of a single matrix.

Let A and B be rectangular complex $m \times n$ matrices. Then there exist unitary matrices $P \in \mathbb{C}^{m\times m}$ and $Q \in \mathbb{C}^{n\times n}$, such that

$$S = P^H AQ = \begin{bmatrix} S_r & \times & \times & \times & \times \\ & S_z & \times & \times & \times \\ & & S_f & \times & \times \\ & & & S_i & \times \\ & & & & S_l \end{bmatrix},$$

$$T = P^H BQ = \begin{bmatrix} T_r & \times & \times & \times & \times \\ & T_z & \times & \times & \times \\ & & T_f & \times & \times \\ & & & T_i & \times \\ & & & & T_l \end{bmatrix}. \tag{4.28}$$

The diagonal blocks of the matrices S and T describe the Kronecker structure of the pencil $A - \lambda B$ as follows:

$S_r - \lambda T_r$ contains all singular blocks of right minimal indices,
$S_z - \lambda T_z$ contains the Jordan structures for the zero eigenvalue,
$S_f - \lambda T_f$ contains blocks with finite, nonzero eigenvalues,
$S_i - \lambda T_i$ contains the Jordan structures for the infinite eigenvalues,
$S_l - \lambda T_l$ contains all singular blocks of left minimal indices.

The regular part of the pencil consists of the parts $A_z - \lambda B_z$, $A_f - \lambda B_f$ and $A_i - \lambda B_i$. The computed GUPTRI form exposes the Jordan structures of the zero eigenvalue ($A_z - \lambda B_z$) and the infinite eigenvalue ($A_i - \lambda B_i$). The singular part of the pencil consists of the right singular blocks $S_r - \lambda T_r$ and left singular blocks $S_l - \lambda T_l$. In addition, the part containing the right singular blocks and the zero eigenvalue blocks as well as the part containing the left singular blocks and the infinite eigenvalue blocks are reduced into the so-called *staircase form*, whose role in finding the Jordan form of a matrix is described further on in Chapter 6. This additional reduction allows finding the right and left minimal indices of the pencil. Thus, the GUPTRI form provides the whole information about the structure of the pencil contained in the Kronecker canonical form being obtained at the same time by using numerically stable equivalence transformations.

Example 4.16. Consider the singular 10×10 pencil with matrices

$$A = \begin{bmatrix} -1 & 1 & -1 & -1 & -1 & 1 & -2 & -2 & -1 & -4 \\ -2 & 0 & 0 & -1 & -1 & 2 & -3 & -5 & 0 & -8 \\ 0 & 0 & 1 & 0 & 1 & -1 & 0 & 0 & 1 & 1 \\ 1 & 0 & -1 & 1 & 0 & -1 & 2 & 3 & -1 & 5 \\ -1 & 0 & 0 & -1 & 1 & 1 & -2 & -3 & 1 & -5 \\ -1 & 0 & 0 & -1 & 0 & 1 & -2 & -3 & 0 & -5 \\ 2 & -1 & 0 & 2 & 0 & 1 & 0 & -1 & 0 & 3 \\ -1 & 3 & -1 & -1 & -1 & 1 & 1 & 0 & -1 & -2 \\ 3 & -4 & 3 & 2 & 3 & -4 & 2 & 4 & 3 & 10 \\ 1 & 1 & 0 & 0 & 0 & -3 & 5 & 9 & -1 & 9 \end{bmatrix}$$

and

$$B = \begin{bmatrix} 0 & 0 & 0 & -1 & 0 & 1 & -1 & -2 & 0 & -3 \\ -1 & 1 & 0 & -1 & 1 & 1 & -2 & -2 & 1 & -4 \\ 1 & 0 & 0 & 2 & 0 & -1 & 3 & 4 & 0 & 7 \\ 0 & -1 & 0 & 0 & 0 & 0 & 0 & -1 & 0 & -1 \\ 0 & 0 & 0 & 0 & 0 & 1 & 0 & 0 & 0 & -1 \\ 0 & 0 & 0 & 1 & 0 & 0 & 1 & 1 & 0 & 2 \\ -1 & 0 & 0 & -1 & 0 & 0 & -1 & -1 & 0 & -3 \\ -1 & -1 & 0 & -1 & 0 & 1 & -3 & -4 & 0 & -6 \\ 2 & 1 & 0 & 3 & -1 & -3 & 6 & 9 & -1 & 14 \\ 2 & -1 & 0 & 1 & -1 & -2 & 2 & 2 & -1 & 6 \end{bmatrix},$$

given in Example 2.24.

The GUPTRI form of this pencil, computed by the function `guptri` from the Matrix Canonical Structure Toolbox

(MCS Toolbox) (MCST), can be represented schematically as

$$
S - \lambda T = \begin{array}{c} \\ 1 \\ 2 \\ 3 \\ 4 \\ 5 \\ 6 \\ 7 \\ 8 \\ 9 \\ 10 \end{array}
\begin{array}{c}
\begin{array}{ccc|cc|cc|ccc} 1 & 2 & 3 & 4 & 5 & 6 & 7 & 8 & 9 & 10 \end{array} \\
\left[\begin{array}{ccc|cc|cc|ccc}
\times & \times & \times & \times & \times & \times & \times & \times & \times & \times \\
\times & \times & \times & \times & \times & \times & \times & \times & \times & \times \\
\hline
 & & & \times & \times & \times & \times & \times & \times & \times \\
 & & & \times & \times & \times & \times & \times & \times & \times \\
\hline
 & & & & & \times & \times & \times & \times & \times \\
 & & & & & \times & \times & \times & \times & \times \\
\hline
 & & & & & & & \times & \times & \times \\
 & & & & & & & \times & \times & \times \\
 & & & & & & & \times & \times & \times \\
 & & & & & & & \times & \times & \times
\end{array}\right]
\end{array}
$$

and the corresponding blocks of the matrices S and T are given in the following.

Right singular blocks of dimension 2×3*:*

$$
\begin{aligned}
S_r &= S_{1:2,1:3} \\
&= \begin{bmatrix} -1.224744871391593 & -0.000000000000005 & -0.000000000000003 \\ 0 & 0.632455532033678 & -0.976187060183959 \end{bmatrix},
\end{aligned}
$$

$$
T_r = T_{1:2,1:3} = \begin{bmatrix} 0 & -0.774596669241481 & 1.195580095479128 \\ 0 & 0 & 0.542326144546643 \end{bmatrix}.
$$

Blocks with finite eigenvalues of order 2:

$$
S_f = S_{3:4,4:5} = \begin{bmatrix} -3.526900110613281 & 2.944840695329253 \\ 0.000000000000003 & -4.295346318982893 \end{bmatrix},
$$

$$
T_f = T_{3:4,4:5} = \begin{bmatrix} 1.175633370204427 & -0.478536612990997 \\ 0 & 1.431782106327628 \end{bmatrix}.
$$

Blocks with infinite eigenvalues of order 2:

$$
S_i = S_{5:6,6:7} = \begin{bmatrix} 1.390443574307605 & 1.411773157513567 \\ 0 & 3.034286827530312 \end{bmatrix},
$$

$$
T_i = T_{5:6,6:7} = \begin{bmatrix} 0 & 2.408318915758422 \\ 0 & 0 \end{bmatrix}.
$$

Left singular blocks of dimension 4×3:

$$S_l = S_{7:10,8:10}$$
$$= \begin{bmatrix} -1.059997880006355 & 0.107503406194865 & 1.374323122465678 \\ 0 & 1.146423008442220 & -0.652753365768232 \\ 0 & 0 & -1.293797914466667 \\ 0 & 0 & 0 \end{bmatrix},$$

$$T_l = T_{7:10,8:10}$$
$$= \begin{bmatrix} 0.090856961143406 & 0.376261921682043 & 0.847579068558971 \\ 0.968904283303602 & -0.348911350395462 & -2.384548009643092 \\ 0 & -0.691564074808127 & 0.352853976672736 \\ 0 & 0 & 0.797724035217477 \end{bmatrix}.$$

In the given case, the block $S_z - \lambda T_z$ that corresponds to zero eigenvalue is missing. The finite eigenvalue -3 is participating in a Jordan block of order 2.

4.4.3 *Notes and References*

4.4.3.1 *Reduction into Generalized Schur Form*

The QZ algorithm for reducing a matrix pencil to triangular form is proposed by Moler and Stewart in Moler and Stewart (1973). Some aspects of the properties of this algorithm are commented in Mastronardi and Van Dooren (2020). Algorithms for reordering the generalized Schur form are proposed in Van Dooren (1982), Kågström (1993) and Kressner (2006). The algorithms for reduction into generalized Schur form with condition and accuracy estimates implemented in the LAPACK routines `xGEES` and `DGEESX` are described in Kågström and Poromaa (1996a, 1996b).

The numerical aspects of finding the Weierstrass canonical form of a regular matrix pencil are discussed in Kalogeropoulos *et al.* (2009) and Malyshev (1989). The numerical solution of the generalized eigenvalue problem in case of a symmetric matrix A and positive definite matrix B is considered, for instance, in Peters and Wilkinson (1970), Wilkinson and Reinsch (1971) and Parlett (1991) and the state of the art programs for solving this problem can be found in LAPACK (Anderson *et al.*, 1999).

Algorithms for computing the generalized eigenvalues and deflating subspaces for Hamiltonian eigenvalue problems are presented in Benner *et al.* (1998, 2002).

4.4.3.2 *GUPTRI Form*

The numerical problems related to the determination of the Kronecker canonical form of a pencil and its application for solving several problems are discussed for the first time in Wilkinson (1979) and Van Dooren (1981). An earlier numerical algorithm for finding this form is presented in Jordan and Godbout (1977). A generalized Schur-staircase form of a singular pencil was developed by Van Dooren in Van Dooren (1979) based on the idea of orthogonal deflation proposed by Kublanovskaya in Kublanovskaya (1966). This algorithm was further improved in Beelen and Van Dooren (1988) and Mastronardi and Van Dooren (2017, 2020). A detailed treatment of different algorithms for finding the Kronecker structure of a pencil can be found in the works of Kublanovskaya (1984, 1989) and her co-researchers Khazanov and Kublanovskaya (1988a, 1988b) and Belyi *et al.* (1988). The algorithm for reducing a pencil to GUPTRI form is developed by Demmel and Kågström (1986, 1987, 1993a, 1993b) based on the RGSVD algorithm of Kågström (1986, 1993). A software implementation of this algorithm is available from MCST.

Chapter 5

Geometry of Jordan Forms

In this chapter, we present the notion of numerical Jordan form of a matrix. In contrast to the "theoretical" Jordan form, the numerical Jordan form preserves its structure when the matrix is subject to small perturbations. This makes the numerical Jordan form a computational object that can be determined reliably in the presence of rounding errors. To find the Jordan structure, it is necessary to know the geometry of the space of matrices and the stratification of the near Jordan forms in this space. This important subject is discussed in simplified form in sections from 5.1 to 5.4. In Section 5.5, we describe the determination of the Jordan form as a solution of an ill-posed numerical problem. A rigorous definition of the numerical Jordan form is given in Section 5.6 and a practical definition, appropriate for computations, is presented in Section 5.7.

5.1 Geometry of Matrix Space

5.1.1 *Orbits and Bundles of Matrices*

The complex square matrix A of dimension $n \times n$ is represented as a point in the complex n^2-dimensional linear space $\mathbb{C}^{n\times n}$. This linear space is called the *matrix space*. The set of all matrices similar to A forms a *manifold* in $\mathbb{C}^{n\times n}$. This manifold, defined formally as

$$\mathcal{O}(A) = \{PAP^{-1} : \det P \neq 0\}, \tag{5.1}$$

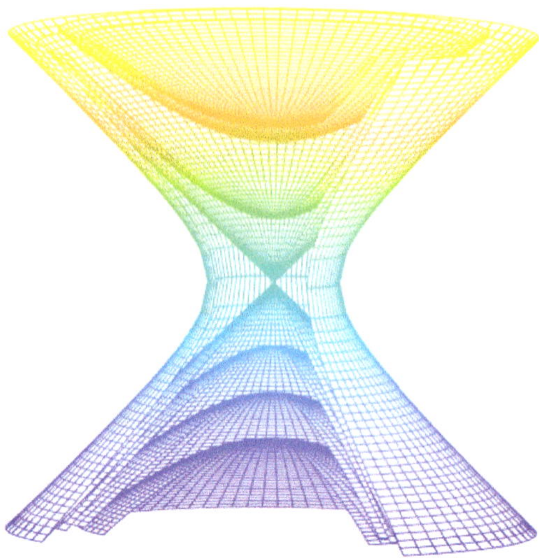

Fig. 5.1. Manifolds in the three-dimensional space.

is called the *orbit* of A. All matrices lying in the same orbit have the same eigenvalues and the same dimensions of the Jordan blocks.

To give an idea of an manifold in a linear space, we represent in Figure 5.1 a division of the three-dimensional space Euclidean space into the strata of manifolds satisfying the equation $x^2 + y^2 - z^2 = C$ for different values of C. Note that the different strata do not intersect with each other.

A *submanifold* of the manifold $\mathcal{M}$ is a subset $\mathcal{S}_{\mathcal{M}}$ which itself has the structure of a manifold. For instance, the sphere in $\mathbb{R}^n$, given by the equation $x^2 + y^2 + z^2 = 1$, represents a submanifold of Euclidean space.

The *bundle* $\mathcal{B}(A)$ of the matrix A is the union of all orbits $\mathcal{O}(A)$. It consists of all matrices with equal Jordan forms which differ only by their eigenvalues but have the same number of distinct eigenvalues and the same dimensions of the corresponding Jordan blocks (Segre characteristics). If two matrices have the same Jordan form except that the distinct eigenvalues are different, then they are in the same bundle. For instance, all diagonal matrices with simple eigenvalues form a bundle. Each bundle is a manifold in the space of matrices whose stratum is an orbit. In a given orbit, only those matrices from the bundle that have the same eigenvalues participate. As an example, consider two matrices with Jordan forms

$$J_1 = \left[\begin{array}{cc|c|c} -2 & 1 & & \\ & -2 & & \\ \hline & & -1 & \\ \hline & & & 6 \end{array}\right], \quad J_2 = \left[\begin{array}{cc|c|c} 5 & 1 & & \\ & 5 & & \\ \hline & & 3 & \\ \hline & & & -4 \end{array}\right]$$

which lie in different orbits but are in the same bundle.

Let $\mathcal{O}(A)$ be the orbit of A. The set of vectors that are tangent to $\mathcal{O}(A)$ at the point x has a structure of a linear space. This linear space is called *tangent space* to $\mathcal{O}(A)$ at x and is denoted by $T_{\mathcal{O}}(x)$. The orthogonal complement to $T_{\mathcal{O}}(x)$ is the *normal space* $N_{\mathcal{O}}(x)$. For instance, if $\mathcal{O}(A)$ is a manifold in the space $\mathbb{C}^3$, then $T_{\mathcal{O}}(x)$ can be represented as a plane that passes at x and $N_{\mathcal{O}}(x)$ is the straight line orthogonal to this plane at the point x (Figure 5.2). Note that the tangent space $T_{\mathcal{O}}(x)$ to $\mathcal{O}(A) \subset \mathbb{C}^{n\times n}$ at $x \in \mathcal{O}(A)$ is a linear subspace of $\mathbb{C}^{n\times n}$, so it must contain 0, but we usually draw the tangent plane passing through the point x instead of through 0.

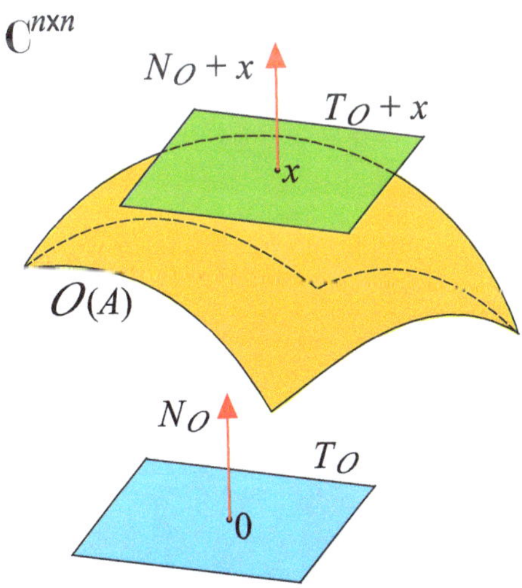

Fig. 5.2. Tangent and normal space of an orbit.

(The same is true for $N_{\mathcal{O}}(x)$.) To get the usual picture of the tangent plane and the normal space, it is necessary to add x to every element of $T_{\mathcal{O}}(0)$ and $N_{\mathcal{O}}(0)$.

The main quantity characterizing the orbits and bundles is their dimension. The dimension of the orbit, $\dim(\mathcal{O}(A))$, is equal to the dimension of its tangent space $T_{\mathcal{O}}(x)$. Usually, it is more convenient to work with the *codimension* of $\mathcal{O}(A)$ denoted by $\operatorname{codim}(\mathcal{O}(A))$, which is equal to the dimension of the normal space $N_{\mathcal{O}}(x)$. Since

$$T_{\mathcal{O}}(x) + N_{\mathcal{O}}(x) = \mathbb{C}^{n\times n},$$

one has that

$$\dim(\mathcal{O}(A)) + \operatorname{codim}(\mathcal{O}(A)) = n^2.$$

Two manifolds $\mathcal{M}$ and $\mathcal{N}$ put in the ambient space $\mathcal{S}$ intersect *transversally*, if at every point x of the intersection their separate tangent spaces $T_{\mathcal{M}}(x)$ and $T_{\mathcal{N}}(x)$ at that point generate the tangent space of the ambient manifold at the same point, i.e.,

$$T_{\mathcal{M}}(x) + T_{\mathcal{N}}(x) = T_{\mathcal{S}}(x),$$

(Figure 5.3). If the manifolds are of complementary dimensions (i.e., their dimensions add up to the dimension of the ambient space), the condition means that the tangent space to the ambient manifold is the direct sum of the two smaller tangent spaces.

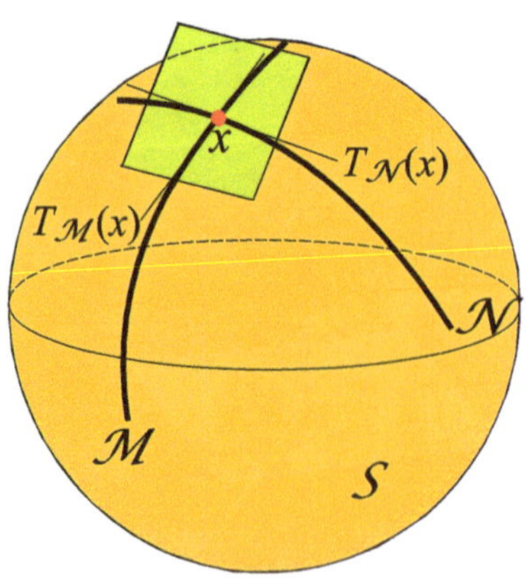

Fig. 5.3. Transverse intersection of two manifolds.

The term frequently used synonymously for transversal is *general position*, i.e., two manifolds which intersect transversally are said to be in general position.

Two linear subspaces $\mathcal{X}$ and $\mathcal{Y}$ of a linear space $\mathcal{L}$ are transverse, if their sum is equal to the whole space, $\mathcal{X} + \mathcal{Y} = \mathcal{L}$. For instance, two planes intersecting at nonzero angles in $\mathbb{R}^3$ are transverse (Figure 5.4).

In three-dimensional space, transverse curves do not intersect (Figure 5.5). Curves transverse to surfaces intersect in points (Figure 5.5), and surfaces transverse to each other intersect in curves. Note that two perpendicular planes in $\mathbb{R}^3$ are transverse, but two perpendicular lines in one and the same plane are not (Figure 5.6). Curves that are tangent to a surface at a point (for instance, curves lying on

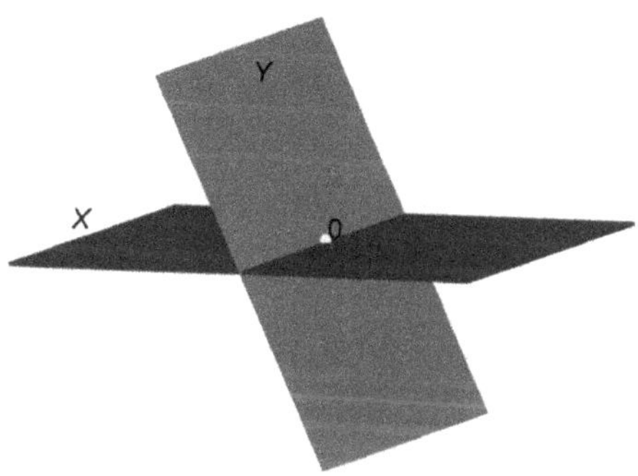

Fig. 5.4. Transverse planes in $\mathbb{R}^3$.

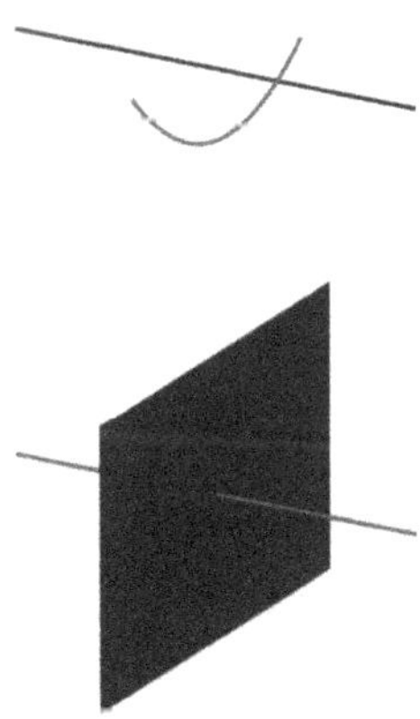

Fig. 5.5. Intersections which are transverse.

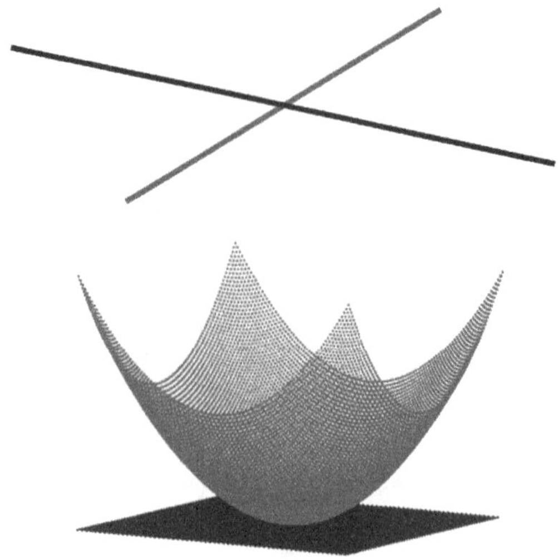

Fig. 5.6. Intersections which are not transverse.

a surface) do not intersect the surface transversally. The same is true for planes that are tangent to a surface at a point (Figure 5.6). If an intersection is transverse, then it is a submanifold whose codimension is equal to the sums of the codimensions of the two manifolds.

It should be emphasized that the transverse intersection is the general case of an intersection of two manifolds, i.e., the transversality is a "typical property".

5.1.2 *Determining the Orbit and Bundle Codimension*

5.1.2.1 *Centralizer Codimension*

Consider how to determine the dimension of the normal space $N_{\mathcal{O}}$, i.e., the codimension of the orbit $\mathcal{O}(A)$. For this aim, we need the following definition. A *centralizer* or *commutator* of the matrix A is called the set of all matrices X commuting with A,

$$Z_A = \{X : [A, X] = 0\}, \quad [A, X] = AX - XA.$$

The centralizer of each nth-order matrix A is a linear subspace of the space $\mathbb{C}^{n\times n}$ of all matrices.

Denoting $\mathcal{L}_A X = [A, X]$, where $\mathcal{L}_A : \mathbb{C}^{n\times n} \to \mathbb{C}^{n\times n}$ is a linear transformation, we have that $Z_A = \{X : \mathcal{L}_A X = 0\}$ which implies

$$Z_A = \mathcal{N}(\mathcal{L}_A).$$

Consider the nonlinear transformation ψ of the set of nonsingular matrices into the space $\mathbb{C}^{n\times n}$ of all matrices, defined by the expression $\psi(P, A) = PAP^{-1}$, and assume that $|X|$ is sufficiently small. Then $P = I + X$ is a nonsingular matrix, and for $|X| \to 0$, we have that $P^{-1} = I - X + O(|X|^2)$. Thus

$$\begin{aligned}\psi(I+X, A) &= (I+X)A(I+X)^{-1}\\ &= A + XA - AX + O(|X|^2)\\ &= \psi(I, A) + \mathcal{L}_A X + O(|X|^2).\end{aligned} \tag{5.2}$$

The tangent vectors to the orbit $\mathcal{O}(A)$ are given by $\psi_* X$, where ψ_* is the derivative of $\psi(P, A)$ in respect to X. It follows from (5.2) that $\psi_* X = \mathcal{L}_A X = AX - XA$. The value of ψ_* at the point $P = I$ is equal to $\mathcal{L}_A$ and hence

$$T_{\mathcal{O}} = \mathcal{R}(\mathcal{L}_A).$$

Since $\psi_* = \mathcal{L}_A$ is a linear transformation of $\mathbb{C}^{n\times n}$ into $\mathbb{C}^{n\times n}$, we obtain that

$$\begin{aligned}\dim(\mathbb{C}^{n\times n}) &= \dim(\mathcal{R}(\mathcal{L}_A)) + \dim(\mathcal{N}(\mathcal{L}_A))\\ &= \dim(T_{\mathcal{O}}) + \dim(Z_A)\\ &= \dim(\mathcal{O}(A)) + \dim(Z_A).\end{aligned}$$

Thus we come to the important result that

$$\dim(Z_A) = \dim(\mathbb{C}^{n\times n}) - \dim(\mathcal{O}(A)) = \operatorname{codim}(\mathcal{O}(A)), \tag{5.3}$$

i.e., the dimension of the centralizer of the matrix A is equal to the dimension of the normal space $N_{\mathcal{O}}$, which in turn is equal to the codimension of the orbit.

The tangent vectors $T = \psi_* X = \mathcal{L}_A X \in \mathbb{C}^{n\times n}$ to the orbit of the matrix A are those matrices that belong to the subspace $T_{\mathcal{O}} = \mathcal{R}(\mathcal{L}_A)$. Using the columnwise vector representations $\operatorname{vec}(T)$

and vec(X) of T and X, respectively, the product $\mathcal{L}_A X \in \mathbb{C}^{n\times n}$ can be represented in $\mathbb{C}^{n^2}$ as (see Example A.2 in the Appendix)

$$\text{vec}(T) = L_A \text{vec}(X),$$

where $L_A = I_n \otimes A - A^T \otimes I_n \in \mathbb{C}^{n^2 \times n^2}$ is the matrix representation of the operator $\mathcal{L}_A$ in $\mathbb{C}^{n^2}$. This shows that the tangent space $T_{\mathcal{O}}$ of the orbit $\mathcal{O}(A)$ in $\mathbb{C}^{n\times n}$ is isomorphic to the image of the matrix L_A in $\mathbb{C}^{n^2}$ and

$$\dim(\mathcal{O}(A)) = \text{rank}(L_A).$$

Since the rank of $L_A = I_n \otimes A - A^T \otimes I_n$ remains unchanged under nonsingular transformations of A, we have that

$$\dim(\mathcal{O}(A)) = \text{rank}(I_n \otimes J - J^T \otimes I_n),$$

where $J = Z^{-1}AZ$ is the Jordan form of A. Exploiting the equality

$$\text{rank}(L_A) + \text{null}(L_A) = n^2$$

and taking into account (5.3), we obtain that

$$\text{codim}(\mathcal{O}(A)) = \text{null}(I_n \otimes J - J^T \otimes I_n). \tag{5.4}$$

Equation (5.4) shows that it is possible to determine the codimension of orbits and bundles of matrices from their Jordan forms. This can be done by determining all matrices X that commute with the matrix A, i.e., by finding the general solution of the equation $AX = XA$ using the Jordan form of A.

5.1.2.2 *Solution of the Matrix Equation $AX = XA$ (Gantmacher, 1959a, Chapter VIII)*

Let the matrix $A \in \mathbb{C}^{n\times n}$ have p distinct eigenvalues $\lambda_1, \lambda_2, \ldots, \lambda_p$ with multiplicities $\nu_1, \nu_2, \ldots, \nu_p$, respectively. With each λ_i is associated the Segre characteristic $\mathcal{S}_i = \{s_{1i}, s_{2i}, \ldots, s_{q_i,i}\}$. Suppose A is reduced into the Jordan form

$$J = Z^{-1}AZ = \text{diag}(J(\mathcal{S}_1, \lambda_1), J(\mathcal{S}_2, \lambda_2), \ldots, J(\mathcal{S}_p, \lambda_p)),$$
$$J(\mathcal{S}_i, \lambda_i) = \text{diag}(J_{s_{1i}}(\lambda_i), J_{s_{2i}}(\lambda_i), \ldots, J_{s_{q_i,i}}(\lambda_i)),$$

where $J(\mathcal{S}_i, \lambda_i) \in \mathbb{C}^{\nu_i \times \nu_i}$, $i = 1, 2, \ldots, p$, and each Jordan block $J_{s_{ki}}(\lambda_i) \in \mathbb{C}^{s_{ki} \times s_{ki}}$, $i = 1, 2, \ldots, p$; $k = 1, 2, \ldots, q_i$, is given by

$$J_{s_{ki}}(\lambda_i) = \begin{bmatrix} \lambda_i & 1 & 0 & \ldots & 0 & 0 \\ 0 & \lambda_i & 1 & \ldots & 0 & 0 \\ 0 & 0 & \lambda_i & \ldots & 0 & 0 \\ \vdots & \vdots & \vdots & \ddots & \vdots & \vdots \\ 0 & 0 & 0 & \ldots & \lambda_i & 1 \\ 0 & 0 & 0 & \ldots & 0 & \lambda_i \end{bmatrix}. \tag{5.5}$$

Note that $\sum_{k=1}^{q_i} s_{ki} = \nu_i$ and the total number of Jordan blocks is equal to $r = q_1 + q_2 + \cdots + q_p$.

Transforming the solution matrix X into $Y = Z^{-1}XZ$, the equation $AX = XA$ is written in the form

$$JY = YJ. \tag{5.6}$$

Partitioning the transformed solution Y according to the Segre characteristics of $\lambda_1, \lambda_2, \ldots, \lambda_p$ as

$$\begin{aligned} Y = [Y_{k\ell}^{(ij)}], \quad & Y_{k\ell}^{(ij)} \in \mathbb{C}^{s_{ki} \times s_{\ell j}}, \\ i = 1, 2, \ldots, p; \quad & k = 1, 2, \ldots, q_i, \\ j = 1, 2, \ldots, p; \quad & \ell = 1, 2, \ldots, q_j, \end{aligned}$$

equation (5.6) is decomposed into r^2 equations

$$\begin{aligned} (\lambda_i I_{s_{ki}} + N_{s_{ki}}) Y_{k\ell}^{(ij)} = Y_{k\ell}^{(ij)} & (\lambda_j I_{s_{\ell j}} + N_{s_{\ell j}}), \\ i = 1, 2, \ldots, p; \quad & k = 1, 2, \ldots, q_i; \\ j = 1, 2, \ldots, p; \quad & \ell = 1, 2, \ldots, q_j, \end{aligned} \tag{5.7}$$

where

$$N_{s_{ki}} = \begin{bmatrix} 0 & 1 & 0 & \ldots & 0 \\ 0 & 0 & 1 & \ldots & 0 \\ \vdots & \vdots & \vdots & \ddots & \vdots \\ 0 & 0 & 0 & \ldots & 1 \\ 0 & 0 & 0 & \ldots & 0 \end{bmatrix} \in \mathbb{C}^{s_{ki} \times s_{ki}},$$

$$N_{s_{\ell j}} = \begin{bmatrix} 0 & 1 & 0 & \dots & 0 \\ 0 & 0 & 1 & \dots & 0 \\ \vdots & \vdots & \vdots & \ddots & \vdots \\ 0 & 0 & 0 & \dots & 1 \\ 0 & 0 & 0 & \dots & 0 \end{bmatrix} \in \mathbb{C}^{s_{\ell j} \times s_{\ell j}}$$

are nilpotent matrices.

Equation (5.7) can be rewritten as

$$(\lambda_j - \lambda_i) Y_{k\ell}^{(ij)} = N_{s_{ki}} Y_{k\ell}^{(ij)} - Y_{k\ell}^{(ij)} N_{s_{\ell j}}, \qquad (5.8)$$
$$i = 1, 2, \dots, p; \quad k = 1, 2, \dots, q_i;$$
$$j = 1, 2, \dots, p; \quad \ell = 1, 2, \dots, q_j.$$

In the solution of (5.8), there are two possible cases:

(1) $\lambda_i \neq \lambda_j$. Using the fact that $N_{s_{ki}}$ and $N_{s_{\ell j}}$ are nilpotent matrices, we can prove that in this case $Y_{k\ell}^{(ij)} = 0$.
(2) $\lambda_i = \lambda_j$. In this case, we have that

$$N_{s_{ki}} Y_{k\ell}^{(ij)} = Y_{k\ell}^{(ij)} N_{s_{\ell j}}, \qquad (5.9)$$
$$i = 1, 2, \dots, p; \quad k = 1, 2, \dots, q_i;$$
$$j = 1, 2, \dots, p; \quad \ell = 1, 2, \dots, q_j.$$

Depending on the values of the Jordan block sizes s_{ki} and $s_{\ell j}$, (5.9) may have one of the following solutions:

$$Y_{k\ell}^{(ij)} = \begin{bmatrix} a & b & c & d \\ 0 & a & b & c \\ 0 & 0 & a & b \\ 0 & 0 & 0 & a \end{bmatrix}, \qquad Y_{k\ell}^{(ij)} = \begin{bmatrix} 0 & 0 & a & b & c \\ 0 & 0 & 0 & a & b \\ 0 & 0 & 0 & 0 & a \end{bmatrix},$$

$(s_{ki} = 4,\ s_{\ell j} = 4)$ $\qquad$ $(s_{ki} = 3,\ s_{\ell j} = 5)$

$$Y_{k\ell}^{(ij)} = \begin{bmatrix} a & b & c \\ 0 & a & b \\ 0 & 0 & a \\ 0 & 0 & 0 \\ 0 & 0 & 0 \end{bmatrix}, \qquad (5.10)$$
$$(s_{ki} = 5,\ s_{\ell j} = 3)$$

where a, b, c, d are arbitrary parameters.

In this way, the solution Y of (5.6) consists of blocks $Y_{k\ell}^{(ij)}$ which are zero matrices if $\lambda_i \neq \lambda_j$ and are matrices in the form (5.10), if $\lambda_i = \lambda_j$. Due to the structure of the nontrivial blocks, the solution corresponding to the blocks associated with a specific λ_i has the form shown in Figure 5.7, where each oblique segment denotes a sequence of identical numbers and the blank entries denote zero. As a result, the solution Y is a block-diagonal matrix where each block corresponding to λ_i has a dimension $\nu_i \times \nu_i$ and a form given in Figure 5.7.

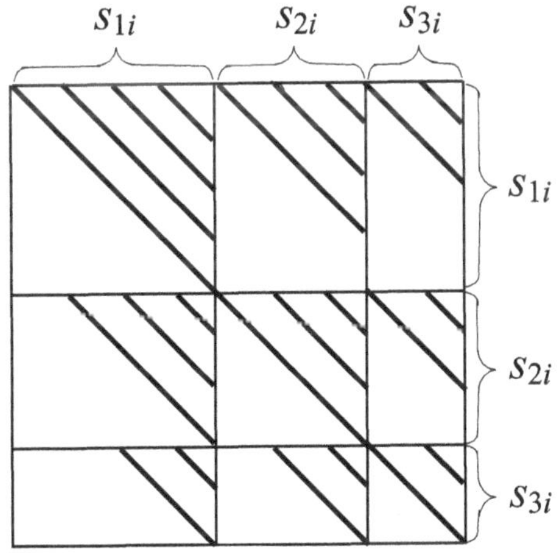

Fig. 5.7. Solution of the equation $JY = YJ$ corresponding to the eigenvalue λ_i.

Example 5.1. Consider the equation $AX = XA$ with the tenth-order matrix

$$A = \begin{bmatrix} 1 & -3 & 5 & -7 & 1 & -4 & 3 & 12 & 3 & -18 \\ 10 & 15 & -7 & 18 & -10 & 7 & 12 & -17 & 2 & 59 \\ 4 & 3 & 4 & 2 & -4 & 0 & 8 & 4 & 3 & 10 \\ 7 & 10 & -6 & 17 & -7 & 6 & 7 & -16 & 0 & 47 \\ -12 & -13 & 2 & -14 & 15 & -4 & -21 & 5 & -7 & -54 \\ -3 & -3 & 0 & -3 & 4 & 3 & -9 & -1 & -5 & -14 \\ 8 & 10 & -1 & 10 & -10 & 1 & 22 & -3 & 9 & 43 \\ -12 & -13 & -1 & -12 & 14 & -1 & -28 & 1 & -12 & -53 \\ -3 & -4 & 0 & -3 & 4 & 1 & -9 & 0 & -3 & -16 \\ -11 & -13 & 3 & -15 & 12 & -4 & -20 & 8 & -7 & -54 \end{bmatrix},$$

given in Example 2.19. The matrix A has three distinct eigenvalues

$$\lambda_1 = -1; \quad \lambda_2 = 2; \quad \lambda_3 = 3$$

with multiplicities $\nu_1 = 1$, $\nu_2 = 5$ and $\nu_3 = 4$, respectively, and associated Segre characteristics

$$\begin{aligned} \mathcal{S}_1 &= \{s_{11}\} = \{1\}, \\ \mathcal{S}_2 &= \{s_{12}, s_{22}\} = \{3, 2\}, \\ \mathcal{S}_3 &= \{s_{13}, s_{23}\} = \{2, 2\}. \end{aligned}$$

In the given case, $q_1 = 1$, $q_2 = 2$, $q_3 = 2$, and the transformed solution matrix looks as

$$Y = \left[\begin{array}{c|ccc|cc|cc|cc} a & & & & & & & & & \\ \hline & b & c & d & e & f & & & & \\ & 0 & b & c & 0 & e & & & & \\ & 0 & 0 & b & 0 & 0 & & & & \\ \hline & 0 & g & h & m & q & & & & \\ & 0 & 0 & g & 0 & m & & & & \\ \hline & & & & & & r & s & t & u \\ & & & & & & 0 & r & 0 & t \\ \hline & & & & & & v & w & y & z \\ & & & & & & 0 & v & 0 & y \end{array}\right],$$

where $a, b, c, d, e, f, g, h, m, q, r, s, t, u, v, w, y, z$ are arbitrary parameters. Note that the number of these parameters is equal to 18.

It is possible to prove that the number of free parameters in the blocks associated with λ_i is

$$D_i = s_{1i} + 3s_{2i} + 5s_{3i} + \cdots + (2q_i - 1)s_{q_i,i}.$$

Summing up these numbers over all the distinct eigenvalues, we obtain that the number of free parameters in the solution of $AX = XB$ is equal to

$$D = \sum_{i=1}^{p} (s_{1i} + 3s_{2i} + 5s_{3i} + \cdots + (2q_i - 1)s_{q_i,i}). \tag{5.11}$$

For the matrix A of Example 5.1, we have that

$$D = s_{11} + s_{12} + 3s_{22} + s_{13} + 3s_{23} = 18.$$

It is easy to show that the number D is equal to the number of linearly independent matrices commuting with A. In fact, the solution matrix Y can be represented as

$$Y = \sum_{k}^{D} c_k Y_k,$$

where $c_1, c_2, \ldots, c_D$ are arbitrary parameters and the matrix Y_k is obtained using expressions of the type (5.10) giving to the parameter c_k the value 1 and to the remaining parameters the value 0. The matrices Y_k are linearly independent, since otherwise for certain values of the free parameters, not all zero, the matrix Y and hence X would be the zero matrix, which is impossible. For instance, the solution matrix $Y_{k\ell}^{(ij)}$ in the case of $s_{kj} = 4$, $s_{\ell j} = 4$ can be represented as

$$Y_{k\ell}^{(ij)} = a \begin{bmatrix} 1 & 0 & 0 & 0 \\ 0 & 1 & 0 & 0 \\ 0 & 0 & 1 & 0 \\ 0 & 0 & 0 & 1 \end{bmatrix} + b \begin{bmatrix} 0 & 1 & 0 & 0 \\ 0 & 0 & 1 & 0 \\ 0 & 0 & 0 & 1 \\ 0 & 0 & 0 & 0 \end{bmatrix} + c \begin{bmatrix} 0 & 0 & 1 & 0 \\ 0 & 0 & 0 & 1 \\ 0 & 0 & 0 & 0 \\ 0 & 0 & 0 & 0 \end{bmatrix}$$

$$+ d \begin{bmatrix} 0 & 0 & 0 & 1 \\ 0 & 0 & 0 & 0 \\ 0 & 0 & 0 & 0 \\ 0 & 0 & 0 & 0 \end{bmatrix}.$$

Clearly, the four matrices participating in the linear combination are linearly independent and $D = 4$.

Previously it was shown that the dimension of the centralizer of A,

$$Z_A = \{X : [A, X] = 0\}, \quad [A, X] = AX - XA,$$

is equal to the codimension of the orbit of A. Since the centralizer is a subspace in the space $\mathbb{C}^{n\times n}$ of all matrices, its dimension is equal to the number of the linearly independent matrices commuting with A, i.e., to the number D of free elements in the solution of the equation $AX = XA$. Hence, we can state the following result.

Theorem 5.2 (Arnold (1971)). *Let $\mathcal{S}_i = \{s_{1i} \geq s_{2i} \geq \cdots, \geq s_{q_i,i}\}$ be the Segre characteristic of A associated with λ_i. Then the codimension of the orbit $\mathcal{O}(A)$ of A is determined from*

$$\begin{aligned}\operatorname{codim}(\mathcal{O}(A)) &= \sum_{i=1}^{p}\sum_{j=1}^{q_i}(2j-1)s_{ji} \\ &= \sum_{i=1}^{p}(s_{1i} + 3s_{2i} + 5s_{3i} + \cdots + (2q_i - 1)s_{q_i,i}),\end{aligned} \tag{5.12}$$

where p is the number of the distinct eigenvalues.

Equation (5.12) can also be expressed in terms of the Weyr characteristic $\mathcal{W}_i = \{w_{i1}, w_{i2}, \ldots, w_{i,h_i}\}$ of A associated with the eigenvalue λ_i as

$$\begin{aligned}\operatorname{codim}(\mathcal{O}(A)) &= \sum_{i=1}^{p}\sum_{j=1}^{h_i} w_{ij}^2 \\ &= \sum_{i=1}^{p}(w_{i1}^2 + w_{i2}^2 + w_{i3}^2 + \cdots + w_{i,h_i}^2).\end{aligned} \tag{5.13}$$

Note that the complex conjugate eigenvalues are considered as two different eigenvalues.

The difference between orbits and bundles is that the eigenvalues of the matrices in a bundle are not specified, i.e., the tangent space of a bundle spans one extra dimension for each distinct eigenvalue compared to the corresponding orbit. Hence the codimension of the bundle $\mathcal{B}(A)$ is given by

$$\begin{aligned}\operatorname{codim}(\mathcal{B}(A)) &= \sum_{i=1}^{p}\left(-1+\sum_{j=1}^{q_i}(2j-1)s_{ji}\right)\\ &= \sum_{i=1}^{p}(-1+s_{1i}+3s_{2i}+5s_{3i}+\cdots+(2q_i-1)s_{q_i,i}).\end{aligned} \tag{5.14}$$

Comparing (5.13) and (5.14), one obtains that

$$\operatorname{codim}(\mathcal{B}(A)) = \operatorname{codim}(\mathcal{O}(A)) - p,$$

i.e., the codimension of a bundle is the same as the codimension of the corresponding orbit minus the number of distinct eigenvalues. This relationship reflects the fact that the bundle has additional degrees of freedom whose number is equal to the number of distinct eigenvalues. Note that in the formula (5.14) simple eigenvalues contribute nothing to the sum. Also, the bundle codimension does not depend on the order n of the matrix but depends only on the sizes of the Jordan blocks corresponding to multiple eigenvalues. Thus, the codimension of a bundle of matrices reflects the dependencies in the space of matrices imposed by the Jordan structure of the matrix.

In Chapter 6, we present Algorithm 6.5, which for a given matrix A determines the dimension of the bundle in which A resides.

5.1.3 *Generic and Nongeneric Bundles*

In this subsection, we consider different types of orbits and bundles determined by the possible Jordan forms of a given matrix.

Example 5.3. Consider first the case of 1×1 matrices. Let $A \in \mathbb{C}^{1\times 1}$, i.e., A is a complex scalar. In such a case, $\mathcal{O}(A) = A$ and the orbit consists of only one matrix. The codimension of $\mathcal{O}(A)$ is equal to 1 (the Jordan form of A consists of a single block of order 1) and the codimension of the bundle $\mathcal{B}(A)$ is equal to zero. Thus in the one-dimensional complex space $\mathbb{C}^1$, we have that $\dim(\mathcal{O}(A)) = 1 - 1 = 0$ and the orbit of A represents a point whose coordinates in the real plane are given by the real and imaginary parts of A. The bundle $\mathcal{B}(A)$ has a dimension equal to $1-0=1$ and coincides with $\mathbb{C}^1$.

Consider then the orbits and bundles of 2×2 matrices. The possible Jordan forms and the codimensions of their orbits and bundles are

$$J_1 = \left[\begin{array}{c|c} \mu_1 & \\ \hline & \mu_2 \end{array}\right], \quad \mu_1 \neq \mu_2;$$

$$s_{11} = 1; \quad s_{21} = 1; \quad p = 2;$$

$$\operatorname{codim}(\mathcal{O}(J_1)) = 1 + 1 = 2;$$

$$\operatorname{codim}(\mathcal{B}(J_1)) = 2 - 2 = 0;$$

$$J_2 = \begin{bmatrix} \mu_1 & 1 \\ & \mu_1 \end{bmatrix}, \quad s_{11} = 2; \quad p = 1;$$

$$\operatorname{codim}(\mathcal{O}(J_2)) = 2;$$

$$\operatorname{codim}(\mathcal{B}(J_2)) = 2 - 1 = 1;$$

$$J_3 = \left[\begin{array}{c|c} \mu_1 & \\ \hline & \mu_1 \end{array}\right], \quad s_{11} = 1; \quad s_{21} = 1; \quad p = 1;$$

$$\operatorname{codim}(\mathcal{O}(J_3)) = 1 + 3 = 4;$$

$$\operatorname{codim}(\mathcal{B}(J_3)) = 4 - 1 = 3.$$

Hence, in the 2×2 case, there are 3 types of possible orbits with codimensions 2, 2, and 4 and 3 types of corresponding bundles with codimensions 0, 1, and 3.

The first type corresponds to matrices with distinct eigenvalues, and according to (5.1), the elements of the orbits are computed as

$$A = PJ_1P^{-1}.$$

Let the similarity transformation matrix be chosen as

$$P = \begin{bmatrix} c & d \\ e & f \end{bmatrix}, \quad P^{-1} = \begin{bmatrix} f & -d \\ -e & c \end{bmatrix} \Big/ (cf - de),$$

where c, d, e, f are free parameters that must satisfy $\det(P) = cf - de \neq 0$. As a result, we obtain

$$A = \begin{bmatrix} cf\mu_1 - de\mu_2 & -cd(\mu_1 - \mu_2) \\ ef(\mu_1 - \mu_2) & -de\mu_1 + cf\mu_2 \end{bmatrix} \Big/ (cf - de).$$

Choosing $cf = 1 + de$ so that $\det(P) = 1$, the entries of the orbit are given by

$$\begin{aligned} a_{11} &= (1 + de)\mu_1 - de\mu_2, \\ a_{12} &= -cd(\mu_1 - \mu_2), \\ a_{21} &= ef(\mu_1 - \mu_2), \\ a_{22} &= -de\mu_1 + (1 + de)\mu_2. \end{aligned}$$

A part of the bundle $\mathcal{B}(J_1)$ is shown in Figure 5.8 in the three-dimensional space of the elements a_{11}, a_{12} and a_{22}, varying the two free parameters $p_1 = cd$ and $p_2 = de$. Each orbit is obtained for a fixed combination of the eigenvalues $1 \leq \mu_1 \leq 7$, $-7 \leq \mu_2 \leq -1$ and represents a plane in the space $(\dim(\mathcal{O}(J_1)) = 2)$. Note that the distances between the planes corresponding to distinct eigenvalues are infinitely small. Thus, the matrices with simple eigenvalues

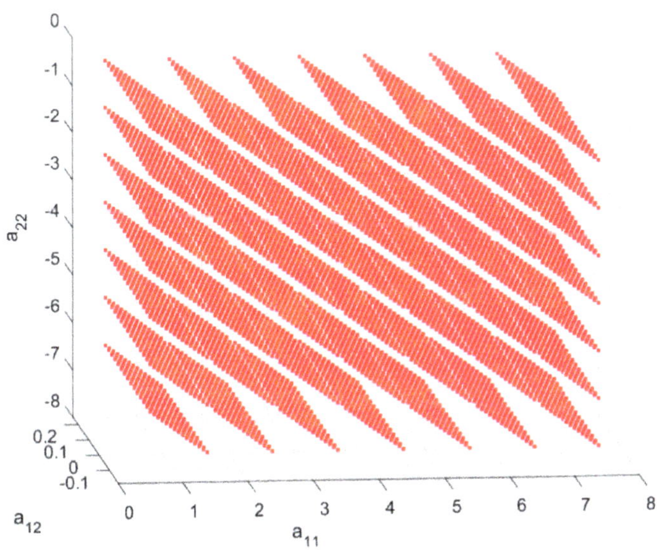

Fig. 5.8. Bundle of 2×2 matrices with codimension equal to 0.

constitute a bundle with dimension equal to 4 and form a *dense set* everywhere in the space of matrices. This type of 2×2 matrices is the most probable one in the sense that a matrix with random entries has distinct (simple) eigenvalues with a probability equal to 1. That is why it is called the *most generic case.*

Consider now the type of matrices with a Jordan form J_2 and double eigenvalue μ_1. In this case, we have

$$A = PJ_2P^{-1} = PJ_0P^{-1} + \mu_1 I_2,$$

where

$$J_0 = \begin{bmatrix} 0 & 1 \\ 0 & 0 \end{bmatrix}.$$

Using a similarity transformation matrix P in the same form as earlier, we obtain

$$A = \begin{bmatrix} -ce & c^2 \\ -e^2 & ce \end{bmatrix} + \mu_1 I_2,$$

where the parameters d and f are chosen so that $\det(P) = cf - de = 1$ and c, e are free parameters. In this way, the entries of the orbit are determined by

$$\begin{aligned} a_{11} &= -ce + \mu_1, \\ a_{12} &= c^2, \\ a_{21} &= -e^2, \\ a_{22} &= ce + \mu_1. \end{aligned}$$

It is interesting that each orbit coincides with the corresponding tangent space

$$T_{\mathcal{O}} = \mathcal{R}(L_A) + \mu_1 I_2,$$

where $L_A = I_2 \otimes J_2 - J_2^T \otimes I_2$. (Since the tangent space passes through 0 for all values of μ_1, to each element of $T_{\mathcal{O}}$, we add the matrix $\mu_1 I_2$ with μ_1 corresponding to the given orbit, see Figure 5.2.)

In Figure 5.9, we show the bundle of orbits of the matrices with Jordan form J_2, which has codimension equal to 1 and dimension

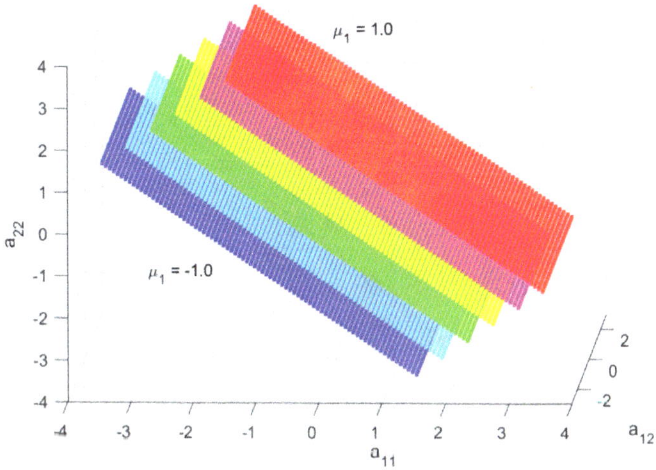

Fig. 5.9. Bundle of orbits of 2×2 matrices with codimension equal to 1.

equal to $2^2 - 1 = 3$. The orbits, corresponding to different values of the eigenvalue μ_1 are represented by parallel planes in the three-dimensional space of the entries a_{11}, a_{12} and a_{22}, varying the free parameters $p_1 = -ce$ and $p_2 = c^2$. The bundle $\mathcal{B}(J_2)$ with $(\mathcal{B}(J_2)) = 1$ is a subspace of dimension three of the whole four-dimensional space.

Consider finally the bundle of matrices with Jordan form J_3, which contains matrices of the form $A = J_3 = \mu_1 I_2$. This is the most degenerate case of 2×2 matrices, since the dimension of the possible orbits is zero (i.e., each orbit consists of only one matrix) and the bundle dimension is equal to 1. In Figure 5.10, we represent this bundle in the two-dimensional space of the entries a_{11} and a_{22} and display seven orbits for different values of μ_1 between -3 and 3. The bundle is a subspace of dimension one which is represented as a straight line in a two-dimensional subspace of the four-dimensional space of 2×2 matrices. This case is called the *least generic* or *most degenerate* one and the probability to meet it in random matrices is equal to zero since the corresponding orbits contain only one matrix.

Example 5.4. Consider next the orbits and bundles of 3×3 matrices. The possible Jordan forms and the codimensions of their orbits

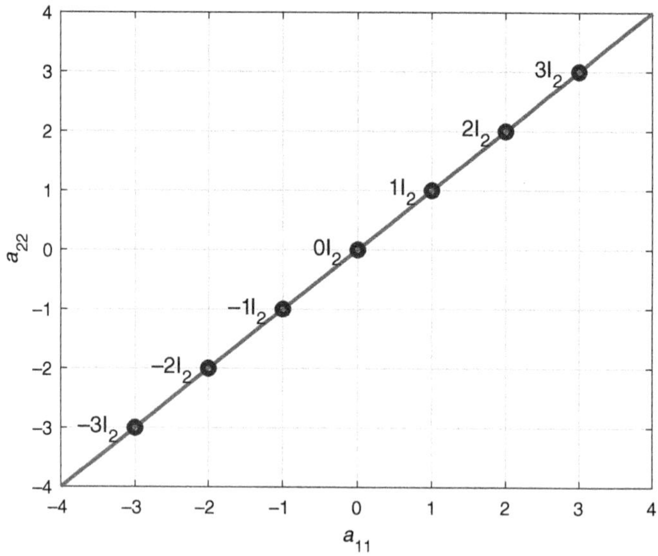

Fig. 5.10. Bundle of orbits of 2 × 2 matrices with codimension equal to 3.

and bundles are

$$J_1 = \left[\begin{array}{c|c|c} \mu_1 & & \\ \hline & \mu_2 & \\ \hline & & \mu_3 \end{array}\right], \qquad \mu_1 \neq \mu_2 \neq \mu_3;$$

$$s_{11} = 1; \quad s_{12} = 1; \quad s_{13} = 1; \quad p = 3;$$
$$\operatorname{codim}(\mathcal{O}(J_1)) = 1 + 1 + 1 = 3;$$
$$\operatorname{codim}(\mathcal{B}(J_1)) = 3 - 3 = 0;$$

$$J_2 = \left[\begin{array}{cc|c} \mu_1 & 1 & \\ \hline & \mu_1 & \\ \hline & & \mu_2 \end{array}\right], \qquad \mu_1 \neq \mu_2;$$

$$s_{11} = 2; \quad s_{12} = 1; \quad p = 2;$$
$$\operatorname{codim}(\mathcal{O}(J_2)) = 2 + 1 = 3;$$
$$\operatorname{codim}(\mathcal{B}(J_2)) = 3 - 2 = 1;$$

$$J_3 = \begin{bmatrix} \mu_1 & 1 & \\ & \mu_1 & 1 \\ & & \mu_1 \end{bmatrix}, \qquad s_{11} = 3; \quad p = 1;$$

$$\operatorname{codim}(\mathcal{O}(J_3)) = 3;$$

$$\operatorname{codim}(\mathcal{B}(J_3)) = 3 - 1 = 2;$$

$$J_4 = \left[\begin{array}{c|c|c} \mu_1 & & \\ \hline & \mu_1 & \\ \hline & & \mu_2 \end{array}\right], \qquad \mu_1 \neq \mu_2;$$

$$s_{11} = 1; \; s_{21} = 1; \; s_{12} = 1; \; p = 2;$$

$$\operatorname{codim}(\mathcal{O}(J_4)) = 1 + 3 + 1 = 5;$$

$$\operatorname{codim}(\mathcal{B}(J_4)) = 5 - 2 = 3;$$

$$J_5 = \left[\begin{array}{cc|c} \mu_1 & 1 & \\ & \mu_1 & \\ \hline & & \mu_1 \end{array}\right], \qquad s_{11} = 2; \quad s_{12} = 1; \quad p = 1;$$

$$\operatorname{codim}(\mathcal{O}(J_5)) = 2 + 3 = 5;$$

$$\operatorname{codim}(\mathcal{B}(J_5)) = 5 - 1 = 4;$$

$$J_6 = \left[\begin{array}{c|c|c} \mu_1 & & \\ \hline & \mu_1 & \\ \hline & & \mu_1 \end{array}\right], \qquad s_{11} = 1; \quad s_{21} = 1; \quad s_{31} = 1; \quad p = 1;$$

$$\operatorname{codim}(\mathcal{O}(J_6)) = 1 + 3 + 5 = 9;$$

$$\operatorname{codim}(\mathcal{B}(J_6)) = 9 - 1 = 8.$$

Hence, in the 3×3 case, there are six different types of orbits and corresponding bundles. The possible Jordan forms are ordered so that the dimensions of the corresponding bundles decrease and their codimensions increase, i.e., they become more and more degenerate. Such ordering of bundles with increasing codimension shows a hierarchy of the Jordan forms of matrices with a given dimension which plays an important role in the next presentation. The first case corresponds to distinct eigenvalues and is the most probable one and

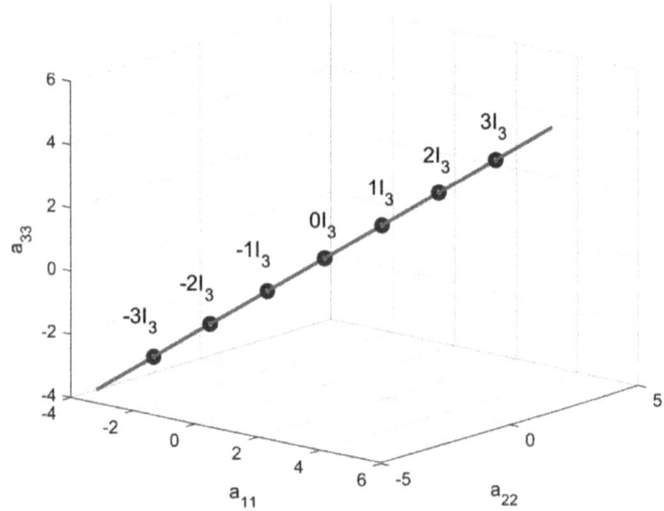

Fig. 5.11. Bundle of orbits of 3×3 matrices with codimension equal to 8.

hence the most generic one. The matrices with simple eigenvalues form a dense set everywhere in the space of matrices representing a bundle with a codimension equal to 0 (Figure 5.8). The last case ($A = J_6 = \mu_1 I_3$) is the most degenerate one since the dimension of the possible orbits is zero and the dimension of the bundle is 1. In Figure 5.11, we represent this bundle in the three-dimensional space of the entries a_{11}, a_{22} and a_{33}, and display seven orbits for different values of μ_1. The bundle is a subspace of dimension one which is represented as a straight line in a three-dimensional subspace of the nine-dimensional space of 3×3 matrices.

In Table 5.1, we show the most generic and most degenerate Jordan forms for the case of a single eigenvalue λ with algebraic multiplicity equal to n. Note that the single nth-order Jordan block associated with such eigenvalue represents the most generic Jordan structure in the case of multiple eigenvalue. The corresponding bundle contains all matrices A which are similar to nth-order companion matrices.

For the general case of $n \times n$ matrices with several distinct eigenvalues, the most generic Jordan form has n one-dimensional blocks corresponding to n distinct eigenvalues (Table 5.2). According to (5.12) and (5.14), in this case, we have that $\text{codim}(\mathcal{O}) = n$ and

Table 5.1. Most generic and most degenerate Jordan forms in case of n-tuple eigenvalue.

Most generic Jordan form

Single nth-order Jordan block

$$J = \begin{bmatrix} \lambda & 1 & & & \\ & \lambda & 1 & & \\ & & \lambda & \ddots & \\ & & & \ddots & 1 \\ & & & & \lambda \end{bmatrix}, \qquad \begin{gathered} \mathcal{S} = \{n\} \\ \dim(\mathcal{O}) = n^2 - n \\ \operatorname{codim}(\mathcal{O}) = n \\ \dim(\mathcal{B}) = n^2 - n + 1 \\ \operatorname{codim}(\mathcal{B}) = n - 1 \end{gathered}$$

Most degenerate Jordan form

n scalar blocks with the same eigenvalue

$$J = \left[\begin{array}{c|c|c|c|c} \lambda & & & & \\ \hline & \lambda & & & \\ \hline & & \lambda & & \\ \hline & & & \ddots & \\ \hline & & & & \lambda \end{array}\right], \qquad \begin{gathered} \mathcal{S} = \{1, 1, \ldots, 1\} \\ \dim(\mathcal{O}) = 0 \\ \operatorname{codim}(\mathcal{O}) = n^2 \\ \dim(\mathcal{B}) = 1 \\ \operatorname{codim}(\mathcal{B}) = n^2 - 1 \end{gathered}$$

Table 5.2. Most generic and most degenerate orbits and bundles of $n \times n$ matrices.

Case	Number of Jordan blocks	Number of distinct eigenvalues	Segre characteristics	Dimensions of orbits and bundles
Most generic	n	n	$\mathcal{S} = [\{1\}, \{1\}, \ldots, \{1\}]$	$\dim(\mathcal{O}) = n^2 - n$ $\operatorname{codim}(\mathcal{O}) = n$ $\dim(\mathcal{B}) = n^2$ $\operatorname{codim}(\mathcal{B}) = 0$
Most degenerate	n	1	$\mathcal{S} = \{1, 1, \ldots, 1\}$	$\dim(\mathcal{O}) = 0$ $\operatorname{codim}(\mathcal{O}) = n^2$ $\dim(\mathcal{B}) = 1$ $\operatorname{codim}(\mathcal{B}) = n^2 - 1$

$\dim(\mathcal{O}) = n^2 - n$. The codimension of the respective bundle is equal to $n - n = 0$ and its dimension is equal to n^2, i.e., the bundle spans the whole matrix space. Otherwise speaking, almost all $n \times n$ matrices have distinct eigenvalues. The least generic (or most degenerate) Jordan form has n one-dimensional (scalar) blocks corresponding to a single eigenvalue of multiplicity n so that the orbits of matrices with such Jordan form have $\text{codim}(\mathcal{O}) = 1 + 3 + \cdots + (2n - 1) = n^2$ and $\dim(\mathcal{O}) = n^2 - n^2 = 0$. The respective bundle has a codimension equal to $n^2 - 1$, dimension equal to 1, and represents a straight line in the matrix space. Hence the orbits corresponding to the least generic case have dimension equal to zero and consists of only one matrix, i.e., the orbit represents a point in the matrix space lying on the bundle line. The cases between the most generic and least generic one are characterized by different Jordan bundles with codimensions between 0 and $n^2 - 1$.

5.2 Stratification of Jordan Bundles

5.2.1 *Matrix Stratification*

Small perturbations may cause a matrix A with given Segre characteristics, i.e., with a given Jordan structure, to move from a bundle of certain codimension to a bundle of smaller codimension. For instance, matrix transitions

$$
\begin{array}{ccccc}
\mathcal{S} = \{5\} & & \mathcal{S} = \{3,2\} & & \mathcal{S} = \{3,1,1\} \\
\begin{bmatrix} \lambda & \eta & & & \\ & \lambda & \zeta & & \\ & & \lambda & \varepsilon & \\ & & & \lambda & \delta \\ & & & & \lambda \end{bmatrix}
& \overset{\varepsilon \to 0}{\longrightarrow} &
\left[\begin{array}{ccc|cc} \lambda & \eta & & & \\ & \lambda & \zeta & & \\ & & \lambda & & \\ \hline & & & \lambda & \delta \\ & & & & \lambda \end{array}\right]
& \overset{\delta \to 0}{\longrightarrow} &
\left[\begin{array}{ccc|c|c} \lambda & \eta & & & \\ & \lambda & \zeta & & \\ & & \lambda & & \\ \hline & & & \lambda & \\ \hline & & & & \lambda \end{array}\right] \\
\text{codim}(\mathcal{B}) = 4 & & \text{codim}(\mathcal{B}) = 8 & & \text{codim}(\mathcal{B}) = 10
\end{array}
$$

$$\Longleftarrow \text{perturbation increasing} \tag{5.15}$$

for some $\eta \neq 0, \zeta \neq 0$ show that a matrix with Segre characteristic $\mathcal{S} = \{3, 1, 1\}$ ($\text{codim}(\mathcal{B}) = 10$) is arbitrarily close to some matrices with Segre characteristic $\mathcal{S} = \{3, 2\}$ ($\text{codim}(\mathcal{B}) = 8$), which in turn are arbitrary near certain matrices with Segre characteristic $\mathcal{S} = \{5\}$ ($\text{codim}(\mathcal{B}) = 4$). In this way, with a small perturbation, one can go from $\mathcal{B}(\{3,1,1\})$ to $\mathcal{B}(\{3,2\})$, i.e., $\mathcal{B}(\{3,1,1\})$ is *in the closure* of $\mathcal{B}(\{3,2\})$ and $\mathcal{B}(\{3,2\})$ *covers* $\mathcal{B}(\{3,1,1\})$. Formally this is written as $\mathcal{B}(\{3,1,1\}) \subset \overline{\mathcal{B}}(\{3,2\})$, where $\overline{\mathcal{B}}(\{3,2\})$ is the closure of $\mathcal{B}(\{3,2\})$. With a small perturbation, one can also go from $\mathcal{B}(\{3,2\})$ to $\mathcal{B}(\{5\})$ which means that $\mathcal{B}(\{3,2\})$ is in the closure of $\mathcal{B}(\{5\})$, i.e., $\mathcal{B}(\{3,2\}) \subset \overline{\mathcal{B}}(\{5\})$. Both closure relations also imply that with a small perturbation, one can go from $\mathcal{B}(\{3,1,1\})$ to $\mathcal{B}(\{5\})$. In the general case, when the matrix A belongs to a given bundle, it can also be in the closure of many other bundles with respect to different Segre characteristics. These bundles have smaller codimension than the bundle in which resides the original matrix. These closure relationships form a hierarchy of stratification of Jordan structures and represent an example of matrix stratification (Figure 5.12). Note that the addition of a perturbation that is in general position to the Jordan form decreases the codimension of the bundle in which moves the perturbed matrix.

5.2.2 *Stratification Rules*

The stratification of Jordan structures is studied by Edelman *et al.* (1999), who show how Jordan and Kronecker forms can be represented as integer partitions such that the closure relations of orbits

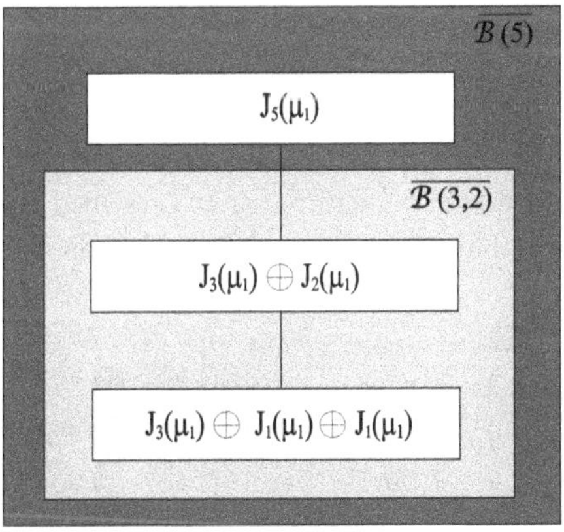

Fig. 5.12. Example of closure hierarchy of Jordan bundles.

and bundles can be revealed by using some simple rules. These rules can be used to test if one structure is more generic than another. The results are presented without proof.

A partition κ of an integer n is a sequence of integers $\{k_1, k_2, k_3, \ldots\}$ such that $k_1 + k_2 + \cdots = n$ and $k_1 \geq k_2 \geq \cdots \geq 0$. If $\mu = \{m_1, m_2, \ldots\}$ is another partition (not necessarily of the same integer n), we say that κ *dominates* μ, i.e., $\kappa \geq \mu$, if

$$k_1 + k_2 + \cdots + k_i \geq m_1 + m_2 + \cdots + m_i$$

for $i = 1, 2, \ldots$. Different partitions of an integer form in this way a *dominance ordering*. We say that $\kappa > \mu$ if and only if $\kappa \geq \mu$ and $\kappa \neq \mu$. If κ and μ are partitions of the same integer n and there is no partition λ such that $\kappa > \lambda > \mu$, then κ is said to *cover* μ.

If, for instance, $\kappa = \{4, 2, 2\}$ and $\mu = \{3, 3, 2\}$, one has that

$$\begin{aligned} 4 &> 3, \\ 4 + 2 &= 3 + 3, \\ 4 + 2 + 2 &= 3 + 3 + 2 \end{aligned}$$

so that κ covers μ.

It is possible to determine easily the covering relationship by placing n coins in a table with k_1 in the first column, k_2 in the second column and so on. A partition κ covers μ if μ may be obtained from κ by moving a coin *rightward one column*, or *downward one row*, under the condition that the partitions remain monotonic, i.e., after the move, it is fulfilled that $k_1 \geq k_2 \geq \cdots \geq 0$ and $m_1 \geq m_2 \geq \cdots \geq 0$ (Figure 5.13).

In a similar way, κ is covered by ν if κ may be obtained from ν by moving a coin *leftward one column*, or *upward one row*, and keeping the monotonicity of the partition (Figure 5.14).

The closure and cover relations for orbits and bundles of matrices can be determined using the integer partitions of the dimension of

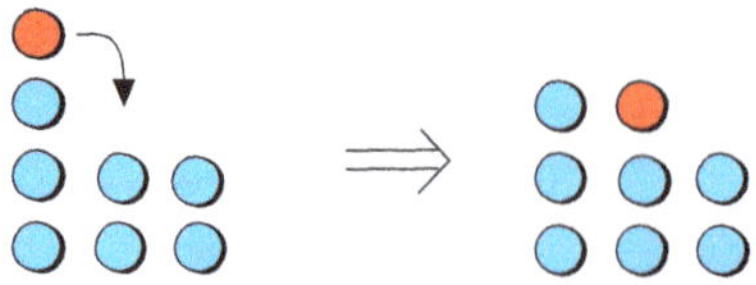

Fig. 5.13. Minimum rightward move illustrates that $\kappa = \{4, 2, 2\}$ covers $\mu = \{3, 3, 2\}$.

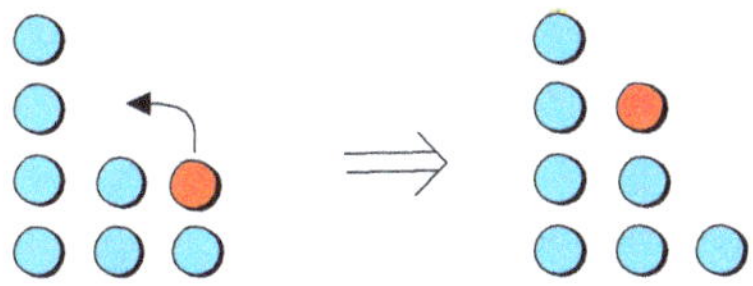

Fig. 5.14. Minimum leftward move illustrates that $\kappa = \{4, 2, 2\}$ is covered by $\nu = \{4, 3, 1\}$.

the matrix space, defined by the Weyr characteristics of the given $n \times n$ matrix. We have the following important results.

Theorem 5.5 (Edelman *et al.* (1999)). *Let A_1 and A_2 be two matrices that have the same eigenvalues μ_i for $i = 1, 2, \ldots, p$, where p is the number of distinct eigenvalues and let $\mathcal{W}(\mu_i)$ be the Weyr characteristic associated with the eigenvalue μ_i, Then $\mathcal{O}(A_1)$ covers $\mathcal{O}(A_2)$, i.e., $\overline{\mathcal{O}}(A_1) \supseteq \overline{\mathcal{O}}(A_2)$, if and only if $\mathcal{W}(A_1, \mu_i) \leq \mathcal{W}(A_2, \mu_i)$.*

From Theorem 5.5 and the definition of covering partition, one gets the following characterization for the orbits of matrices A_1 and A_2.

Corollary 5.6. *$\mathcal{O}(A_1)$ covers $\mathcal{O}(A_2)$ if and only if some $\mathcal{W}(A_2, \mu_i)$ can be obtained from $\mathcal{W}(A_1, \mu_i)$ by a minimum leftward coin move and $\mathcal{W}(A_2, \mu_j) = \mathcal{W}(A_1, \mu_j)$ for all $m_j \neq \mu_i$.*

In Theorem 5.5 and Corollary 5.6, one assumes that the number of eigenvalues and their multiplicities are the same for all orbits in the closure hierarchy. This is not fulfilled if we consider the hierarchy of bundles when some eigenvalues can coalesce or split apart. In such case, the coalescing of two eigenvalues is equivalent to taking the union of their Weyr characteristics and the splitting of an eigenvalue into two eigenvalues is equivalent to the division of its Weyr characteristic into two parts. Thus the conditions for covering relations of bundles are given by the following theorems:

Theorem 5.7. *Suppose that we have two bundles $\mathcal{B}(A_1)$ and $\mathcal{B}(A_2)$ such that the former has at least as many distinct eigenvalues as the latter. Then $\overline{\mathcal{B}}(A_1) \supseteq \overline{\mathcal{B}}(A_1)$ if and only if it is possible to coalesce eigenvalues and apply the dominance ordering coin moves to the structure integer partitions of the bundle defined by A_1 to reach that of A_2.*

Table 5.3. Stratification rules of bundles of a matrix.

A.	$\mathcal{B}(A_1)$ covers $\mathcal{B}(A_2)$	**B.**	$\mathcal{B}(A_1)$ is covered by $\mathcal{B}(A_2)$
(1)	Minimum leftward coin move in any $\mathcal{W}(\mu_i)$.	(1)	Minimum rightward coin move in any $\mathcal{W}(\mu_i)$.
(2)	Make any pair of eigenvalues to coalesce, i.e., take the of their sets of coins union.	(2)	For any $\mathcal{W}(\mu_i)$, divide the set of coins into two new sets so that their union is $\mathcal{W}(\mu_i)$.

Theorem 5.8. *$\mathcal{B}(A_1)$ covers $\mathcal{B}(A_2)$ if and only if some $\mathcal{W}(A_2, \mu_i)$ either can be obtained from $\mathcal{W}(A_1, \mu_i))$ by a minimum leftward coin move or by coalescing the partitions from two distinct eigenvalues and $\mathcal{W}(A_2, \mu_j) = \mathcal{W}(A_1, \mu_j)$ for all $m_j \neq \mu_i$.*

The stratification rules for bundles are summarized in Table 5.3.

Example 5.9. Consider a 3×3 matrix A with Jordan canonical form $J = J_2(\mu_1) \oplus J_1(\mu_2)$ which corresponds to a bundle $\mathcal{B}$ with codimension equal to $2 + 1 - 2 = 1$. Using the stratification rules A. and B. from Table 5.3, we show how to derive all nearest neighbors to the matrix J in a bundle stratification.

The Jordan canonical form of A is represented by the Weyr characteristics and their corresponding sets of coins as shown in Figure 5.15.

First, rule A.1 (minimal leftward coin move) is applied to all sets of coins to which it is feasible. In the given case, this is only the set $\mathcal{W}(\mu_1)$.

As a result, one obtains the set $\mathcal{W}(\mu_1) = \{2\}$ and clearly it covers the set $\mathcal{W}(\mu_1) = \{1, 1\}$. The set $\mathcal{W}(\mu_2) = \{1\}$ remains unchanged so

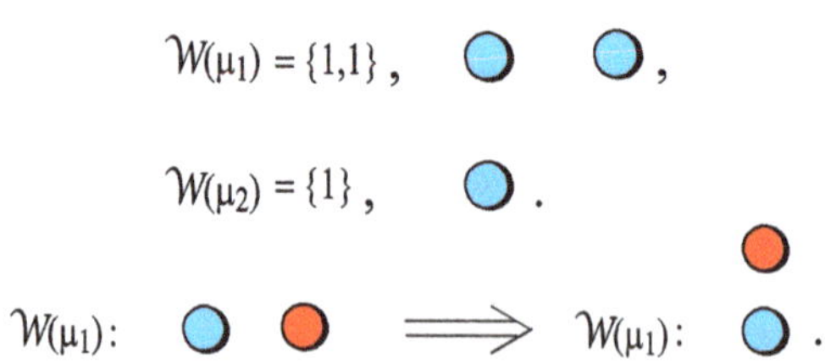

Fig. 5.15. Applying rule A.1 for bundle stratification.

that the Weyr characteristics of the covered bundle $\widetilde{\mathcal{B}}$ are $\mathcal{W}(\mu_1) = \{2\}, \mathcal{W}(\mu_2) = \{1\}$. The associated Segre characteristics of the bundle $\widetilde{\mathcal{B}}$ are $\mathcal{S}(\mu_1) = \{1, 1\}, \mathcal{S}(\mu_2) = \{1\}$. Hence, the bundle $\mathcal{B}$ covers the matrices with Jordan canonical form $J_1(\mu_1) \oplus J_1(\mu_1) \oplus J_1(\mu_2)$. Its codimension is equal to $1 + 3 + 1 - 2 = 3$.

Next, one applies rule A.2 (take the union of two sets) to all the sets of coins. For the given Jordan form, one has only the sets $\mathcal{W}(\mu_1)$ and $\mathcal{W}(\mu_2)$.

The result is a bundle with Weyr characteristic $\mathcal{W}(\mu_1) = \{1,\ 1,\ 1\}$, which corresponds to the Segre characteristic $\mathcal{S}(\mu_1) = \{3\}$ (Figure 5.16). Thus the bundle $\mathcal{B}$ covers the matrices with a single Jordan block $J_3(\mu_1)$.

Consider now how to find the matrix bundles which cover the given bundle $\mathcal{B}$. The rule B.1 (minimum rightward coin move) can not be applied in the given case. The rule B.2 (divide one set of coins into two) can be applied only to the set $\mathcal{W}(\mu_1)$ (Figure 5.17).

As a result, one obtains the sets $\mathcal{W}(\mu_1) = \{1\}, \mathcal{W}(\mu_1) = \{1\}, \mathcal{W}(\mu_3) = \{1\}$. This shows that the bundle $\mathcal{B}$ is covered by a bundle corresponding to matrices with Jordan form $J_1(\mu_1) \oplus J_1(\mu_2) \oplus J_1(\mu_3)$, i.e., all 3×3 matrices with distinct eigenvalues. The codimension of this bundle is equal to $1 + 1 + 1 - 3 = 0$, i.e., this is the most generic bundle.

The nearest neighbors to the bundle of the matrices with Jordan canonical form $J_2(\mu_1) \oplus J_1(\mu_2)$ are shown in Figure 5.18.

The procedure for bundle stratification of all 3×3 matrices is shown in Figure 5.19. The stratification is done from the top to the bottom using the covering relations A.1 and A.2. The same result can be obtained if the relations B.1 and B.2 are used from the bottom to the top.

Fig. 5.16. Applying rule A.2.

Fig. 5.17. Applying rule B.2.

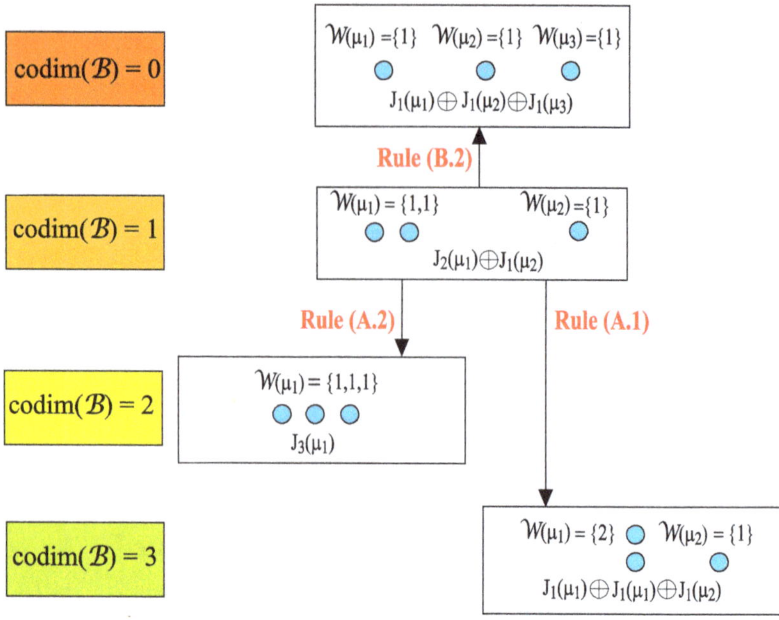

Fig. 5.18. All nearest neighbors in the closure hierarchy to the bundle of 3×3 matrices with Jordan canonical form $J_2(\mu_1) \oplus J_1(\mu_2)$.

The hierarchical closure relationship between matrix bundles can be represented as a stratification graph in which nodes represent bundles of different canonical structures and edges between pairs of nodes indicate the relations between their canonical forms. An example of such a stratification graph is given in Figure 5.20 displaying the stratification of 3×3 bundles. Note that a structure can never be covered by a less or equally generic structure. This allows ordering the structures within the closure hierarchy by their dimension or their codimension. (Further on we shall order the structures by their codimensions as shown in Figure 5.20.) Several bundles in a closure hierarchy can have the same codimension.

Note that the most degenerate Jordan structure $J_1(\mu_1) \oplus J_1(\mu_1) \oplus J_1(\mu_1)$ corresponds to a bundle of codimension equal to 8, i.e., the bundle dimension is equal to one. This bundle contains orbits of dimension 0 (points in the nine-dimensional space) which lie on a straight line representing the bundle as it is shown in Figure 5.11.

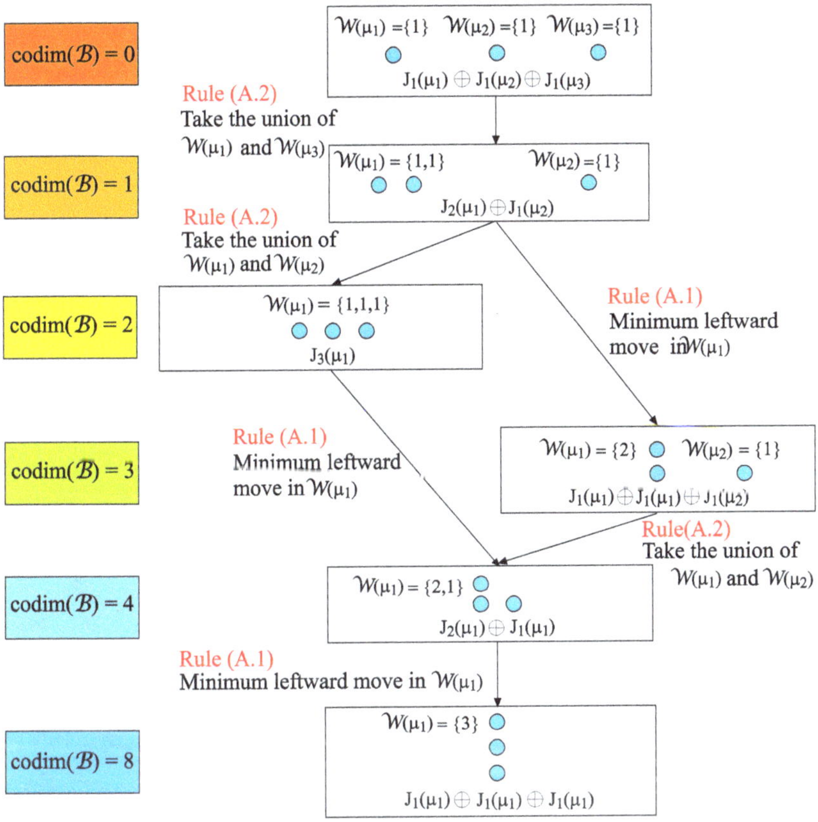

Fig. 5.19. The bundle stratification of the set of 3×3 matrices, illustrated by coin tables.

The most generic bundle, containing all matrices with distinct eigenvalues, is with codimension 0 and dimension equal to 9.

The stratification of the Jordan forms of higher dimension can be done easily using the StratiGraph tool (Elmroth *et al.*, 2001; Johansson, 2003; Kågström *et al.*, 2012; StratiGraph, 2019) developed by a group of researchers from the Department of Computing Science of the Umeå University in Sweden. This tool allows determining and visualizing the stratification graph of a given matrix of arbitrary dimension n. It should be taken into account, however, that deciding whether a bundle is in the closure of another bundle is

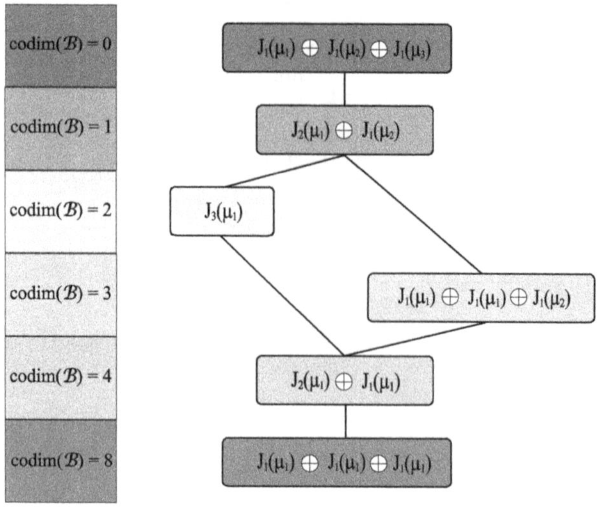

Fig. 5.20. The stratification graph representing the hierarchy of the Jordan structures of 3×3 matrices.

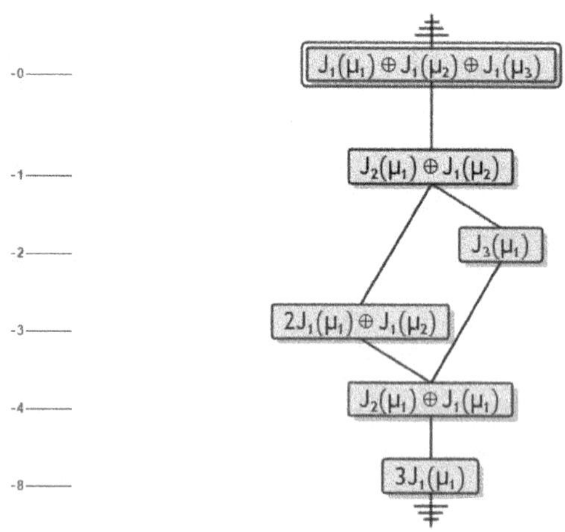

Fig. 5.21. The stratification graph for 3×3 matrices obtained by StratiGraph.

an NP-complete problem (Edelman *et al.*, 1999) so that the volume of the computational work increases exponentially with the increase in n. In Figure 5.21, we show the bundle stratification graph for 3×3 matrices obtained by using the StratiGraph tool.

The stratification graph for 4×4 matrices is shown in Figure 5.22.

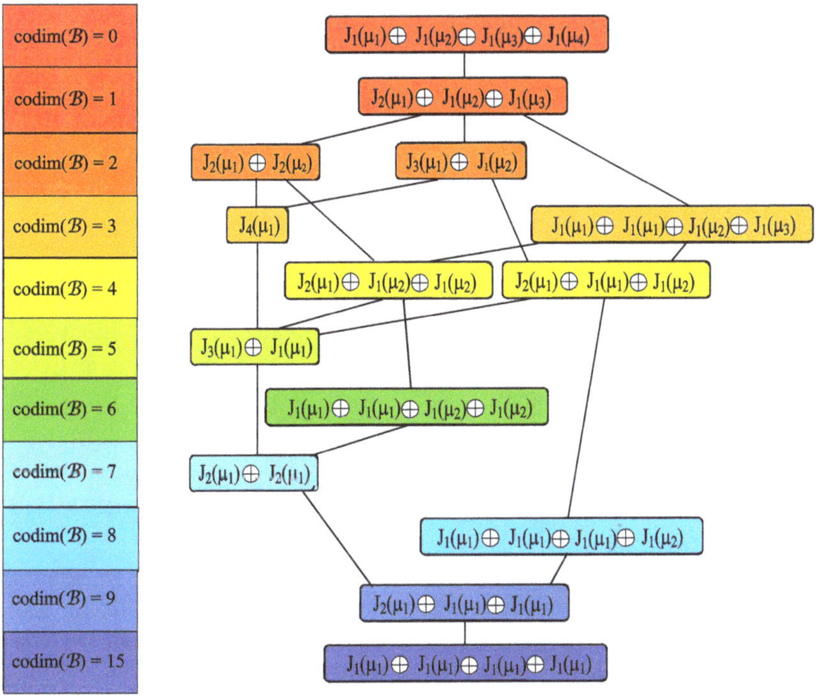

Fig. 5.22. The stratification graph for 4×4 matrices.

5.3 Normal Forms of Matrices Depending on Parameters

This section is devoted to the definition of the notion of a versal family and its application to a problem related to the Jordan canonical form of matrices.

5.3.1 *Matrix Deformations*

In finding the Jordan form of a matrix, instead of the exact matrix A we compute the Jordan form of a close matrix $\widetilde{A}$, whose entries are contaminated by small errors of various nature. In this subsection, we find the number of the independent parameters determining the errors of the entries of $\widetilde{A}$ and represent the matrix $\widetilde{A}$ as a function of these parameters.

Let the entries of the matrix A depend on some parameters. It is supposed that these parameters belong to a linear space $\mathcal{P}^k$ called *parameter space* with $\dim(\mathcal{P}^k) = k$, where k is the number of independent parameters. Assume $A_0 \in \mathbb{C}^{n\times n}$ is fixed. A family of matrices $A(\delta)$, $\delta \in \Delta \subseteq \mathcal{P}^k$, where Δ is a neighborhood of the origin in $\mathcal{P}^k$, is called a *deformation* of A_0 if $A : \Delta \to \mathbb{C}^{n\times n}$ is such a mapping that the entries of $A(\delta)$ are convergent power series of δ_i and $A(0) = A_0$. A deformation is also called *matrix family* and the subset Δ of the parameter space is referred to as a *base* of the family. Note that a function that is locally given by a convergent power series is called *analytic function* and the complex analytic function is referred to as *holomorphic function.* A holomorphic function has derivatives of every order (i.e., it is infinitely differentiable) at each point in its domain so that it is a continuous function. A mapping $A : \Delta \to \mathbb{C}^{n\times n}$ whose entries are holomorphic functions is referred to as *holomorphic mapping.* Since their entries depend smoothly on the parameters, the implementation of such mappings is convenient from computational point of view.

To determine a canonical form of the matrix family $A(\delta)$, $\delta \in \Delta$, we may try to find the Jordan canonical form of each $A(\delta)$ by using a similarity transformation depending on δ. However, this process is unstable since under a small change of the elements of the matrix, both the Jordan form and the similarity transformation leading to it change discontinuously, in general. That is why it is better to derive a canonical (or *normal*) form for a family of matrices that depend smoothly on parameters, implementing a similarity transformation also depending smoothly on parameters.

Example 5.10. A one-parameter deformation of the Jordan block

$$A_0 = \begin{bmatrix} -1 & 1 \\ 0 & -1 \end{bmatrix}$$

may look as

$$A(\delta) = \begin{bmatrix} -1 & 1 \\ 0 & -1 \end{bmatrix} + \begin{bmatrix} 3\delta^2 & \delta - 4\delta^2 \\ 0 & 22\delta^2 \end{bmatrix}.$$

In Figure 5.23, we show the matrix family $A(\delta)$ for small changes of the parameter δ.

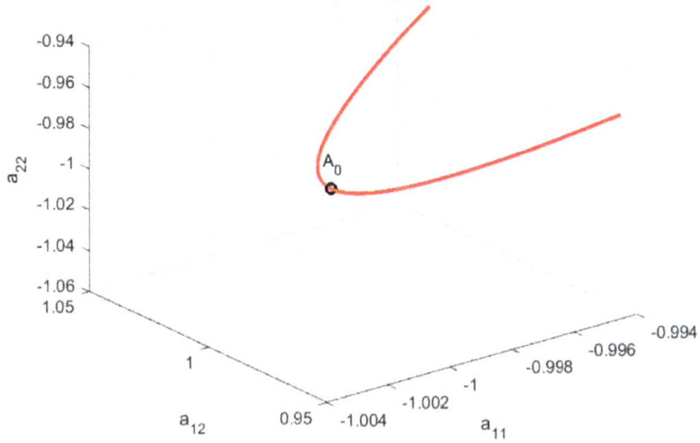

Fig. 5.23. One-parameter matrix deformation.

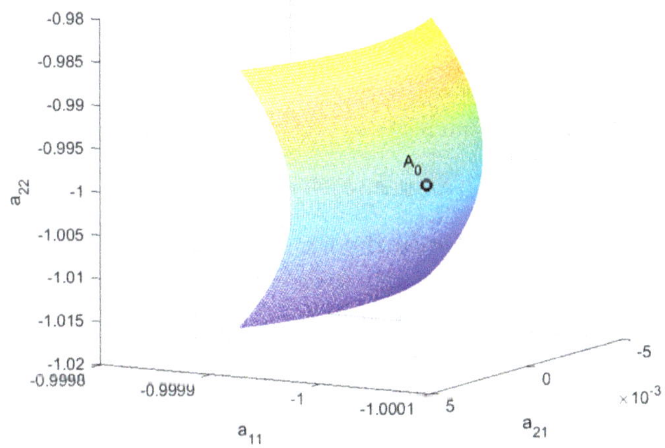

Fig. 5.24. Two-parameter matrix deformation.

In Figure 5.24, we show a two-parameter deformation

$$A(\delta) = \begin{bmatrix} -1 & 1 \\ 0 & -1 \end{bmatrix} + \begin{bmatrix} \delta_1^2 + 2\delta_2^2 & 0 \\ \delta_1 - 4\delta_2^2 & 10\delta_1^2 - 3\delta_2 + 15\delta_2^2 \end{bmatrix},$$

of the same matrix A_0.

Further on we shall use the following terminology. By a *map $B(\eta) \to A(\delta)$ of one family into another family*, we mean that to each value of the parameter η of the family $B(\eta)$ there corresponds a definite value $\delta = \phi(\eta)$ of the parameter of the family $A(\delta)$.

A *versal family* of matrices is one into which we can map every other family of matrices by means of a suitable mapping.

A *universal family* is a versal family $A(\delta)$ such that in mapping a family $B(\eta)$ into it, the "change of parameters" $\delta = \phi(\eta)$ is uniquely determined by the family $B(\eta)$.

A *miniversal family* is a versal family depending on the minimum possible number of parameters. Clearly, such families are the most interesting from numerical point of view.

Note that the term "versal" is formed from the word "universal" by dropping the prefix "uni", which denotes the uniqueness of the map ϕ. As Arnold remarks (Arnol'd, 1972), "versal" is the intersection of the two words "universal" and "transversal".

To define formally the earlier notation, we need the following definitions.

Two deformations $A(\delta)$ and $A'(\delta)$ of the matrix A_0 are called *equivalent* if there exists a deformation $P(\delta)$ of the identity matrix I_n with the same base such that

$$A'(\delta) = P(\delta)A(\delta)P^{-1}(\delta).$$

Let both $A(\delta)$ and $A'(\eta)$ be deformations of A_0, where $\delta \in \Delta \subseteq \mathcal{P}^k$ and $\eta \in \mathrm{H} \subseteq \mathcal{P}^\ell$. If there exist a deformation $P(\eta)$ of the identity matrix with $\eta \in \widetilde{\mathrm{H}} \subseteq \mathrm{H}$ and a holomorphic mapping $\phi : \widetilde{\mathrm{H}} \to \Delta$ with $\phi(0) = 0$ such that

$$A'(\eta) = P(\eta)A(\phi(\eta))P^{-1}(\eta), \quad \eta \in \widetilde{\mathrm{H}},$$

then we say that $A'(\eta)$ is *induced* from the deformation $A(\delta)$ by $P(\eta)$ and $\phi(\eta)$ (see Figure 5.25).

A deformation $A(\delta)$ of a matrix A_0 is said to be versal, if every other deformation $A'(\eta)$ of A_0 is equivalent to a deformation induced from $A(\delta)$ under a suitable change of the parameters, i.e.,

$$A'(\eta) = P(\eta)A(\phi(\eta))P^{-1}(\eta), \quad P(0) = I_n, \quad \phi(0) = 0. \qquad (5.16)$$

A versal deformation of A_0 is called universal if the map ϕ inducing A' is uniquely determined by A' and a versal deformation of A_0 is

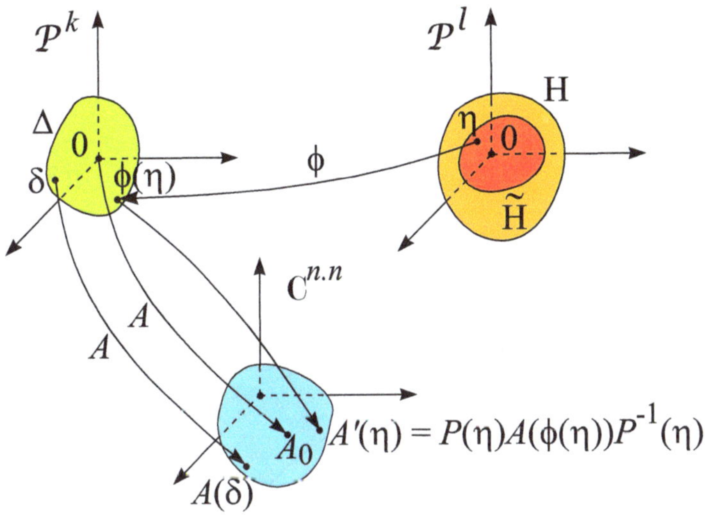

Fig. 5.25. Versal deformation.

called miniversal deformation if the dimension of its parameter space is the smallest among all versal deformations of A_0.

The following lemma gives a condition under which a matrix deformation is versal.

Lemma 5.11 (Arnold (1971, 1988, Chapter 6, Section 30)). *A deformation $A(\delta)$ of A_0 is a versal deformation if and only if $A(\delta)$ is transversal to the orbit of A_0 at $\delta = 0$.*

We will prove that the versal deformation $A(\delta)$ is transversal. Let $A'(\eta)$ be arbitrary deformation of A_0. Then by the versatility of A we have that

$$A'(\eta) = P(\eta)A(\phi(\eta))P^{-1}(\eta), \quad \eta \in \widetilde{\mathrm{H}}.$$

Taking into account that

$$\frac{\partial P^{-1}}{\partial \eta} = -P^{-1}\frac{\partial P}{\partial \eta}P^{-1},$$

and $P(0) = P^{-1}(0) = I_n$, we get

$$A'_* = A_*\phi_* + P_*A_0 - A_0P_* = A_*\phi_* + [P_*, A_0],$$

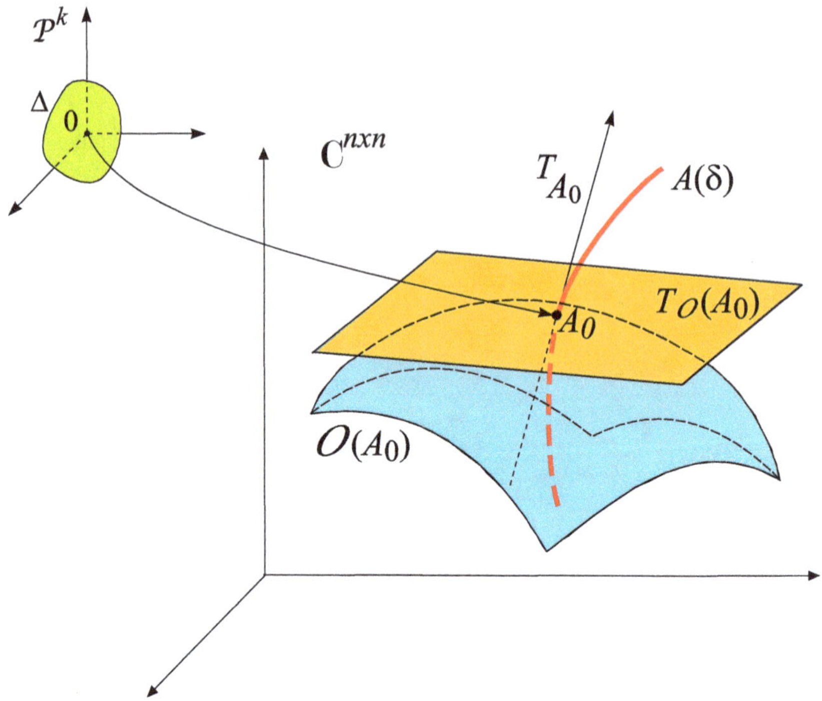

Fig. 5.26. The versal deformation $A(\delta)$ is transversal to $\mathcal{O}(A_0)$.

where the subscript (*) denotes the derivative in respect to η at $\eta = 0$. Consequently, for every tangent to the base of A' at the origin vector $v \in \mathcal{P}^{\ell}$, we have that

$$A'_* v = A_* \phi_* v + [P_* v, A_0].$$

Since $A_* \phi_* v \in \mathcal{R}(A_*)$ and $[P_* v, A_0] \in T_{\mathcal{O}}(A_0)$, any vector in the space T_{A_0}, tangent to $A(\delta)$ at A_0, can be represented as a sum of a vector in the image of A_* and a vector tangent to the orbit of A_0. Thus, $A(\delta)$ is transversal to the orbit of A_0 (Figure 5.26).

A proof that a deformation which is transversal to the orbit is versal can be found in Chow *et al.* (1994), Section 2.9.

According to Lemma 5.11, we have that

$$T_{A_0} + T_{\mathcal{O}}(A_0) = C^{n \times n}.$$

That is why the dimension of the parameter space of a versal deformation of A_0 is equal to $n^2 - \dim(\mathcal{O}(A_0))$, i.e., to the codimension of the orbit $\mathcal{O}(A_0)$. In its turn, $\text{codim}(\mathcal{O}(A_0))$ is equal to the dimension of the centralizer Z_{A_0}. Thus, we come to the conclusion that if A_0 is in Jordan normal form with p distinct eigenvalues $\lambda_1, \lambda_2, \ldots, \lambda_p$ with associated Segre characteristics

$$\mathcal{S}_i = \{s_{1i} \geq s_{2i} \geq \cdots, \geq s_{q_i,i},\ i = 1, 2, \ldots, p\},$$

then according to Theorem 5.2 the dimension of the miniversal deformation of A_0 is given by

$$D = \sum_{i=1}^{p} (s_{1i} + 3s_{2i} + 5s_{3i} + \cdots + (2q_i - 1)s_{q_i,i}).$$

As we see latter on, although this dimension is unique, the miniversal deformation itself is not unique, since the mapping $A(\delta)$ which is transversal to $\mathcal{O}(A_0)$ is not unique.

Equation (5.17) concerns a particular case of a general situation which can be described as follows. Let $N \subset M$ be a smooth submanifold of a manifold M. We consider a mapping $A : \Delta \to M$ of another manifold Δ into M and let δ be a point in Δ such that $A(\delta) \in N$. The mapping $A : \Delta \to M$ is transversal to $N \subset M$ at δ if the tangent space to M at $A(\delta)$ is the sum

$$TM_{A(\delta)} = A_*(T_0\Delta) + TN_{A(\delta)}, \tag{5.17}$$

where $TN_{A(\delta)}$ is the tangent space to N.

Equation (5.17) gives a condition for transverse intersection of two manifolds.

Example 5.12. (1) Let

$$A_0 = \begin{bmatrix} \lambda_1 & & & \\ & \lambda_2 & & \\ & & \ddots & \\ & & & \lambda_n \end{bmatrix}, \quad \lambda_i \neq \lambda_j.$$

Then the codimension of the orbit $\mathcal{O}(A_0)$ is equal to n and an n-parameter versal deformation of A_0 is

$$A(\delta) = \begin{bmatrix} \lambda_1 + \delta_1 & & & \\ & \lambda_2 + \delta_2 & & \\ & & \ddots & \\ & & & \lambda_n + \delta_n \end{bmatrix},$$

where δ_i, $i = 1, 2, \ldots, n$, are arbitrary parameters. This deformation is also universal and miniversal.

(2) Let $A_0 = 0$. Then $\text{codim}(\mathcal{O}(A_0)) = n^2$ and an n^2-parameter versal deformation of A_0 is

$$A(\delta) = \begin{bmatrix} \delta_{11} & \delta_{12} & \ldots & \delta_{1n} \\ \delta_{21} & \delta_{22} & \ldots & \delta_{2n} \\ \vdots & \vdots & \ddots & \vdots \\ \delta_{n1} & \delta_{n2} & \ldots & \delta_{nn} \end{bmatrix},$$

i.e., the family of all matrices. This deformation is also miniversal.

Equation (5.16) may be viewed as an approximation in a neighborhood of 0 of an arbitrary matrix family $A(\delta)$ by a versal family $A'(\eta)$. The versal deformation of a matrix A_0 is a "normal form" into which not only A_0 but also any family of nearby matrices can be put. Of course, this new normal form must itself contain parameters; its advantage is that both the elements of the normal form and the similarity transformation leading to it can be chosen depending smoothly on the elements of the original matrix (varying in a neighborhood of A_0). Note that the versal deformations leave the matrices of the family $A(\delta)$ in the same bundle determined by the Jordan form of A_0 provided the second order errors in the linear approximation (5.2) are sufficiently small.

5.3.2 *The Construction of Versal Deformations*

Since the versal deformation $A(\delta)$ is transversal to the orbit of A_0, the problem for finding a miniversal deformation of A_0 can be reduced to the algebraic problem for finding a complementary subspace $N_{\mathcal{O}} = \mathcal{N}(\mathcal{L}_{A_0})$ to the tangent subspace $T_{\mathcal{O}} = \mathcal{R}(\mathcal{L}_{A_0})$. Remember that according to the presentation in Section 5.1.2, $\mathcal{L}_{A_0}X = [A_0, X] = A_0X - XA_0$ and $\dim(T_{\mathcal{O}}) + \dim(N_{\mathcal{O}}) = n^2$.

We introduce in the space of matrices $\mathbb{C}^{n^2}$ the Hermitian scalar product $\langle A, B\rangle = \text{tr}(AB^H)$, where B^H is a matrix obtained from B by transposition and complex conjugation. The trace of the corresponding scalar product represents the sum of the squares of the moduli of the product elements.

We have the following result.

Lemma 5.13 (Arnold (1971)). *The vector B from the tangent space to $\mathbb{C}^{n^2}$ at the point A_0 is perpendicular to the orbit of the matrix A_0, i.e., lies in the normal space to the orbit, if and only if $[B^H, A_0] = B^H A_0 - A_0 B^H = 0$.*

In fact, the tangent vectors to the orbit of A_0 are the matrices that can be represented in the form $[X, A_0]$. Since B is orthogonal to the orbit for any X, we have that $\langle [X, A_0], B\rangle = 0$. Otherwise speaking, for any X we have that

$$\begin{aligned} 0 &= \operatorname{tr}([X, A_0], B^H) = \operatorname{tr}(A_0 B^H - A_0 X B^H) \\ &= \operatorname{tr}([A_0, B^H], X) = \langle [A_0, B^H], X\rangle. \end{aligned}$$

Since X is arbitrary, this condition is equivalent to $[A_0, B^H] = 0$. In this way, the orthogonal complement to the orbit of A_0 is the Hermitian transpose of its centralizer and the determining of the miniversal deformation reduces to the solution of the matrix equation $B^H A_0 - A_0 B^H = 0$ considered in Section 5.1.2. Assuming that A_0 has one eigenvalue λ with Segre characteristic $\mathcal{S} = \{s_{1i},\ s_{2i}, \ldots, s_{qi}\}$ and taking into account that the solution of $XA_0 - A_0 X = 0$ has the form shown in Figure 5.7, we obtain the following result.

Lemma 5.14. *The matrix B that satisfies $B^H A_0 - A_0 B^H = 0$ has the form shown in Figure* 5.27 *for $q_i = 3$ in which all entries that are not on the black segments are equal to zero.*

If the Jordan form of A_0 has several distinct eigenvalues, we divide it into blocks corresponding to these eigenvalues. Then matrices commuting with A_0 are block-diagonal, where a block corresponding to an eigenvalues λ_i has the form described in Figure 5.27.

Hence, for a transversal deformation of a matrix A_0, we may choose the family of the form $A(\delta) = A_0 + B$ in which the number of nonzero entries in B is minimal and equal to the number of parameters, i.e., $A(\delta)$ is a miniversal deformation.

It follows from the earlier results that a basis $(B_1, B_2, \ldots, B_k)$ of a complementary subspace $N_{\mathcal{O}}$ to $\mathcal{R}(\mathcal{L}_{A_0})$ can be chosen such that each B_i is a matrix with only one entry on one of the oblique segments described in Lemma 5.13 equal to 1 and all other entries being zero.

Thus

$$A(\delta) = A_0 + \sum_{i=1}^{k} \delta_i B_i \tag{5.18}$$

is a miniversal deformation of A_0.

5.3.3 *Selection of the Normal Form*

Since the matrix B has many nonzero elements, it is possible to use other forms of the versal family $A_0 + B$ in which the number of the nonzero entries of B is minimal and equal to the number of parameters. To obtain a family transversal to the orbit, it is sufficient to take a matrix B for which on each oblique segment in Figure 5.27 one entry as an independent parameter and all the other entries are taken as zeros. The nonzero entry can be chosen on each oblique segment at an arbitrary place. In Figure 5.28, we present two types of normal forms containing the minimum possible number of parameters. Other forms are also possible which confirms the observation that the miniversal deformation is not unique. Note that although the corresponding families remain transversal to the orbit, the forms in Figure 5.28 may not satisfy the orthogonality condition $B^H A_0 - A_0 B^H = 0$, differently to the form presented in Figure 5.27.

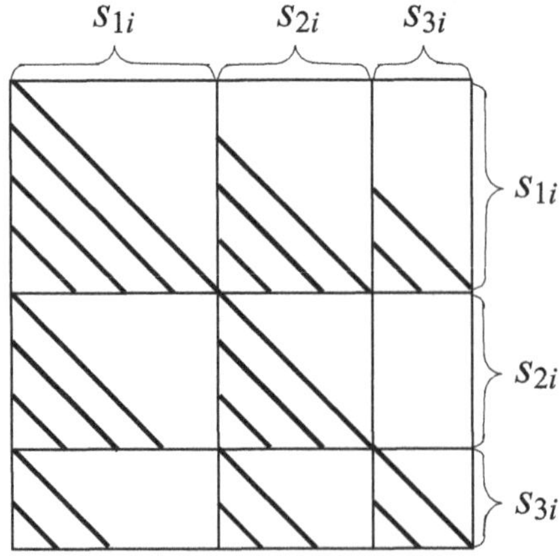

Fig. 5.27. A vector orthogonal to the orbit.

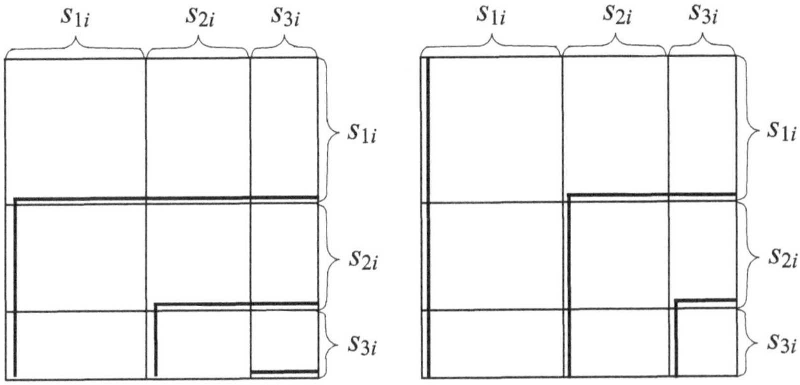

Fig. 5.28. Two types of normal forms.

The earlier considerations can be summarized by the following theorem.

Theorem 5.15. *If A_0 is in Jordan normal form with p distinct eigenvalues $\lambda_1, \lambda_2, \ldots, \lambda_p$ with associated Segre characteristics*

$$\mathcal{S}_i = \{s_{1i} \geq s_{2i} \geq \cdots, \geq s_{q_i,i},\ i = 1, 2, \ldots, p\},$$

then the dimension of the parameter space of a miniversal deformation of A_0 is

$$D = \sum_{i=1}^{p} (s_{1i} + 3s_{2i} + 5s_{3i} + \cdots + (2q_i - 1)s_{q_i,i})$$

and a miniversal deformation of A_0 is block-diagonal, each of the blocks being a miniversal deformation described by (5.18) *for the block of A_0 corresponding to a distinct eigenvalue.*

If the dimension k of the base Δ of $A(\delta)$ is greater than the minimum D (i.e., the codimension of the orbit), we can replace the base by a submanifold whose dimension is equal to the codimension of the orbit and such that the restriction of A_0 to this subspace is still transversal to the orbit. It may be proved that the resulting deformation is versal (Arnold, 1971).

In conclusion, any complex matrix close to the given matrix A_0 can be reduced to the D-parameter normal form $A_0 + B$, where A_0 is the Jordan form of the given matrix, so that the reduced mapping and the parameters of the normal form depend holomorphically on the elements of the given matrix.

5.3.4 *Examples*

Example 5.16. Consider several low-order examples for constructing of versal deformations of matrices.

(1) A versal and universal two-parameter deformation of a 2×2 Jordan block:

$$A(\delta) = \begin{bmatrix} \lambda & 1 \\ 0 & \lambda \end{bmatrix} + \begin{bmatrix} 0 & 0 \\ \delta_1 & \delta_2 \end{bmatrix}.$$

(2) A versal but not universal four-parameter deformation of a 2×2 matrix with semisimple eigenvalues:

$$A(\delta) = \begin{bmatrix} \lambda & 0 \\ 0 & \lambda \end{bmatrix} + \begin{bmatrix} \delta_1 & \delta_2 \\ \delta_3 & \delta_4 \end{bmatrix}.$$

(3) A versal and universal three-parameter deformation of a 3×3 Jordan block:

$$A(\delta) = \begin{bmatrix} \lambda & 1 & 0 \\ 0 & \lambda & 1 \\ 0 & 0 & \lambda \end{bmatrix} + \begin{bmatrix} \delta_1 & 0 & 0 \\ \delta_2 & \delta_1 & 0 \\ \delta_3 & \delta_2 & \delta_1 \end{bmatrix}.$$

Other versions of the deformation are also possible, for instance,

$$A(\delta) = \begin{bmatrix} \lambda & 1 & 0 \\ 0 & \lambda & 1 \\ 0 & 0 & \lambda \end{bmatrix} + \begin{bmatrix} 0 & 0 & 0 \\ 0 & 0 & 0 \\ \delta_1 & \delta_2 & \delta_3 \end{bmatrix}$$

or

$$A(\delta) = \begin{bmatrix} \lambda & 1 & 0 \\ 0 & \lambda & 1 \\ 0 & 0 & \lambda \end{bmatrix} + \begin{bmatrix} \delta_1 & 0 & 0 \\ \delta_2 & 0 & 0 \\ \delta_3 & 0 & 0 \end{bmatrix}.$$

(4) A versal five-parameter deformation of a matrix with a multiple eigenvalue participating in 2×2 and 1×1 Jordan blocks (the dimension of the family is $\operatorname{codim}(\mathcal{O}(A_0)) = 2 + 3.1 = 5$):

$$A(\delta) = \left[\begin{array}{cc|c} \lambda & 1 & 0 \\ 0 & \lambda & 0 \\ \hline 0 & 0 & \lambda \end{array}\right] + \left[\begin{array}{cc|c} 0 & 0 & 0 \\ \delta_1 & \delta_2 & \delta_3 \\ \hline \delta_4 & 0 & \delta_5 \end{array}\right].$$

Other possible deformation is

$$A(\delta) = \left[\begin{array}{cc|c} \lambda & 1 & 0 \\ 0 & \lambda & 0 \\ \hline 0 & 0 & \lambda \end{array}\right] + \left[\begin{array}{cc|c} \delta_1 & 0 & 0 \\ \delta_2 & 0 & \delta_4 \\ \hline \delta_3 & 0 & \delta_5 \end{array}\right].$$

(5) A miniversal three-parameter deformation of a matrix A_0 with two distinct eigenvalues participating in 2×2 and 1×1 Jordan blocks, respectively ($\operatorname{codim}(\mathcal{O}(A_0)) = 2 + 1 = 3$),

$$A(\delta) = \left[\begin{array}{cc|c} \lambda_1 & 1 & 0 \\ 0 & \lambda_1 & 0 \\ \hline 0 & 0 & \lambda_2 \end{array}\right] + \left[\begin{array}{cc|c} \delta_1 & 0 & 0 \\ \delta_2 & \delta_1 & 0 \\ \hline 0 & 0 & \delta_3 \end{array}\right].$$

(6) A miniversal deformation of a 5×5 matrix A_0 with with a single multiple eigenvalue participating in 3×3 and 2×2 Jordan blocks ($\operatorname{codim}(\mathcal{O}(A_0)) = 3 + 3.2 = 9$):

$$A(\delta) = \left[\begin{array}{ccc|cc} \lambda & 1 & 0 & 0 & 0 \\ 0 & \lambda & 1 & 0 & 0 \\ 0 & 0 & \lambda & 0 & 0 \\ \hline 0 & 0 & 0 & \lambda & 1 \\ 0 & 0 & 0 & 0 & \lambda \end{array}\right] + \left[\begin{array}{ccc|cc} 0 & 0 & 0 & 0 & 0 \\ 0 & 0 & 0 & 0 & 0 \\ \delta_1 & \delta_2 & \delta_3 & \delta_4 & \delta_5 \\ \hline \delta_6 & 0 & 0 & 0 & 0 \\ \delta_7 & 0 & 0 & \delta_8 & \lambda_9 \end{array}\right].$$

(7) A miniversal deformation of a 5×5 matrix A_0 with two multiple eigenvalues λ_1 and λ_2 with Segre characteristics $\mathcal{S}_1 = \{2,1\}$ and $\mathcal{S}_2 = \{2\}$, respectively $(\operatorname{codim}(\mathcal{O}(A_0)) = 2 + 3.1 + 2 = 7)$:

$$A(\delta) = \left[\begin{array}{ccc|cc} \lambda_1 & 1 & 0 & 0 & 0 \\ 0 & \lambda_1 & 0 & 0 & 0 \\ 0 & 0 & \lambda_1 & 0 & 0 \\ \hline 0 & 0 & 0 & \lambda_2 & 1 \\ 0 & 0 & 0 & 0 & \lambda_2 \end{array}\right] + \left[\begin{array}{ccc|cc} 0 & 0 & 0 & 0 & 0 \\ \delta_1 & \delta_2 & \delta_3 & 0 & 0 \\ \delta_4 & 0 & \delta_5 & 0 & 0 \\ \hline 0 & 0 & 0 & 0 & 0 \\ 0 & 0 & 0 & \delta_6 & \lambda_7 \end{array}\right].$$

(8) A miniversal deformation of a Jordan block A_0 of size n with an eigenvalue λ (the dimension of the family is $\operatorname{codim}(\mathcal{O}(A_0)) = n$):

$$A(\delta) = \begin{bmatrix} \lambda & 1 & & & \\ & \lambda & \ddots & & \\ & & \ddots & 1 & \\ & & & \lambda & 1 \\ & & & & \lambda \end{bmatrix} + \begin{bmatrix} \delta_1 & 1 & & & \\ \delta_2 & \delta_1 & \ddots & & \\ \vdots & \vdots & \ddots & 1 & \\ \delta_{n-1} & \delta_{n-2} & \dots & \delta_1 & 1 \\ \delta_n & \delta_{n-1} & \dots & \delta_2 & \delta_1 \end{bmatrix}.$$

Other deformations with the same dimension are also possible.

(9) An universal deformation of the $n \times n$ companion matrix

$$A_0 = \begin{bmatrix} 0 & 1 & & & \\ & 0 & \ddots & & \\ & & & 1 & \\ & & & 0 & 1 \\ -p_n & -p_{n-1} & \dots & -p_2 & -p_1 \end{bmatrix}.$$

The companion matrix has one Jordan block for each distinct eigenvalue (it is a nonderogatory matrix) so that

$\text{codim}(\mathcal{O}(A_0)) = n$). The elements $p_1, p_2, \ldots, p_n$ are the coefficients of the characteristic polynomial

$$\det(\lambda I_n - A_0) = \lambda^n + p_1\lambda^{n-1} + \cdots + p_n$$

which are uniquely determined. Since the dimension of the deformation is n, the matrices in the corresponding family $A(\delta)$ can be chosen in companion form like A_0. Such a family is called *Sylvester family.* It follows that if a matrix A_0 can be reduced to a companion form, then also matrices close to A_0 can be reduced to companion form without loosing their smooth dependence on the parameters.

Example 5.17. Consider the tenth-order matrix A presented in Example 5.1, which has three distinct eigenvalues

$$\lambda_1 = -1, \quad \lambda_2 = 2, \quad \lambda_3 = 3$$

with multiplicities $\nu_1 = 1$, $\nu_2 = 5$ and $\nu_3 = 4$, respectively, and associated Segre characteristics

$$\begin{aligned} \mathcal{S}_1 &= \{s_{11}\} = \{1\}, \\ \mathcal{S}_2 &= \{s_{12}, s_{22}\} = \{3, 2\}, \\ \mathcal{S}_3 &= \{s_{13}, s_{23}\} = \{2, 2\}. \end{aligned}$$

This matrix lies in a bundle with codimension equal to $\text{codim}(\mathcal{B}(A)) = 1 + 3 + 3.2 + 2 + 3.2 - 3 = 15$. Taking into account the form of the solution Y of the matrix equation $JY = YJ$, where $J = Z^{-1}AZ$, the versal deformation of J can be chosen in the form

$$B = \left[\begin{array}{c|ccc|cc|cc|cc}
\delta_1 & & & & & & & & & \\ \hline
 & \delta_2 & 0 & 0 & 0 & 0 & & & & \\
 & \delta_3 & \delta_2 & 0 & \delta_5 & 0 & & & & \\
 & \delta_4 & \delta_3 & \delta_2 & \delta_6 & \delta_5 & & & & \\ \hline
 & \delta_7 & 0 & 0 & \delta_9 & 0 & & & & \\
 & \delta_8 & \delta_7 & 0 & \delta_{10} & \delta_9 & & & & \\ \hline
 & & & & & & \delta_{11} & 0 & \delta_{13} & 0 \\
 & & & & & & \delta_{12} & \delta_{11} & \delta_{14} & \delta_{13} \\ \hline
 & & & & & & \delta_{15} & 0 & \delta_{17} & 0 \\
 & & & & & & \delta_{16} & \delta_{15} & \delta_{18} & \delta_{17}
\end{array}\right]$$

or

$$C = \left[\begin{array}{c|ccc|cc|cc|cc}
\delta_1 & & & & & & & & & \\ \hline
 & & & & & & & & & \\
 & & & & & & & & & \\
 & \delta_4 & \delta_3 & \delta_2 & \delta_6 & \delta_5 & 0 & 0 & 0 & 0 \\ \hline
 & \delta_7 & & & & & & & & \\
 & \delta_8 & & & \delta_{10} & \delta_9 & 0 & 0 & 0 & 0 \\ \hline
 & 0 & & & 0 & & & & & \\
 & 0 & & & 0 & & \delta_{12} & \delta_{11} & \delta_{14} & \delta_{13} \\ \hline
 & 0 & & & 0 & & \delta_{15} & & & \\
 & 0 & & & 0 & & \delta_{16} & & \delta_{18} & \delta_{17}
\end{array}\right],$$

where $\delta_1, \delta_2, \ldots, \delta_{18}$ are free parameters. Since $A = ZJZ^{-1}$, the corresponding perturbations of the matrix A are found as $\Delta A_B = ZBZ^{-1}$ and $\delta A_C = ZCZ^{-1}$. To find numerically the Jordan form of A, the free parameters are chosen as shown in Table 5.4, where s is a variable scaling factor.

The computation of the Jordan forms of the matrices $A + \Delta A_B$ and $A + \delta A_C$ is done by the Algorithm 7.5 presented in Chapter 7 for different values of s. The numerical experiments show that these matrices remain in a bundle with codimension equal to 15 for perturbations with norms up to $\|\Delta A_B\|_F = 1.00059 \cdot 10^{-6}$ and $\|\Delta A_C\|_F = 4.28826 \cdot 10^{-7}$, which correspond to scaling factors $s = 7 \cdot 10^{-11}$ and $s = 3 \cdot 10^{-11}$, respectively. This means that the Jordan form of any matrix $A + \Delta A_B$ remains the same as the Jordan form of A for

$$\|\Delta A_B\|_F \leq 1.00059 \cdot 10^{-6} \tag{5.19}$$

Table 5.4. Parameter values.

Parameter	δ_1	δ_2	δ_3	δ_4	δ_5	δ_6	δ_7	δ_8	δ_9
Value $s\times$	7	5	2	−3	4	−7	−2	8	3
Parameter	δ_{10}	δ_{11}	δ_{12}	δ_{13}	δ_{14}	δ_{15}	δ_{16}	δ_{17}	δ_{18}
Value $s\times$	5	−6	2	−8	4	−5	−3	−1	9

provided that the versal perturbation ΔA_B satisfy (5.19). Note that $\|JB^H - B^H J\|_2 = 2.58 \cdot 10^{-26}$, while $\|JC^H - C^H J\|_2 = 6.08 \cdot 10^{-10}$ which results from the nonorthogonality of C to the orbit of J.

For perturbations with larger norms, both matrices have Jordan forms with Segre characteristics

$$\begin{aligned} \mathcal{S}_1 &= \{s_{11}\} = \{1\}, \\ \mathcal{S}_2 &= \{s_{12}, s_{22}\} = \{3, 2\}, \\ \mathcal{S}_3 &= \{s_{13}, s_{23}\} = \{3, 1\}, \end{aligned}$$

which correspond to a bundle with codimension equal to $1+3+3.2+3+3.1-3 = 13$. These perturbations are not versal, since the second-order errors in the linear approximation (5.2) are no longer negligible. Note that if we try to compute the Jordan form of the matrix $A+\Delta A$, where ΔA is a random matrix with $\|\Delta A\|_F = 1.00059 \cdot 10^{-6}$, we obtain the Segre characteristics

$$\begin{aligned} \mathcal{S}_1 &= \{s_{11}\} = \{1\}, \\ \mathcal{S}_2 &= \{s_{12}, s_{22}\} = \{5\}, \\ \mathcal{S}_3 &= \{s_{13}, s_{23}\} = \{4\}, \end{aligned}$$

which shows that the perturbed matrix goes in a bundle with codimension equal to $1 + 5 + 4 - 3 = 7$. The reason for this transition is that although with the same norm as ΔA_B and ΔA_C, the random perturbation is not versal.

The construction of versal forms with the minimal number of parameters is closely related to the analysis of the bifurcation diagrams of a family of matrices, considered in the following section.

5.4 Bifurcation Diagrams

The word *bifurcation* means division into two and is generally used to indicate every qualitative, topological change in the geometrical picture of a mathematical object under the variation of parameters on which the object depends. As noted previously, if a matrix depends on parameters, we say that it is given a family. If we are interested in the family locally, i.e., for small changes of the parameters in the vicinity of fixed values, then we speak about deformations of

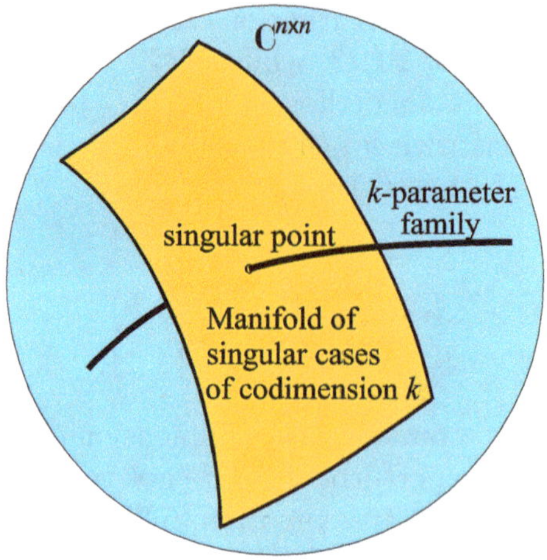

Fig. 5.29. A generic k-parameter family and the manifold of singular cases.

the matrix corresponding to these small changes of the parameters. Matrix families which are transverse to all bundles are called *generic families.*

In Figure 5.29, we represent symbolically a k-parameter generic matrix family which intersects transversally the hypersurface of *singular cases* (the intersection occurs at "nonzero angle"). A family with k parameters is represented by a k-dimensional manifold in the matrix space. For example, a one-parameter family is a curve in the three-dimensional matrix space with a codimension equal to 2 and the manifold of singular cases has a codimension equal to 1. The manifold of singular cases consists of all matrices which are defective or/and derogatory. It follows from the transversality that in the general case the manifold of singular cases has dimension $n^2 - k$ and a codimension equal to k. Hence, the codimension of the manifold of singular cases is equal to the number of parameters that determine the matrix family. The one-dimensional family contains matrices with simple eigenvalues except in the *singular points*, where the curve representing the one-parameter family intersects the manifold of singular cases. Since the singular points have dimensions equal to zero, in the generic families, almost all matrices have simple eigenvalues. The singular point corresponds to a *bifurcation value* of the parameter. As the parameter changes, at the moment it passes through a bifurcation value, the number and type of singular points (the Jordan structure of the matrix) change — the object undergoes a "qualitative change" or bifurcation. Note that a singular point can be moved to the domain of generic cases by an arbitrarily small perturbation of the matrix elements but cannot be removed at all from the family.

Corresponding to the partition of the matrix space in bundles, the parameter space can be partitioned into subsets. The exceptional values of the parameters to which correspond matrices with multiple eigenvalues (the singular cases) constitute a subset in the parameter space. This subset is called *bifurcation diagram*. The bifurcation diagram of a generic family of matrices represents a finite union of manifolds — to each bundle of orbits corresponds its own manifold in the parameter space $\mathcal{P}$. The codimension k of a manifold in the parameter space of a generic family is equal to the codimension of the corresponding bundle in the space of all matrices, i.e., to the number of parameters determining the matrix family. Therefore, the bifurcation diagram of a family of matrices is a partition of the parameter space $\mathcal{P}$ according to Jordan types of matrices. The partition consists of grouping together the matrices with the same dimensions of the Jordan blocks differing only in the eigenvalues. Thus, the bifurcation diagrams allow studying the partition of space of matrices into matrices with Jordan forms of distinct types. This makes them a useful tool to analyze the qualitative metamorphosis (or "catastrophes") of a matrix family.

The partition of the space in a finite number of manifolds set by algebraic equations and inequalities is called *stratification*. The partition of the space of matrices into matrices with the same dimensions of the Jordan blocks groups the matrices into bundles of equal codimensions and represents a finite stratification of the space of matrices. In the space of families of matrices of order n, the families transversal to the stratification into Jordan types constitute an everywhere dense set. Clearly, such families consist of matrices with simple eigenvalues.

In this section, we represent the bifurcation diagrams of matrix families with a small number of parameters ($k \leq 3$). From the

Table 5.5. Dimensions of objects in different parameter spaces.

Parameter space	Codimension of the bundle in the parameter space	Dimension of the bundle in the parameter space	Type of the object in the parameter space
$\mathbb{C}^1$	1	0	Point
	0	1	Line
$\mathbb{C}^2$	2	0	Point
	1	1	Curve
	0	2	Plane
$\mathbb{C}^3$	3	0	Point
	2	1	Curve
	1	2	Surface
	0	3	3D-space

considerations given earlier, it follows that the number of parameters is equal to the codimension $\mathrm{codim}(\mathcal{B})$ of the bundle $\mathcal{B}$ associated with the matrix family.

In constructing the bifurcation diagrams, it is useful to know the dimensions of the distinct objects in the parameter spaces $\mathbb{C}^1, \mathbb{C}^2$ and $\mathbb{C}^3$. For a bundle with a given codimension in the matrix space, the data in Table 5.5 allow determining the dimension and type of the corresponding manifold in the parameter space. Note that objects with the same dimension in the different parameter spaces have different codimensions in these spaces.

5.4.1 *One-Parameter Matrix Families*

In a generic one-parameter family of matrices, there are only matrices with a simple eigenvalue α ($\mathrm{codim}(\mathcal{B}) = 0$) and matrices with a single Jordan block with $\mathrm{codim}(\mathcal{B}) = 1$. In such a family becomes unremovable a degeneration of codimension one, corresponding to a hypersurface (a submanifold given by one equation) of the matrix space under consideration. From $\mathrm{codim}(\mathcal{B}) = 1$, it follows that the corresponding bundle of matrices has only one eigenvalue α of multiplicity 2 participating in a Jordan block of order 2:

$$J_2(\alpha) = \begin{bmatrix} \alpha & 1 \\ 0 & \alpha \end{bmatrix}.$$

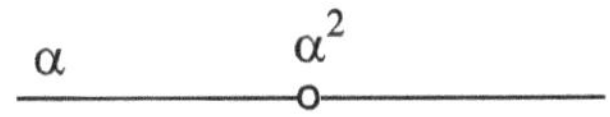

Fig. 5.30. Bifurcation diagram for α^2.

Matrices with such Jordan form are defective but nonderogatory and are similar to the companion matrix

$$A = \begin{bmatrix} 0 & 1 \\ -\alpha^2 & 2\alpha \end{bmatrix}.$$

For simplicity, a stratum of this type will be denoted by α^2. Hence, in a generic family of matrices depending on one parameter, there are only matrices with a simple eigenvalue α, and for some isolated values of the parameter, there are matrices of type α^2 (with one Jordan block of order 2 and $\operatorname{codim}(\mathcal{B}) = 1$). The bifurcation diagram of a generic family of matrices depending on one parameter is shown in Figure 5.30. If in a family there are matrices with a more complicated Jordan structure, then they can be removed by an arbitrarily small perturbation of the family. This follows from the more general fact that in the study of generic k-parameter families, the only unremovable degeneracies are those of codimension not exceeding k. All other degenerate objects form a set of codimension greater than k in the matrix space, and we can get rid of them by an arbitrarily small deformation of the k-parameter family (Arnol'd, 1994, Part II, Section 4.3).

5.4.2 *Two-Parameter Matrix Families*

In generic two-parameter families, one unavoidably finds degenerations of codimension two (but not of codimension three). Codimensions equal to two have the bundles of matrices with Jordan type α^3 (one Jordan block of size 3) and $\alpha^2\beta^2$ (two Jordan blocks $J_2(\alpha)$ and $J_2(\beta)$ of order 2 with distinct eigenvalues α and β). Matrices of this type are defective and nonderogatory.

Further on we shall call the bundles whose matrices have only one multiple eigenvalue, *elementary*. The rest bundles are in fact transverse intersections of the elementary. In particular, the codimension of a complex singularity corresponding to several multiple eigenvalues is equal to the sum of the codimensions of the elementary components. In the given case, this allows considering separately the bundles α^2 and β^2.

Consider first the case of a family of third-order companion matrices

$$A = \begin{bmatrix} 0 & 1 & 0 \\ 0 & 0 & 1 \\ -a_3 & -a_2 & -a_1 \end{bmatrix}.$$

The triple eigenvalue α is obtained as the root of the third-degree characteristic equation

$$z^3 + a_1 z^2 + a_2 z + a_3 = 0. \tag{5.20}$$

Using the substitution

$$z = t - a/3,$$

equation (5.20) is reduced to the depressed cubic equation

$$t^3 + ct + d = 0, \tag{5.21}$$

where

$$c = a_2 - a_1^2/3, \quad d = 2a_1^3/27 - a_1 a_2/3 + a_3.$$

The representation (5.21) confirms that the parameter space corresponding to the bundle α^3 has dimension equal to 2 (the same as the bundle codimension). The roots of (5.21) are given by

$$\begin{aligned} t_1 &= -\frac{1}{2}(S+T) + \frac{1}{2} i_o \sqrt{3}(S-T), \\ t_2 &= -\frac{1}{2}(S+T) - \frac{1}{2} i_o \sqrt{3}(S-T), \\ t_3 &= S+T, \end{aligned} \tag{5.22}$$

where

$$\begin{aligned} Q &= \frac{c}{3}, \\ R &= -\frac{d}{2}, \end{aligned}$$

$$S = \sqrt[3]{R + \sqrt{Q^3 + R^2}},$$
$$T = \sqrt[3]{R - \sqrt{Q^3 + R^2}}.$$

The roots of the original equation (5.20) are linked to t_1, t_2, t_3 as

$$\alpha_i = t_i - a_1/3, \quad i = 1, 2, 3.$$

The case of double eigenvalues $\alpha = \alpha_1 = \alpha_2 = -R^{1/3} - \frac{1}{3}a_1$ corresponds to zero discriminant $D = Q^3 + R^2 = 0$, i.e., $R^2 = -Q^3$ and $S = T = R$. In this way, we obtain two curves in the plane of parameters c and d, described by the equations

$$Q = \frac{c}{3},$$
$$R = \pm\sqrt{Q^3},$$
$$d = -2R.$$

The case of triple root is obtained for $R = 0$ which implies $Q = 0$, $c = 0$, $d = 0$ and $t_1 = t_2 = t_3 = 0$. Hence the triple root of (5.20) is equal to $\alpha = \alpha_1 = \alpha_2 = \alpha_3 = -\frac{1}{3}a_1$.

The bifurcation diagram of the bundle of matrices with Jordan block α^3 represents a semicubical parabola — a curve with a singular point in the form of a cusp corresponding to the triple eigenvalue α (Figure 5.31).

Consider now the case of an elementary bundle of matrices of type β^2. In this case, the eigenvalue β is determined as the double root

$$\beta = -a_1/2 \tag{5.23}$$

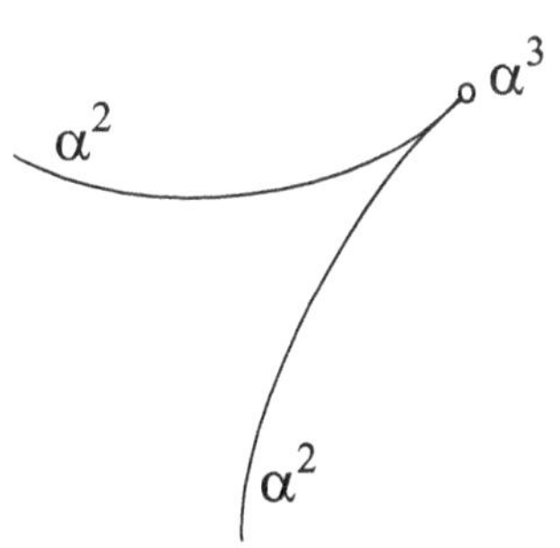

Fig. 5.31. Bifurcation diagram for α^3.

of the quadratic equation $z^2 + a_1 z + a_2 = 0$. Such eigenvalue is obtained for values of the parameters a_1 and a_2 that satisfy

$$D = a_1^2 - 4a_2 = 0. \tag{5.24}$$

Equation (5.24) is parameterized as

$$c^2 + d^2 = r^2, \tag{5.25}$$

where $c = r\cos(\Theta), d = r\sin(\Theta), 0 \leq \Theta \leq 2\pi$ and $r = a_1 = 2\sqrt{a_2}$.

Equation (5.25) describes a circle in the plane of parameters c and d with a radius equal to $2|\beta|$.

The earlier considerations indicate that the bifurcation diagram of a bundle of matrices of type $\alpha^2\beta^2$ (two Jordan blocks of order 2 with distinct eigenvalues) has a singular point in the form of self-intersection (Figure 5.32). Thus, the bifurcation diagram of a generic two-parameter family of matrices represents a curve whose singularities are cusps and points of self-intersection. Matrices with one Jordan block of order 2 correspond to the regular points of the curve and matrices with simple eigenvalues correspond to points outside the curve. Similar to the case of one-parameter families, if a family contains matrices of more complicated Jordan forms or the bifurcation diagram has more complicated singularities, then we can remove them by an arbitrarily small perturbation of the family.

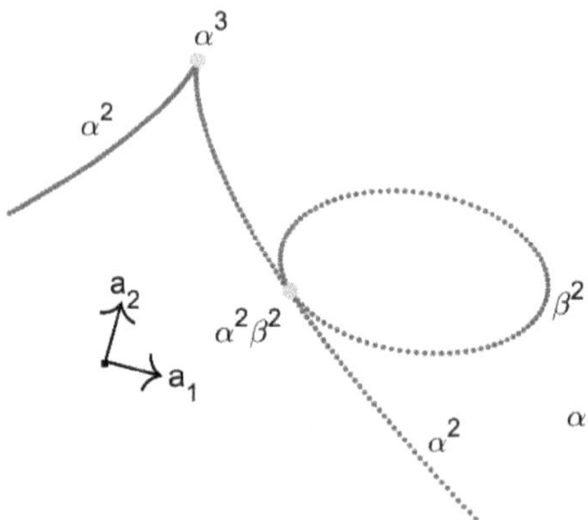

Fig. 5.32. Bifurcation diagram for α^3 and $\alpha^2\beta^2$.

5.4.3 *Three-Parameter Matrix Families*

Codimension equal to 3 have four types of bundles: $\alpha\alpha$ (two Jordan blocks of order 1 with the same eigenvalues), $\alpha^3\beta^2$ (two blocks of orders 3 and 2 with distinct eigenvalues), α^4 (one 4-by-4 Jordan block) and $\alpha^2\beta^2\gamma^2$ (three 2-by-2 blocks with distinct eigenvalues). All more complex singularities in the three-dimensional space resolve into these types of bundles on small parameter variations.

5.4.3.1 *The Case $\alpha\alpha$*

The degenerate matrices pertaining to this family have Jordan form

$$J = \begin{bmatrix} \alpha & 0 \\ 0 & \alpha \end{bmatrix}$$

consisting of two 1×1 blocks. The matrices $A = PJP^{-1} = \alpha I_2$ are nondefective but derogatory. In the given case, the eigenvalue $\alpha = -a_1/2$ is determined as the double root of the quadratic equation $z^2 + a_1 z + a_2 = 0$ for values of the parameters a_1 and a_2 satisfying

$$D = a_1^2 - 4a_2 = 0. \tag{5.26}$$

Equation (5.26) is parameterized as

$$\sqrt{x_1^2 + x_2^2} = \pm r, \tag{5.27}$$

where $x_1 = r\cos(\Theta), x_2 = r\sin(\Theta), 0 \leq \Theta \leq 2\pi, r = 2\sqrt{a_2}$.

The expression (5.27) describes a manifold in $\mathbb{C}^3$ in the form of a cone with center at $r = 0$ corresponding to the double semisimple eigenvalue $\alpha\alpha$ (Figure 5.33). The point $r = 0$ is a singular point of the manifold and the cone surfaces for $r \neq 0$ correspond to bundles of the type α^2 with codimension 1 (single Jordan block of order 2). The points $\alpha\beta$ outside the bifurcation diagram represent matrices with two simple eigenvalues.

It is necessary to note that the presentation in this section is relevant to the complex case and the surfaces shown in the corresponding figures should be considered complex.

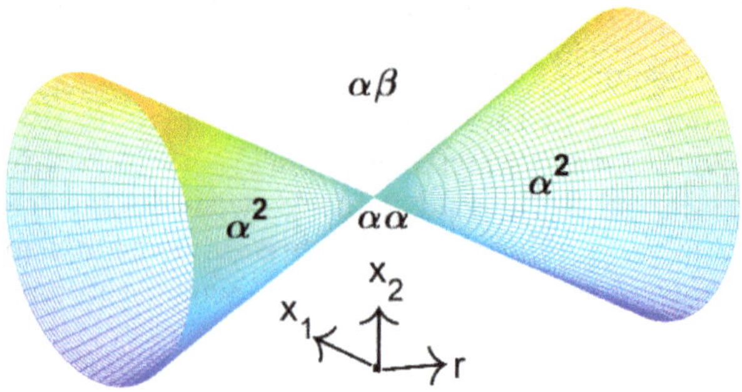

Fig. 5.33. Bifurcation diagram for $\alpha\alpha$.

5.4.3.2 *The Case* $\alpha^3\beta^2$

This bundle contains matrices that are defective but nonderogatory. The triple eigenvalue α is obtained as a root of the third-degree equation

$$z^3 + a_1 z^2 + a_2 z + a_3 = 0, \tag{5.28}$$

which was already considered in connection with the bundle α^3. From the condition for a double eigenvalue $\alpha = R^{1/3} - \frac{1}{3}a_1$, we obtain two surfaces in the space of parameters a_1, a_2 and a_3, described by the equations

$$\begin{aligned} Q &= \frac{3a_2 - a_1^2}{9}, \\ R &= \pm\sqrt{Q^3}, \\ a_3 &= -2a_1^3/27 + a_1 a_2/3 - 2R. \end{aligned}$$

The case of triple eigenvalue $\alpha = \alpha_1 = \alpha_2 = \alpha_3 = -\frac{1}{3}a_1$ is obtained for $R = 0$ which implies $Q = 0$ and $a_2 = a_1^2/3$, $a_3 = -2a_1^3/27 + a_1 a_2/3$.

In Figure 5.34, we show the bifurcation diagram of a generic matrix family containing a third-order Jordan block α^3 in the three-dimensional space of parameters (a_1, a_2, a_3). The two surfaces representing matrices with double eigenvalues intersect in a curve with a codimension equal to 2 corresponding to the Jordan block $J_3(\alpha)$. The points $\alpha\beta\gamma$ outside the bifurcation diagram represent matrices with simple eigenvalues.

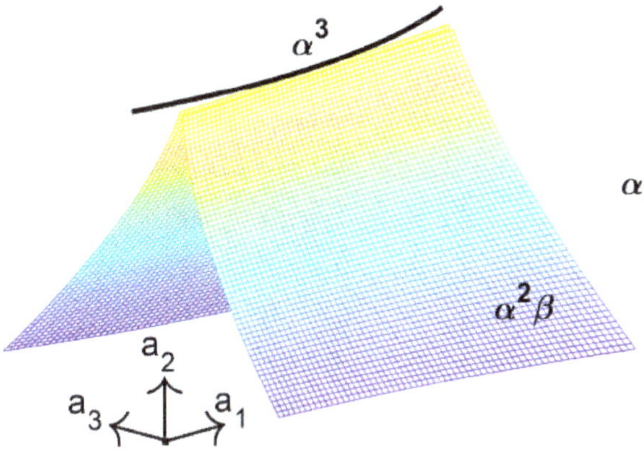

Fig. 5.34. Bifurcation diagram of a matrix family containing a third-order Jordan block.

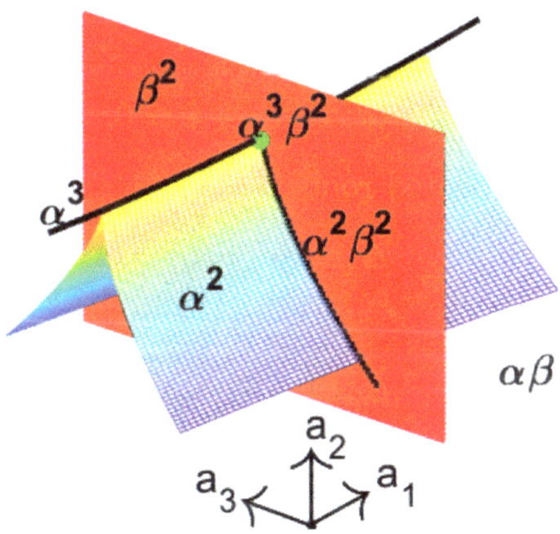

Fig. 5.35. Bifurcation diagram for $\alpha^3\beta^2$.

The bifurcation diagram for the bundle of matrices of type $\alpha^3\beta^2$ is shown in Figure 5.35. The bundle of the type β^2 (one Jordan block of order 2) with a codimension equal to 1 corresponds to a plane in $\mathbb{C}^3$ which is transverse to the line α^3. (The changes of β^2 are independent on the changes of α^3.) The bundle $\alpha^2\beta^2$ with codimension equal to 2 is represented by a curve which is the intersection of the surfaces α^2 and the plane β^2. Finally, the bundle $\alpha^3\beta^2$ of codimension equal to 3 is represented by a point where the line α^3 intersects the plane β^2.

5.4.3.3 *The Case* α^4

This case illustrates the complicated mutual disposition of the Jordan forms bundles in the parameter space. The family under consideration consists of 4×4 companion matrices of the form

$$A = \begin{bmatrix} 0 & 1 & 0 & 0 \\ 0 & 0 & 1 & 0 \\ 0 & 0 & 0 & 1 \\ -e & -d & -c & 0 \end{bmatrix},$$

where the parameters c, d and e determine the matrix family. The singular cases of this family consist of matrices that are defective and nonderogatory. Note that a fourth-order companion matrix with a 4-tuple eigenvalue has only one Jordan block in the Jordan form. The characteristic equation of the matrix A has the form of a depressed quartic equation

$$z^4 + cz^2 + dz + e = 0.$$

This equation has repeated roots if

$$\Delta(c, d, e) = 0, \tag{5.29}$$

where

$$\Delta(c, d, e) = 16c^4e - 4c^3d^2 - 128c^2e^2 + 144cd^2e - 27d^4 + 256e^3$$

is the discriminant of the quartic polynomial $z^4 + cz^2 + dz + e$.

Equation (5.29) can be represented as a biquadratic equation $a_1d^4 + a_2d^2 + a_3 = 0$ with respect to the parameter d, where

$$\begin{aligned} a_1 &= -27, \\ a_2 &= 144ce - 4c^3, \\ a_3 &= 16c^4e - 128c^2e^2 + 256e^3. \end{aligned}$$

This equation has four roots given by

$$d = \pm\sqrt{\left(-a_2 \pm \sqrt{a_2^2 - 4a_1a_3}\right) / (2a_1)}.$$

Thus, the bifurcation diagram corresponding to repeated roots depends on the two free parameters c and e.

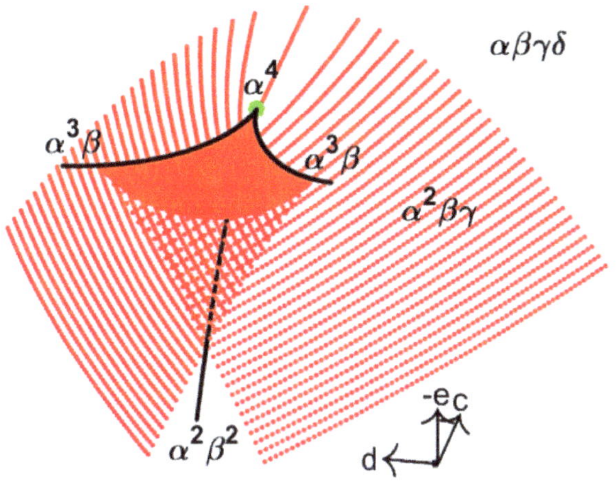

Fig. 5.36. Swallowtail diagram for matrices of type α^4.

Table 5.6. Bundles of Jordan forms represented by the swallowtail diagram.

Notation	Segre characteristics	codim($\mathcal{B}$)	dim($\mathcal{B}$)	Representation in $\mathbb{C}^3$
α^4	$\{4\}$	3	0	Point
$\alpha^3\beta$	$\{3\}, \{1\}$	2	1	Curve
$\alpha^2\beta^2$	$\{2\}, \{2\}$	2	1	Curve
$\alpha^2\beta\gamma$	$\{2\}, \{1\}, \{1\}$	1	2	Surface
$\alpha\beta\gamma\delta$	$\{1\}, \{1\}, \{1\}, \{1\}$	0	3	All points of $\mathbb{C}^3$ outside the swallowtail

The surface representing the points (c, d, e), shown in Figure 5.36, is called *swallowtail*.

According to the data given in Table 5.6, the point α^4 (the swallowtail point) represents the 4×4 Jordan blocks $J_4(\alpha)$ of defective matrices (the most degenerate companion matrices A), the curve $\alpha^3\beta$ (the two cuspidal edges coming out from the swallowtail point) corresponds to the Jordan forms $J_3(\alpha) \oplus J_1(\beta)$, the curve $\alpha^2\beta^2$

(the intersection of the swallowtail wings) represents the Jordan forms $J_2(\alpha) \oplus J_2(\beta)$, the surface $\alpha^2\beta\gamma$ (the swallowtail surface) represents the Jordan forms $J_2(\alpha) \oplus J_1(\beta) \oplus J_1(\gamma)$ and the points $\alpha\beta\gamma\delta$ (everything outside the swallowtail) correspond to the diagonal Jordan forms $J_1(\alpha) \oplus J_1(\beta) \oplus J_1(\gamma) \oplus J_1(\delta)$ of all matrices with simple eigenvalues. The latest case is the most generic one, hence it is the most probable.

Figure 5.36 shows that the numerical determination of the Jordan form has to deal with the complicated geometry of the parameter space and supports Arnold's conclusion that the larger the codimension of degeneracy, the more difficult it is to study this degeneracy (Arnold, 1988, Chapter 6, Section 30).

5.4.3.4 *Case* $\alpha^2\beta^2\gamma^2$

The bifurcation diagram for this case is given in Figure 5.37. The bundles α^2, β^2 and γ^2 consist of matrices that are defective and nonderogatory. These bundles have codimensions equal to 1 and are represented by planes in $\mathbb{C}^3$ which are transverse to each other since the changes of the bundles are independent. The three orthogonal axes represent the bundles $\alpha^2\beta^2$, $\alpha^2\gamma^2$ and $\beta^2\gamma^2$ which have codimensions

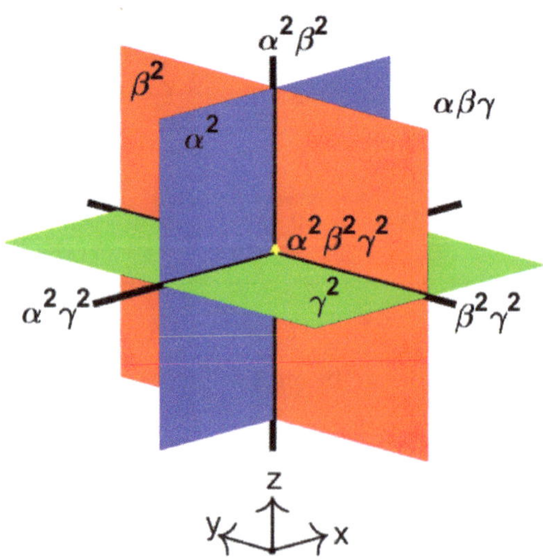

Fig. 5.37. Bifurcation diagram for $\alpha^2\beta^2\gamma^2$.

Table 5.7. List of all Jordan bundles for $k \leq 4$.

k	Bundles of Jordan forms
1	α^2
2	$\alpha^3, \alpha^2\beta^2$
3	$\alpha^4, \alpha\alpha, \alpha^3\beta^2, \alpha^2\beta^2\gamma^2$
4	$\alpha^5, \alpha^2\alpha, \alpha^4\beta^2, \alpha\alpha\beta^2, \alpha^3\beta^2\gamma^2, \alpha^3\beta^3, \alpha^2\beta^2\gamma^2\delta^2$

equal to 2. The intersection point of these axes represent the bundle $\alpha^2\beta^2\gamma^2$ which has a codimension equal to 3.

All cases of Jordan bundles for $k = \text{codim}(\mathcal{B}) \leq 4$ are summarized in Table 5.7. By increasing the number of parameters k, the number of possible Jordan forms increases quickly which makes the usage of the bifurcation diagrams more difficult. It should be noted that the bifurcation diagram of a matrix family in the case of several distinct eigenvalues can be determined as a direct product of diagrams, corresponding to the cases of one multiple eigenvalue.

5.5 Reduction into Jordan Form as an Ill-Posed Problem

Determining the Jordan canonical form of a square nonsymmetric matrix A with defective eigenvalues by the use of a computer is one of the most challenging problems of numerical linear algebra. This is due to two causes. First, deciding which of the eigenvalues are multiple in the presence of rounding errors is associated with serious difficulties. Second, the determination of the sizes of the Jordan blocks associated with a given multiple eigenvalue (the Segre characteristic) is related to the determination of the rank of a matrix and this can be a difficult task working with finite-precision machine arithmetic.

The Jordan canonical form is *structurally unstable* in the sense that it is not a continuous function of the matrix elements. The following example of a 2×2 matrix illustrates the instability of this form.

Example 5.18. Let

$$A(\varepsilon) = \begin{bmatrix} 0 & 1 \\ \varepsilon & 0 \end{bmatrix},$$

where ε is a small positive number. This matrix has the Jordan canonical form

$$J = \begin{bmatrix} 0 & 1 \\ 0 & 0 \end{bmatrix}$$

for $\varepsilon = 0$ and

$$J = \begin{bmatrix} -\sqrt{\varepsilon} & 0 \\ 0 & \sqrt{\varepsilon} \end{bmatrix}$$

for $\varepsilon > 0$. Clearly, the Jordan canonical form of $A(\varepsilon)$ change its structure by jumping at $\varepsilon = 0$, i.e., it is not continuous at $\varepsilon = 0$. For small ε, the diagonalizing nonsingular transformation matrix

$$V = \begin{bmatrix} 1 & 1 \\ -\sqrt{\varepsilon} & \sqrt{\varepsilon} \end{bmatrix}$$

is ill conditioned. In fact, we may represent V as

$$V = \sqrt{2} D \cdot F_2,$$

where

$$D = \begin{bmatrix} 1 & 0 \\ 0 & -\sqrt{\varepsilon} \end{bmatrix}$$

and the 2×2 *Fourier matrix*

$$F_2 = \frac{1}{\sqrt{2}} \begin{bmatrix} 1 & 1 \\ 1 & -1 \end{bmatrix}$$

is orthogonal. Hence

$$\text{cond}_2(V) = \text{cond}_2(D) = \frac{1}{\sqrt{\varepsilon}}.$$

For instance, if $\varepsilon = 10^{-8}$, then the condition number of V with respect to inversion is $\text{cond}_2(V) = 10^4$.

Mathematical problems for which infinitely small perturbations of the parameters lead to large changes of the solution are called *ill-posed problems.*

The eigenvalue problem for a matrix A with distinct eigenvalues λ_i, $i = 1, 2, \ldots, n$, is a *well-posed computational problem* since for a sufficiently small perturbation ΔA, the eigenvalues $\widetilde{\lambda}_i$, $i = 1, 2, \ldots, n$, of the perturbed matrix $A + \Delta A$ satisfy the inequality

$$|\widetilde{\lambda}_i - \lambda_i| \leq c_{\lambda_i} \|\Delta A\|_2, \tag{5.30}$$

where

$$c_{\lambda_i} = \frac{\|x_i\|_2 \|y_i\|_2}{|y_i^H x_i|}$$

is the condition number of λ_i computed by using of the corresponding right x_i and left y_i^H eigenvectors (Section 3.3). Note that $c_{\lambda_i} = \|P_i\|_2$, where

$$P_i = \frac{x_i y_i^H}{y_i^H x_i}$$

is the spectral projection on the invariant subspace spanned by x_i. If c_{λ_i} is large, then λ_i is ill conditioned. The ill-conditioned eigenvalues are computed with large errors due to their high sensitivity to perturbations of the matrix. By decreasing the differences between the eigenvalues, their sensitivity increases, and in the limit case of defective eigenvalues, their conditioning becomes infinitely large. In such case, the linear estimate (5.30) is not valid and the eigenvalue problem becomes *ill posed.*

We note that the eigenvalue sensitivity estimates available in the linear algebra package LAPACK (Anderson *et al.*, 1999) and the program system MATLAB® (MATLAB) produce meaningful results only for well-posed problems when the matrices have simple eigenvalues, i.e., eigenvalues to which correspond linearly independent eigenvectors.

The determination of the Jordan form of a matrix with defective eigenvalues is an ill-posed computational problem, since in contrast

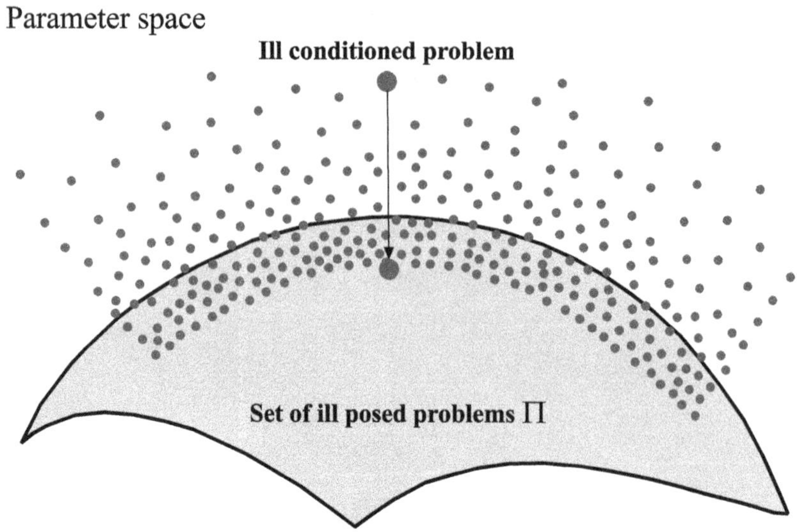

Fig. 5.38. Regularization of an ill-conditioned problem.

with the linear estimate (5.30), a perturbation of size ε in a Jordan block $J_{s_{ji}}(\lambda_i) \in \mathbb{C}^{s_{ji} \times s_{ji}}$ of order $s_{ji} > 1$ can change its eigenvalues by the quantity $\varepsilon^{1/s_{ji}}$ whose derivative at $\varepsilon = 0$ is infinite (Section 3.4). The set of matrices with defective eigenvalues form a low-dimensional surface Π in the n^2 parameter space corresponding to the matrix elements. It is typical for many problems of matrix computations that the conditioning of the problems worsens with the decrease in the distance to the set of ill-posed problems (Demmel, 1987a). That is why the matrices that are in the vicinity of Π have ill-conditioned eigenvalue problems (Figure 5.38). While approaching Π, the conditioning of the problem is worsening which is illustrated symbolically by the larger density of the points representing the computational problems. When the problem lies on Π, it is usually well conditioned since the sensitivity of the defective eigenvalues is finite (see the discussion after Theorem 3.12). That is why the solution of ill-conditioned problems is done by projecting them on Π and solving as ill-posed problems. Such an approach is called *regularization.*

The use of regularization in the solution of the eigenvalue problem is illustrated by the following example.

Example 5.19. Given is the matrix

$$A = \begin{bmatrix} -109 & -126 & 675 \\ -45 & -52 & 279 \\ -26 & -30 & 161 \end{bmatrix}.$$

Using the function `eig` in MATLAB®, one obtains the eigenvalues

$$\begin{aligned} 1.999999999999170 &+ j_o 0.000000000000000, \\ -0.999999999999574 &+ j_o 0.000001024409445, \\ -0.999999999999574 &- j_o 0.000001024409445. \end{aligned}$$

The corresponding eigenvalue condition numbers, determined by the function `condeig` of MATLAB®, are

$$\begin{aligned} &1.6689517668 \cdot 10^2, \\ &1.3224701370 \cdot 10^8, \\ &1.3224701370 \cdot 10^8, \end{aligned}$$

which shows that the last two eigenvalues are very sensitive. For this reason, the eigenvector matrix V is also ill conditioned with condition number equal to

$$\text{cond}(V) = 3.24 \cdot 10^8.$$

The eigenvalue problem can be regularized in the following way. The matrix A is reduced into Jordan form, using the algorithm presented in Chapter 7. In this algorithm, the last two eigenvalues are recognized as multiple, taking as an eigenvalue the mean value

$$(\lambda_2 + \lambda_3)/2 = -0.999999999999574.$$

This approach is justified by the fact that as opposed to the individual eigenvalues which can be very sensitive, their mean value is not. In the given case, the errors in the computed second and third eigenvalues are of the order $\sqrt{\mathbf{u}\|A\|_2} \approx 1 \cdot 10^{-6}$, while the mean value is changed with the quantity $4.26 \cdot 10^{-13}$, which is of order of the backward error $\mathbf{u}\|A\|_2$, where $\mathbf{u}$ is the unit roundoff (Section 1.1.2). In this way, the ill-conditioned eigenvalue problem is projected onto the set of ill-posed problems which includes the matrices with defective multiple eigenvalues. As a result of the reduction of A to Jordan form J, one obtains

$$J = Z^{-1}AZ$$
$$= \begin{bmatrix} 1.999999999999214 & 0 & 0 \\ 0 & -0.999999999999611 & 15.248813586295865 \\ 0 & 0 & -0.999999999999611 \end{bmatrix},$$

where

$$Z = \begin{bmatrix} -0.886460226183731 & 0.928279121632659 & 0.334637240705078 \\ -0.406294270334237 & 0.309426373877600 & -0.936984273974103 \\ -0.221615056545937 & 0.206284249251711 & -0.100391172211498 \end{bmatrix}.$$

Clearly, there is a quadratic elementary divisor corresponding to the eigenvalue -1.0, i.e., this eigenvalue participates in a Jordan block of size 2. At the same time, the matrix Z of the transformation to Jordan form has condition number cond(Z), which is equal only to 333.79.

To summarize, as a result of the rounding errors and the properties of the method for finding the eigenvalues, they are obtained as simple but ill conditioned. The algorithm for reduction into Jordan form recognizes these eigenvalues as multiple and found them with maximum possible accuracy. In this way, the ill-conditioned eigenvalue problem is reduced to a problem of determining multiple eigenvalues which is solved with high accuracy reducing the matrix into Jordan form.

Example 5.19 confirms that the computation of defective multiple eigenvalues of a matrix is an ill-posed problem. This problem can be regularized using an appropriate criterion to determine the dimensions of the null subspaces (the elements of the Weyr characteristics) in finding the Jordan structure of the matrix. This allows finding the exact canonical form $\widehat{J}$ of a near matrix $A + \Delta A$, where the size of ΔA is an upper bound on the distance of A to the regularized problem with computed structure $\widehat{J}$.

5.6 The Numerical Jordan Form: Rigorous Definition

Following the presentation in Zeng (2005, 2021) and Zeng and Li (2021), the problem for finding the Jordan form of a matrix in presence of errors can be formalized in the following way.

Consider for definiteness the case of complete stratification of the Jordan structures of 3×3 matrices using the Segre characteristics, as shown in Figure 5.39. Let $\operatorname{dist}(A, \Pi) = \min\{\|B - A\|_F : B \in \Pi\}$ be the *distance* from the matrix A to a bundle Π. While approaching Π, the conditioning of the problem is worsening, and for this reason, Π is called *pejorative*[1] *manifold* by Kahan (1972). When $A \in \mathbb{C}^{n\times n}$

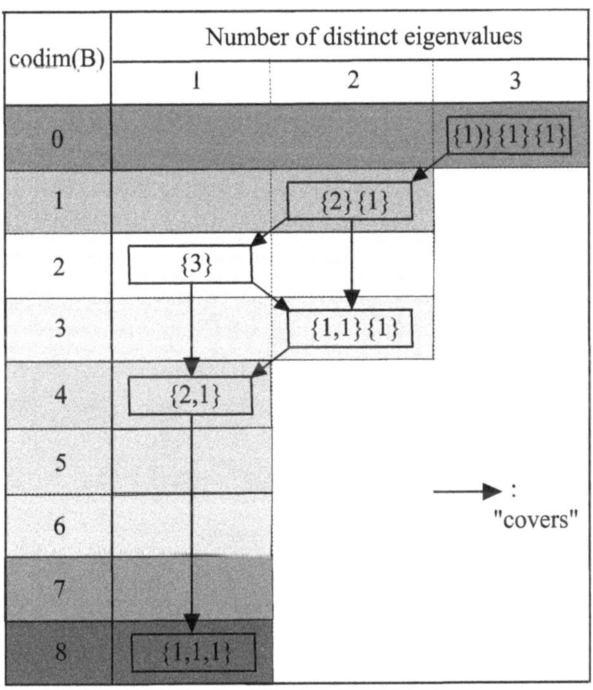

Fig. 5.39. Stratification of the Jordan bundles of 3×3 matrices using the Segre characteristics.

[1]From the Latin word *pejorare* — to make worse.

is near a bundle Π, say $\Pi = \mathcal{B}(\{2,1\})$ listed in Figure 5.39, then A is close to four other bundles with lower codimension above $\mathcal{B}(\{2,1\})$ following the hierarchy. This means that small perturbations can move A from the bundle Π with codimension codim(Π) to a bundle with lower dimension. In the actual computations, the matrix A is obtained from imperfect data or is a subject of rounding errors during the computations. That is why A generically resides in the bundle $\mathcal{B}(\{1\},\{1\},\ldots,\{1\}$ of codimension 0 and the information about the exact Jordan form of A is lost. Fortunately, the original bundle Π where A resides has the distinctive future that it is of highest codimension among all the bundles lying in the ε-neighborhood of A provided that $\varepsilon < \rho$, where ρ is the smallest distance from A to a bundle Π' with codimension $\text{codim}(\Pi') > \text{codim}(\Pi)$. Therefore, to recover the true Jordan structure of A, it is necessary to determine the matrix bundle Π of highest codimension in the neighborhood of A.

As mentioned previously, the addition of a perturbation of general position to the Jordan form decreases the codimension of the bundle in which moves the perturbed matrix. However, it should be emphasized that larger distance to the given matrix does not always mean smaller codimension of the bundle in which resides the perturbed matrix. Let ΔA_1 and ΔA_2 be two perturbations of A. The following example shows that the inequality

$$\text{dist}(A, \mathcal{B}(A + \Delta A_2)) > \text{dist}(A, \mathcal{B}(A + \Delta A_2))$$

does not necessarily mean that

$$\text{codim}(\mathcal{B}(A + \Delta A_2) < \text{codim}(\mathcal{B}(A + \Delta_1).$$

Example 5.20. Consider the matrix

$$A = \begin{bmatrix} -1 & 0 & 0 \\ 0 & -1 & 0 \\ 0 & 0 & -1 \end{bmatrix}$$

which is in a bundle $\mathcal{B}(\mathcal{A})$ with codimension equal to 8. If A is perturbed by

$$\Delta A_1 = \begin{bmatrix} 0 & \delta & \eta \\ 0 & 0 & 0 \\ 0 & 0 & 0 \end{bmatrix},$$

where $\delta \neq 0,\ \eta \neq 0$, we obtain the matrix

$$A_1 = A + \Delta A_1 = \begin{bmatrix} -1 & \delta & \eta \\ 0 & -1 & 0 \\ 0 & 0 & -1 \end{bmatrix}.$$

Using a plane rotation

$$R = \begin{bmatrix} 1 & 0 & 0 \\ 0 & c & s \\ 0 & -s & c \end{bmatrix},$$

where $c = \delta/\sqrt{\delta^2 + \eta^2},\ s = -\eta/\sqrt{\delta^2 + \eta^2}$, it is easy to obtain that

$$R^T A_1 R = \begin{bmatrix} -1 & \sqrt{\delta^2 + \eta^2} & 0 \\ 0 & -1 & 0 \\ 0 & 0 & -1 \end{bmatrix},$$

which shows that $\operatorname{codim}(\mathcal{B}(A_1)) = 4$.

If we now perturb A by

$$\Delta A_2 = \begin{bmatrix} 0 & \delta & 0 \\ 0 & 0 & \zeta \\ 0 & 0 & 0 \end{bmatrix},$$

where $\zeta \neq 0$, then we obtain the matrix

$$A_2 = \begin{bmatrix} -1 & \delta & 0 \\ 0 & -1 & \zeta \\ 0 & 0 & -1 \end{bmatrix}$$

which is an bundle of dimension 2. Hence, a perturbation ΔA_1 with $\|\Delta A_1\|_F = \sqrt{\delta^2 + \eta^2}$ moves A to a bundle with codimension 4, while a perturbation ΔA_2 with $\|\Delta A_2\|_F = \sqrt{\delta^2 + \zeta^2}$ moves A to a bundle with codimension 2. If $|\eta| > |\zeta|$, we see that

$$\operatorname{dist}(A, \mathcal{B}(A_1)) > \operatorname{dist}(A, \mathcal{B}(A_2)),$$

but

$$\operatorname{codim}(\mathcal{B}(A_2)) < \operatorname{codim}(\mathcal{B}(A_1)).$$

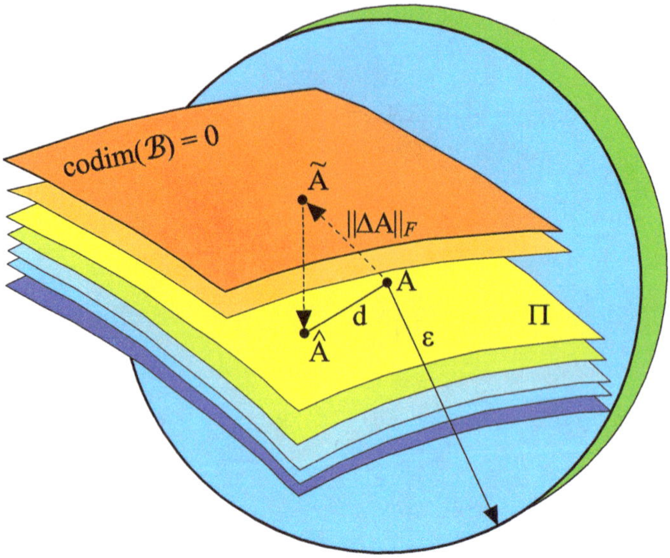

Fig. 5.40. Determining the numerical Jordan form.

Hence, in the given case, A is moved to a bundle with smaller codimension by a smaller instead of by a greater perturbation. This shows that the ordering of the bundles in the matrix space is generally different from the ordering imposed by the stratification hierarchy.

The determination of the Jordan form of a matrix with an ill-conditioned eigenvalue problem may be illustrated in a simplified setting as shown in Figure 5.40 where the space $\mathbb{R}^3$ is used as a substitution of the n^2 dimensional parameter space. (Note that the ordering of the strata is illustrative.) The objective is to find the Jordan normal form of the matrix A which is represented as a point on the manifold Π. In practice, we don't know the exact matrix A and we use its approximation $\widetilde{A} = A + \Delta A$ which is contaminated by empirical or/and rounding errors ΔA such that $\|\Delta A\| < \varepsilon$. With respect to the Frobenius norm, the point $\widetilde{A}$ is inside a sphere with center A and radius equal to ε. Theoretically, the matrix $\widetilde{A}$ has distinct eigenvalues and it is used by the numerical method to find an approximation of the eigenstructure. However, since $\widetilde{A}$ is outside Π, its eigenvalue problem is ill conditioned and the corresponding numerical results contain large errors. To regularize the problem, the point $\widetilde{A}$ is projected onto Π which gives the new problem $\widehat{A}$ whose eigenvalues are defective but well conditioned. Note that there are several manifolds that do not intersect, each of them corresponding to a different Jordan form. It may be proved rigorously (Zeng, 2021) that the best regularization results are obtained if the distance $d = \|\widehat{A} - \widetilde{A}\|$ between $\widetilde{A}$ and Π is minimal which corresponds to the orthogonal projection of $\widetilde{A}$ on the closest pejorative manifold Π. (This shows that the numerical determination of the Jordan form may be recast as a least squares problem.) As a result, the Jordan form J of $\widehat{A}$ is taken as the Jordan canonical form of A. The value of $d = \|\widehat{A} - A\|$ characterizes the backward error in finding J.

Thus, we come to the following rigorous definition of the notion of numerical Jordan form.

Definition 5.21. Suppose that for $A \in \mathbb{C}^{n\times n}$ and $\varepsilon > 0$ we are given the approximation $\widetilde{A} = A + \Delta A$ of A, where $\|\delta A\| < \varepsilon$. Let $\Pi \subset \mathbb{C}^{n\times n}$ be the matrix bundle such that

$$\operatorname{codim}(\Pi) = \max\{\operatorname{codim}(\Pi') : \operatorname{dist}(\widetilde{A}, \Pi') < \varepsilon\} \tag{5.31}$$

and let the matrix $\widehat{A} \in \Pi$ satisfy $\|\widehat{A} - A\|_F = \min_{B\in\Pi}\|B - A\|_F$ with exact Jordan decomposition $\widehat{A} = ZJZ^{-1}$. Then J is called the *numerical Jordan canonical form* of A within ε, and ZJZ^{-1} is called the *numerical Jordan decomposition* of A within ε.

Example 5.22. Given is the sixth-order matrix

$$A = \begin{bmatrix} -670 & -1302 & 4610 & 11257 & 895 & -16559 \\ -1481 & -2740 & 9987 & 24831 & 3390 & -40853 \\ 820 & 1652 & -5758 & -13884 & -556 & 18800 \\ -792 & -1525 & 5441 & 13336 & 1225 & -20142 \\ -320 & -574 & 2127 & 5345 & 905 & -9323 \\ -184 & -340 & 1240 & 3084 & 424 & -5083 \end{bmatrix}$$

whose exact Jordan form is

$$J = \operatorname{diag}\left(\begin{bmatrix} -2 & 1 & & \\ & -2 & 1 & \\ & & -2 & 1 \\ & & & -2 \end{bmatrix}, \begin{bmatrix} -1 & 1 \\ & -1 \end{bmatrix}\right).$$

This matrix is defective but nonderogatory. It is in a bundle Π in the 36-dimensional space whose codimension is equal to $4 + 2 - 2 = 4$.

The diagonal elements of the Schur form computed by the function `schur` of MATLAB are

$$\begin{aligned}
&-2.005581021025717 + j_o 0.000000000000000,\\
&-2.000018276197160 + j_o 0.005599556627086,\\
&-2.000018276197160 - j_o 0.005599556627086,\\
&-1.994382426681778 + j_o 0.000000000000000,\\
&-0.999999999949297 + j_o 0.000018939381468,\\
&-0.999999999949297 - j_o 0.000018939381468.
\end{aligned}$$

These eigenvalues are distinct so that they are the exact eigenvalues of a matrix $\widetilde{A} + \Delta A$ which is in a bundle of codimension $\text{codim}(\mathcal{B}) = 0$. Using a numerical algorithm, one computes a Jordan form $\widehat{J}$ with Segre characteristics $s_{11} = 4$ and $s_{12} = 2$ associated with the eigenvalues

$$\begin{aligned}
&-2.000000000025454,\\
&-2.000000000025454,\\
&-2.000000000025454,\\
&-2.000000000025454,
\end{aligned}$$

and

$$\begin{aligned}
&-0.999999999949297,\\
&-0.999999999949297,
\end{aligned}$$

respectively. The eigenvalues of A are computed exactly up to eleven decimal digits. The computed Segre characteristics show that the matrix

$$\widehat{A} = Z\widehat{J}Z^{-1} = 10^4$$

$$\cdot \begin{bmatrix}
-0.066999999999759 & -0.130200000000348 & 0.461000000000600\\
-0.148100000000135 & -0.274000000000015 & 0.998700000000243\\
0.082000000000064 & 0.165199999999896 & -0.575799999999809\\
-0.079199999999947 & -0.152499999999883 & 0.544099999999599\\
-0.031999999999793 & -0.057399999999809 & 0.212699999999166\\
-0.018399999999934 & -0.033999999999923 & 0.123999999999690
\end{bmatrix} \cdots \rightarrow$$

$$\leftarrow \cdots \begin{bmatrix}
1.125699999999035 & 0.089499999992746 & -1.655899999978640\\
2.483100000001330 & 0.339000000002283 & -4.085300000008000\\
-1.388400000000210 & -0.055600000002015 & 1.880000000005861\\
1.333599999999064 & 0.122500000000053 & -2.014199999998984\\
0.534499999997335 & 0.090499999997931 & -0.932299999990937\\
0.308399999999091 & 0.042399999999476 & -0.508299999997412
\end{bmatrix},$$

whose exact Jordan form is $\widehat{J}$, is in the same bundle Π as the matrix A with codimension equal to 4. The relative distance between matrices $\widehat{A}$ and A, characterizing the backward error in computing $\widehat{J}$, is

$$\text{dist}(\widehat{A} - A) = \|\widehat{A} - A\|_F / \|A\|_F = 4.16 \cdot 10^{-12}.$$

Note that sometimes the matrix $\widehat{A}$ can appear in a different bundle than A, if the Weyr characteristics of A are not computed correctly.

5.7 Numerical Structure of a Matrix

The rigorous definition of the numerical Jordan form, given in the previous section, is not constructive from a computational point of view, since it requires knowing all bundles of possible Jordan forms as well as the distances between them. That is why in this section, we present another definition, which may be easily computed. This definition is given by Ruhe and Kågström (1980b) in terms of the Weyr canonical form and makes use of the notion of a numerically multiple eigenvalue. This notion in turn is based on the notion of the numerical rank of a matrix (Definition 1.19).

Definition 5.23 (Ruhe (1970a) and Kågström and Ruhe (1980b). The matrix A has *numerically multiple eigenvalue* λ with Weyr characteristic

$$\mathcal{W}(A, \lambda) = \{w_1, \ w_2, \ldots, w_h\},$$

if

$$\begin{aligned} \sigma_{n_k+1}(N^k) &\geq gap \cdot tol\|N\|_2^k \geq tol\|N\|_2^k \geq \sigma_{n_k}(N^k), \\ &\qquad k = 1, \ 2, \ldots, \ h-1, \\ tol\|N\|_2^h &\geq \sigma_{n_h}(N^h), \\ &\qquad w_k = n_k - n_{k-1}, \ \ k = 1, 2, \ldots, h, \ \ n_0 = 0 \end{aligned} \tag{5.32}$$

for some small *tol* and sufficiently large *gap*, where $N = A - \lambda I$, the singular values of N^k are in increasing order, i.e., $0 \leq \sigma_1 \leq \cdots \leq \sigma_n$ and $\sum_{j=1}^k w_j = n_k, \ k = 1, 2, \ldots, h$.

The second inequality of (5.32) is considered separately, since the matrix $N^h = (A - \lambda I)^h$ should be theoretically a zero matrix and all its singular values must be equal to zero.

Definition 5.23 makes use of the inequality $\|N^k\| \le \|N\|^k$. Clearly, it is fulfilled that

$$\frac{\sigma_{n_k+1}(N^k)}{\sigma_{n_k}(N^k)} \ge \text{gap}.$$

In Figure 5.41, for a given k, we give an interpretation of (5.32).

The structure $\mathcal{W}(A, \lambda) = \{w_1, w_2, \ldots, w_h\}$ defined by using (5.32) or the corresponding Segre characteristic is referred to as the *numerical structure* of the matrix A. Since the Weyr and Segre characteristics are determined with respect to some tolerances, depending on the size of the uncertainty in A, it is reasonable to call them the *numerical Weyr characteristic* and the *numerical Segre characteristic*, respectively.

The structure $\mathcal{W}(A, \lambda)$ is *well conditioned*, if the number *tol* is sufficiently small and the number *gap* is sufficiently large in the ideal case *tol* being of the order of the machine precision and *gap* close to one. On the other hand, if such a difference in the sizes of the singular values does not exists, one says that the structure is *ill conditioned.*

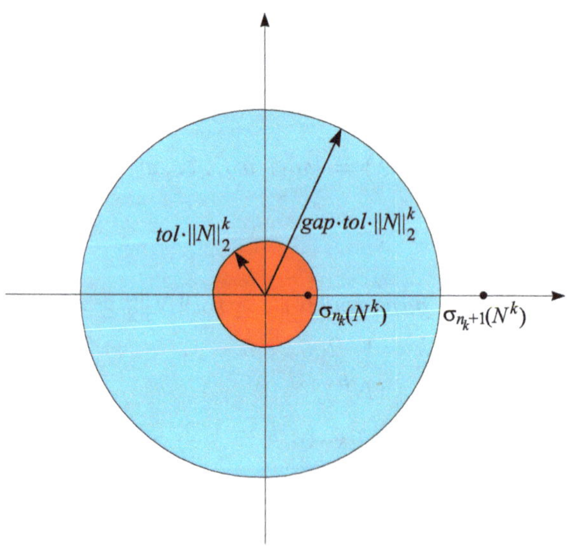

Fig. 5.41. A geometrical interpretation of Definition 5.23.

Then a change of *gap* or *tol* will change n_k and will produce a new structure $\mathcal{W}(A, \lambda)$. The most frequent ill conditioning of $\mathcal{W}(A, \lambda)$ is present when w_1 (the geometric multiplicity of λ) can be augmented by a small perturbation of A. As a consequence of this, h or w_k, $k = 2, \ldots, h$, may decrease.

Example 5.24. Consider the matrix

$$A = \begin{bmatrix} 6 & 0 & 3 & -6 & -5 & -11 & 14 & 33 \\ -2 & -10 & 0 & -3 & 9 & 10 & -2 & 4 \\ 5 & 7 & 4 & -2 & -10 & -16 & 13 & 25 \\ 1 & -6 & 0 & 0 & 2 & 1 & 4 & 11 \\ -1 & -3 & -1 & 1 & 3 & 4 & -2 & -4 \\ 1 & -6 & 0 & -3 & 3 & 3 & 3 & 11 \\ 0 & 6 & -2 & 4 & -5 & -2 & -2 & -13 \\ -1 & -7 & 0 & -2 & 6 & 6 & -1 & 4 \end{bmatrix}.$$

This matrix has an eigenvalue λ equal to 1 with multiplicity eight, which participate in two Jordan blocks — one of size 6 and the other with size 2 so that the Segre characteristic is $\mathcal{S}(A, \lambda) = \{6, 2\}$. The Weyr characteristic of this matrix is $\mathcal{W}(A, \lambda) = \{2, 2, 1, 1, 1, 1\}$ and the index of nilpotence is 6.

In Table 5.8, we give the parameters, characterizing the numerical Jordan structure determined according to Definition 5.23. These parameters are obtained by singular value decompositions of the matrices $N^k, N = A - \lambda I_8$, $k = 1, 2, \ldots, 6$. In computing the parameters, we used the tolerances $gap = 1000$ and $tol = 10^{-12}$.

Table 5.8. Parameters, characterizing the numerical structure of a matrix.

k	n_k	w_k	$\sigma_{n_k+1}(N^k)$	$gap \cdot tol\|N\|_2^k$	$tol\|N\|_2^k$	$\sigma_{n_k}(N^k)$
1	2	2	$1.939 \cdot 10^{-1}$	$5.466 \cdot 10^{-8}$	$5.466 \cdot 10^{-11}$	$9.856 \cdot 10^{-16}$
2	4	2	$7.552 \cdot 10^{-1}$	$2.987 \cdot 10^{-6}$	$2.987 \cdot 10^{-9}$	$2.665 \cdot 10^{-15}$
3	5	1	$2.775 \cdot 10^{0}$	$1.633 \cdot 10^{-4}$	$1.633 \cdot 10^{-7}$	$7.675 \cdot 10^{-15}$
4	6	1	$2.338 \cdot 10^{1}$	$8.925 \cdot 10^{-3}$	$8.925 \cdot 10^{-6}$	$2.441 \cdot 10^{-15}$
5	7	1	$4.699 \cdot 10^{1}$	$4.878 \cdot 10^{-1}$	$4.878 \cdot 10^{-4}$	$1.608 \cdot 10^{-15}$
k	n_k	w_k			$tol\|N\|_2^k$	$\sigma_{n_k}(N^k)$
6	8	1			$2.666 \cdot 10^{-2}$	0

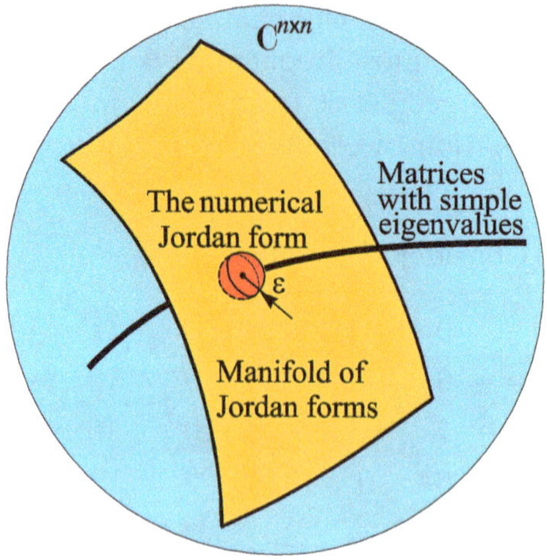

Fig. 5.42. A matrix family and the ball of parameters corresponding to the numerical Jordan form.

It is seen from the table that the numerical structure of the matrix under consideration is well conditioned since the singular value $\sigma_{n_k+1}(N^k)$ is larger than $gap \cdot tol\|N\|_2^k$ and considerably larger than $\sigma_{n_k}(N^k)$ for each k.

Figure 5.42 illustrates symbolically the difference between the "pure" Jordan form and its numerical counterpart defined by (5.32). In contrast to the "theoretical case" (Figure 5.29), the numerical Jordan structure remains the same inside a certain set of parameter values defined by the inequality

$$\|E\|_F \leq \varepsilon, \tag{5.33}$$

where $\varepsilon > 0$ depends on the values of $\sigma_{n_k+1}(N^k),\ k = 1, 2, \ldots, h-1$. The parameter set determined by (5.33) is represented as a ball with a radius ε and a center coinciding with the singular point relevant to the theoretical case. Thus, differently from the pure Jordan form that is defined for a single combination of parameters at the singular point, the numerical Jordan form remains the same for all parameter values contained in the ball with radius ε. The trajectory of the matrix family leaves the ball when the conditions (5.32) are violated as the parameters change.

It should be emphasized that Definitions 5.21 and 5.23 are relevant not only to the Jordan form but they also hold for other matrix forms which possess Jordan structure, like the Jordan–Schur form present in Chapter 6 and the block-diagonal staircase form considered in Chapter 7.

5.8 Notes and References

The presentation in Sections 5.1 and 5.2 is based on the works of Arnold (1971, 1972, 1988, Chapter 6, Section 30), Gantmacher (1959a), Chapter VIII, Demmel and Edelman (1995), Edelman *et al.* (1997, 1999), Johansson (2006), Johansson (2009) and Zeng and Li (2021).

5.8.1 *Geometry of Matrix Space*

Versal deformations were introduced by the great French mathematician Henri Poincaré in his Thesis Poincaré (1928), Lemme IV, in connection with his studies of equilibria and cycles in dynamical systems.

The investigation of abrupt changes which take place in families of objects depending smoothly on parameters is a subject of the singularity theory (Arnold *et al.*, 1998, 1993; Guckenheimer, 2010). As the French mathematician Bernard Teissier says, "*I claim that Singularity Theory sits inside Mathematics much as Mathematics sits inside the general scientific culture.... But singularity theory is also, again in a role played by Mathematics in general science, a crucible where different types of mathematical problems interact and surprising connections are born*" (Teissier, 2020). The singularity theory lies in the basis of the famous Catastrophe theory (Thom, 1977; Zeeman, 1979; Arnold, 1992, 1994; Golubitsky, 1978; Guckenheimer, 1978). The first one who studied the perturbations of Jordan canonical forms using the methods of elementary Catastrophe theory was the Russian mathematician Vladimir Arnold. In his works (Arnold, 1971, 1972, 1988), Arnold developed the method for determining the miniversal deformations and normal forms of matrices subject to perturbations that are characterized by a minimum number of independent perturbations. More on this subject can be found in Gilmore (1993), Chapter 14, Chow *et al.* (1994), Section 2.9, Mailybaev (2001) and Stolovitch (1993). In Gilmore (1993), Chapter 14, Gilmore refers to the normal form of a matrix family as the *Jordan–Arnold form.*

5.8.2 *Structural Stability and Genericity*

In general terms, a mathematical object is *structurally stable*, if its structure remains invariant under perturbations of the object parameters. If the object is structurally stable for large variations of parameters, it is called *robust.* The Jordan canonical form is an example of an object, which is structurally unstable. Properties that hold for "typical" examples are called *generic properties.* Specifically, a generic property of a class of objects is one that is true for "almost all" of those objects. For instance, a generic square matrix has simple eigenvalues or a generic property of a space is a property that holds at "almost all" points of the space. The notions of structural stability, genericity and transversality in application to the dynamical systems, are discussed rigorously in Hirsh and Smale (1974), Chapter 16, Palis and de Melo (1982), Chapter 4, Wiggins (2003), Chapter 12, and Arrowsmith and Place (1990), Chapter 3.

5.8.3 *Stratification of Jordan Bundles*

The hierarchy and stratification of Jordan bundles and their relation to the Weyr characteristics are discussed very thoroughly in Edelman *et al.* (1997, 1999). In these papers, one can find the proof of Theorem 5.5 and the rules for stratification of orbits and bundles.

5.8.4 *Bifurcation*

The name "bifurcation" was introduced by Poincaré in 1885 in a paper that marks the beginning of the bifurcation theory (Poincaré, 1885). A clear introduction to the bifurcation theory of dynamical systems is given in Glendinning (2015) and deep presentations of this theory can be found in Crawford (1991), Guckenheimer (2010) and Kuznetsov (2004). Bifurcation analysis of eigenvalues and generalized eigenvalues is presented in Seyranian and Mailybaev (2003),

Chapter 2. The computations in the neighborhood of algebraic singularities are discussed in Chaitin-Chatelin *et al.* (1995).

5.8.5 *Well-Posed and Ill-Posed Problems*

According to the French mathematician Jacques Hadamard (1923), the mathematical models of physical phenomena should possess the following properties:

(1) a solution exists,
(2) the solution is unique,
(3) the solution depends continuously on the data.

A mathematical problem satisfying all three requirements is called *well posed*. Otherwise, it is called *ill posed*.

The geometry of ill conditioning in numerical computations is considered in depth by Demmel (1987a, 1988, 1993) and Smale (1990).

5.8.6 *Finding the Jordan Form as a Regularization Problem*

The methods for obtaining stable approximations of ill-posed problems are called *regularization methods*. We refer the reader to the book Björck (2015), Section 2.6, for a very good introduction to the regularization methods and to the books Chaitin-Chatelin and Frayssé (1996) and Hansen (1998) for the application of these methods to the solution of ill-posed computational problems in finite precision arithmetic. The idea to "curb" the ill conditioning of an eigenvalue problem by projecting the problem onto the set of ill-posed problems is proposed by Kahan (1972). The determination of the size of the Jordan block associated with multiple eigenvalues is considered in the papers of Ruhe (1970a) and Kågström and Ruhe (1980a, 1980b).

Definition 5.21 is a slight modification of the definition of the numerical Jordan form given in Zeng and Li (2021).

The numerical difficulties in finding the Jordan form may be removed to some extent if the matrix A is reduced to a block-diagonal form

$$B = X^{-1}AX = \operatorname{diag}(B_1 B_2, \ldots, B_p),$$

where each B_i is square matrix of order ν_i. Such an algorithm is proposed in Bavely and Stewart (1979) and attempts to find B using only well-conditioned transformations. If a group of eigenvalues cannot be split off into a block by a transformation whose condition is limited by a tolerance prescribed by the user, the block is enlarged until a well-conditioned reducing transformation can be found. Unfortunately, as shown by Gu (1995), the problem of finding a similarity transformation X such that $X^{-1}AX$ is block-diagonal with the condition number of X being at most τ is nonpolynomial (NP)-hard.

The structure and properties of multiple eigenvalues and polynomial root problems as well as their regularization are studied in depth by Zeng (2005, 2016, 2021) and by Zeng and Li (2021).

5.8.7 *Geometry of Matrix Pencils*

The geometry of the Kronecker forms of matrix pencils is similar to the geometry of Jordan forms. Early papers on this subject are Waterhouse (1984), Pokrzywa (1986) and Berg and Kwatny (1995). Basic results about the codimension of matrix pencils and hierarchy of the Kronecker forms are obtained in Edelman *et al.* (1997) and Edelman *et al.* (1999). New results on this topic can be found in Elmroth *et al.* (2001, 2003), Johansson (2009), Dmytryshyn *et al.* (2017) and De Terán and Dopico (2023).

Another ill-posed problem, closely related to the determining of the Jordan form, is the numerical computation of the Drazin inverse arising in the solution of singular differential and difference equations (Campbell and Meyer, 2009). The problems related to the determination of the numerical counterpart of the Drazin inverse are discussed in Campbell and Meyer (2009), Chapter 12, Wei and Wang (1997) and Wei and Li (2003).

Chapter 6

Reduction into Jordan–Schur Form

In this chapter, we present numerical algorithms intended to obtain the *Jordan–Schur form* of a square matrix. The Jordan–Schur form is an intermediate step in determining the Weyr and Jordan forms and has an independent applications. This chapter begins with a short review of the available algorithms for determining the Jordan structure of a matrix. After a brief description of the staircase form and the Jordan–Schur form, we discuss three basic algorithms whose sequential application allows reducing the Schur form of a matrix into the Jordan–Schur form. The numerical properties of these algorithms are evaluated and some information about their implementation using LAPACK and MATLAB® is given. The numerical computation of the Jordan–Schur form is in the basis of the algorithm for finding the Jordan form, considered in the following chapter.

6.1 Algorithms for Reduction into Jordan Form

In this section, we discuss in brief of the available methods for finding the Jordan form of a matrix. The comparison of these methods motivates the presentation in this and the following chapter.

First we present the remarkable algorithm of Kublanovskaya–Ruhe–Kågström, which is the cornerstone of the contemporary methods for finding the numerical Jordan and Weyr canonical forms of nonsymmetric square matrices.

6.1.1 *Algorithm of Kublanovskaya–Ruhe–Kågström*

The basis of one of the best algorithms for numerical computation of the Jordan form at this moment is a method proposed by the Russian mathematician Vera Kublanovskaya. In her paper published in Russian in 1966 and translated into English in 1968 (Kublanovskaya, 1966), Kublanovskaya shows how with the aid of a sequence of unitary transformations it is possible to find the dimensions of the null spaces null$((A - \lambda I)^j)$ (respectively, the Weyr characteristic) without computing the matrix powers $(A - \lambda I)^j$ and, consequently, the sizes of the Jordan blocks associated with a multiple eigenvalue. As a result, the matrix is reduced to *staircase form* and the corresponding algorithm is called *staircase algorithm* because of the shape of the resulting matrix. The Kublanovskaya method contains the basic idea which makes possible the development of efficient algorithms for obtaining canonical forms of matrices and matrix pencils. The idea of Kublanovskaya was extended and refined by Axel Ruhe who proposed a working algorithm for the reduction of a matrix to Jordan form in 1970 (Ruhe, 1970a). One of the important additions done by Ruhe is the clustering of multiple eigenvalues with the aid of the Gershgorin theorem (see Theorem 3.2). This clustering is essential since it determines the conditioning of the nonsingular transformations used in the block-diagonalization of the matrix. Ruhe proposed for the first time to order the eigenvalues along the diagonal of the Schur form using numerically stable plane rotations. Also, he proposed to find the dimensions of the null spaces by the singular value decomposition which is the most reliable tool for determining the numerical rank of a matrix (Section 1.6.1). Ruhe introduced the notion of numerically multiple eigenvalue which in turn can be used to define the numerical structure of a matrix. In contrast to the "pure" Jordan structure, the numerical structure is determined by using some tolerances which makes it insensitive to small changes (perturbations) of the matrix elements. In his paper, Ruhe gave also an upper bound on the condition number of the nonsingular transformation which reduces the initial matrix into Jordan form. The algorithm for finding the numerical Jordan form was refined and presented in its final form by Kågström and Ruhe in (1980b) and the corresponding Fortran program JNF was given in Kågström and Ruhe (1980a). Additional information about this remarkable algorithm can be found in Kågström (1981) and Kågström and Wiberg (2000).

The problem of finding the Jordan form has an interesting development. Hardly in 1999 it was realized (Shapiro, 1999) that the algorithm of Kublanovskaya, Kågström and Ruhe actually finds the Weyr characteristics of the given matrix. In this way, the algorithm of Kågström and Ruhe determines first the Weyr canonical form thus avoiding to some extent the numerical problems related to the finding of the Jordan form.

It should be noted that the history of the Kublanovskaya–Ruhe–Kågström algorithm is an inseparable part of the rich history of eigenvalue computation, see Golub and van der Vorst (2000) and Golub and Uhlig (2009).

6.1.2 *Algorithm Using the Chain Relations*

An alternative to the algorithm of Kågström and Ruhe is determining the chain relations (see equation (2.18)) as proposed by Golub and Wilkinson in Golub and Wilkinson (1976). This algorithm finds successively the principal vectors associated with a multiple eigenvalue in order to construct a nonsingular basis of the transformation to Jordan form. For this aim, it is necessary to use nonorthogonal bases for some intermediate calculations which can create numerical difficulties in some cases. The comparison done in Kågström (1981) shows that the method of chain relations usually performs worse than the Kublanovskaya–Ruhe–Kågström algorithm.

6.1.3 *Algorithm Based on the Generalized Eigenvalue Problem*

The reduction into Jordan form can be done reformulating the ordinary eigenvalue problem $Ax = \lambda x$ as a generalized eigenvalue problem $Ax = \lambda I x$ and using the corresponding algorithms for reduction into GUPTRI form (Section 4.4.2). In fact, the GUPTRI algorithm is a generalization of the algorithm of Kågström and Ruhe intended for the problem $Ax = \lambda Bx$ and can be used successfully in the particular case of $B = I$. However, there is one significant obstacle. As it is shown in Section 3.8, the sensitivity of a deflation subspace associated with the generalized eigenvalue problem $Ax = \lambda I x$ is measured by the quantity

$$1/\mathrm{dif}(A_{11}, I; A_{22}, I), \tag{6.1}$$

where $\mathrm{dif}(A_{11}, I; A_{22}, I)$ is the difference between the spectra of the two pencils $A_{11} - \lambda I$ and $A_{22} - \lambda I$, while the sensitivity of the corresponding invariant subspace associated with the ordinary eigenvalue problem $Ax = \lambda x$ is measured by

$$1/\mathrm{sep}(A_{11}, A_{22}), \tag{6.2}$$

where $\mathrm{sep}(A_{11}, A_{22})$ is the separation between A_{11} and A_{22}. The difficulty comes from the fact that the quantity (6.1) is always greater or equal to (6.2) so that the sensitivity of the pencil reformulation is greater or at best equal to the sensitivity of the corresponding ordinary eigenvalue problem. This is a drawback of the generalized eigenvalue problem formulation which restricts to some extent its using in finding the Jordan structure of a matrix.

As mentioned earlier, a software implementation of the GUPTRI algorithm is available from MCST.

6.1.4 *Algorithm of Zeng and Li*

A completely different and numerically reliable alternative to the algorithm of Kågström and Ruhe is proposed by Zeng and Li (2021). Further on this algorithm will be referred to as NumJCF. This is a two-stage algorithm based on the powerful method of Zeng (2005, 2016) for finding the multiple roots of inexact polynomials. First, the algorithm finds the Jordan structure in terms of the Segre/Weyr characteristics at each distinct eigenvalue. With this structure as a constraint, the problem of computing the Jordan canonical form is reformulated as a least squares problem. Then, the accurate eigenvalues and the staircase decomposition of the matrix are determined iteratively, and the Jordan decomposition can follow as an option. The NumJCF algorithm is a probabilistic algorithm in the sense that there are several steps in it that require parameters generated at random. Consequently, failures are possible from time to time (in 0.1% of the subsequent runs according to the authors). These failures are verifiable, and in case of a failure, one should run the code again that may be inconvenient in some cases. The algorithm is implemented by the function `NumericalJordanForm` in the

MATLAB® toolbox for numerical algebraic computation NAClab (Zeng and Li, 2013).

6.1.5 *Symbolic Computation*

The reduction into Jordan form can also be done using symbolic computations. For such aim, one can use the algorithm described in Li *et al.* (1997) or the function `jordan` from the Symbolic Math Toolbox of MATLAB® (Symbolic Toolbox). Since this function determines the pure Jordan form, the matrix A must be known exactly, so its elements must be integers or ratios of small integers. Clearly, the symbolic reduction into Jordan form is appropriate for low order matrices.

Other methods for finding the Jordan form or for improving the existing methods are proposed in Sridhar and Jordan (1973), Fletcher and Sorensen (1983), Coleman and Sorensen (1984), Beelen and Van Dooren (1990), Mailybaev (2006), Guglielmi *et al.* (2015) and Mastronardi and Van Dooren (2017). Aspects of the numerical determination of the Jordan form are discussed in Steer (1982) and Demmel (1983b).

6.2 The Jordan–Schur Form

As already mentioned, the first efficient numerical algorithm to compute the Jordan structure associated with a given eigenvalue λ_i of a non-Hermitian matrix A is proposed by Kublanovskaya (1966). The Jordan structure of the multiple eigenvalue λ_i is determined by a special Schur form which has the form

$$Q^H A Q = S = \begin{bmatrix} S_{11} & S_{12} \\ 0 & S_{22} \end{bmatrix}, \tag{6.3}$$

where Q is unitary and S is an upper triangular matrix. The block S_{11} has eigenvalues equal to λ_i and the block S_{22} corresponds to the other eigenvalues, if any. The block S_{11} has the form

$$S_{11} = \lambda_i I + N,$$

where N is a nilpotent matrix with a block structure in a *staircase form*, revealing the Jordan structure associated with λ_i:

$$N = \begin{bmatrix} 0_{w_1\times w_1} & N_{12} & \dots & N_{1,h_i-1} & N_{1,h_i} \\ 0 & 0_{w_2\times w_2} & \dots & N_{2,h_i-1} & N_{2,h_i} \\ \vdots & \vdots & \ddots & \vdots & \vdots \\ 0 & 0 & \dots & 0_{w_{h_i-1}\times w_{h_i-1}} & N_{h_i-1,h_i} \\ 0 & 0 & \dots & 0 & 0_{w_{h_i}\times w_{h_i}} \end{bmatrix}. \qquad (6.4)$$

The sizes $w_1 \geq w_2 \ldots \geq w_{h_i} \geq 1$ of the diagonal blocks, where h_i is the corresponding index of nilpotence, constitute the set of nonzero consecutive differences

$$\begin{aligned} w_j &= \text{null}((A-\lambda I)^j) - \text{null}((A-\lambda I)^{j-1}) \\ &= \dim(\mathcal{N}((A-\lambda I)^j)) - \dim(\mathcal{N}((A-\lambda I)^{j-1})), \\ & \quad j = 1,\ 2, \ldots, h_i \end{aligned}$$

of the dimensions of the null spaces (the nullities) of the matrices $(A-\lambda_i I)^j$, i.e., the Weyr characteristic associated with λ_i (see Section 2.2.3). The number w_j is the dimension of the subspace spanned by the principal vectors of grade j and the corresponding columns of Q constitute a basis of this subspace. The integer h_i is the height of the longest chain of principal vectors corresponding to λ_i.

A numerical algorithm for finding the staircase form of a matrix is presented latter on in Section 6.5.

Example 6.1. Given is the matrix

$$A = \begin{bmatrix} 1 & -6 & 4 & -8 & 2 & -8 \\ -1 & -5 & 2 & -4 & 6 & 0 \\ 2 & -1 & -1 & -2 & -2 & -4 \\ 2 & 0 & -1 & -1 & -3 & -3 \\ 0 & -4 & 2 & -4 & 3 & -2 \\ 1 & 1 & 0 & 0 & -3 & -3 \end{bmatrix},$$

which has a 6-tuple eigenvalue $\lambda = -1$, participating in two Jordan blocks with dimensions 3×3. The Weyr characteristic associated with this eigenvalue is $\mathcal{W} = \{2,\ 2,\ 2\}$. The staircase form, obtained with the aid of a unitary similarity transformation, is (the matrix entries are displayed up to eight decimal digits)

$$S = \left[\begin{array}{cc|cc}
-1 & 0 & 3.2083304 - j_o 2.5964029 & -0.3880774 + j_o 1.6119868 \\
0 & -1 & 0.3750128 - j_o 0.0650310 & 0.7374406 - j_o 1.0917652 \\
\hline
0 & 0 & -1 & 0 \\
0 & 0 & 0 & -1 \\
\hline
0 & 0 & 0 & 0 \\
0 & 0 & 0 & 0
\end{array}\right. \cdots \rightarrow$$

$$\leftarrow \cdots \left.\begin{array}{cc}
-6.8781261 + j_o 11.9136047 & -3.6188798 + j_o 6.2682700 \\
2.6643870 - j_o 4.6149178 & -2.3644500 + j_o 4.0954223 \\
\hline
-1.6755549 + j_o 0.8767155 & -3.7435045 + j_o 0.6542011 \\
-0.9679524 + j_o 0.2782862 & 7.1323581 + j_o 0.2076560 \\
\hline
-1 & 0 \\
0 & -1
\end{array}\right].$$

An important application of the staircase algorithm is the obtaining of a special Schur form of a square matrix under the action of unitary similarity transformations. This special form is referred to as the *Jordan–Schur form* of a matrix.

Assume that the matrix A has p distinct eigenvalues λ_1, $\lambda_2, \ldots, \lambda_p$ with multiplicities $\nu_1,\ \nu_2, \ldots, \nu_p$, respectively, and associated Weyr characteristics $\mathcal{W}_1,\ \mathcal{W}_2, \ldots, \mathcal{W}_p$. Then the matrix A can be transformed into the upper block-triangular form

$$J_S = U_S^H A U_S = \begin{bmatrix} S(\mathcal{W}_1, \lambda_1) & T_{12} & \cdots & T_{1p} \\ & S(\mathcal{W}_2, \lambda_2) & \cdots & T_{2p} \\ & & \ddots & \vdots \\ & & & S(\mathcal{W}_p, \lambda_p) \end{bmatrix}, \quad (6.5)$$

where U_S is an unitary similarity transformation matrix. The diagonal blocks of J_S have the form

$$S(\mathcal{W}_i, \lambda_i) = \begin{bmatrix} \lambda_i I_{w_{i1}} & N_{12}^{(i)} & \cdots & N_{1,h_i-1}^{(i)} & N_{1,h_i}^{(i)} \\ & \lambda_i I_{w_{i2}} & \cdots & N_{2,h_i-1}^{(i)} & N_{2,h_i}^{(i)} \\ & & \ddots & \vdots & \vdots \\ & & & \lambda_i I_{w_{i,h_i-1}} & N_{h_i-1,h_i}^{(i)} \\ & & & & \lambda_i I_{w_{i,h_i}} \end{bmatrix} \tag{6.6}$$

$$\in \mathbb{C}^{\nu_i \times \nu_i}, \quad i = 1, 2, \ldots, p,$$

where $S(\mathcal{W}_i, \lambda_i) - \lambda_i I_{\nu_i}$ is in staircase form. The associated Weyr characteristics can be used to find easily by Algorithm 2.1 the corresponding Segre characteristics of A. The off-diagonal blocks T_{ij} of J_S have no special structure.

The matrix J_S in (6.5) is the Jordan–Schur form of A. The shape of the Jordan–Schur form is shown in Figure 6.1 for a matrix with Weyr characteristics

$$\mathcal{W}_1 = \{3,\ 2,\ 1,\ 1\}, \quad \mathcal{W}_2 = \{3,\ 2,\ 1\}, \ldots, \quad \mathcal{W}_p = \{2,\ 2,\ 1\}.$$

The Jordan–Schur form lays in bundles in the same way as the Jordan form. The codimension of such a bundle is determined by the Weyr (or Segre) characteristics associated with the corresponding eigenvalues using (5.12) and (5.13).

The Jordan–Schur form is an intermediate step in determining the Weyr and Jordan form of a matrix. An important advantage of this form is that it can be obtained in a numerically stable way using only unitary similarity transformations of the given matrix providing at the same time information about the Weyr (and hence the Jordan) structure of the matrix. An algorithm for reducing a matrix to Jordan–Schur form is present in Section 6.6.

To give an impression for the Jordan–Schur form and the subsequent reduction into Weyr and Jordan forms, we consider the following example.

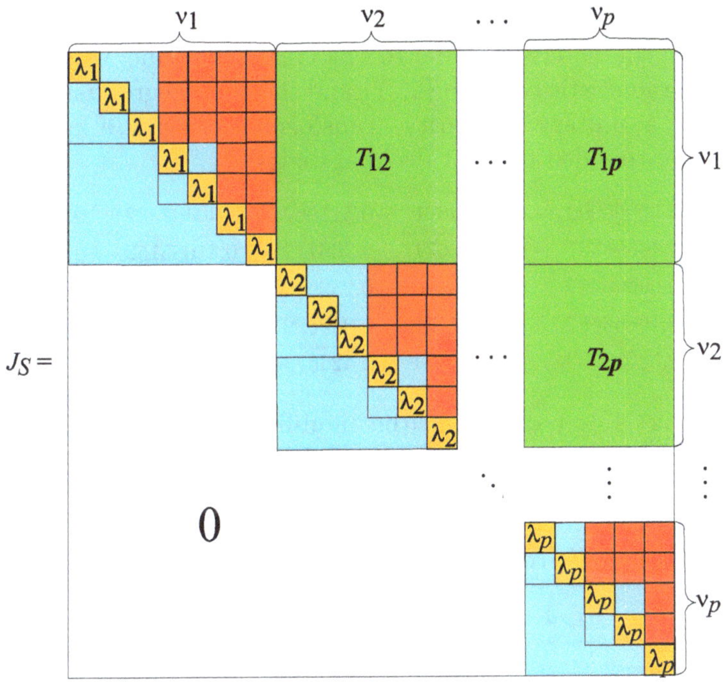

Fig. 6.1. The Jordan–Schur form.

Example 6.2. Consider the matrix

$$A = \begin{bmatrix} 1 & -5 & 7 & -10 & 9 \\ 0 & -1 & -1 & 2 & -2 \\ -3 & 6 & -10 & 12 & -11 \\ 0 & -2 & 3 & -6 & 5 \\ 2 & -5 & 7 & -10 & 8 \end{bmatrix}.$$

This matrix has the Jordan form

$$J = \operatorname{diag}\left(\begin{bmatrix} -1 & 1 \\ & -1 \end{bmatrix}, \left[\begin{array}{cc|c} -2 & 1 & \\ & -2 & \\ \hline & & -2 \end{array}\right] \right).$$

The eigenvalue $\lambda_1 = -1$ has a Weyr characteristic equal to $\{1,\ 1\}$ and Segre characteristic equal to $\{2\}$. The eigenvalue $\lambda_2 = -2$ has a Weyr characteristic equal to $\{2,\ 1\}$ and the same Segre characteristic.

Using a unitary similarity transformation with a matrix (all results are rounded to eight decimal digits)

$$U_S = \begin{bmatrix} 0.50000000 & -0.56694671 & -0.51506910 & 0.40407332 & 0.00000000 \\ 0.16666667 & -0.56694671 & 0.70612281 & 0.31085152 & 0.23570226 \\ 0.66666667 & 0.00000000 & -0.23261998 & 0.52841393 & -0.47140452 \\ -0.16666667 & 0.56694671 & 0.04982915 & 0.65275454 & 0.47140452 \\ -0.50000000 & 0.18898224 & 0.42367369 & 0.18651091 & -0.70710678 \end{bmatrix},$$

the matrix A is reduced by the modification of the algorithm of Kågström and Ruhe to the Jordan–Schur form

$$J_S = U_S^H A U_S$$

$$= \left[\begin{array}{cc|cc|c} -1 & 4.53557368 & 5.37066484 & 3.31051818 & 30.05203820 \\ \hline & -1 & 1.24183998 & -0.28780802 & -0.13363062 \\ \hline & & -2 & 0 & -1.40973734 \\ & & & -2 & -3.29693547 \\ \hline & & & & -2 \end{array}\right].$$

Note that the position of the horizontal and vertical lines corresponds to the Weyr characteristics associated with the two eigenvalues. The reduction is done with a relative backward error

$$\|U_S \cdot J_S \cdot U_S^H - A\|_F / \|A\|_F = 8.57 \cdot 10^{-16}.$$

It is convenient to represent the above Jordan–Schur form as

$$J_S = \left[\begin{array}{cc|cc|c} -1 & \times & \times & \times & \times \\ \hline & -1 & \times & \times & \times \\ \hline & & -2 & 0 & \times \\ & & & -2 & \times \\ \hline & & & & -2 \end{array}\right],$$

where $(\times)$ is arbitrary nonzero entry. Such representation reveals clearly the Weyr characteristics associated with the eigenvalues when the values of the nonzero elements are not of interest.

Zeroing the corresponding superdiagonal elements using elementary similarity transformations, one obtains the Weyr canonical form

$$W = X^{-1}J_S X$$

$$= \operatorname{diag}\left(\begin{bmatrix} -1 & 4.53557368 \\ & -1 \end{bmatrix}, \left[\begin{array}{cc|c} -2 & & 0 \\ & -2 & 3.52679178 \\ \hline & & -2 \end{array}\right]\right).$$

The Weyr form is obtained in un-normalized form, i.e., the nonzero elements of the superdiagonal blocks (the elements which "carry" the ranks of the corresponding nilpotent blocks) are different from 1. The reduction into normalized form requires applying additional diagonal similarity transformations which may increase the condition number of the transformation matrix with respect to inversion and should be avoided.

The Jordan–Schur form is reduced into Weyr form with the aid of nonsingular similarity transformations that may introduce large rounding errors in case of ill-conditioned eigenvalue problems. In the given case, the transformation matrix into Weyr form,

$$X = \begin{bmatrix} -0.50000000 & -0.56694671 & 0.18312953 & -0.56708763 & -0.25000000 \\ 0.16666667 & -0.56694671 & 0.88169795 & 0.00000000 & -0.25000000 \\ 0.66666667 & 0.00000000 & -0.18312953 & 0.56708763 & -0.75000000 \\ -0.16666667 & 0.56694671 & -0.34928421 & -0.28354382 & -0.50000000 \\ -0.50000000 & 0.18898224 & 0.18312953 & -0.56708763 & -0.25000000 \end{bmatrix},$$

has a condition number with respect to the 2-norm equal to 41.60

Using permutation similarity transformation with the matrix

$$P_J = \begin{bmatrix} 1 & 0 & 0 & 0 & 0 \\ 0 & 1 & 0 & 0 & 0 \\ 0 & 0 & 0 & 0 & 1 \\ 0 & 0 & 1 & 0 & 0 \\ 0 & 0 & 0 & 1 & 0 \end{bmatrix},$$

($P_J^{-1} = P_J^T$) the Weyr form is reduced into the Jordan form

$$J = P_J^T W P_J = \operatorname{diag}\left(\begin{bmatrix} -1 & 4.53557368 \\ & 1 \end{bmatrix}, \left[\begin{array}{cc|c} -2 & 3.52679178 & \\ & -2 & \\ \hline & & -2 \end{array}\right]\right).$$

Hence, the determination of the Jordan form from the Weyr form requires only permutation of the columns of the transformation matrix and thus it is free from rounding errors. Obviously, the obtained Jordan form is also un-normalized. The similarity transformation matrix into Jordan form $J = Z^{-1}AZ$ is equal to $Z = XP_J$ so that $\mathrm{cond}_2(Z) = \mathrm{cond}_2(X)$.

In the following three sections, we present algorithms whose sequential implementation leads to the reduction of a matrix into the Jordan–Schur form.

6.3 Reduction into Ordered Schur Form

The reduction into the Jordan–Schur form of the matrix $A \in \mathbb{C}^{n\times n}$ is done in three steps. The first and second steps are the reduction into the upper triangular (Schur) form and the reordering of the diagonal elements (eigenvalues), respectively. The algorithm for this aim assumes that the matrix A has complex entries.

The reduction into the Schur form

$$T = U^H AU$$

is done by the QR algorithm using iteratively unitary similarity transformations accumulated in the matrix $U \in \mathbb{C}^{n\times n}$ (see Section 4.1). In the given case, we prefer to use complex unitary transformations even for real matrices to obtain the Schur form in a triangular form which significantly simplifies the reduction into Jordan–Schur form. This step is implemented by using the LAPACK routine `xGEESX` or MATLAB® function `schur` implementing the option `'complex'`. It requires about $25n^3$ flops in the complex case and is backward stable.

If A has one or more numerically multiple eigenvalues, then the Schur form will have close diagonal elements. Since these elements can be obtained at different positions on the diagonal of T, usually it is necessary to reorder the diagonal elements so that the close eigenvalues appear in adjacent positions. This is done in order to group the numerically multiple eigenvalues in clusters whose Jordan structure is determined later on. The reordering is implemented by the following sorting operation.

Denote the eigenvalue approximations found at the first step as $\lambda_k,\ k = 1,\ 2, \ldots, n$. Sort them into λ_k' so that

$$\lambda_n' = \lambda_n,$$
$$|\lambda_{k-1}' - \lambda_k'| = \min_{j<k}|\lambda_j' - \lambda_k'|, \quad k = n, \ldots, 2.$$

Each time when $\lambda_k' \neq \lambda_k$, the Schur form must be reordered by a sequence of unitary transformations (usually complex plane rotations), each of them exchanging two adjacent diagonal elements.

Let

$$T = \begin{bmatrix} t_{11} & t_{12} \\ 0 & t_{22} \end{bmatrix}.$$

The elements t_{11} and t_{22} are exchanged by using the complex plane rotation

$$R = \begin{bmatrix} c & -\bar{s} \\ s & \bar{c} \end{bmatrix}.$$

The numbers c and s are chosen so that the transformed matrix has the form

$$T' = R^H T R = \begin{bmatrix} t_{22} & \bar{t}_{12} \\ 0 & t_{11} \end{bmatrix}.$$

From here, one finds

$$\xi = \sqrt{|t_{22} - t_{11}|^2 + |t_{12}|^2},$$
$$c = t_{12}/\zeta,$$
$$s = (t_{22} - t_{11})/\xi.$$

Note that when the matrix T is $n \times n$ with $n > 2$, then after the modification of the 2×2 diagonal block of T, it is necessary to multiply by R^H and R the corresponding two rows and two columns, respectively, of the Schur form (Figure 6.2).

After the reordering, the close eigenvalues will appear together since each time the chosen next eigenvalue is the one that is closest to the already reordered eigenvalues. The maximum number of exchanges is $n(n-1)/2$, but in reality, this number rarely exceeds n.

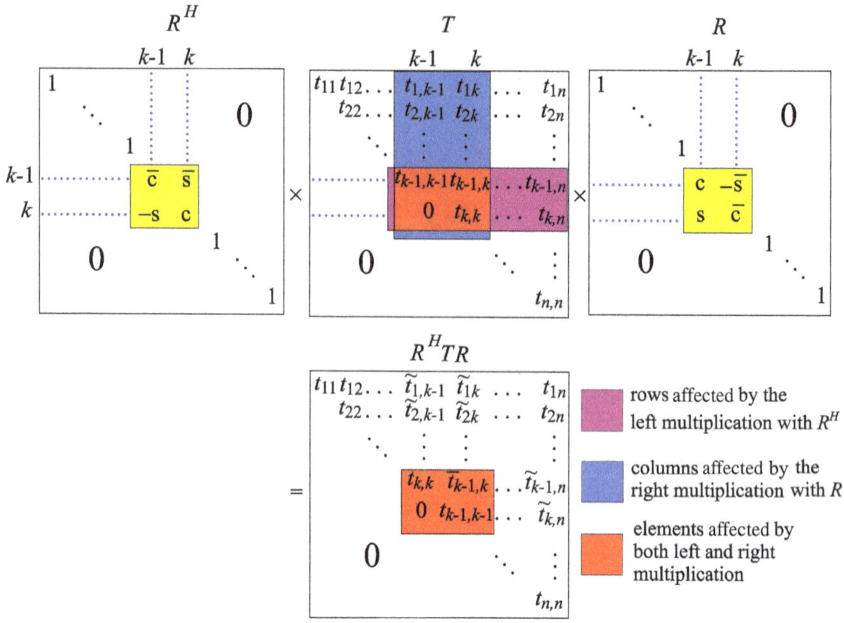

Fig. 6.2. Exchanging of two adjacent diagonal elements of the Schur form.

Due to the use of unitary transformations in each step, the algorithm described is backward stable (Bai and Demmel, 1993).

This step of the reduction is realized by Algorithm 6.1. The algorithm can be implemented by using the LAPACK routine `xTREXC` or MATLAB® function `ordschur`. It requires at most $6n^2$ flops and is backward stable.

Example 6.3. Given is the 6×6 matrix

$$A = \begin{bmatrix} -3 & -6 & 5 & -9 & 9 & 4 \\ 0 & -5 & 3 & -3 & 2 & -1 \\ -1 & -5 & 2 & -6 & 7 & 3 \\ 0 & -1 & 1 & -3 & 1 & 0 \\ 2 & 0 & 1 & 1 & -6 & -6 \\ -2 & -3 & 2 & -5 & 7 & 4 \end{bmatrix},$$

which has a simple eigenvalue $\lambda_1 = -3$, double eigenvalue $\lambda_2 = -1$ and a triple eigenvalue $\lambda_3 = -2$.

Algorithm 6.1. Reordering of the Schur form
Reorders the Schur form by the aid of unitary similarity transformations

```
function [T_2, V] = reorder(T)
Input: The matrix T ∈ ℂ^{n×n} in Schur form.
Output: The matrix T_2 in Schur form, reordered so that the close
        eigenvalues are located in adjacent positions; the matrix V
        of the unitary transformation, which reduces T into the
        ordered Schur form T_2.
Set V = I_n
for i = 2,...,n
  k = n − i + 2
  d_min = |t_kk − t_{k−1,k−1}|
  Set k' = k, k_2 = k − 2
  if k_2 ≥ 1
    Find the closest to t_kk diagonal element
    for j_1 = 1,...,k_2
      j = k_2 − j_1 + 1
      s_k = |t_jj − t_kk|
      if s_k < d_min
        k' = j
        d_min = s_k
      end
    end
  end
  if k' ≠ k and k' ≤ k_2
    for j = k',...,k_2
      Compute a complex rotation R which exchanges t_jj and t_{j+1,j+1}
      for ℓ = j,...,n
        [t_jℓ; t_{j+1,ℓ}] = R^H [t_jℓ; t_{j+1,ℓ}]
      end
      for ℓ = 1,...,j
        [t_ℓj t_{ℓ,j+1}] = [t_ℓj t_{ℓ,j+1}]R
      end
      for ℓ = 1,...,n
        [v_ℓj v_{ℓ,j+1}] = [v_ℓj v_{ℓ,j+1}]R
      end
    end
  end
end
```

Using the function `schur` of MATLAB® with option `'real'`, one obtains the unordered Schur form

$$T = \begin{bmatrix} -3.0000 & -2.7450 & 0.2522 & -2.6080 & 1.4435 & 21.1723 \\ 0 & -1.0000 & -3.7427 & 1.4781 & -7.3518 & -6.8512 \\ 0 & 0 & -1.0000 & -0.2728 & 1.5633 & 0.6786 \\ 0 & 0 & 0 & -2.0000 & 2.0226 & 0.0000 \\ 0 & 0 & 0 & 0 & -2.0000 & -2.2361 \\ 0 & 0 & 0 & 0 & 0 & -2.0000 \end{bmatrix}.$$

Implementing an orthogonal similarity transformation with the matrix

$$V = \begin{bmatrix} -0.8082 & -0.4835 & 0.3362 & 0 & 0 & 0 \\ 0.5889 & -0.6636 & 0.4614 & 0 & 0 & 0 \\ 0 & 0.5709 & 0.8210 & 0 & 0 & 0 \\ 0 & 0 & 0 & 1.0000 & 0 & 0 \\ 0 & 0 & 0 & 0 & 1.0000 & 0 \\ 0 & 0 & 0 & 0 & 0 & 1.0000 \end{bmatrix},$$

one obtains the ordered Schur form

$$T_2 = V^H T V$$
$$= \begin{bmatrix} -1.0000 & -3.6283 & -0.4098 & 2.9782 & -5.4959 & -21.1465 \\ 0 & -1.0000 & 2.8764 & 0.1244 & 5.0731 & -5.3028 \\ 0 & 0 & -3.0000 & -0.4187 & -1.6233 & 4.5136 \\ 0 & 0 & 0 & -2.0000 & 2.0226 & 0.0000 \\ 0 & 0 & 0 & 0 & -2.0000 & -2.2361 \\ 0 & 0 & 0 & 0 & 0 & -2.0000 \end{bmatrix}.$$

In the given case are only exchanged the $(1,1)$-element of the matrix T with the 2×2-block, containing the eigenvalues -1. This is reflected by the nontrivial leading 3×3-block of the matrix V.

6.4 Clustering of Multiple Eigenvalues

The third step of the reduction into the Jordan–Schur form is to decide which of the computed eigenvalue approximations constitute groups corresponding to numerically multiple eigenvalues. This problem is difficult, since due to the high sensitivity of the multiple eigenvalues their computed approximations may differ significantly from the actual values. A reliable clustering of the numerically multiple eigenvalues can be done with the aid of the Gershgorin theorem (Section 3.2). In the following, we consider the algorithm proposed by Ruhe in (1970a).

The close eigenvalues are grouped in p clusters corresponding to numerically multiple eigenvalues. This is done by using a Gershgorin heuristic to construct a diagonal similarity transformation which isolates groups of intersecting disks containing the eigenvalues of the matrix $T + E, \|E\| \leq ein$, where ein is an appropriate tolerance reflecting the uncertainty of A.

The eigenvalue clustering is based on the important fact that when a given matrix is a perturbation of a matrix with a multiple eigenvalue, then the mean of the corresponding group of eigenvalues is perturbed by the same amount as the matrix itself, while the individual eigenvalues are perturbed by fractional powers of the perturbation (Ruhe, 1970b). Thus, if we may recognize a group of eigenvalues as multiple, then the eigenvalue of the corresponding block can be taken as the mean value of the eigenvalues in the group.

It is assumed that the matrix $A \in \mathbb{C}^{n\times n}$ is preliminarily reduced to Schur form T with ordered diagonal elements. Let the diagonal scaling matrix be given as

$$D = \mathrm{diag}(d_1,\ d_2, \ldots, d_n), \quad 1 \leq d_1 \leq \cdots \leq d_n.$$

The matrix DTD^{-1} has the form

$$\begin{bmatrix} t_{11} & t_{12}\frac{d_1}{d_2} & \ldots & t_{1n}\frac{d_1}{d_n} \\ 0 & t_{22} & \ldots & t_{2n}\frac{d_2}{d_n} \\ \vdots & \vdots & \ddots & \vdots \\ 0 & 0 & \ldots & t_{nn} \end{bmatrix}.$$

According to the Gershgorin theorem (Theorem 3.2), the region of the complex plane, determined by the n disks

$$|z - t_{kk}| \leq \sum_{j=k+1}^{n} \left|\frac{d_k}{d_j}\right| |t_{kj}|, \quad k = 1,\ 2, \ldots, n,$$

contains the eigenvalues of DTD^{-1}. Moreover, if m of the disks intersect, then the union of these disks contains exactly m eigenvalues.

With respect to the perturbed matrix $\widetilde{T} = T + E$, the disks are

$$|z - \widetilde{t}_{kk}| \leq \sum_{j=k+1}^{n} \left|\frac{d_k}{d_j}\right| |\widetilde{t}_{kj}|, \quad \widetilde{t}_{kj} = t_{kj} + e_{kj}, \quad k = 1, 2, \ldots, n.$$

Hence

$$|z - \widetilde{t}_{kk}| \leq |d_k| \left(\sum_{\substack{j \neq k \\ j=1}}^{n} \frac{|e_{kj}|}{|d_j|} + \sum_{j=k+1}^{n} \frac{|t_{kj}|}{|d_j|} \right).$$

Under the condition that $|d_j| \geq 1$ for all j, it follows that

$$\sum_{\substack{j \neq k \\ j=1}}^{n} \frac{|e_{kj}|}{|d_j|} \leq \|E\|_\infty.$$

In this way, one obtains the upper bound on the disk radius,

$$|r_k| = |z - t_{kk}| \leq |d_k| \left(\|E\|_\infty + \sum_{j=k+1}^{n} \frac{|t_{kj}|}{|d_j|} \right), \quad k = 1, 2, \ldots, n.$$

For a given $\|E\|_\infty$, this bound describes a set of disks whose centers are located at the computed eigenvalues. If the disks' radii are known, then from them it is possible to find recursively the diagonal elements of the scaling matrix according to the expression

$$d_k = \frac{r_k}{\|E\|_\infty + \sum_{j=k+1}^{n} |t_{kj}|/d_j}, \quad k = n, n-1, \ldots \tag{6.7}$$

The idea is to construct disks centered at each eigenvalue with radii which are as larger as possible without intersecting whichever disk from another group.

The algorithm for clustering numerically multiple eigenvalues performs in the following way. Starting with $s = 1$, the last s eigenvalues are grouped and one plot a disk in the complex plane which contains only $\lambda_{n-s+1}, \ldots, \lambda_n$. Fixing this disk, one plot disks around all

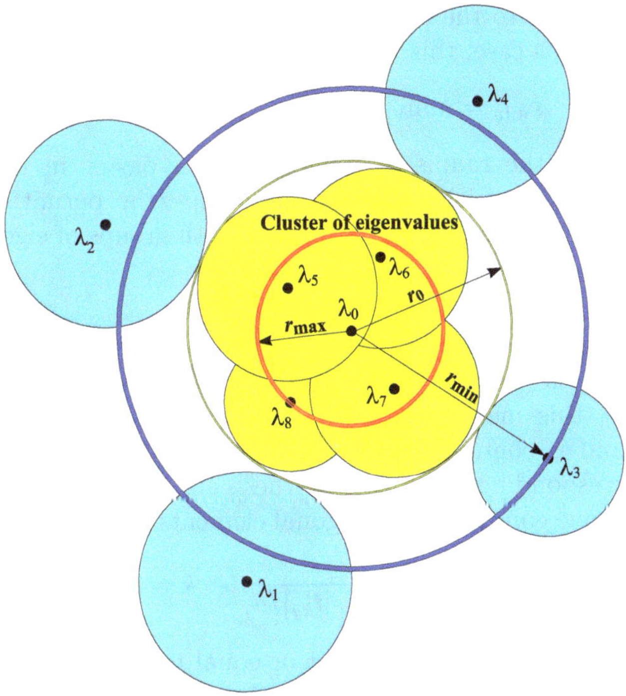

Fig. 6.3. Use of Gershgorin disks to cluster multiple eigenvalues.

n eigenvalues which touch the first disk, and thus we obtain upper bounds on the radii of the permitted Gershgorin disks. Beginning with $k = n$, the diagonal elements d_k are computed from (6.7) for given radii of the disks. In case when $d_1 \geq 1$, we have succeeded to isolate the last s eigenvalues from the rest, otherwise we increase s by one and continue again with another set of disks.

This strategy of clustering is illustrated in Figure 6.3 for the case of eight eigenvalues. Suppose that the eigenvalues λ_8, λ_7 and λ_6 are already clustered and we want to check if λ_5 enters the group, i.e., if these 4 eigenvalues come from one 4-tuple eigenvalue. For this group, we compute the mean

$$\lambda_0 = \sum_{i=5}^{8} \lambda_i \Big/ 4,$$

the maximum deviation of the eigenvalues from the mean,

$$r_{\max} = \max_k |t_{kk} - \lambda_0|, \quad k = 5, \ldots, 8,$$

and the distance to the nearest eigenvalue which is not accepted as multiple (in this case, this is λ_3),

$$r_{\min} = \min_k |t_{kk} - \lambda_0|, \quad k = 1, \ldots, 4.$$

Now compute the radius of the disk which passes in the middle between the disks with radii $r_{\max}$ and $r_{\min}$ (the permitted extensions of the disks are taken the same for both groups of eigenvalues),

$$r_0 = (r_{\max} + r_{\min})/2.$$

Around each eigenvalue for $k = 1, \ldots, 8$, we plot a disk that is tangent to the disk with center λ_0 and radius r_0. If the existence of such disks is possible, this means that the group $\lambda_5, \ldots, \lambda_8$ is isolated from the rest and is contained in a region obtained by the intersection of the corresponding disks around these eigenvalues. To check this possibility, one computes the diagonal elements of the scaling matrix,

$$d_k = \frac{r_k}{\|E\|_\infty + \sum_{j=k+1}^{n} |t_{kj}|/d_j}, \quad k = 8, 7, \ldots, 1.$$

If all diagonal elements are greater or equal to 1, i.e.,

$$d_k \geq 1, \quad k = 8, \ldots, 1,$$

this means that the disks shown in Figure 6.3 may exist, i.e., the four eigenvalues in the group can be considered as multiple. If, however, there is at least one k, for which $d_k < 1$, this means that the four eigenvalues cannot be isolated from the rest. In this case, it is necessary to enlarge the group with a further eigenvalue, in the given case λ_4. After that, the procedure is repeated. As a result, one finds a cluster of eigenvalues that can be accepted as multiple and they are included in a block whose eigenvalue is the mean value of all eigenvalues in the group.

At this step, the upper triangular form takes the block-triangular form (Figure 6.4)

$$T_3 = \begin{bmatrix} T_{11} & T_{12} & \ldots & T_{1p} \\ & T_{22} & \ldots & T_{2p} \\ & & \ddots & \vdots \\ & & & T_{pp} \end{bmatrix}, \tag{6.8}$$

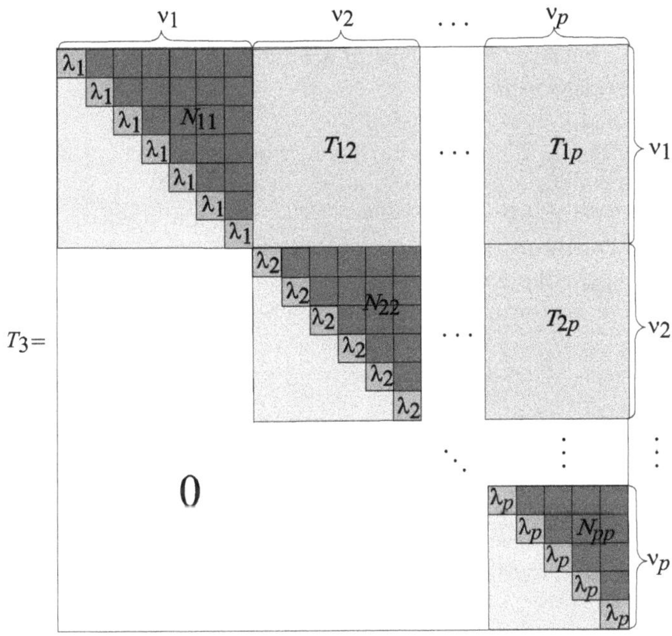

Fig. 6.4. Clustering of multiple eigenvalues.

where each diagonal block

$$T_{ii} = \lambda_i I_{\nu_i} + N_{ii} \in \mathbb{C}^{\nu_i \times \nu_i}$$

corresponds to a distinct eigenvalue λ_i with multiplicity ν_i and N_{ii} is a strictly upper triangular and hence, a nilpotent matrix.

This strategy may be overly conservative in its construction of the disks, but in most cases, it works reliably.

The clustering done at this step represents a regularization procedure in the sense that if A has ill-conditioned eigenvalues, we seek in the *ein*-vicinity of A a matrix with one or several Jordan blocks associated with a near eigenvalue. In this way, the ill-conditioned eigenvalue problem is projected onto the surface (the bundle) of the nearest ill-posed problems. In the case of ill-conditioned eigenproblem, this strategy allows improving significantly the conditioning of the matrix Z of the similarity transformation into the Jordan form J. The tolerance parameter *ein*, used at this step, plays the role of a regularization parameter and its increasing reduces cond(Z) but increases the backward error $E = ZJZ^{-1} - A$.

Algorithm 6.2. Clustering of multiple eigenvalues
Grouping the multiple eigenvalues of a matrix in Schur form using the Gershgorin theorem.
function $[\lambda_{\text{cluster}}, s] = \text{cluster}(A, ein)$
Input: Matrix $T \in \mathbb{C}^{n\times n}$ in Schur form, whose close eigenvalues are located in adjacent positions; the perturbation size $ein \approx \|E\|_\infty$, which influences the clustering.
Output: λ_{cluster}, the mean values of s clustered eigenvalues in the bottom right corner of the Schur form

$s = 0$
repeat
  $s = s + 1$
  $m_p = \sum_{k=n-s+1}^{n} t_{kk}/s$
  $r_{\max} = \max_{n-s+1\le k\le n} |t_{kk} - m_p|$
  $r_{\min} = \min_{1\le k\le n-s} |t_{kk} - m_p|$
  $r_0 = (r_{\max} + r_{\min})/2$
  $rad_k = ||t_{kk} - m_p| - r_0|, \quad k = 1, 2, \ldots, n$
  $k = n$
  repeat
    $\rho = ein + \sum_{i=k+1}^{n} |t_{ki}|/d_i$
    $d_k = rad_k/\rho$
    $k = k - 1$
  until $d_k < 1$ **or** $k = 0$
  if $s = n - 1$ **and** $d_k < 1$ **then** $s = n$
until $d_1 \ge 1$ **or** $s = n$
$\lambda_{\text{cluster}} = m_p$

The earlier considerations are summarized in Algorithm 6.2. Algorithm 6.2 requires at most $6n^3$ flops.

Example 6.4. Given is the matrix

$$A = \begin{bmatrix} 3 & -1 & 1 & -4 & -4 & -7 \\ 0 & -4 & 2 & -3 & 2 & 0 \\ 3 & -1 & 0 & -2 & -3 & -6 \\ 2 & 1 & 0 & 0 & -4 & -5 \\ 0 & -2 & 1 & -1 & 0 & 0 \\ 2 & 1 & 0 & -1 & -3 & -5 \end{bmatrix},$$

which is similar to a Jordan block of size 6 with an eigenvalue $\lambda = -1$.

The clustering of multiple eigenvalues is done by Algorithm 6.2 using a tolerance $ein = 10^{-12}$.

After reordering the Schur form by Algorithm 6.1, its diagonal elements are obtained as

$$\begin{aligned}
&-1.002038236632993 + \jmath_o 0.003525899443428,\\
&-0.997964321125235 + \jmath_o 0.003530329729337,\\
&-0.995923526278148 + \jmath_o 0.000000000000000,\\
&-0.997964321125235 - \jmath_o 0.003530329729337,\\
&-1.002038236632993 - \jmath_o 0.003525899443428,\\
&-1.004071358205400 + \jmath_o 0.000000000000000.
\end{aligned}$$

The computed eigenvalues are accurate up to at most the third digit after the decimal point. Also, the six real eigenvalues are converted into two pairs of complex conjugate eigenvalues and two real eigenvalues.

The results of the tests for clustering of groups of $s = 1,\ 2, \ldots, 6$ eigenvalues are given in Table 6.1.

The procedure for testing the isolation of an eigenvalue from the rest eigenvalues begins with the last (6th) eigenvalue. The computed Gershgorin disks for $s = 1,\ 2, \ldots, 5$ are shown respectively in Figures 6.5–6.9.

It is seen in Figure 6.5 that to isolate the eigenvalue

$$\lambda_6 = -1.004071358205400 + \jmath_o 0.000000000000000$$

from the rest eigenvalues ($s = 1$), it is necessary that the Gershgorin disks touch the circle with center λ_6 and radius equal to half of the sum of $r_{\max} = 0$ and $r_{\min} = 4.070 \cdot 10^{-3}$. However, it is seen from Table 6.1 that in this case the third diagonal element of the scaling matrix D is obtained equal to $d_3 = 7.920 \cdot 10^{-1} < 1$, which means that the shown sizes and locations of the disks cannot be realized, i.e., λ_6 cannot be isolated from the rest eigenvalues.

Similarly, for $s = 2$ (Figure 6.6), the Gershgorin disks should touch the circle, drawn at the half distance between $r_{\max} = 3.710 \cdot 10^{-3}$ and $r_{\min} = 5.386 \cdot 10^{-3}$. In this case, the third element $d_3 = 1.069 \cdot 10^{-2}$ of the diagonal scaling matrix D is also smaller than 1, which shows that λ_6 and λ_5 cannot be isolated from the rest eigenvalues.

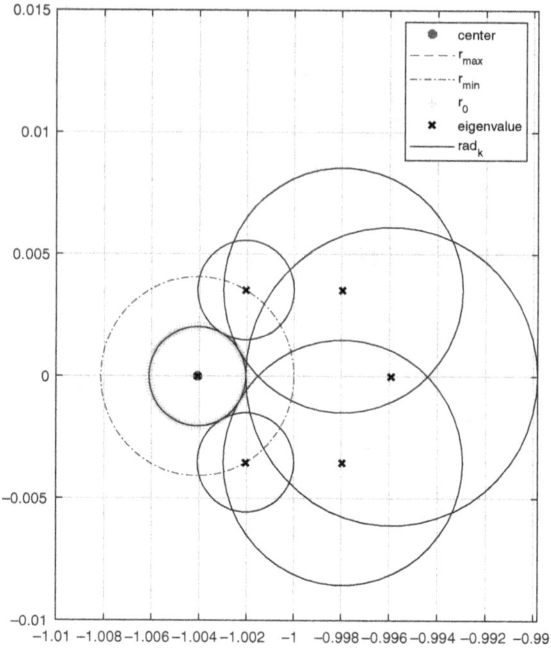

Fig. 6.5. Gershgorin disks for $s = 1$.

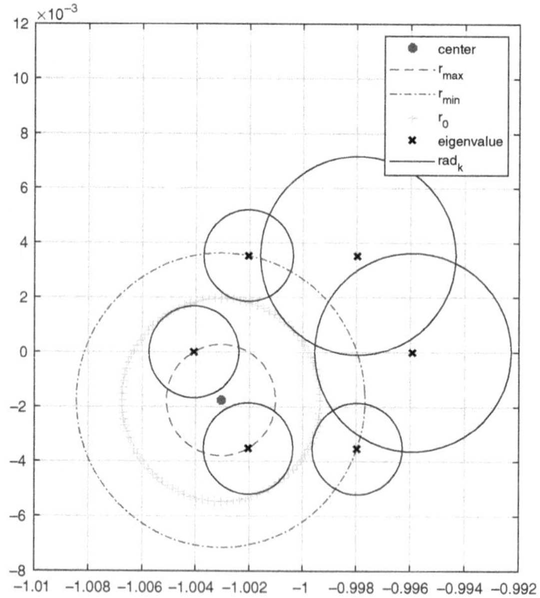

Fig. 6.6. Gershgorin disks for $s = 2$.

Table 6.1. Clustering of eigenvalues.

s	mp	$r_{\max}$	$r_{\min}$	diag(D)
1	-1.004071358205400	0	$4.070 \cdot 10^{-3}$	$d_6 = 7.373 \cdot 10^6$ $d_5 = 5.528 \cdot 10^3$ $d_4 = 2.309 \cdot 10^1$ $d_3 = 7.920 \cdot 10^{-2}$
2	-1.003054797419196 $-j_o 0.001762949721714$	$2.035 \cdot 10^{-3}$	$5.386 \cdot 10^{-3}$	$d_6 = 6.070 \cdot 10^6$ $d_5 = 3.747 \cdot 10^3$ $d_4 = 5.235 \cdot 10^0$ $d_3 = 1.069 \cdot 10^{-2}$
3	-1.001357971987876 $-j_o 0.002352076390922$	$3.592 \cdot 10^{-3}$	$5.917 \cdot 10^{-3}$	$d_6 = 4.217 \cdot 10^6$ $d_5 = 5.281 \cdot 10^3$ $d_4 = 5.099 \cdot 10^0$ $d_3 = 3.344 \cdot 10^{-3}$
4	-0.999999360560444 $-j_o 0.0017640572931918$	$4.441 \cdot 10^{-3}$	$5.609 \cdot 10^{-3}$	$d_6 = 2.238 \cdot 10^6$ $d_5 = 1.947 \cdot 10^3$ $d_4 = 3.823 \cdot 10^0$ $d_3 = 1.319 \cdot 10^{-3}$
5	-0.999592352673402 $-j_o 0.000705179888686$	$4.538 \cdot 10^{-3}$	$4.887 \cdot 10^{-3}$	$d_6 = 6.457 \cdot 10^5$ $d_5 = 2.330 \cdot 10^2$ $d_4 = 2.819 \cdot 10^{-1}$
6	-1.000000000000001 $-j_o 0.000000000000000$			

The same result produces also the attempts to cluster the eigenvalues in groups of 3, 4 and 5 eigenvalues. In Figure 6.7, we show the Gershgorin disks when in one group are clustered λ_6, λ_5 and λ_4 and in Figure 6.8 are shown the corresponding disks in the case when $\lambda_6, \lambda_5, \lambda_4$ and λ_3 in one group are clustered. In both cases, the third element of D is again smaller than 1, which shows the impossibility to isolate this group from the rest eigenvalues.

Finally, in Figure 6.9, we show the Gershgorin disks for the case when one attempts to combine in one group all eigenvalues except λ_1. In this case, one obtains that the element d_4 is smaller than one. Hence, we reach the conclusion that all 6 eigenvalues should participate in one group, i.e., there is an eigenvalue with algebraic multiplicity equal to 6. At this stage, it is not possible to say whether the eigenvalues participate in one or several Jordan blocks, i.e., what their geometric multiplicity is.

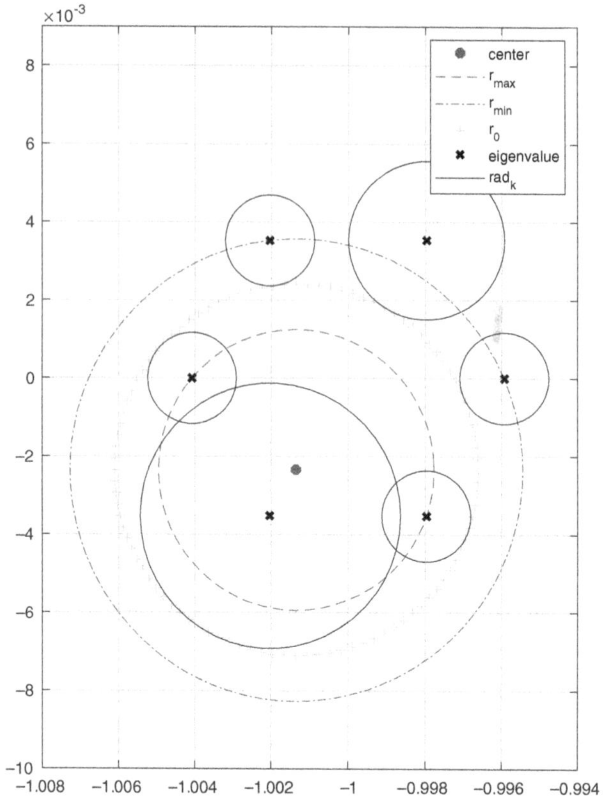

Fig. 6.7. Gershgorin disks for $s = 3$.

Note that after combining the 6 multiple eigenvalues in one group, their mean value

$$m_p = -1.000000000000001 - j_o 0.000000000000000$$

turns out to be equal to the true eigenvalue $\lambda = -1$ up to the 14th decimal digit inclusive (compare with the results obtained from the Schur form). In this respect, Algorithm 6.2, based on the Gershgorin theorem, itself is a powerful tool for the exact computation of the multiple eigenvalues of a matrix. Hence, it is appropriate to use this algorithm jointly with the standard algorithms for computing the eigenvalues in cases of multiple eigenvalues.

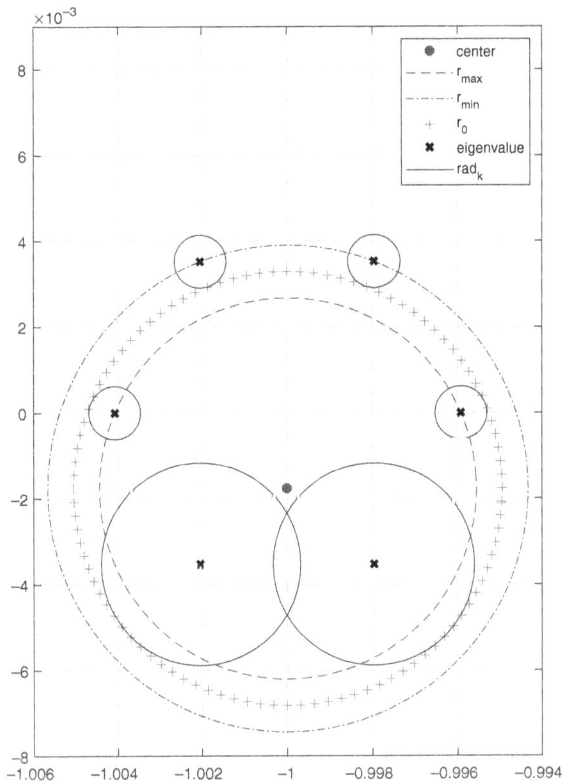

Fig. 6.8. Gershgorin disks for $s = 4$.

6.5 Reduction into Staircase Form

The heart of Kublanovskaya–Ruhe–Kågström algorithm is the unitary reduction of a square matrix into staircase form.

Consider the complex matrix $A \in \mathbb{C}^{n \times n}$ which has a single eigenvalue λ participating in one or several Jordan blocks.

With the aid of the shifting

$$A \leftarrow A - \lambda I_n, \tag{6.9}$$

the matrix A is transformed into a matrix that has only zero eigenvalues, i.e., this matrix is nilpotent. The structure of this matrix and its Weyr characteristic, in particular, is the same as that of A.

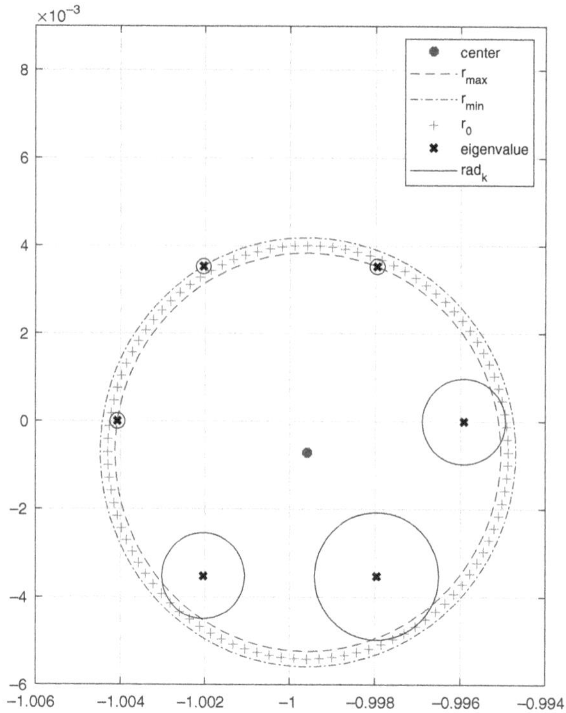

Fig. 6.9. Gershgorin disks for $s = 5$.

In this section, we show how a nilpotent square matrix is reduced to a staircase form using unitary similarity transformations. The reduction into staircase form consists of consecutive determination of the numerical null spaces of matrices with decreasing dimension. (For the definition of numerical null space, see Section 1.6.2.) This is done by the staircase algorithm, which at each step of the reduction uses the criterion for a numerical rank of a matrix (Section 1.6.1). For an $n \times n$ matrix A with singular values $\sigma_1 \leq \sigma_2 \leq \cdots \leq \sigma_n$ that are in increasing order, the numerical rank is the largest index k, such that $\sigma_{n-k+1} > tol \cdot ||A||_F$ and $\sigma_{n-k+1} > gap \cdot tol \cdot ||A||_F$, where tol is a tolerance reflecting the uncertainty in A and gap is a sufficiently large number usually taken as 1000. This numerical rank remains invariant to perturbations of A of size up to $gap \cdot tol \cdot ||A||_F$. After determining the rank, the current matrix is transformed and replaced by a near rank deficient matrix, whose exact rank is equal to the numerical rank of A.

The reduction of the nilpotent matrix $N^{(1)} := A$ in staircase form is done in h steps, where the index of nilpotence h is not known *a priori* and is found during the reduction process.

6.5.1 *First Step*

In the first step, one performs a singular value decomposition of the matrix A,

$$N^{(1)} = A = U\Sigma V^H = [U_1 \quad U_2]\begin{bmatrix}\Sigma_1 & 0\\ 0 & \Sigma_2\end{bmatrix}\begin{bmatrix}V_1^H\\ V_2^H\end{bmatrix},$$

where the matrices are separated according to the numerical rank k_1 of A. Hence, the diagonal matrix Σ_1 corresponds to "zero" singular values, whose number is $w_1 = n - k_1$ (the first element of the Weyr characteristic). Then, the numerical null space $\mathcal{N}(N^{(1)}) = \text{span}(V_1)$ is deflated by using unitary similarity transformation as

$$A^{(1)} = V^H AV = \begin{bmatrix}V_1^H\\ V_2^H\end{bmatrix} A\,[V_1 \quad V_2] = \begin{bmatrix}V_1^H U_1\Sigma_1 & V_1^H U_2\Sigma_2\\ V_2^H U_1\Sigma_1 & V_2^H U_2\Sigma_2\end{bmatrix}.$$

In this way,

$$A^{(1)} = Q^{(1)^H} AQ^{(1)}, \quad Q^{(1)} = V.$$

In accordance with the criterion for a numerical rank (see Section 1.6.1), we accept the leading w_1 columns of $A^{(1)}$ as zeros and replace this matrix with a matrix $\widetilde{A}^{(1)}$ which is rank deficient, i.e., strictly incomplete with respect to the rank (Figure 6.10),

$$\widetilde{A}^{(1)} = \begin{bmatrix}0_{w_1\times w_1} & \widetilde{A}_{12}\\ 0_{k_1\times w_1} & \widetilde{A}_{22}\end{bmatrix} = \begin{bmatrix}0_{w_1\times w_1} & V_1^H U_2\Sigma_2\\ 0_{k_1\times w_1} & V_2^H U_2\Sigma_2\end{bmatrix}, \tag{6.10}$$

where $\widetilde{A}_{12} \in \mathbb{C}^{w_1\times n_b}$, $\widetilde{A}_{22} \in \mathbb{C}^{n_b\times n_b}$, $n_b = n - w_1 = k_1$. We have that

$$\widetilde{A}^{(1)} = Q^{(1)^H}(A + E_1)Q^{(1)}, \tag{6.11}$$

where the perturbation $E_1 = -U_1\Sigma_1 V_1^H$ reflects the neglecting of the diagonal matrix Σ_1. Due to the unitary invariance of the Frobenius norm, one has the relationship

$$\|E_1\|_F^2 = \|\Sigma_1\|_F^2 = \sigma_1^2 + \cdots + \sigma_{w_1}^2.$$

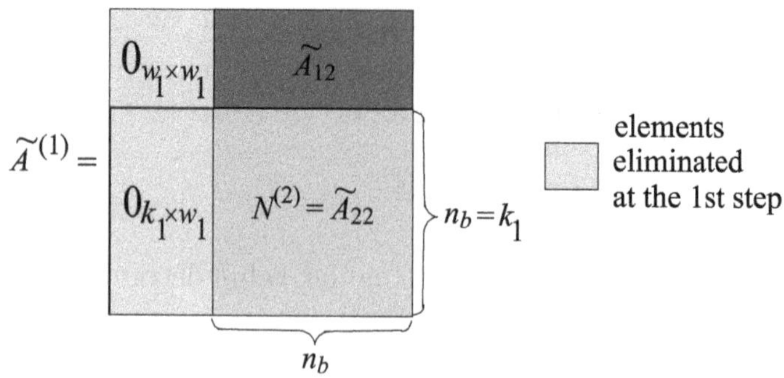

Fig. 6.10. The reduction into staircase form after the first step.

The equation

$$(A + E_1)V_1 = 0_{n \times w_1}$$

implies that V_1, the first w_1 columns of V, are the exact eigenvectors of the matrix $A + E_1$.

Note that this step corresponds to the determining of the numerical null space $\mathcal{N}(N^{(1)}) = \text{span}(V_1)$ of A, as defined in Section 1.6.2. We have that $\text{null}(N^{(1)}) = \dim(\mathcal{N}(N^{(1)})) = w_1$. This gives reason to call the quantity $\dim(\mathcal{N}(N^{(1)})$ *dimension of the numerical nullspace* or *numerical nullity*. The quantity w_1 is the first element of the numerical Weyr characteristics.

6.5.2 *Next Steps*

It follows from (6.10) that if we take the vector

$$z = \begin{bmatrix} 0_{w_1} \\ x \end{bmatrix},$$

where $x \in \mathbb{C}^{k_1}$ is in the null space of $\widetilde{A}_{22}$, i.e., $\widetilde{A}_{22}x = 0$, then

$$\widetilde{A}^{(1)}z = \widetilde{A}^{(1)} \begin{bmatrix} 0_{w_1} \\ x \end{bmatrix} = \begin{bmatrix} \widetilde{A}_{12}x \\ \widetilde{A}_{22}x \end{bmatrix} = \begin{bmatrix} \widetilde{A}_{12}x \\ 0 \end{bmatrix}.$$

Since

$$(\widetilde{A}^{(1)})^2 = \begin{bmatrix} 0 & \widetilde{A}_{12}\widetilde{A}_{22} \\ 0 & \widetilde{A}_{22}^2 \end{bmatrix},$$

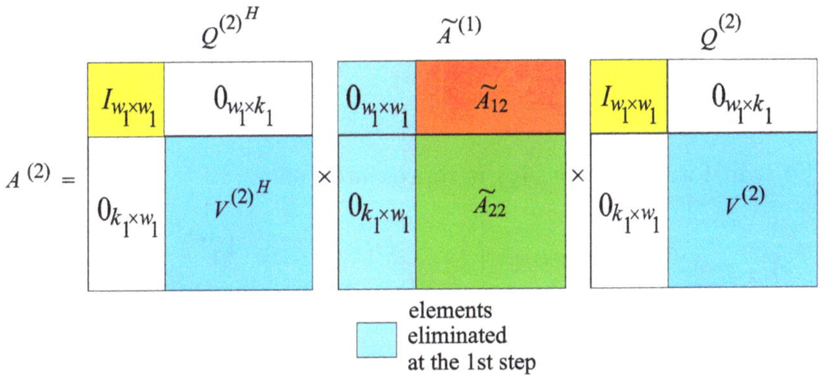

Fig. 6.11. The similarity transformation at the second step.

we obtain that the vector

$$\widetilde{A}^{(1)} z = \begin{bmatrix} \widetilde{A}_{12} x \\ 0 \end{bmatrix}$$

is in the null space of $(\widetilde{A}^{(1)})^2$, i.e., z is a principal vector of $\widetilde{A}^{(1)}$ with grade equal to 2 (see Section 2.2.1). The principal vectors of grade 3 and higher are computed by recursive application of the described procedure to the block $\widetilde{A}_{22}$. Thus, at the second step of the reduction, one performs a singular value decomposition of the nilpotent matrix $N^{(2)} = \widetilde{A}_{22} \in \mathbb{C}^{n_b \times n_b}$,

$$\widetilde{A}_{22} = U^{(2)} \Sigma^{(2)} V^{(2)^H} = \begin{bmatrix} U_1^{(2)} U_2^{(2)} \end{bmatrix} \begin{bmatrix} \Sigma_1^{(2)} & 0 \\ 0 & \Sigma_2^{(2)} \end{bmatrix} \begin{bmatrix} V_1^{(2)^H} \\ V_2^{(2)^H} \end{bmatrix}.$$

At this step, a unitary similarity transformation is done (Figure 6.11) so that

$$A^{(2)} = \begin{bmatrix} I_{w_1} & 0 \\ 0 & V^{(2)^H} \end{bmatrix} \begin{bmatrix} 0_{w_1 \times w_1} & \widetilde{A}_{12} \\ 0_{k_1 \times w_1} & \widetilde{A}_{22} \end{bmatrix} \begin{bmatrix} I_{w_1} & 0 \\ 0 & V^{(2)} \end{bmatrix}$$

$$= \begin{bmatrix} 0_{w_1 \times w_1} & \widetilde{A}_{12} V^{(2)} \\ 0_{k_1 \times w_1} & V^{(2)^H} \widetilde{A}_{22} V^{(2)} \end{bmatrix}.$$

Thus,

$$A^{(2)} = Q^{(2)^H} \widetilde{A}^{(1)} Q^{(2)},$$

where

$$Q^{(2)} = \begin{bmatrix} I_{w_1} & 0 \\ 0 & V^{(2)} \end{bmatrix}.$$

The transformed block A_{22} is represented as

$$A_{22}^{(2)} = V^{(2)^H} \widetilde{A}_{22} V^{(2)} = \begin{bmatrix} V_1^{(2)^H} U_1^{(2)} \Sigma_1^{(2)} & V_1^{(2)^H} U_2^{(2)} \Sigma_2^{(2)} \\ V_2^{(2)^H} U_1^{(2)} \Sigma_1^{(2)} & V_2^{(2)^H} U_2^{(2)} \Sigma_2^{(2)} \end{bmatrix},$$

where the matrix $\Sigma_2^{(2)}$ has numerical rank $k_2 < k_1$. Accepting as zeros the singular values of $\Sigma_1^{(2)}$ whose number is $w_2 = n_b - k_2$, one obtains

$$\widetilde{A}_{22}^{(2)} = \begin{bmatrix} 0_{w_2 \times w_2} & \widetilde{A}_{12}^{(3)} \\ 0_{k_2 \times w_2} & \widetilde{A}_{22}^{(3)} \end{bmatrix} = \begin{bmatrix} 0_{w_2 \times w_2} & V_1^{(2)^H} U_2^{(2)} \Sigma_2^{(2)} \\ 0_{k_2 \times w_2} & V_2^{(2)^H} U_2^{(2)} \Sigma_2^{(2)} \end{bmatrix}.$$

Thus, we determine the matrix

$$\widetilde{A}^{(2)} = \begin{bmatrix} 0_{w_1 \times w_1} & \widetilde{A}_{12} V^{(2)} \\ 0_{k_1 \times w_1} & \widetilde{A}_{22}^{(2)} \end{bmatrix} = \begin{bmatrix} 0_{w_1 \times w_1} & \widetilde{A}_{12} V_1^{(2)} & \widetilde{A}_{12} V_2^{(2)} \\ 0_{w_2 \times w_1} & 0_{w_2 \times w_2} & \widetilde{A}_{12}^{(3)} \\ 0_{k_2 \times w_1} & 0_{k_2 \times w_2} & \widetilde{A}_{22}^{(3)} \end{bmatrix}.$$

This step corresponds to the determining of the numerical null space $\mathcal{N}(N^{(2)}) = \operatorname{span}(V_1^{(2)})$ of $\widetilde{A}_{22}$. Clearly, $\operatorname{null}(N^{(2)}) = \dim(\mathcal{N}(N^{(2)}) = w_2$.

The transformed matrix A after the second step of the reduction is shown in Figure 6.12.

We have that

$$\widetilde{A}^{(2)} = Q^{(2)^H} (\widetilde{A}^{(1)} + E_2) Q^{(2)}, \tag{6.12}$$

where the perturbation E_2 reflects the neglecting of the matrix $\Sigma_1^{(2)}$. The norm of E_2 is found from

$$\|E_2\|_F^2 = \left(\sigma_1^{(2)}\right)^2 + \cdots + \left(\sigma_{w_2}^{(2)}\right)^2,$$

where $\sigma_\ell^{(2)}$, $\ell = 1, \ldots, w_2$, are the singular values of $\Sigma_1^{(2)}$.

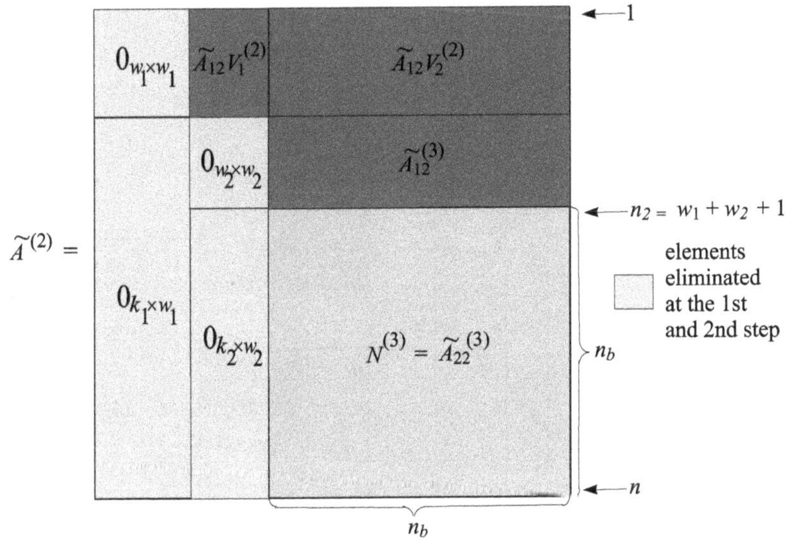

Fig. 6.12. The reduction into staircase form after the second step.

Let

$$z = \begin{bmatrix} 0_{w_1} \\ 0_{w_2} \\ x \end{bmatrix},$$

where $x \in \mathbb{C}^{n-w_1-w_2}$ belongs to the numerical null space $\mathcal{N}(N^{(3)})$ of $N^{(3)} = \widetilde{A}_{22}^{(3)}$, i.e., $\widetilde{A}_{22}^{(3)}x = 0$.

Then it is possible to show that the vector

$$\widetilde{A}^{(2)}z = \widetilde{A}^{(2)} \begin{bmatrix} 0_{w_1} \\ 0_{w_2} \\ x \end{bmatrix} = \begin{bmatrix} \widetilde{A}_{12}V_2^{(2)}x \\ \widetilde{A}_{12}^{(3)}x \\ \widetilde{A}_{22}^{(3)}x \end{bmatrix} = \begin{bmatrix} \widetilde{A}_{12}V_2^{(2)}x \\ \widetilde{A}_{12}^{(3)}x \\ 0 \end{bmatrix}$$

belongs to the null space of $(\widetilde{A}^{(2)})^3$, i.e., z is a principal vector of $\widetilde{A}^{(2)}$ of grade 3.

The transformed matrix A after the ith step of the reduction is shown in Figure 6.13.

At the last hth step of the reduction, one obtains that $k_h = 0$. Note that at this step the numerical rank is obtained using only the

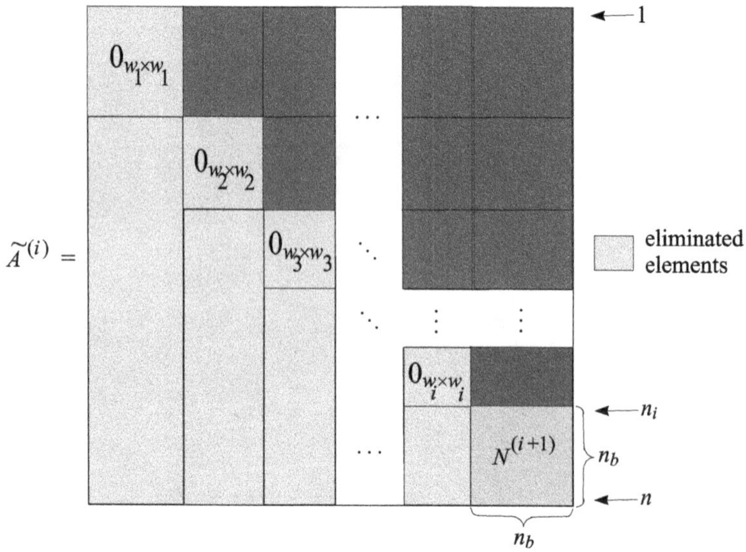

Fig. 6.13. The reduction into staircase form at step i.

condition $\sigma_{n_b-k}\left(N^{(h)}\right) < tol\|A\|_F$. Hence, after this step, we have determined all elements of the numerical Weyr characteristic

$$\begin{aligned}
w_1 &= \text{null}(A-\lambda I_n)^1 - \text{null}(A-\lambda I_n)^0 = \dim(\mathcal{N}(N^{(1)})),\\
w_2 &= \text{null}(A-\lambda I_n)^2 - \text{null}(A-\lambda I_n)^1 = \dim(\mathcal{N}(N^{(2)})),\\
\vdots &\quad \vdots\\
w_h &= \text{null}(A-\lambda I_n)^h - \text{null}(A-\lambda I_n)^{h-1} = \dim(\mathcal{N}(N^{(h)})).
\end{aligned}$$

After the last step, the matrix A is reduced into the *staircase form*

$$\begin{aligned}
S &= \widetilde{A}^{(h)} = Q^H(A+E)Q\\
&= \begin{bmatrix} 0_{w_1\times w_1} & N_{12} & \cdots & N_{1,h-1} & N_{1h}\\ 0 & 0_{w_2\times w_2} & \cdots & N_{2,h-1} & N_{2h}\\ \vdots & \vdots & \ddots & \ddots & \vdots\\ 0 & 0 & \cdots & 0_{w_{h-1}\times w_{h-1}} & N_{h-1,h}\\ 0 & 0 & \cdots & 0 & 0_{w_h\times w_h}\end{bmatrix}, \quad (6.13)
\end{aligned}$$

where

$$Q = Q^{(1)}Q^{(2)}\ldots Q^{(h)},$$

and the perturbation E reflects the effect of all neglected elements of the transformed matrices. The nilpotent matrix S has the Weyr structure $\mathcal{W} = \{w_1, w_2, \dots, w_h\}$, and the matrix E has a norm that can be found from

$$\begin{aligned}\|E\|_F^2 &= \|E_1\|_F^2 + \|E_2\|_F^2 + \cdots + \|E_h\|_F^2\\ &= \|\Sigma_1\|_F^2 + \|\Sigma_1^{(2)}\|_F^2 + \cdots + \|\Sigma^{(h)}\|_F^2\\ &= \sigma_1^2 + \cdots + \sigma_{w_1}^2 + \left(\sigma_1^{(2)}\right)^2 + \cdots + \left(\sigma_{w_2}^{(2)}\right)^2\\ &\quad + \cdots + \left(\sigma_1^{(h)}\right)^2 + \cdots + \left(\sigma_{w_h}^{(h)}\right)^2. \end{aligned} \tag{6.14}$$

The following estimate holds

$$\|E\|_F = O(\text{tol} \cdot \|A\|_F).$$

The number $\|E\|_F$ can be interpreted as the distance of the matrix $A - \lambda I_n$ to the computed nilpotent matrix S. In this way, the relative distance from $A - \lambda I_n$ to the matrix S, which has the numerical structure $\mathcal{W}(A, \lambda) = \{w_1, w_2, \dots, w_h\}$, is given by the quantity $\|E\|_F / \|A\|_F$.

One has that

$$\nu = \sum_{i=1}^{h} w_i,$$

i.e., the algebraic multiplicity of the eigenvalue λ is equal to the sum of the elements of the Weyr characteristic.

In Figure 6.14, we show the shape of the staircase form of a 12×12 matrix A which has a multiple eigenvalue λ with associated Weyr characteristic $\mathcal{W} = \{5, 3, 2, 1, 1\}$.

The reduction considered earlier is summarized in Algorithm 6.3 (see also Figure 6.13).

The reduction performed by Algorithm 6.3 is done by implementing unitary transformations determined using only the singular value decomposition. It can be realized using the LAPACK routine `xGESVD` or MATLAB® function `svd`. The function `staircase` can also be used independently, since it does not require the matrix A to be preliminary reduced to Schur form. A version of the algorithm requiring

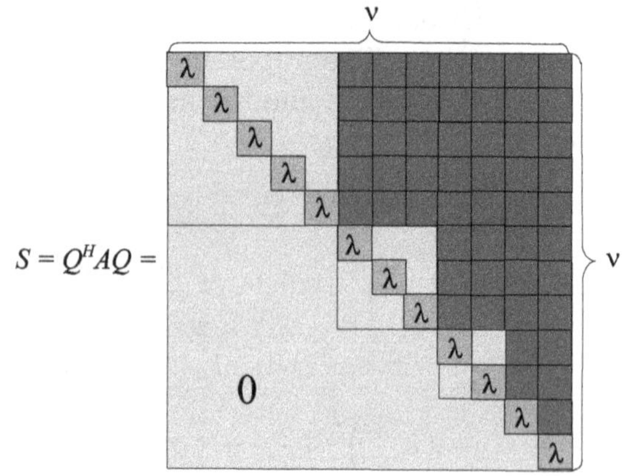

Fig. 6.14. The staircase form of a matrix A with Weyr characteristic $\mathcal{W} = \{5,\ 3,\ 2,\ 1,\ 1\}$.

less computational work can be realized by using rank revealing QR decomposition, but it is not so reliable with respect to the determination of the Jordan structure.

Consider the numerical properties of the algorithm. Algorithm 6.3 is backward stable and produces accurate results for the Weyr and Segre characteristics, if the numerical structure of the matrix is well conditioned. According to the definition, given in Section 5.7, the original matrix A has a well-conditioned numerical structure at $\lambda = 0$, if the separation between the numerical images and null spaces in the staircase algorithm is well defined with respect to the deflation tolerances *tol* and *gap*. If A has the zero as an exact multiple eigenvalue, then the ratio $\|E\|_F/\|A\|_F$ will have an order $O(n \cdot \mathbf{u})$, where $\mathbf{u}$ is the unit rounding error of the arithmetic used. Note that the perturbation $E = QSQ^H - A$ can be easily computed from the result obtained.

6.5.3 *Strategy for Choosing a Bundle*

If we interpret the staircase algorithm from the point of view of the stratification of Jordan bundles, considered in Chapter 5, it is easy to see that the strategy used in this algorithm consists of finding at each step a numerical null space of the corresponding transformed matrix of maximum possible dimension with respect to the given

Algorithm 6.3. Reduction into staircase form
Reduces a nilpotent matrix into staircase form and determines the Weyr characteristic associated with the zero eigenvalue.
function $[S, Q, \mathcal{W}] = \text{staircase}(A, tol, gap)$
Input: The matrix $A \in \mathbb{C}^{n \times n}$ with a single zero eigenvalue; tolerance parameters tol and gap used in the deflation.
Output: A staircase matrix $S \in \mathbb{C}^{n \times n}$ and a unitary matrix $Q \in \mathbb{C}^{n \times n}$, so that $Q^H(A+E)Q = S$, where $||E||_F = O(tol||A||_F)$. The zero eigenvalue of $A+E$ has the associated Weyr characteristic $\mathcal{W} = \{w_1, w_2, \ldots, w_h\}$.
Set $S = A,\ Q = I_n$
Let $U\Sigma V^H$ be the SVD of A with singular values in increasing order. Find the largest $k > 0$, such that $\sigma_{n-k+1} > tol \cdot ||A||_F$ and $\sigma_{n-k+1} > gap \cdot tol \cdot ||A||_F$. Partition conformably $U = [U_1,\ U_2]$, $\Sigma = \text{diag}(\Sigma_1, \Sigma_2)$ and $V = [V_1,\ V_2]$.
Set $i = 1,\ n_i = 1;\ n_b = n,\ w_1 = n - k$
while $w_i > 0$ **do**
  $Q(:, n_i : n) = Q(:, n_i : n)V$
  if $i = 1$

$$S = \begin{bmatrix} 0_{w_1 \times w_1} & V_1^H U_2 \Sigma_2 \\ 0_{n_b \times w_1} & V_2^H U_2 \Sigma_2 \end{bmatrix}$$

  else

$$S(1:n, n_i:n) = \begin{bmatrix} S(1:n_i-1, n_i:n)V \\ 0_{w_i \times w_i} \quad V_1^H U_2 \Sigma_2 \\ 0_{n_b \times w_i} \quad V_2^H U_2 \Sigma_2 \end{bmatrix}$$

  end
  $i = i + 1,\ n_i = n_i + w_i,\ n_b = n_b - w_i$
  Let $U\Sigma V^H$ be the SVD of the $n_b \times n_b$ matrix $N^{(i+1)} = S(n_i{:}n, n_i{:}n)$.
  Find the largest index k, such that $\sigma_{n_b-k+1} > tol \cdot ||A||_F$ and $\sigma_{n_b-k+1} > gap \cdot tol \cdot ||A||_F$.
  $U = [U_1,\ U_2],\ \Sigma = \text{diag}(\Sigma_1, \Sigma_2),\ V = [V_1,\ V_2],\ U_1, V_1 \in \mathbb{C}^{nb \times k}$.
  $w_i = n - n_i - k + 1$
end

deflation criterion. This is equivalent to finding the *maximum possible elements of the Weyr characteristic* associated with the zero eigenvalue with respect to the deflation criterion. According to the discussion in Section 5.1, for a zero eigenvalue λ of multiplicity n, the most generic Jordan form of an $n \times n$ nilpotent matrix is a $J_n(0)$

Table 6.2. Jordan and Weyr forms corresponding to an eigenvalue with multiplicity n.

Most generic forms

$\operatorname{codim}(\mathcal{B}) = n - 1$

$\mathcal{S} = \{n\}$ $\qquad$ $\mathcal{W} = \{1, 1, \ldots, 1\}$

$$J = \begin{bmatrix} \lambda & 1 & & & \\ & \lambda & 1 & & \\ & & \lambda & \ddots & \\ & & & \ddots & 1 \\ & & & & \lambda \end{bmatrix}, \qquad W = \left[\begin{array}{c|c|c|c|c} \lambda & 1 & & & \\ \hline & \lambda & 1 & & \\ \hline & & \lambda & \ddots & \\ \hline & & & \ddots & 1 \\ \hline & & & & \lambda \end{array}\right]$$

Intermediate cases

$n - 1 \le \operatorname{codim}(\mathcal{B}) \le n^2 - 1$

$\operatorname{codim}(\mathcal{B}) = s_1 + 3s_2 + \cdots + (2q - 1)s_q$

$\mathcal{S} = \{s_1, s_2, \ldots, s_q\}$ $\qquad$ $\mathcal{W} = \{w_1, w_2, \ldots, w_h\}$

$s_1 \ge s_2 \ge \cdots \ge s_q \ge 1$ $\qquad$ $w_1 \ge w_2 \ge \cdots \ge w_h \ge 1$

$$J = \left[\begin{array}{c|c|c|c|c} J_{s_1}(\lambda) & & & & \\ \hline & J_{s_2}(\lambda) & & & \\ \hline & & J_{s_3}(\lambda) & & \\ \hline & & & \ddots & \\ \hline & & & & J_{s_q}(\lambda) \end{array}\right], \quad W = \left[\begin{array}{c|c|c|c|c} \lambda I_{w1} & G_{w_1,w_2} & & & \\ \hline & \lambda I_{w_2} & G_{w_2,w_3} & & \\ \hline & & \lambda I_{w_3} & \ddots & \\ \hline & & & \ddots & G_{w_{h-1},w_h} \\ \hline & & & & \lambda I_{w_h} \end{array}\right]$$

Most degenerate forms

$\operatorname{codim}(\mathcal{B}) = n^2 - 1$

$\mathcal{S} = \{1, 1, \ldots, 1\}$ $\qquad$ $\mathcal{W} = \{n\}$

$$J = \left[\begin{array}{c|c|c|c|c} \lambda & & & & \\ \hline & \lambda & & & \\ \hline & & \lambda & & \\ \hline & & & \ddots & \\ \hline & & & & \lambda \end{array}\right], \qquad W = \begin{bmatrix} \lambda & & & & \\ & \lambda & & & \\ & & \lambda & & \\ & & & \ddots & \\ & & & & \lambda \end{bmatrix}$$

block (defective but nonderogatory matrix) and the least generic (most degenerate) Jordan form consists of n Jordan blocks $J_1(0)$ (derogatory but nondefective matrices) (see Table 6.2). The latter form corresponds to the maximum possible value equal to n of the single element of the Weyr characteristic. All other possible Jordan structures under small perturbations are given by the closure hierarchy of the nilpotent orbits of $n \times n$ matrices with zero eigenvalue. In this way, we come to the conclusion that *Algorithm 6.2 determines the least generic (i.e., most degenerate) numerical structure of the given matrix concerning the deflation criterion.* This property of the algorithm satisfies the condition (5.31) imposed in the definition of the numerical Jordan form (Definition 5.21).

In case of an ill-conditioned numerical structure, the algorithm may produce a wrong nearby Jordan structure. For this reason, it is useful to know the stratification of the Jordan structures that shows which structures are nearby other structures in the hierarchy of Jordan forms (see Section 5.4).

If Algorithm 6.3 produces an unacceptable result from the point of view of the size of the perturbation $\|E\|_F$, the computations should be repeated for other values of *tol* and *gap*, taking into account the considerations given latter on in Section 6.6.

Example 6.5. Consider the matrix

$$A = \begin{bmatrix} 3 & -4 & 2 & -2 & -4 & -10 & 17 \\ 1 & -5 & 1 & -4 & 2 & 0 & 8 \\ 2 & 0 & -1 & 2 & -4 & -6 & 5 \\ 1 & -2 & 0 & -2 & 0 & -1 & 4 \\ 1 & -3 & 1 & -3 & 0 & -1 & 6 \\ 1 & -1 & 0 & 0 & -1 & -3 & 4 \\ 0 & -2 & 0 & -2 & 2 & 2 & 1 \end{bmatrix},$$

given in Example 2.14 which has 7-tuple eigenvalue $\lambda = -1$. Further on in this example we shall use the nilpotent matrix

$$N = A - \lambda I_7 = \begin{bmatrix} 4 & -4 & 2 & -2 & -4 & -10 & 17 \\ 1 & -4 & 1 & -4 & 2 & 0 & 8 \\ 2 & 0 & 0 & 2 & -4 & -6 & 5 \\ 1 & -2 & 0 & -1 & 0 & -1 & 4 \\ 1 & -3 & 1 & -3 & 1 & -1 & 6 \\ 1 & -1 & 0 & 0 & -1 & -2 & 4 \\ 0 & -2 & 0 & -2 & 2 & 2 & 2 \end{bmatrix}.$$

Since this matrix has Jordan form

$$J = \operatorname{diag}\left(\begin{bmatrix} 0 & 1 & 0 \\ 0 & 0 & 1 \\ 0 & 0 & 0 \end{bmatrix}, \begin{bmatrix} 0 & 1 \\ 0 & 0 \end{bmatrix}, \begin{bmatrix} 0 & 1 \\ 0 & 0 \end{bmatrix} \right),$$

with blocks of orders $s_1 = 3,\ s_2 = 2$ and $s_3 = 2$ and zero eigenvalues, the following conditions on the ranks and nullities of the powers of N hold:

$$\begin{aligned} \text{rank}(N^0) &= 7, & \text{null}(N^0) &= 0, \\ \text{rank}(N^1) &= 4, & \text{null}(N^1) &= 3, \\ \text{rank}(N^2) &= 1, & \text{null}(N^2) &= 6, \\ \text{rank}(N^3) &= 0, & \text{null}(N^3) &= 7 \end{aligned}$$

and the Weyr characteristic has elements

$$\begin{aligned} w_1 &= \text{null}(N^1) - \text{null}(N^0) = 3, \\ w_2 &= \text{null}(N^2) - \text{null}(N^1) = 3, \\ w_3 &= \text{null}(N^3) - \text{null}(N^2) = 1. \end{aligned}$$

Thus, the matrix A is in a bundle with codimension equal to 18.

The reduction into staircase form of the matrix A is done by Algorithm 6.3 and the computations produce the following results.

In the first step, the singular values of the matrix $N = A - \lambda I_7$ are

$$\begin{array}{c|l} & \sigma_1 = 0.000000000000000, \\ \Sigma_1 & \sigma_2 = 0.000000000000000, \\ & \sigma_3 = 0.000000000000001, \\ \hline & \sigma_4 = 0.550048116530568, \\ \Sigma_2 & \sigma_5 = 1.317436171248185, \\ & \sigma_6 = 10.256796098960987, \\ & \sigma_7 = 25.470766435043270. \end{array}$$

In the given case, in the unitary deflation, we will use tolerances equal to $tol = 10^{-12}$ and $gap = 1000$. This means that in determining the numerical rank of a matrix block, the smallest singular value σ_{n-k+1}, which is accepted as different from zero, should satisfy the conditions $\sigma_{n-k+1} \geq tol \cdot \|A\|_F = 2.76 \cdot 10^{-11}$ and $\sigma_{n-k+1} > 1000 \cdot tol \cdot \|A\|_F = 2.76 \cdot 10^{-8}$. The application of these conditions produce numerical rank of $A - \lambda I_7$ equal to $k = 4$, nullity of $A - \lambda I_7$ equal to 3 and element of the Weyr characteristic equal to $w_1 = 3$.

As a result from the transformation at the first step, one obtains (up to four decimal digits)

$$S_1 = V^H N V$$

$$= \left[\begin{array}{ccc|cccc} 0 & 0 & 0 & 0.2388 & -0.2093 & 1.6458 & 22.2117 \\ 0 & 0 & 0 & -0.3335 & -0.8438 & 4.7240 & 2.5584 \\ 0 & 0 & 0 & 0.3532 & -0.4054 & 5.3361 & -12.0100 \\ \hline 0 & 0 & 0 & 0.0942 & -0.8705 & -6.9309 & -2.0705 \\ 0 & 0 & 0 & -0.0227 & 0.2103 & 1.6745 & 0.5002 \\ 0 & 0 & 0 & 0.0072 & -0.0668 & -0.5319 & -0.1589 \\ 0 & 0 & 0 & -0.0103 & 0.0956 & 0.7614 & 0.2275 \end{array}\right].$$

At the second step, one transforms the fourth-order matrix

$$N^{(2)} = \begin{bmatrix} 0.0942 & -0.8705 & -6.9309 & -2.0705 \\ -0.0227 & 0.2103 & 1.6745 & 0.5002 \\ 0.0072 & -0.0668 & -0.5319 & -0.1589 \\ -0.0103 & 0.0956 & 0.7614 & 0.2275 \end{bmatrix}.$$

Its singular values are

$$\begin{array}{c|l} & \sigma_1 = 0.000000000000000, \\ \Sigma_1 & \sigma_2 = 0.000000000000001, \\ & \sigma_3 = 0.000000000000003, \\ \hline \Sigma_2 & \sigma_4 = 7.559289460184546 \end{array}$$

which gives $\text{null}(N^{(2)}) = 3$ and element of the Weyr characteristic equal to $w_2 = 3$. As a result of the transformation at this step, one obtains the matrix

$$\left[\begin{array}{ccc|c} 0 & 0 & 0 & -6.2276 \\ 0 & 0 & 0 & -4.2654 \\ 0 & 0 & 0 & 0.4062 \\ \hline 0 & 0 & 0 & 3.5930 \cdot 10^{-15} \end{array}\right].$$

At the 3rd step, we deal with the first-order matrix

$$N^{(3)} = 3.5930 \cdot 10^{-15}$$

whose singular value is

$$\sigma_1 = 3.5930 \cdot 10^{-15}$$

and can be accepted as zero. Hence, null$(N^{(3)})$ is equal to 1 and the corresponding element of the Weyr characteristic is $w_3 = 1$. (Note that in case of a scalar last block, the corresponding element of the Weyr characteristic by necessity is equal to one.)

As a result, one obtains the staircase form

$$S = \left[\begin{array}{ccc|ccc|c} 0 & 0 & 0 & -0.0124 & -1.4595 & 20.7950 & 7.8492 \\ 0 & 0 & 0 & -0.7511 & 1.1341 & 1.2563 & 5.1240 \\ 0 & 0 & 0 & 0.1035 & 2.3499 & -12.8405 & 1.6100 \\ \hline 0 & 0 & 0 & 0 & 0 & 0 & -6.2276 \\ 0 & 0 & 0 & 0 & 0 & 0 & -4.2655 \\ 0 & 0 & 0 & 0 & 0 & 0 & 0.4062 \\ \hline 0 & 0 & 0 & 0 & 0 & 0 & 0 \end{array}\right],$$

which has the Weyr characteristic

$$\mathcal{W} = \{3,\ 3,\ 1\}.$$

Hence, the eigenvalue $\lambda = -1$ of A has a Segre characteristic, which is $\mathrm{S} = \{3,\ 2,\ 2\}$, i.e., the Jordan canonical form of A has one block of order 3 and two blocks of order 2.

The unitary transformation matrix into staircase form (in the case under consideration this matrix is orthogonal) is

$$Q = \begin{bmatrix} -0.6616 & -0.3048 & 0.5051 & -0.4043 & 0.1230 & -0.1888 & 0.0000 \\ -0.2283 & -0.4607 & 0.0594 & 0.7148 & -0.0972 & 0.1082 & 0.4472 \\ -0.2050 & 0.6166 & 0.3863 & 0.2414 & -0.5909 & -0.1453 & 0.0000 \\ -0.2940 & 0.5567 & -0.1200 & 0.0691 & 0.6167 & 0.0647 & 0.4472 \\ -0.6114 & 0.0361 & -0.5669 & 0.0939 & -0.1488 & 0.2695 & -0.4472 \\ 0.0890 & 0.0599 & 0.5063 & 0.2568 & 0.3191 & 0.6036 & -0.4472 \\ 0.0000 & 0.0000 & 0.0000 & 0.4333 & 0.3492 & -0.7002 & -0.4472 \end{bmatrix}.$$

The vector containing the norms of the matrices neglected during the individual deflation steps is

$$[5.0564 \cdot 10^{-15}\ 0\ 0\ 0\ 0\ 0\ 0].$$

This shows that the staircase form obtained is accurate for a matrix which is different from the matrix A by a perturbation E, satisfying

$$\|E\|_F = 5.0564 \cdot 10^{-15}.$$

6.6 Reduction into Jordan–Schur Form

Algorithm 6.3, presented in the previous section, can reduce into staircase form only one diagonal block of the Schur form corresponding to a given numerically multiple eigenvalue. To reduce the whole $n \times n$ matrix into Jordan–Schur form, it is necessary to transform consecutively each diagonal block T_{ii} of the block-triangular form (6.8) of size $\nu_i \times \nu_i$ corresponding to distinct eigenvalue and to apply the transformations on the off-diagonal blocks. In Figure 6.15, we represent part of the Jordan–Schur form, obtained after the reduction into staircase form of the first three diagonal blocks. Note the off-diagonal rows and columns transformed at the third step by the unitary similarity transformation.

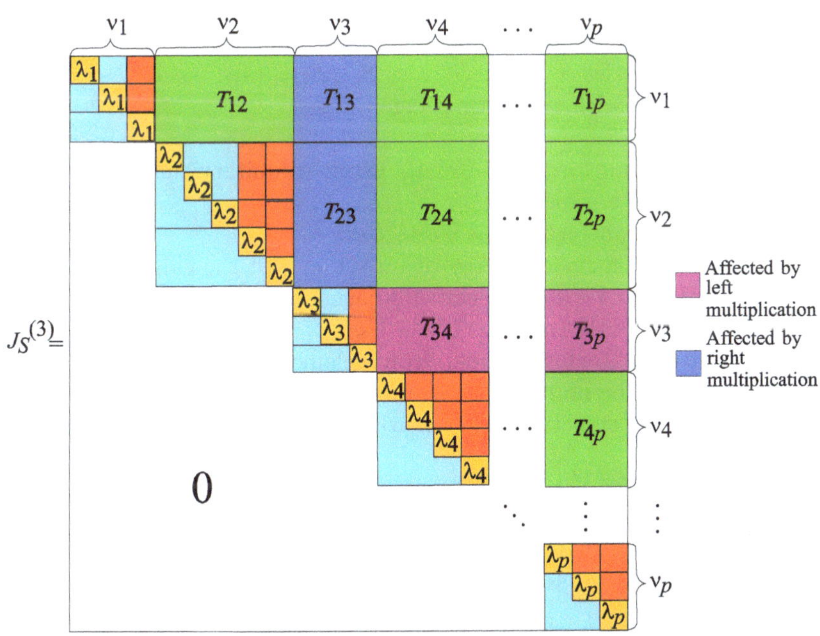

Fig. 6.15. The Jordan–Schur form after transforming the third diagonal block.

Algorithm 6.4 reduces a square complex matrix A into Jordan–Schur form. The algorithm makes use of Algorithms 6.1 and 6.2 that reduce A to upper block-triangular form and Algorithm 6.3 that reduces a nilpotent block into staircase form.

An example of the reduction into Jordan–Schur form is given at the beginning of this chapter (Example 6.1).

It should be emphasized that the reduction into the Jordan–Schur form by Algorithm 6.4 is more reliable than the reduction

Algorithm 6.4. Reduction into Jordan–Schur form
Reduces a square complex matrix into Jordan–Schur form by using unitary similarity transformations
function $[J_S, U_S, \mathcal{W}, \mathcal{S}] = \text{jordan\_schur}(A, ein, tol, gap)$
Input: The matrix $A \in \mathbb{C}^{n\times n}$; tolerance parameter ein used in the clustering of the eigenvalues; tolerance parameters tol and gap used in the deflation.
Output: The Jordan–Schur form $J_S \in \mathbb{C}^{n\times n}$ and a unitary matrix $U_S \in \mathbb{C}^{n\times n}$, such that $U_S^H(A+E)U_S = J_S$, where $||E||_F = O(tol||A||_F)$; the numerical Weyr characteristics $\mathcal{W} = [\mathcal{W}_1,\ \mathcal{W}_2, \dots,\ \mathcal{W}_p]$, $\mathcal{W}_i = \{w_{i1},\ w_{i2}, \dots,\ w_{i,h_i}\}$ associated with the distinct eigenvalues; the numerical Segre characteristics $\mathcal{S} = [\mathcal{S}_1,\ \mathcal{S}_2, \dots,\ \mathcal{S}_p]$, $\mathcal{S}_i = \{s_{1i},\ s_{2i}, \dots,\ s_{q_i,i}\}$ associated with the distinct eigenvalues.
Set $U_S = I_n$
Reduce A into Schur form $T = U^H AU$ using an unitary similarity transformation with a matrix U.
Accumulate the transformations $U_S \leftarrow U_S U$
Reorder the Schur form T by Algorithm 6.1 so that the close eigenvalues of $T_2 = V^H TV$ appear in adjacent positions
Accumulate the transformations $U_S \leftarrow U_S V$
Using Algorithm 6.2, cluster the diagonal elements of T_2 in groups corresponding to the multiple eigenvalues λ_i, $i = 1, 2, \dots, p$ and determine the structure of the matrix

$$T_3 = \begin{bmatrix} T_{11} & T_{12} & \dots & T_{1p} \\ & T_{22} & \dots & T_{2p} \\ & & \ddots & \vdots \\ & & & T_{pp} \end{bmatrix}, \quad T_{ii} \in \mathbb{C}^{\nu_i\times\nu_i}, \quad i = 1, 2, \dots, p$$

for $i = 1, 2, \ldots, p$
  Make the diagonal block nilpotent
  $T_{ii} \leftarrow T_{ii} - \lambda_i I_{\nu_i}$
  Reduce by Algorithm 6.3 the block T_{ii} into the staircase form
  $S_i = Q_i^H T_{ii} Q_i$ and determine the Weyr characteristic $\mathcal{W}_i$ associated with λ_i
  $T_{ii} \leftarrow S_i + \lambda_i I_{\nu_i}$
  Transform the block-triangular matrix T_3
  for $j = i + 1 : p$
    $T_{ij} \leftarrow Q_i^H T_{ij}$
  end
  for $k = 1 : i - 1$
    $T_{ki} \leftarrow T_{ki} Q_i$
  end
  Accumulate the transformations
  $Y_i = [I_{\nu_1+\cdots+\nu_{i-1}}, Q_i, I_{\nu_{i+1}+\cdots+\nu_p}]$
  $U_S \leftarrow U_S Y_i$
  Using Algorithm 2.1, determine from $\mathcal{W}_i$ the Segre characteristic
  $\mathcal{S}_i$ associated with λ_i
end
$J_S = T_3$

into Jordan form. This follows from the fact that the computation of the Weyr and Jordan forms at the final step of the full algorithm for reduction into Jordan form requires using of additional nonunitary transformations which can be ill conditioned in some cases. From this point of view, the Jordan–Schur form represents a good compromise between the Schur and Jordan forms since it contains information about the Jordan structure which is obtained using only unitary similarity transformations.

Using the information about the Jordan structure of a matrix A, provided by Algorithm 6.4, it is possible to determine the codimension of the bundle $\mathcal{B}(A)$ in which A resides. This can be done exploiting the relationship (5.12) or (5.12). Algorithm 6.5 determines the bundle codimension $\text{codim}(\mathcal{B}(A))$ using the Segre characteristics of A with respect to the tolerances *ein* and *tol* implemented in clustering the eigenvalues and finding the Weyr characteristics, respectively.

Algorithm 6.5. Determining the codimension of the bundle in which resides a given matrix

Determines the codimension of the bundle $\mathcal{B}(A)$ in which resides the matrix A with respect to the tolerances ein and tol.

function $[codim, ev, \mathcal{W}, \mathcal{S}] = \text{bundle\_dimension}(A, ein, tol)$

Input: The matrix $A \in \mathbb{C}^{n\times n}$; tolerance parameter ein used in the clustering of the eigenvalues; tolerance parameter tol used in the reduction into staircase form.

Output: The codimension $codim$ of the bundle $\mathcal{B}(A)$ in which resides the matrix A; the distinct eigenvalues $ev = [\lambda_1,\ \lambda_2, \dots,\ \lambda_p]$ of A; the Weyr characteristics $\mathcal{W} = [\mathcal{W}_1,\ \mathcal{W}_2, \dots,\ \mathcal{W}_p]$, $\mathcal{W}_i = \{w_{i1},\ w_{i2}, \dots,\ w_{i,h_i}\}$ associated with the distinct eigenvalues; the Segre characteristics $\mathcal{S} = [\mathcal{S}_1,\ \mathcal{S}_2, \dots,\ \mathcal{S}_p]$, $\mathcal{S}_i = \{s_{1i},\ s_{2i}, \dots,\ s_{q_i,i}\}$ associated with the distinct eigenvalues.

Using Algorithm 6.4 determine the distinct eigenvalues $\lambda_1,\ \lambda_2, \dots,\ \lambda_p$ of the matrix A and the associated with them Weyr $\mathcal{W}$ and Segre $\mathcal{S}$ characteristics.

Set $codim = 0$

for $i = 1, 2, \dots, p$

  Determine the number of the Jordan blocks corresponding to λ_i

  $q_i = w_{i1}$

  for $j = 1, 2, \dots, q_i$

    Determine the size of the Jordan block

    $nb = s_{ji}$

    $codim \leftarrow codim + (2j-1)nb$

  end

end

$codim \leftarrow codim - p$

6.7 Numerical Issues

6.7.1 *Operation Count*

With respect to the computational cost, the reduction into the Jordan–Schur form requires more operations than the reduction into Schur form. This is mainly due to the reduction of the diagonal blocks to staircase form which needs a large volume of operations if these blocks have a high index of nilpotence, i.e., if their Jordan forms involve high-order blocks.

An approximate operation count for Algorithm 6.4 can be done as follows. As mentioned in Section 4.1.3, the reduction of a complex matrix into Hessenberg form and the subsequent reduction into Schur form by the explicitly shifted QR algorithm require

$$\frac{20}{3}n^3 + 12kn^3 \quad \text{flops},$$

where k is the average number of iterations necessary to find an eigenvalue (usually k does not exceed 2). According to the estimates given in Kågström and Ruhe (1980b), under the assumption of equal eigenvalue multiplicities $\nu = n/p$, the reduction of the diagonal blocks to staircase form and the transformation of the superdiagonal blocks require

$$2\left(\frac{1}{3}\frac{h}{p}n^3 + \frac{2h}{p^2}n^3 + \frac{p-1}{p^2}n^3\right) \quad \text{flops},$$

where h is the average index of nilpotence of the blocks.

The approximate total volume of work is given in Table 6.3 and depends on the structure of the matrix. For $p = 1,\ h = n$ (nonderogatory but defective matrix with a single Jordan block), this volume of work is of order

$$4\frac{2}{3}n^4 + 42\frac{2}{3}n^3 \quad \text{flops}$$

Table 6.3. Operation count in flops for the reduction into Jordan–Schur form.

Step	Operation	Average	Maximum	Maximum for
1	Reduction into Schur form	$12kn^3 + \frac{20}{3}n^3$	$30\frac{2}{3}n^3$	2 iterations per eigenvalue
2	Grouping of close eigenvalues	$O(n^2)$	$6n^3$	Totally unordered Schur form
3	Clustering of multiple eigenvalues	$2n^3$	$6n^3$	No clustering possible
4	Reduction into staircase form	$2\left(\frac{1}{3}\frac{n^3h}{p} + \frac{2n^3h}{p^2}\right.$	$4\frac{2}{3}n^4$	$p = 1,\ h = n$
		$\left.+\frac{(p-1)n^3}{p^2}\right)$	$\frac{2}{3}(2n^2 + n)$	$p = n,\ h = 1$
	Total	$32\frac{2}{3}n^3 + \frac{2}{3}\frac{n^3h}{p}$	$4\frac{2}{3}n^4 + 42\frac{2}{3}n^3$	$p = 1,\ h = n$
		$+4\frac{n^3h}{p^2} + 2\frac{(p-1)n^3}{p^2}$	$42\frac{2}{3}n^3 + \frac{2}{3}(2n^2 + n)$	$p = n,\ h = 1$

and for $p = n$, $h = 1$ (derogatory but non-defective matrix with n scalar Jordan blocks), it is approximately

$$42\frac{2}{3}n^3 \quad \text{flops.}$$

In this way, for matrices with large Jordan blocks, the reduction to Jordan–Schur form requires a volume of work of order n^4. Note that this operation count does not include the accumulation of the unitary similarity transformation matrix.

6.7.2 *Well-Conditioned Numerical Structures*

The numerical experiments with Algorithm 6.4 show that it usually finds accurately the Weyr characteristics of the given matrix. This is because the matrices arising in practical computations usually have well-conditioned numerical structures. (The ill-conditioned Jordan structures belong to a restricted set and hence are low probable.)

For matrix eigenvalues that are well conditioned in accordance with the definition (5.23) for a numerically multiple eigenvalue, the relative backward error $\|U_SJ_SU_S^H - A\|_F/\|A\|_F$ of the computed Jordan–Schur form is usually of size $\mathbf{u}$. This is due to the numerical stability of Algorithm 6.4 which implements only unitary similarity transformations.

Example 6.6. Consider the matrix

$$A = \begin{bmatrix} 6 & 0 & 3 & -6 & -5 & -11 & 14 & 33 \\ -2 & -10 & 0 & -3 & 9 & 10 & -2 & 4 \\ 5 & 7 & 4 & -2 & -10 & -16 & 13 & 25 \\ 1 & -6 & 0 & 0 & 2 & 1 & 4 & 11 \\ -1 & -3 & -1 & 1 & 3 & 4 & -2 & -4 \\ 1 & -6 & 0 & -3 & 3 & 3 & 3 & 11 \\ 0 & 6 & -2 & 4 & -5 & -2 & -2 & -13 \\ -1 & -7 & 0 & -2 & 6 & 6 & -1 & 4 \end{bmatrix}$$

used in Example 5.24. This matrix has an eigenvalue λ equal to 1 with multiplicity eight which participate in two Jordan blocks — one of size 6 and the other of size 2 so that the Segre characteristic is $\mathcal{S}(A,\lambda) = \{6, 2\}$. The Weyr characteristic of this matrix is $\mathcal{W}(A,\lambda) = \{2,\ 2,\ 1,\ 1,\ 1,\ 1\}$ and the index of nilpotence is 6.

Using Algorithm 6.4, we obtain the Jordan–Schur form

$$J_S = \left[\begin{array}{cc|cc|c|c|c|c}
1 & 0 & \times & \times & \times & \times & \times & \times \\
 & 1 & \times & \times & \times & \times & \times & \times \\
\hline
 & & 1 & 0 & \times & \times & \times & \times \\
 & & & 1 & \times & \times & \times & \times \\
\hline
 & & & & 1 & \times & \times & \times \\
\hline
 & & & & & 1 & \times & \times \\
\hline
 & & & & & & 1 & \times \\
\hline
 & & & & & & & 1
\end{array}\right],$$

where the position of the lines corresponds to the values of the elements of the Weyr characteristic. Since the Jordan structure of the matrix is well conditioned, the eigenvalues of A and the associated Weyr and Segre characteristics are determined accurately. The relative backward error in determining the Jordan–Schur form is $2.17 \cdot 10^{-16}$.

As mentioned earlier, the backward stability of the algorithm does not guarantee that the Jordan–Schur form and the associated numerical structure of the matrix are always determined correctly. Despite the proven performance of Algorithm 6.4, there are some cases when the algorithm can fail. This happens when the numerical structure of the given matrix is ill conditioned.

6.7.3 *Ill-Conditioned Numerical Structures*

In case of ill-conditioned numerically multiple eigenvalues, the numerical structure of the matrix and its Jordan–Schur form can be determined with errors. Due to the importance of this issue, we shall consider in detail three ill-conditioned examples for reduction into Jordan–Schur form.

Example 6.7. The matrix

$$A = \begin{bmatrix} 1153 & 4009 & 9132 & -14860 & -36532 & 109632 \\ -1308 & -4567 & -10409 & 16966 & 41686 & -125094 \\ 722 & 2522 & 5749 & -9372 & -23028 & 69102 \\ -617 & -2147 & -4891 & 7961 & 19570 & -58728 \\ -408 & -1416 & -3226 & 5248 & 12911 & -38739 \\ -244 & -848 & -1932 & 3144 & 7732 & -23201 \end{bmatrix}$$

has a 6-tuple eigenvalue equal to 1, which participates in two Jordan blocks of sizes 4 and 2. The Weyr and Segre characteristics associated with this eigenvalue are $\mathcal{W} = \{2,\ 2,\ 1,\ 1\}$ and $\mathcal{S} = \{4,\ 2\}$, respectively. Thus, the matrix A lays in a bundle with codimension equal to 9.

The diagonal elements of the Schur form, obtained after reordering and clustering of the eigenvalues, are

$$\begin{aligned} \lambda_1 &= 0.999966571817327 + \jmath_o 0.017257290374225, \\ \lambda_2 &= 1.017292598344664 + \jmath_o 0.000000000000000, \\ \lambda_3 &= 0.999966571817327 - \jmath_o 0.017257290374225, \\ \lambda_4 &= 0.982774258015524 + \jmath_o 0.000000000000000, \\ \lambda_5 &= 1.000000000000913 + \jmath_o 0.000002654149934, \\ \lambda_6 &= 1.000000000000913 - \jmath_o 0.000002654149934. \end{aligned}$$

The first four eigenvalues are computed with larger errors of order 10^{-2} since they participate in the larger Jordan block ($s_1 = 4$). The condition number of the eigenvector matrix with respect to 2-norm is $2.94 \cdot 10^{11}$ which shows that the eigensystem is ill conditioned.

The parameters characterizing the numerically multiple eigenvalues of the matrix for $tol = 10^{-12}$ and $gap = 1000$ are given in Table 6.4. Note that

$$\sum_{j=1}^{k} w_j = n_k, \quad k = 1, 2, \ldots, h$$

and $N = A - \lambda I_6$, where $\lambda = 0.999999999999444$ is the average of the cluster eigenvalues.

The requirements for well-conditioned numerical structure (Section 5.7) are not satisfied since for $k = 2$ and $k = 3$ the quantity

Table 6.4. Parameters, characterizing the numerical structure of matrix A.

k	n_k	w_k	$\sigma_{n_k+1}(N^k)$	$gap \cdot tol\|N\|_2^k$	$tol\|N\|_2^k$	$\sigma_{n_k}(N^k)$
1	2	2	$2.106 \cdot 10^{-1}$	$2.077 \cdot 10^{-4}$	$2.077 \cdot 10^{-7}$	$4.669 \cdot 10^{-13}$
2	4	2	$4.307 \cdot 10^{1}$	$4.316 \cdot 10^{1}$	$4.316 \cdot 10^{-2}$	$1.549 \cdot 10^{-12}$
3	5	1	$2.754 \cdot 10^{5}$	$8.966 \cdot 10^{6}$	$8.966 \cdot 10^{3}$	$5.031 \cdot 10^{-12}$
k	n_k	w_k			$tol\|N\|_2^k$	$\sigma_{n_k}(N^k)$
4	6	1			$1.863 \cdot 10^{9}$	0.000

$\sigma_{n_k+1}(N^k)$ is smaller than $gap \cdot tol\|N\|_2^k$. The ill conditioning of this structure implies that in the vicinity of the given matrix, there exist matrices with different Jordan structures.

Let us first reduce the matrix A into Jordan–Schur form using tolerances $ein = 10^{-12}$ and $tol = 10^{-12}$. At the first step of the deflation procedure, the singular values of the matrix $N = A - \lambda I_6$ are

$$
\begin{array}{c|l}
\Sigma_1 & \sigma_1 = 1.961841248248860 \cdot 10^{-13}, \\
 & \sigma_2 = 1.436574640779701 \cdot 10^{-12}, \\
\hline
 & \sigma_3 = 2.106477676415382 \cdot 10^{-1}, \\
\Sigma_2 & \sigma_4 = 3.496529810049557 \cdot 10^{0}, \\
 & \sigma_5 = 1.449173453287902 \cdot 10^{1}, \\
 & \sigma_6 = 2.077444445652385 \cdot 10^{5}.
\end{array}
$$

Since we use tolerances equal to $tol = 10^{-12}$ and $gap = 1000$, the smallest singular value σ_{n-k+1}, which is accepted as larger than zero, should satisfy the conditions

$$\sigma_{n-k+1} \geq tol \cdot \|A\|_F = 2.08 \cdot 10^{-7}$$

and

$$\sigma_{n-k+1} > gap \cdot tol \cdot \|A\|_F = 2.08 \cdot 10^{-4}.$$

The application of these conditions produces numerical rank of $A - \lambda I_6$ equal to $k = 4$, nullity of $A - \lambda I_6$ equal to 2 and element of the Weyr characteristic equal to $w_1 = 2$.

At the second step, we obtain that the singular values of the block $N^{(2)}$ are

$$\begin{array}{c|l} \Sigma_1 & \sigma_1 = 1.109912961190676 \cdot 10^{-12}, \\ & \sigma_2 = 1.598165641462267 \cdot 10^{-11}, \\ \hline \Sigma_2 & \sigma_3 = 8.117382912081016 \cdot 10^{0}, \\ & \sigma_4 = 2.768307861730003 \cdot 10^{2}. \end{array}$$

Using the same tolerances $tol \cdot \|A\|_F$ and $gap \cdot tol \cdot \|A\|_F$, we found that $\text{null}(N^{(2)}) = 2$ and consequently $w_2 = 2$.

The singular values of the block $N^{(3)}$ at the third step are

$$\begin{array}{c|l} \Sigma_1 & \sigma_1 = 1.601707967897870 \cdot 10^{-12}, \\ \hline \Sigma_2 & \sigma_2 = 1.051266676695842 \cdot 10^{2}. \end{array}$$

Thus, $\text{null}(N^{(3)}) = 1$ and $w_3 = 1$. Finally, $\sigma(N^{(4)}) = 8.084636423113301 \cdot 10^{-12}$ and $w_4 = 1$. As a result, Algorithm 6.4 produces the Jordan–Schur form

$$J_S = \left[\begin{array}{cc|cc|c|c} 1 & 0 & \times & \times & \times & \times \\ 0 & 1 & \times & \times & \times & \times \\ \hline 0 & 0 & 1 & 0 & \times & \times \\ 0 & 0 & 0 & 1 & \times & \times \\ \hline 0 & 0 & 0 & 0 & 1 & \times \\ \hline 0 & 0 & 0 & 0 & 0 & 1 \end{array}\right]$$

and determines correctly the Weyr characteristic as $\mathcal{W} = \{2,\ 2,\ 1,\ 1\}$ and the Segre characteristic as $\mathcal{S} = \{4,\ 2\}$. The eigenvalues are computed exactly up to 12 digits. The relative backward error in computing the Jordan–Schur form is

$$\|U_S J_S U_S^H - A\|_F / \|A\|_F = 1.30 \cdot 10^{-15}.$$

Let us increase now the tolerance tol to 10^{-6}. It is instructive to see how the numerical null space dimensions (the numerical nullities) are affected by this choice. As shown earlier, at the first step of the deflation procedure, the singular values of the matrix $N = A - \lambda I_6$ are

$$
\begin{array}{c|l}
 & \sigma_1 = 1.961841248248860 \cdot 10^{-13}, \\
 & \sigma_2 = 1.436574640779701 \cdot 10^{-12}, \\
\Sigma_1 & \sigma_3 = 2.106477676415382 \cdot 10^{-1}, \\
 & \sigma_4 = 3.496529810049557 \cdot 10^{0}, \\
 & \sigma_5 = 1.449173453287902 \cdot 10^{1}, \\
\hline
\Sigma_2 & \sigma_6 = 2.077444445652385 \cdot 10^{5}.
\end{array}
$$

Due to the change of the tolerances *tol* and *gap*, the smallest singular value σ_{n-k+1}, which is accepted as larger than zero, should satisfy

$$\sigma_{n-k+1} \geq \text{tol} \cdot \|A\|_F = 2.08 \cdot 10^{-1}$$

and

$$\sigma_{n-k+1} > \text{gap} \cdot \text{tol} \cdot \|A\|_F = 2.08 \cdot 10^{2}.$$

These conditions produce numerical rank of $A - \lambda I_6$ equal to $k = 1$, numerical nullity of $A - \lambda I_6$ equal to 5 and element of the numerical Weyr characteristic equal to $w_1 = 5$. Hence, at the second step of the deflation, we obtain by necessity that $w_2 = 1$. In this way, the numerical Segre characteristic of the computed Jordan–Schur form is $\mathcal{S} = \{2,\ 1,\ 1,\ 1,\ 1\}$. Thus, the first two eigenvalues participate in one 2×2 Jordan block, while the rest eigenvalues participate in 4 single Jordan blocks. The codimension of the bundle in which resides the Jordan–Schur form is equal to 25. (Remember that the exact matrix A lays in a bundle with codimension 9.) This error in the determination of the sizes of the Jordan blocks is an example of a failure of the staircase algorithm due to the ill conditioning of the numerical structure.

As a result, we obtain the Jordan–Schur form

$$
J_S = \left[\begin{array}{ccccc|c}
\lambda_1 & 0 & 0 & 0 & 0 & \times \\
0 & \lambda_2 & 0 & 0 & 0 & \times \\
0 & 0 & \lambda_3 & 0 & 0 & \times \\
0 & 0 & 0 & \lambda_4 & 0 & \times \\
0 & 0 & 0 & 0 & \lambda_5 & \times \\
\hline
0 & 0 & 0 & 0 & 0 & \lambda_6
\end{array}\right]
$$

with a numerical Weyr characteristic $\mathcal{W} = \{5,\ 1\}$ and eigenvalues

$$\begin{aligned}
\lambda_1 &= 0.999999999999444,\\
\lambda_2 &= 0.999999999999444,\\
\lambda_3 &= 0.999205415857171,\\
\lambda_4 &= 0.838828579634766,\\
\lambda_5 &= 0.574331135880728,\\
\lambda_6 &= 1.587634868624911.
\end{aligned}$$

The Jordan–Schur form is computed with relative backward error $1.45 \cdot 10^{-15}$. In the given case, the problem is regularized to a Jordan–Schur form with two blocks at the price of larger errors in the eigenvalues. This confirms that the quantities *ein* and *tol* can be used as regularization parameters for obtaining Jordan–Schur forms with different numerical structures.

The following example reveals the behavior of Algorithm 6.4 in the reduction into Jordan–Schur form of a family of matrices whose eigenvalue conditioning increases gradually.

Example 6.8. Given is a family of $n \times n$ matrices constructed as shown in Example 4.4,

$$A = QJQ^{-1},$$

where J is the desired Jordan form of A,

$$Q = H_2 \Sigma H_1,$$

H_1 and H_2 are elementary reflections of the form

$$\begin{aligned}
H_1 &= I_n - \frac{2}{n} e e^T, \quad e = [1,\ 1, \ldots,\ 1]^T,\\
H_2 &= I_n - \frac{2}{n} f f^T, \quad f = [1,\ -1, \ldots,\ (-1)^{n-1}]^T,\\
\Sigma &= \operatorname{diag}(1, \sigma, \sigma^2, \ldots, \sigma^{n-1})
\end{aligned}$$

and $\sigma > 0$ is a given parameter. The condition number of the nonsingular transformation matrix Q is equal to the condition number of the matrix Σ given by $\operatorname{cond}(\Sigma) = \sigma^{n-1}$.

Table 6.5. Results from the reduction into Jordan–Schur form for the matrix A of Example 6.8.

σ	cond(Q)	λ_1, λ_2	$\mathcal{W}(\lambda)$	codim($\mathcal{B}(A)$)	*berr*
1	$1.00 \cdot 10^0$	1.000000000000000	{1,1,1,1}	3	$2.62 \cdot 10^{-15}$
		0.800000000000000	{1}		
2	$1.60 \cdot 10^1$	0.999999999999999	{1,1,1,1}	3	$3.05 \cdot 10^{-15}$
		0.800000000000001	{1}		
3	$8.10 \cdot 10^1$	0.999999999999977	{1,1,1,1}	3	$1.60 \cdot 10^{-14}$
		0.800000000000072	{1}		
4	$2.56 \cdot 10^2$	0.999999999999947	{1,1,1,1}	3	$1.94 \cdot 10^{-14}$
		0.800000000000164	{1}		
6	$1.30 \cdot 10^3$	0.999999999998177	{1,1,1,1}	3	$3.59 \cdot 10^{-13}$
		0.800000000007105	{1}		
8	$4.10 \cdot 10^3$	1.000000000018368	{1,1,1,1}	3	$1.05 \cdot 10^{-12}$
		0.799999999926848	{1}		
10	$1.00 \cdot 10^4$	0.999999999074030	{1,1,1,1}	3	$2.96 \cdot 10^{-11}$
		0.800000001301106	{1}		
12	$2.07 \cdot 10^4$	0.959999999996148	{1,1,1,1,1}	4	$5.92 \cdot 10^{-12}$

In Table 6.5, we present the results from the computation of the Jordan–Schur form for a fifth-order matrices with prescribed Jordan form

$$J = \left[\begin{array}{cccc|c} 1 & 1 & 0 & 0 & 0 \\ 0 & 1 & 1 & 0 & 0 \\ 0 & 0 & 1 & 1 & 0 \\ 0 & 0 & 0 & 1 & 0 \\ \hline 0 & 0 & 0 & 0 & 0.8 \end{array}\right]$$

and for different transformation matrices corresponding to different values of σ. The tolerances used in the reduction are the same for all σ and equal to $ein = tol = 10^{-12}$. In the last column, we give the relative backward error

$$berr = \|U_S J_S U_S^T - A\|_F / \|A\|_F$$

in computing the Jordan–Schur form. With the increase in the condition number of the similarity transformation matrix Q, the conditioning of the Jordan structure gets worse, and for $\sigma = 12$ instead of in the bundle with codimension 3 corresponding to Segre characteristics $\{4\}$ and $\{1\}$, the algorithm finds a Jordan structure in a bundle with a codimension 4 corresponding to numerical Segre characteristic equal to $\{5\}$.

The following example shows that the ill conditioning of the Jordan structure can be related to the presence of small coupling elements in the Jordan form of the matrix.

Example 6.9. Consider the problem of finding the Jordan structure of a family of matrices in the form $A = QJQ^{-1}$ where

$$J = \left(\begin{bmatrix} -1 & 1 & \\ & -1 & 10^{2-j} \\ & & -1 \end{bmatrix}, \begin{bmatrix} -1 & 1 & \\ & -1 & 1 \\ & & -1 \end{bmatrix} \right),$$

$$Q = \begin{bmatrix} -2 & -1 & 0 & -2 & 2 & 1 \\ 1 & 0 & 1 & 0 & -1 & -1 \\ -2 & 0 & 0 & -1 & 1 & 1 \\ -1 & 0 & 0 & 0 & 1 & 0 \\ 0 & -1 & 1 & 0 & 0 & -1 \\ -1 & 0 & -1 & -1 & 1 & 1 \end{bmatrix}$$

and j is a variable integer. For all values of j, the exact Weyr characteristic is $\mathcal{W} = \{2, 2, 2\}$. By increasing j, the coupling element J_{23} decreases and the 3×3 Jordan block gets closer to the direct sum of a 2×2 and 1×1 Jordan block. This leads to the worsening of the numerical problem for finding the Jordan structure of A.

In Table 6.6, we give the results obtained by using Algorithm 6.4 for six values of the integer j as well as the codimension of the bundle $\mathcal{B}(A)$ determined by Algorithm 6.5. In the reduction into Jordan–Schur form, we use tolerances $tol = 5 \cdot 10^{-13}$ and $gap = 1000$. For each step i of the deflation procedure, we show the numerical rank

$$k = \operatorname{rank}(N^{(i)})$$

of the corresponding nilpotent matrix $N^{(i)} \in \mathbb{C}^{n_b \times n_b}$ as well as the singular value $\sigma_{n_b-k+1}(N^{(i)})$ (or $\sigma_{n_b-k}(N^{(i)})$ at the last step (i) determining the numerical rank of $N^{(i)}$. At each step i, the difference $n_b - k = \operatorname{null}(N^{(i)})$ determines the corresponding element of the

Table 6.6. Determining the numerical ranks during the reduction into Jordan–Schur form.

j	J_{23}	i	n_b	k	$\sigma(N^{(i)})$	$w_i = n_b - k$	codim($\mathcal{B}(A)$)
1	10^1	1	6	4	$\sigma_{n_b-k+1}(N^{(i)}) = 3.59 \cdot 10^{-1}$	{2}	11
		2	4	2	$\sigma_{n_b-k+1}(N^{(i)}) = 1.06 \cdot 10^{0}$	{2}	
		3	2	0	$\sigma_{n_b-k}(N^{(i)}) = 3.08 \cdot 10^{-16}$	{2}	
2	10^0	1	6	4	$\sigma_{n_b-k+1}(N^{(i)}) = 3.68 \cdot 10^{-1}$	{2}	11
		2	4	2	$\sigma_{n_b-k+1}(N^{(i)}) = 9.88 \cdot 10^{-1}$	{2}	
		3	2	0	$\sigma_{n_b-k}(N^{(i)}) = 2.47 \cdot 10^{-15}$	{2}	
3	10^{-1}	1	6	4	$\sigma_{n_b-k+1}(N^{(i)}) = 7.01 \cdot 10^{-2}$	{2}	11
		2	4	2	$\sigma_{n_b-k+1}(N^{(i)}) = 1.37 \cdot 10^{-1}$	{2}	
		3	2	0	$\sigma_{n_b-k}(N^{(i)}) = 3.00 \cdot 10^{-16}$	{2}	
4	10^{-2}	1	6	4	$\sigma_{n_b-k+1}(N^{(i)}) = 6.84 \cdot 10^{-3}$	{2}	11
		2	4	2	$\sigma_{n_b-k+1}(N^{(i)}) = 1.35 \cdot 10^{-2}$	{2}	
		3	2	0	$\sigma_{n_b-k}(N^{(i)}) = 1.94 \cdot 10^{-16}$	{2}	
5	10^{-3}	1	6	4	$\sigma_{n_b-k+1}(N^{(i)}) = 6.81 \cdot 10^{-4}$	{2}	9
		2	4	2	$\sigma_{n_b-k+1}(N^{(i)}) = 1.34 \cdot 10^{-3}$	{2}	
		3	2	1	$\sigma_{n_b-k+1}(N^{(i)}) = 1.55 \cdot 10^{-10}$	{1}	
		4	1	0	$\sigma_{n_b-k}(N^{(i)}) = 1.27 \cdot 10^{-15}$	{1}	
6	10^{-4}	1	6	4	$\sigma_{n_b-k+1}(N^{(i)}) = 6.81 \cdot 10^{-5}$	{2}	9
		2	4	2	$\sigma_{n_b-k+1}(N^{(i)}) = 1.34 \cdot 10^{-4}$	{2}	
		3	2	1	$\sigma_{n_b-k+1}(N^{(i)}) = 6.10 \cdot 10^{-8}$	{1}	
		4	1	0	$\sigma_{n_b-k}(N^{(i)}) = 8.73 \cdot 10^{-16}$	{1}	

numerical Weyr characteristic w_i associated with $\lambda = -1$. The results show that with the increase in j, the singular value $\sigma_{n_b-k+1}(N^{(i)})$ generally decreases for the respective i and for $j = 5$ and $j = 6$, the numerical rank of $N^{(i)}$ becomes different from the "true" rank which leads to errors in the determining the Weyr characteristic and bundle codimension.

It is appropriate to make the following commentary in connection with the poor performance of Algorithm 6.4 in case of ill-conditioned Jordan structures. Matrices with badly conditioned numerical structure may appear in practice due to two reasons. First, such matrices may arise from poor mathematical models in which the quantities are badly scaled or which do not describe adequately the corresponding physical phenomenon. Second, matrices with badly conditioned structure may arise as a result from numerical algorithms in which there is no control on the properties of the intermediate quantities. In both cases, it is not possible to expect reliable results from the computations, and in such cases, it is not correct to blame the algorithm for determining the Jordan form. Finding the correct Jordan structure is a decision-making procedure and as such some errors in case of badly conditioned examples seem inevitable.

6.7.4 *Stratification of the Jordan–Schur Forms*

As mentioned previously, the tolerances *ein* and *tol* used in Algorithm 6.4 play the role of regularization parameters which allow achieving a Jordan structure by the use of an well-conditioned similarity transformation. The tolerance *ein* influences the clustering of the eigenvalues in groups of numerically multiple eigenvalues, while the tolerance *tol* in its turn affects the determination of the dimensions of the null spaces and consequently the Weyr characteristics of the associated eigenvalues. Thus choosing different *ein* and *tol* one may obtain Jordan–Schur forms that reside in bundles with different codimensions. The choice of appropriate tolerances can be done to minimize the distance between $\widehat{A}_k = U_S \cdot J_S \cdot U_S^H$ and A or to choose a specific Jordan structure. As the following example demonstrates, there are some matrices for which the minimum distance to the given matrix is not a guarantee for finding the best solution.

Example 6.10. Consider the matrix

$$A = \begin{bmatrix} -111 & -240 & -572 & 1752 & -4272 \\ -149 & -335 & -778 & 2340 & -5688 \\ 132 & 288 & 657 & -1980 & 4744 \\ -38 & -84 & -201 & 613 & -1504 \\ -22 & -48 & -112 & 340 & -823 \end{bmatrix}$$

which is obtained as $A = QJQ^{-1}$, where

$$J = \left(\begin{bmatrix} 1 & 1 \\ & 1 \end{bmatrix}, 1, \begin{bmatrix} -1 & 1 \\ & -1 \end{bmatrix} \right)$$

and

$$Q = \begin{bmatrix} 56 & 187 & 64 & -8 & 6 \\ 109 & 261 & 90 & -16 & 6 \\ -100 & -234 & -80 & 14 & -5 \\ 20 & 64 & 22 & -3 & 2 \\ 14 & 38 & 13 & -2 & 1 \end{bmatrix}.$$

The matrix of the similarity transformation into Jordan form is relatively ill conditioned with condition number $\text{cond}(Q) = 1.34 \cdot 10^5$. Since the Segre characteristics associated with the eigenvalues $\lambda_1 = 1$ and $\lambda_2 = -1$ are $\mathcal{S}_1 = \{2, 1\}$ and $\mathcal{S}_2 = \{2\}$, respectively, the matrix A is in a bundle $\mathcal{B}(A)$ with codimension equal to 5.

In Table 6.7, we give the Weyr characteristics and bundle codimensions of the six matrices $\widehat{A}_k = U_S \widehat{J_S} U_S^H$ corresponding to the Jordan–Schur forms $\widehat{J}_{S,k}$, $k = 1, 2, \ldots, 6$, computed for different tolerances *ein* and *tol*. The last column contains the relative distances

$$\text{dist}(\widehat{A}_k, A) = \frac{\|\widehat{A}_k - A\|_F}{\|A\|_F}, \quad k = 1, 2, \ldots, 6$$

between the matrices $\widehat{A}_k$ and the original matrix A in the space $\mathbb{C}^{25}$. By increasing the tolerances, the first element of the Weyr characteristic (the number of Jordan blocks corresponding to the mean eigenvalue) increases and the Jordan structure is moving toward the most degenerate case ($\mathcal{W} = \{5\}$). The matrix $\widehat{A}_1$ is in the same bundle as A and its distance to A is very small, so it is reasonable to accept that its exact Jordan form is the numerical Jordan

Table 6.7. Jordan–Schur forms of A computed for different tolerances.

k	ein, tol	Matrix	Weyr characteristics	$\text{codim}(\mathcal{B}(A))$	$\text{dist}(\widehat{A}_k, A)$
1	10^{-9}	$\widehat{A}_1$	{1,1},{2,1}	5	$8.35 \cdot 10^{-16}$
2	10^{-8}	$\widehat{A}_2$	{2,1,1,1}	6	$5.33 \cdot 10^{-7}$
3	10^{-7}	$\widehat{A}_3$	{2,2,1}	8	$5.29 \cdot 10^{-7}$
4	10^{-6}	$\widehat{A}_4$	{3,1,1}	10	$3.07 \cdot 10^{-5}$
5	10^{-5}	$\widehat{A}_5$	{4,1}	16	$1.86 \cdot 10^{-15}$
6	1	$\widehat{A}_6$	{5}	24	$7.88 \cdot 10^{-16}$

form of A. Note, however, that the bundles with large codimensions (16 and 24) are also close to the bundle with codimension 5, where resides the matrix A, so that it is impossible to recognize the true Jordan–Schur form only from the distance $\text{dist}(\widehat{A}_k, A)$ (the relative backward error in $\widehat{A}_k$). These results confirm the conclusion made in Section 5.5 that the ordering of the bundles in the matrix space is generally different from the ordering imposed by the bundle stratification hierarchy.

In Figure 6.16, we show the distances between all computed matrices $\widehat{A}_k$ and between them and the given matrix A, as well as a graph illustrating the position of the matrices $\widehat{A}_k$ with respect to A (the subscript 0 refers to the matrix A).

6.7.5 *Choice of Appropriate Tolerances*

The choice of appropriate tolerances *ein* and *tol* in the reduction to the Jordan–Schur form depends on several factors.

The choice of the tolerance *ein* affects the clustering of numerically multiple eigenvalues in groups and thus influences the determination of the eigenvalue multiplicities. Decreasing *ein* generally decreases the dimension of the cluster containing numerically multiple eigenvalues (the algebraic multiplicity ν) and vice versa. If the tolerance *ein* is appropriately chosen, then the norms of the spectral projections on the invariant subspaces are relatively small. Large norms of the projectors indicate that the clusters are not separated well and hence *ein* should be changed. Changing this tolerance, one can move the Jordan structure to different bundles trying to find better numerical structure of the matrix. Such numerical structure should be well conditioned and should correspond to a small relative backward error $\|U_S J_S U_S^H - A\|/\|A\|$.

The choice of *tol* determines the tolerance in finding the numerical rank of the matrices during the transformation to staircase form. In this way, *tol* affects determining the sizes of null spaces (the nullities) of the corresponding matrices and hence the Weyr characteristic associated with the eigenvalue. The appropriate value of this tolerance depends on the size of uncertainty in the given matrix and changing of this tolerance also moves the Jordan structure to different bundles. In particular, increasing *tol* generally increases the dimensions of the numerical null spaces thus moving the numerical

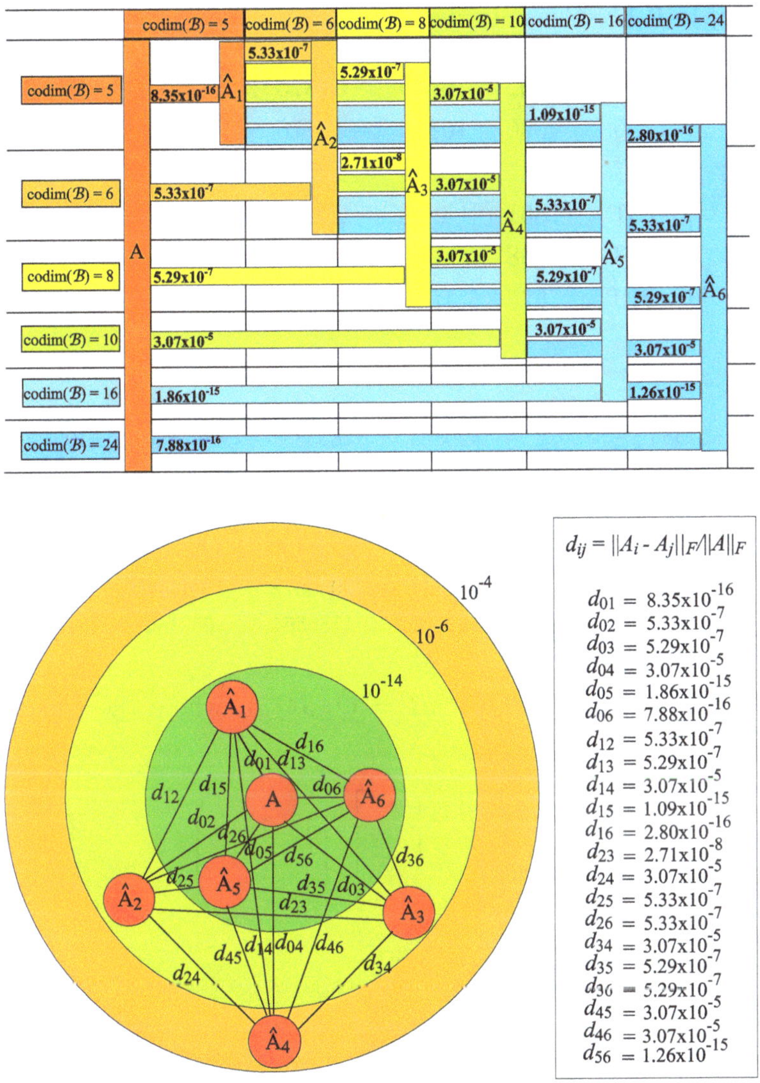

Fig. 6.16. Distances between the computed matrices $\widehat{A}_k$ and the matrix A.

Jordan structure toward to the most degenerate bundle containing scalar Jordan blocks with the same eigenvalues.

The following example shows how the choice of appropriate values of *ein* and *tol* is affected by the eigenvalue conditioning of the matrix under consideration.

Example 6.11. Consider the matrices

$$A = \begin{bmatrix} 2126 & 6302 & 16529 & -68873 & 42565 & -231570 \\ 1355 & 4096 & 10816 & -44776 & 27324 & -149698 \\ -897 & -2611 & -6796 & 28518 & -17855 & 96448 \\ -631 & -1851 & -4833 & 20223 & -12595 & 68229 \\ 163 & 498 & 1320 & -5447 & 3303 & -18156 \\ 210 & 625 & 1641 & -6831 & 4213 & -22945 \end{bmatrix}$$

and

$$B = \begin{bmatrix} -64585 & -388012 & -1420823 & 12855386 & 48638485 & -129640575 \\ -98012 & -588739 & -2155696 & 19504777 & 73796230 & -196695514 \\ 84680 & 508608 & 1862227 & -16849641 & -63750412 & 169919368 \\ -61344 & -368477 & -1349188 & 12207497 & 46186991 & -123106191 \\ -5444 & -32700 & -119736 & 1083372 & 4098940 & -10925258 \\ -8728 & -52426 & -191960 & 1736861 & 6571405 & -17515333 \end{bmatrix},$$

obtained as $A = Q_A J Q_A^{-1}$, $B = Q_B J Q_B^{-1}$, where the prescribed Jordan form is

$$J = \left(\begin{bmatrix} 1 & 1 & \\ & 1 & 1 \\ & & 1 \end{bmatrix}, \begin{bmatrix} 1 & 1 \\ & 1 \end{bmatrix}, 2 \right)$$

and

$$Q_A = \begin{bmatrix} -1121 & -273 & -9 & 42 & -13 & 8 \\ -751 & -202 & -15 & 26 & -7 & 5 \\ 442 & 94 & -2 & -17 & 7 & -4 \\ 317 & 71 & 0 & -12 & 4 & -3 \\ -90 & -25 & -2 & 3 & 0 & 1 \\ -110 & -27 & -1 & 4 & -1 & 1 \end{bmatrix},$$

Table 6.8. Jordan–Schur forms of A and B computed for different tolerances.

k	Matrix	ein, tol	Weyr characteristics	$\text{codim}(\mathcal{B}(A))$ $\text{codim}(\mathcal{B}(B))$	$\text{dist}(\widehat{A}_k, A)$ $\text{dist}(\widehat{B}_k, B)$
1	A	10^{-14}	{1,1,1,1,1,1}	5	$4.02 \cdot 10^{-9}$
	B	10^{-15}	{1,1,1,1,1,1}	5	$1.99 \cdot 10^{-12}$
2	A	10^{-11}	{2,1,1,1,1}	7	$1.42 \cdot 10^{-9}$
	B	10^{-14}	{2,1,1,1,1}	7	$6.31 \cdot 10^{-13}$
3	A	10^{-9}	{3,2,1}	13	$3.28 \cdot 10^{-8}$
	B	10^{-11}	{3,2,1}	13	$7.39 \cdot 10^{-11}$
4	A	10^{-8}	{4,1,1}	17	$4.02 \cdot 10^{-8}$
	B	10^{-10}	{4,1,1}	17	$2.43 \cdot 10^{-13}$
5	A	10^{-5}	{5,1}	25	$5.12 \cdot 10^{-16}$
	B	10^{-9}	{5,1}	25	$1.68 \cdot 10^{-15}$
6	A	1	{6}	35	$4.67 \cdot 10^{-16}$
	B	1	{6}	35	$1.06 \cdot 10^{-15}$

$$Q_B = \begin{bmatrix} 14904 & 3173 & 579 & -69 & -2 & 12 \\ 22537 & 4760 & 868 & -99 & -15 & 10 \\ -19437 & -4090 & -742 & 85 & 14 & -6 \\ 14102 & 2977 & 542 & -62 & -9 & 6 \\ 1254 & 266 & 48 & -6 & 0 & 1 \\ 2007 & 424 & 77 & -9 & -1 & 1 \end{bmatrix}.$$

The matrices A and B are in a bundle with codimension equal to 8. The similarity transformation matrices have condition numbers

$$\text{cond}(Q_A) = 2.30 \cdot 10^6 \text{ and } \text{cond}(Q_B) = 1.73 \cdot 10^9,$$

respectively. Clearly, the conditioning of the eigenvalue problem of B is worse than the conditioning of the eigenvalue problem of A.

In Table 6.8, we compare the results of the reduction into the Jordan–Schur form for the two matrices using different values of the tolerances $ein = tol$. Due to the greater eigensystem sensitivity of the matrix B, the transition of its Jordan structure to a specific bundle $\mathcal{B}(B)$ could be done by using smaller tolerances. This indicates that for matrices with ill-conditioned eigensystems the tolerances can be chosen smaller.

The choice of appropriate tolerances may be facilitated by using a diagram showing the distances between different Jordan–Schur forms, like the one displayed in Figure 6.16.

6.8 Notes and References

The presentation in this chapter is based on the millstone papers of Kågström and Ruhe (1980a, 1980b) and the reach surveys of Kågström and Wiberg (2000) and Sapiro (1999). In Bai *et al.* (2000), p. 24, James Demmel proposed to call the matrix J_S in (6.5) the Jordan–Schur form of A and Elmroth *et al.* (2003) called it the *Schur-staircase form*. With a risk to introduce additional confusion into the names, one could propose the name *Schur–Weyr form* as a more appropriate one.

A detailed history of the staircase algorithm is given in the paper Edelman and Ma (2000). The occasional failures of the staircase algorithm are explained in this paper in terms of the matrix space geometry. Algorithms for exchanging diagonal blocks and reordering of Schur form are proposed by several authors, see Bai and Demmel (1993), Bojanczyk and Van Dooren (1993), Brandts (2001), Granat *et al.* (2009), Kågström and Poromaa (1996a), Kressner (2006), Stewart (1976) and Van Dooren (1982).

The idea of using the Gershgorin disks to isolate matrix eigenvalues is utilized in several sources, see the book of Varga (2004) and his papers with co-workers (Householder *et al.*, 1970; Levinger and Varga, 1966; Medley and Varga, 1968; Varga, 1965). As noted in Section 6.4, the Gershgorin clustering strategy may be overly conservative in its construction of the isolating disks. That is why it can be replaced by the sharper heuristic proposed in Kågström and Wiberg (2000) which allows for larger radii of the Gershgorin disks for some of the clustered eigenvalues. The experiments show that in most cases both algorithm produce the same results. Since the clustering is a key algorithm in finding the Jordan structure, it seems appropriate to perform it using a multiple precision arithmetic.

The algorithm of orthogonal deflation is refined to reduce the number of computational operations in several papers, see Beelen and Van Dooren (1990), Guglielmi *et al.* (2015) and Mastronardi and Van Dooren (2017). One of the obstacles to wider implementation of the algorithm for determining the Jordan structure is the necessity to use the tolerances *ein* and *tol* which require some adjustment thus preventing the automatic ("blind") usage of the algorithm. For this reason in the latter development of the algorithm intended to find the Kronecker structure of a matrix pencil, implemented in the routine `guptri` (StratiGraph 2019), a procedure for automatic adjustment of these tolerances is used.

The idea of orthogonal deflation was further implemented to find the Kronecker and Weierstrass canonical forms of matrix pencils (Beelen and Van Dooren, 1988; Kågström, 1986; Van Dooren, 1979; Demmel and Kågström, 1987, 1993a, 1993b; Mastronardi and Van Dooren, 2017, 2020).

The unitary reduction into staircase form is implemented also to solve other matrix problems, for instance, to find the Drazin inverse (Campbell and Meyer, 2009, Chapter 12).

Chapter 7

Reduction into Weyr and Jordan Forms

This chapter presents the full algorithm of the reduction of a matrix into numerical Jordan form. An intermediate step of the reduction is the determining of the Weyr characteristics and the Weyr form of the matrix.

The first three steps of the reduction into numerical Weyr and Jordan forms coincide with the corresponding steps of the reduction into the Jordan–Schur form present in Chapter 6. Then the reduction continues further on with additional transformations leading ultimately to the Jordan form of the matrix.

The reduction of a complex matrix into numerical Jordan form by the algorithm of Kågström–Ruhe can be summarized as follows:

(1) Reduction of A into upper triangular (Schur) form $T = U^H AU$ using unitary similarity transformations.
(2) Reordering the Schur form by unitary transformations such that close eigenvalues appear in adjacent positions. This is done by using Algorithm 6.1 presented in Chapter 6.
(3) Clustering of multiple eigenvalues in p groups corresponding to distinct numerically multiple eigenvalues. The clustering is performed by Algorithm 6.2.

(4) Reduction of the upper block-triangular form

$$\begin{bmatrix} T_{11} & T_{12} & \dots & T_{1p} \\ & T_{22} & \dots & T_{2p} \\ & & \ddots & \vdots \\ & & & T_{pp} \end{bmatrix},$$

into the block-diagonal form

$$B = \text{diag}(T_{11}, T_{22}, \dots, T_{pp}),$$

where each diagonal block

$$T_{ii} = \lambda_i I_{\nu_i} + N_{ii} \in \mathbb{C}^{\nu_i \times \nu_i}$$

corresponds to a distinct numerically multiple eigenvalue λ_i with multiplicity ν_i and N_{ii} is a nilpotent matrix. This reduction is done by using well-conditioned nonsingular transformations and decomposes the problem of finding the Jordan form into subproblems of smaller sizes with unique numerically multiple eigenvalues.

(5) Orthonormalization of the vectors which constitute a basis of the invariant subspace corresponding to each diagonal block. This allows reducing the condition number of the matrix of the similarity transformation into the block-diagonal form.

(6) Transforming the diagonal blocks into staircase form. In this step, each block T_{ii} is reduced by using unitary similarity transformations into the form

$$S(\mathcal{W}_i, \lambda_i) = \begin{bmatrix} \lambda_i I_{w_{i1}} & N_{12}^{(i)} & \dots & N_{1,h_i-1}^{(i)} & N_{1,h_i}^{(i)} \\ & \lambda_i I_{w_{i2}} & \dots & N_{2,h_i-1}^{(i)} & N_{2,h_i}^{(i)} \\ & & \ddots & \vdots & \vdots \\ & & & \lambda_i I_{w_{i,h_i-1}} & N_{h_i-1,h_i}^{(i)} \\ & & & & \lambda_i I_{w_{i,h_i}} \end{bmatrix} \in \mathbb{C}^{\nu_i \times \nu_i},$$

where $S(\mathcal{W}_i, \lambda_i) - \lambda_i I_{\nu_i}$ is in staircase form and $w_{i1}, w_{i2}, \dots, w_{i,h_i}$ are the elements of the Weyr characteristic $\mathcal{W}_i$ associated with λ_i, $i = 1, 2, \dots p$. At this step, the Jordan structure of the matrix is determined completely. This reduction is done with the aid of Algorithm 6.3, present in Chapter 6.

(7) Determining the coupling elements and reduction into numerical Weyr form

$W = X_W^{-1} A X_W = \text{diag}(W(\mathcal{W}_1, \lambda_1), W(\mathcal{W}_2, \lambda_2), \ldots, W(\mathcal{W}_p, \lambda_p))$,

where each Weyr block $W(\mathcal{W}_i, \lambda_i)$ corresponding to a distinct eigenvalue λ_i is in the form

$$W(\mathcal{W}_i, \lambda_i) = \begin{bmatrix} \lambda_i I_{w_{i1}} & G_{w_{i1}, w_{i2}} & & \\ & \lambda_i I_{w_{i2}} & \ddots & \\ & & \ddots & G_{w_{i,h_i-1}, w_{i,h_i}} \\ & & & \lambda_i I_{w_{i,h_i}} \end{bmatrix}, \quad i = 1, 2, \ldots, p.$$

The rectangular matrices $G_{w_{ij}, w_{i,j+1}}$ participating in the Weyr blocks are in diagonal form and carry the coupling elements of the Weyr and Jordan forms. Note that in contrast to the "pure" Jordan form, the coupling elements are generally different from one which reduces the conditioning of the transformation matrix.

(8) The numerical Jordan form is determined by transforming the Weyr form with permutation similarity transformations with a matrix P_J.

The nomenclature of the various forms and the transformations between them, used in this chapter, are shown in Figure 7.1.

Next, we describe in detail steps 4–8 of the full algorithm.

7.1 Block-Diagonalization

Let by the aid of Algorithm 6.1 the matrix A be reduced to the Schur form

$$T = U^H A U = \begin{bmatrix} T_{11} & T_{12} & T_{13} & \cdots & T_{1p} \\ & T_{22} & T_{23} & \cdots & T_{2p} \\ & & T_{33} & \cdots & T_{3p} \\ & & & \ddots & \vdots \\ & & & & T_{pp} \end{bmatrix}, \quad (7.1)$$

where each diagonal block T_{kk}, $k = 1, \ldots, p$, corresponds to distinct numerically multiple eigenvalue with multiplicity ν_k.

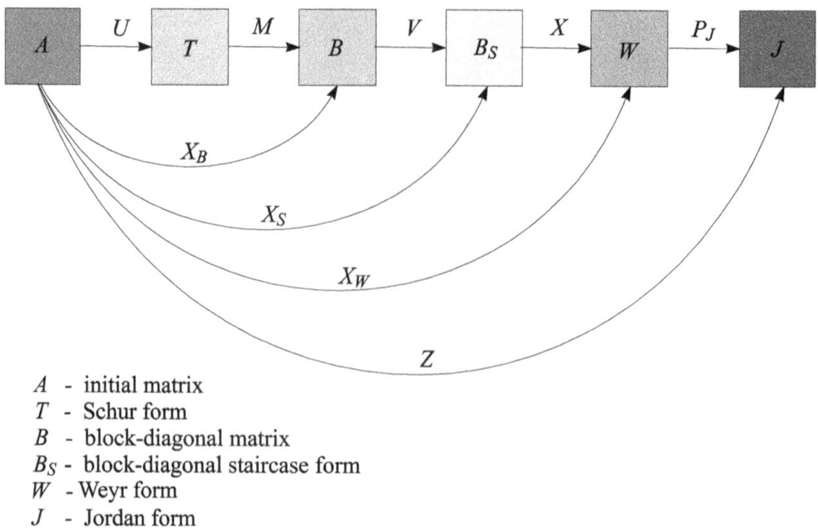

Fig. 7.1. Nomenclature of different forms and transformations.

Using a sequence of nonsingular similarity transformations, the Schur form can be reduced into the block-diagonal form (Figure 7.2)

$$B = M^{-1}TM = \begin{bmatrix} T_{11} & & & & \\ & T_{22} & & & \\ & & T_{33} & & \\ & & & \ddots & \\ & & & & T_{pp} \end{bmatrix}.$$

Hence,

$$B = X_B^{-1} A X_B = \operatorname{diag}(T_{11}, T_{22}, \dots, T_{pp}), \tag{7.2}$$

where $X_B = U \cdot M$ and $\operatorname{cond}_2(X_B) = \operatorname{cond}_2(M)$. In particular, this reduction can be done by a block-diagonalization method based on the solution of the Sylvester equation (see Section 2.1.6). In this method, the zeroing of the superdiagonal block T_{ij} is performed by

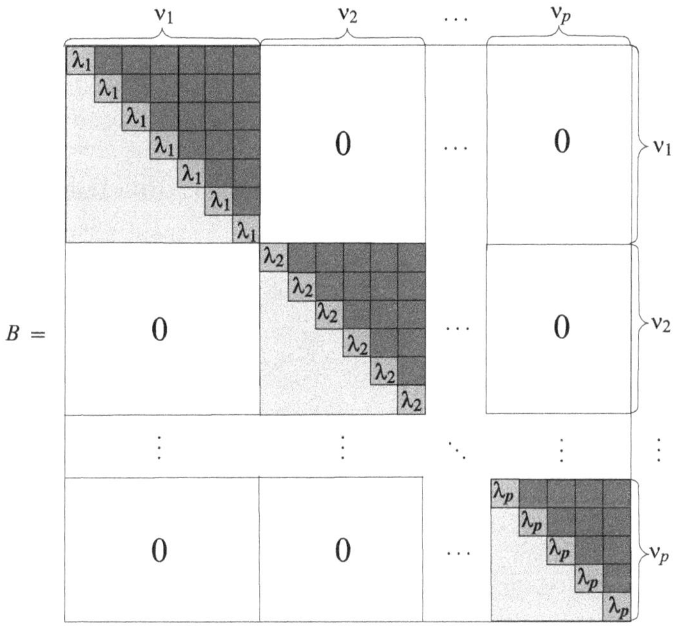

Fig. 7.2. The block-diagonal form B.

$p(p-1)/2$ similarity transformations of the form

$$M_k = \begin{bmatrix} I_{\nu_1\times\nu_1} & & & & & & \\ & \ddots & & & & & \\ & & I_{\nu_i\times\nu_i} & \cdots & X_{ij} & & \\ & & & \ddots & \vdots & & \\ & & & & I_{\nu_j\times\nu_j} & & \\ & & & & & \ddots & \\ & & & & & & I_{\nu_p\times\nu_p} \end{bmatrix},$$

where the matrix X_{ij} is the solution of the Sylvester matrix equation

$$T_{ii}X_{ij} - X_{ij}T_{jj} + T_{ij} = 0. \tag{7.3}$$

The matrices T_{ii} and T_{jj} in the Sylvester equation (7.3) are in triangular form so that it can be solved efficiently by the Bartels–Stewart algorithm (see Algorithm 4.1 in Section 4.2.3). Note that this equation is ill conditioned if the eigenvalues of T_{ii} and T_{jj} are not well separated. The solution of the Sylvester equation can be done by using the LAPACK routine **xTRSYL** or MATLAB® function **lyap**. The solution of each Sylvetser equation requires $O(\nu_k^3)$ operations.

The upper block-triangular similarity transformation matrix

$$M = M_1 M_2 \dots M_{p(p-1)/2} = \begin{bmatrix} I_{\nu_1\times\nu_1} & M_{12} & M_{13} & \dots & M_{1p} \\ & I_{\nu_2\times\nu_2} & M_{23} & \dots & M_{2p} \\ & & I_{\nu_3\times\nu_3} & \dots & M_{3p} \\ & & & \ddots & \vdots \\ & & & & I_{\nu_p\times\nu_p} \end{bmatrix}$$

is obtained as the product of the nonsingular transformations used at each step. Since the matrix

$$M_k^{-1} = \begin{bmatrix} I_{\nu_1\times\nu_1} & & & & & \\ & \ddots & & & & \\ & & I_{\nu_i\times\nu_i} & \dots & -X_{ij} & \\ & & & \ddots & \vdots & \\ & & & & I_{\nu_j\times\nu_j} & \\ & & & & & \ddots \\ & & & & & & I_{\nu_p\times\nu_p} \end{bmatrix}$$

differs from M_k only by the sign of the block X_{ij}, it is also possible to compute simultaneously with the transformation matrix M its inverse M^{-1}.

Let the block-diagonal form $B = M^{-1}T_3M$ of T_3 be determined and define for the ith diagonal block T_{ii} the corresponding spectral projection

$$P_k = M_k \cdot (M^{-1})_k \in \mathbb{C}^{n\times n}$$

on the invariant subspace on which T_{kk} acts, where $M_k \in \mathbb{C}^{n\times\nu_k}$ are the columns of M corresponding to T_{kk} and $(M^{-1})_k \in \mathbb{C}^{\nu_k\times n}$ are the corresponding rows of the matrix M^{-1}. The norm $\|P_k\|_2$ of the spectral projection is a measure of the sensitivity of the eigenvalues of T_{kk}.

If the eigenvalues of the blocks are well separated, then the norms of the spectral projections are small and the block-diagonalization is done precisely. Thus the spectral projections can be used to control the accuracy of eigenvalues clustering.

The reduction into block-diagonal form can be used to estimate the norms of the spectral projections corresponding to groups of multiple eigenvalues. The spectral projection, corresponding to the kth group of eigenvalues is given by

$$P_k = \begin{bmatrix} M_{1k} \\ \vdots \\ M_{k-1,k} \\ I_{n_k \times n_k} \\ 0 \end{bmatrix} \cdot \begin{bmatrix} 0 & I_{n_k \times n_k} & (M^{-1})_{k,k+1} \dots (M^{-1})_{kp}, \end{bmatrix} \tag{7.4}$$

where $(M^{-1})_{kj}$ are the corresponding blocks of M^{-1}. Hence,

$$\|P_k\|_2 \le \left(1 + \left(\sum_{j=1}^{k-1} \|M_{jk}\|_F^2\right)^{1/2}\right) \times \left(1 + \left(\sum_{j=k+1}^{p} \|(M^{-1})_{kj}\|_F^2\right)^{1/2}\right). \tag{7.5}$$

A large norm of the spectral projection implies that the eigenvalues in the respective group are not well separated from the eigenvalues in the other groups.

The reduction into block-diagonal form is done by Algorithm 7.1.

The initial matrix is reduced into Schur form by using the QR algorithm (Section 4.1), the reordering of the eigenvalues is done by Algorithm 6.1 and the grouping of the multiple eigenvalues is implemented by Algorithm 6.2.

The solution of the Sylvester equation (7.3) is performed by the Bartels–Stewart algorithm (Algorithm 4.1). The block-diagonalization requires

$$\frac{p(p-1)n^3}{p^3} + 2\frac{(p-1)n^3}{p} \text{ flops},$$

Algorithm 7.1. Block-diagonalization

Block-diagonalization of a matrix in Schur form whose multiple eigenvalues are clustered in groups. The blocks of the nonsingular similarity transformation are computed by solving Sylvester equations.

function $[B, X_B] = \text{block\_diag}(T, U)$

Input: Matrix $T \in \mathbb{C}^{n \times n}$ in Schur form, whose p diagonal blocks $T_{ii} \in \mathbb{C}^{\nu_i \times \nu_i}$, $i = 1, \ldots, p$ correspond to distinct numerically multiple eigenvalues; unitary matrix $U \in \mathbb{C}^{n \times n}$ of the transformation of A into Schur form T.

Output: The block-diagonal form $B = M^{-1}TM = Y^{-1}AY$; $X_B = U \cdot M$ contains the matrix of the nonsingular transformation which reduces A into B.

Set $M = I_n$
for $j = 2, \ldots, p$
  for $i = 1, \ldots, j-1$
    Solve the Sylvester equation for X_{ij}
      $T_{ii}X_{ij} + X_{ij}T_{jj} + T_{ij} = 0$
    Left multiplication by the transformation matrix M_k^{-1}
    for $k = j, \ldots, p$
      $T_{ik} = T_{ik} - X_{ij}T_{jk}$
    end
    Right multiplication by the transformation matrix M_k
    for $k = i, \ldots, p$
      $T_{kj} = T_{kj} - T_{ki}X_{ij}$
    end
    Accumulation of the transformations in M
    for $k = 1, \ldots, p$
      $M_{kj} = M_{kj} - M_{ki}X_{ij}$
    end
  end
end
$X_B = U \cdot M$

where the first term corresponds to the solution of the Sylvester equations used in the zeroing of the $p(p-1)/2$ superdiagonal blocks and the second term corresponds to the transformation of the block Schur form. In this count, it is assumed that the solution of each

Sylvester equation by the Bartels–Stewart algorithm requires $(n/p)^3$ flops. The accuracy of block-diagonalization depends on how well are separated the eigenvalues in the distinct groups. For close eigenvalues, the norms of the blocks X_{ij} become large and the accuracy of reduction decreases.

Example 7.1. Given is the 6×6 matrix

$$A = \begin{bmatrix} -3 & -6 & 5 & -9 & 9 & 4 \\ 0 & -5 & 3 & -3 & 2 & -1 \\ -1 & -5 & 2 & -6 & 7 & 3 \\ 0 & -1 & 1 & -3 & 1 & 0 \\ 2 & 0 & 1 & 1 & -6 & -6 \\ -2 & -3 & 2 & -5 & 7 & 4 \end{bmatrix},$$

which has a double eigenvalue $\lambda_1 = -1$, a simple eigenvalue $\lambda_2 = -3$ and a triple eigenvalue $\lambda_3 = -2$. The Schur form in which the multiple eigenvalues are grouped has the form (to obtain results with real numbers, in this example the function `schur` is used with the option 'real')

$$T = U^H A U = \left[\begin{array}{cc|c|ccc} -1.0000 & -3.6283 & -0.4098 & 2.9782 & -5.4960 & -21.1465 \\ 0 & -1.0000 & 2.8764 & 0.1244 & 5.0731 & -5.3029 \\ \hline 0 & 0 & -3.0000 & -0.4187 & -1.6233 & 4.5136 \\ \hline 0 & 0 & 0 & -2.0000 & 2.0226 & 0 \\ 0 & 0 & 0 & 0 & -2.0000 & -2.2361 \\ 0 & 0 & 0 & 0 & 0 & -2.0000 \end{array}\right].$$

Using the transformation

$$M_1 = \left[\begin{array}{cc|c|ccc} 1.0000 & 0 & -2.4042 & 0 & 0 & 0 \\ 0 & 1.0000 & -1.4382 & 0 & 0 & 0 \\ \hline 0 & 0 & 1.0000 & 0 & 0 & 0 \\ \hline 0 & 0 & 0 & 1.0000 & 0 & 0 \\ 0 & 0 & 0 & 0 & 1.0000 & 0 \\ 0 & 0 & 0 & 0 & 0 & 1.0000 \end{array}\right],$$

one obtains

$$T_2 = M_1^{-1} T M_1$$

$$= \left[\begin{array}{cc|c|ccc} -1.0000 & -3.6283 & 0 & 1.9715 & -9.3986 & -10.2949 \\ 0 & -1.0000 & 0 & -0.4779 & 2.7385 & 1.1886 \\ \hline 0 & 0 & -3.0000 & -0.4187 & -1.6233 & 4.5136 \\ \hline 0 & 0 & 0 & -2.0000 & 2.0226 & 0 \\ 0 & 0 & 0 & 0 & -2.0000 & -2.2361 \\ 0 & 0 & 0 & 0 & 0 & -2.0000 \end{array}\right].$$

The next transformation with the matrix

$$M_2 = \left[\begin{array}{cc|c|ccc} 1.0000 & 0 & 0 & -0.2377 & 2.4888 & 14.7933 \\ 0 & 1.0000 & 0 & 0.4779 & -1.7719 & 2.7735 \\ \hline 0 & 0 & 1.0000 & 0 & 0 & 0 \\ \hline 0 & 0 & 0 & 1.0000 & 0 & 0 \\ 0 & 0 & 0 & 0 & 1.0000 & 0 \\ 0 & 0 & 0 & 0 & 0 & 1.0000 \end{array}\right]$$

leads to

$$T_3 = M_2^{-1} T_2 M_2$$

$$= \left[\begin{array}{cc|c|ccc} -1.0000 & -3.6283 & 0 & 0 & 0 & 0 \\ 0 & -1.0000 & 0 & 0 & 0 & 0 \\ \hline 0 & 0 & -3.0000 & -0.4187 & -1.6233 & 4.5136 \\ \hline 0 & 0 & 0 & -2.0000 & 2.0226 & 0 \\ 0 & 0 & 0 & 0 & -2.0000 & -2.2361 \\ 0 & 0 & 0 & 0 & 0 & -2.0000 \end{array}\right].$$

The final transformation with matrix

$$M_3 = \left[\begin{array}{cc|c|ccc} 1.0000 & 0 & 0 & 0 & 0 & 0 \\ 0 & 1.0000 & 0 & 0 & 0 & 0 \\ \hline 0 & 0 & 1.0000 & -0.4187 & -0.7763 & 2.7776 \\ \hline 0 & 0 & 0 & 1.0000 & 0 & 0 \\ 0 & 0 & 0 & 0 & 1.0000 & 0 \\ 0 & 0 & 0 & 0 & 0 & 1.0000 \end{array}\right]$$

produces the block-diagonal form

$$B = M_3^{-1} T_3 M_3 = \left[\begin{array}{cc|c|ccc} -1 & -3.6283 & 0 & 0 & 0 & 0 \\ 0 & -1 & 0 & 0 & 0 & 0 \\ \hline 0 & 0 & -3 & 0 & 0 & 0 \\ \hline 0 & 0 & 0 & -2 & 2.0226 & 0 \\ 0 & 0 & 0 & 0 & -2 & -2.2361 \\ 0 & 0 & 0 & 0 & 0 & -2 \end{array}\right].$$

The matrix of the similarity transformation of T to B is

$$M = M_1 M_2 M_3$$
$$= \left[\begin{array}{cc|c|ccc} 1.0000 & 0 & -2.4042 & 0.7690 & 4.3553 & 8.1154 \\ 0 & 1.0000 & -1.4382 & 1.0801 & -0.6554 & -1.2212 \\ \hline 0 & 0 & 1.0000 & -0.4187 & -0.7763 & 2.7776 \\ \hline 0 & 0 & 0 & 1.0000 & 0 & 0 \\ 0 & 0 & 0 & 0 & 1.0000 & 0 \\ 0 & 0 & 0 & 0 & 0 & 1.0000 \end{array}\right].$$

The condition number of the matrix of the similarity transformation into block-diagonal form is $\text{cond}(M) = 155.19$.

In the given case, the separations between the diagonal blocks,

$$\text{sep}(T_{11}, T_{22}) = 0.8861, \quad \text{sep}(T_{11}, T_{33}) = 0.0183,$$
$$\text{sep}(T_{22}, T_{33}) = 0.1763,$$

are large and the Sylvester equations at the individual steps are solved with high accuracy.

For the estimates of the norms of the spectral projections, one obtains

$$\|P_1\|_2 \leq 15.4588, \quad \|P_2\|_2 \leq 9.1652, \quad \|P_3\|_2 \leq 9.6186,$$

which confirms that the eigenvalues in the three groups are well separated.

7.2 Orthonormalization

Let the matrix $Y \in \mathbb{C}^{n\times m}$ have linearly independent columns $y_1, y_2, \ldots, y_m$. The *classical Gram–Schmidt orthogonalization (CGS)* (Björck, 2015, Section 2.3.4; Golub and Van Loan, 2013, Section 5.2.7) is a process of constructing an orthonormal set of vectors $q_1, q_2, \ldots, q_m$ by determining q_k for $k = 1, 2, \ldots, m$ as

$$q_k = \left(y_k - \sum_{i=1}^{k-1} r_{ik} q_i\right) \Big/ r_{kk}, \tag{7.6}$$

$$r_{ik} = q_i^H y_k, \quad i = 1, 2, \ldots, k-1.$$

The quantities r_{kk} are nonzero due to the linear independence of the vectors y_k. As a result, the vectors q_k fulfill $q_k^H q_k = 1$ and

$$\operatorname{span}(q_1, q_2, \ldots, q_k) = \operatorname{span}(y_1, y_2, \ldots, y_k), \quad 1 \le k \le m.$$

The linearly independent complex vectors $y_1, y_2, \ldots, y_m$ are orthonormalized by the following CGS algorithm.

Classical Gram–Schmidt algorithm

Set $r_{11} = \|y_1\|_2$, $q_1 = y_1/r_{11}$
for $k = 2, 3, \ldots, m$
  for $i = 1, 2, \ldots, k-1$
    $r_{ik} = q_i^H y_k$
  end
  $z_k = y_k - \sum_{i=1}^{k-1} r_{ik} q_i$
  $r_{kk} = \|z\|_2$
  $q_k = z_k / r_{kk}$
end

The CGS algorithm produces the QR decomposition (see Section 1.5.4) of the matrix Y as

$$Y = Q \cdot R,$$

where $Q = [q_1, q_2, \ldots, q_m] \in \mathbb{C}^{n\times m}$ has orthonormal columns and

$$R = \begin{bmatrix} r_{11} & r_{12} & \cdots & r_{1m} \\ & r_{22} & \cdots & r_{2m} \\ & & \ddots & \vdots \\ & & & r_{mm} \end{bmatrix} \in \mathbb{C}^{m\times m}.$$

The CGS algorithm has poor numerical properties which may lead to loss of orthogonality among the computed vectors q_i. That is why it is preferable to implement the so-called *modified Gram–Schmidt algorithm (MGS)* in which the calculations are rearranged. Skipping the details presented in Björck (2015), Section 2.3.4, and Golub and Van Loan (2013), Section 5.2.8, the MGS algorithm is described as follows:

Modified Gram–Schmidt algorithm

```
for k = 1, 2, ..., m
    r_kk = ||y_k||_2
    q_k = y_k / r_kk
    for j = k + 1, k + 2, ..., m
        r_kj = q_k^H y_j
        y_j = y_j - q_k r_kj
    end
end
```

Consider now how to orthonormalize the basis vectors of the invariant subspaces, corresponding to the blocks of the block-diagonal form (7.2). According to (7.2), we have

$$AY_i = Y_iT_{ii}, \quad Y_i \in \mathbb{C}^{n\times\nu_i}, \quad i = 1, 2, \dots, p, \tag{7.7}$$

where to simplify the notation we set $Y = X_B$. The columns of the matrix Y_i are basis vectors for the invariant subspace in which acts the restriction T_{ii}.

At this step of the reduction to Jordan form, we perform an orthonormalization of the columns of Y_i by the modified Gram–Schmidt process so that

$$Y_i = Q_iR_{ii}, \tag{7.8}$$

where the matrix $Q_i \in \mathbb{C}^{n\times\nu_i}$ has orthonormal columns and $R_{ii} \in \mathbb{C}^{\nu_i\times\nu_i}$ is an upper triangular matrix. Equation (7.7) is then rewritten as

$$AQ_i = Q_iT'_{ii}, \quad i = 1, 2, \dots, p, \tag{7.9}$$

where $T'_{ii} = R_{ii}T_{ii}R_{ii}^{-1}$ is again in upper triangular form. Thus

$$B' = Q^{-1}AQ = \operatorname{diag}(T'_{11}, T'_{22}, \dots, T'_{pp}), \quad Q = [Q_1, Q_2, \dots, Q_p].$$

Note that the matrix Q itself is not unitary.

Algorithm 7.2. Orthonormalization of invariant subspace bases
Orthonormalization of the columns of the nonsingular transformation matrix that are bases of invariant subspaces.
function $[B, X_B] = \text{orthonormal}(B, X_B)$
Input: Block-diagonal matrix $B \in \mathbb{C}^{n \times n}$, whose p diagonal blocks $T_{ii} \in \mathbb{C}^{\nu_i \times \nu_i}$, $i = 1, \ldots, p$ correspond to distinct numerically multiple eigenvalues; X_B — the matrix of the transformation of A into block-diagonal form B.
Output: B contains the updated block-diagonal form; X_B contains the matrix of the transformation into updated block-diagonal form.

```
Set Q = 0_{n×n}, Y = X_B
τ_1 = 0
for i = 1, 2, ..., p
    if i < p
        τ_{i+1} = ν_{i+1}
    else
        τ_{i+1} = 0
    end
    if τ_i < τ_{i+1}
        for k = τ_i + 1, τ_i + 2, ..., τ_{i+1}
            r_kk = ||y_{1:n,k}||_2
            q_{1:n,k} = y_{1:n,k} / r_kk
            if k < τ_{i+1}
                for j = k + 1, k + 2, ..., τ_{i+1}
                    r_kj = q^H_{1:n,k} y_{1:n,j}
                    y_{1:n,j} = y_{1:n,j} − q_{1:n,k} r_kj
                end
            end
        end
        B_ii = R T_ii R^{-1}
    end
end
X_B = Q
```

The orthonormalization makes it possible to obtain unitary basis Q_i of the ν_i-dimensional invariant subspace in which acts the restriction T'_{ii}. This allows for reducing the condition number of the similarity transformation matrix Q. The transformations done at this step require $2n\nu_i^2$ computational operations for each block Y_i and are backward stable when the vectors to be orthonormalized are fairly independent (Golub and Van Loan, 2013, Section 5.2.9). As an alternative, it is possible to use the Householder triangularization (see Algorithm 1.8 in Section 1.5.4) which requires twice the number of computational operations required by the Gram–Schmidt orthogonalization but has better numerical properties.

The orthonormalization of the basis vectors of the invariant subspaces corresponding to multiple eigenvalues is done by Algorithm 7.2.

Example 7.2. Using Algorithm 7.2 to orthonormalize the columns of the transformation matrix, computed in Example 7.1, one obtains a new matrix whose condition number is equal to 33.18. This is almost five times less than the initial condition number equal to 155.19. The last three columns of the transformation matrix to a block-diagonal form corresponding to the triple eigenvalue at -2 are

$$y_1 = \begin{bmatrix} -0.8563328 \\ 0.8563568 \\ 0.0000080 \\ 0.0000000 \\ 0.8563488 \\ -0.8563408 \end{bmatrix}, \quad y_2 = \begin{bmatrix} -3.4641180 \\ -1.7320506 \\ -1.7320589 \\ -0.0000081 \\ 0.0000002 \\ -1.7320591 \end{bmatrix}, \quad y_3 = \begin{bmatrix} -5.9385672 \\ -1.8073886 \\ -3.8729797 \\ -3.8729833 \\ -1.8073922 \\ -2.0655875 \end{bmatrix}.$$

The angles between these vectors are (in degrees)

$$\angle(y_1, y_2) = 67.8, \quad \angle(y_2, y_3) = 31.7, \quad \angle(y_3, y_1) = 75.4,$$

which shows that they are far from orthogonality.

As a result of Algorithm 7.2, we obtain the orthonormal vectors

$$q_1 = \begin{bmatrix} -0.4999930 \\ 0.5000070 \\ 0.0000047 \\ 0.0000000 \\ 0.5000023 \\ -0.4999977 \end{bmatrix}, \quad q_2 = \begin{bmatrix} -0.6123784 \\ -0.6123660 \\ -0.4082488 \\ -0.0000019 \\ -0.2041192 \\ -0.2041296 \end{bmatrix}, \quad q_3 = \begin{bmatrix} -0.1065990 \\ 0.3198024 \\ -0.2131998 \\ -0.8528029 \\ -0.3198006 \\ 0.1066008 \end{bmatrix}.$$

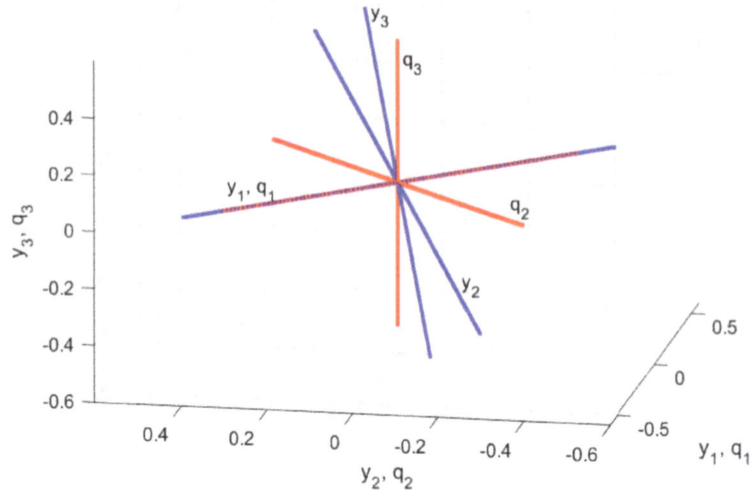

Fig. 7.3. Orthonormalization of the columns corresponding to the triple eigenvalue at -2.

To measure the deviation from orthonormality, we use the quantity

$$\|I_3 - [q_1,\ q_2,\ q_3]^H [q_1,\ q_2,\ q_3]\|_2^{1/2} = 2.10 \cdot 10^{-8},$$

which is of order $\mathbf{u}^{1/2}$.

The mutual disposition of the vectors y_1, y_2, y_3 and the corresponding orthonormal vectors q_1, q_2, q_3 is illustrated in Figure 7.3.

7.3 Reduction into Block-Diagonal Staircase Form

After the block-diagonalization of the Schur form and the orthonormalization of the basis vectors of the invariant subspaces, each diagonal upper triangular block is reduced into staircase form using Algorithm 6.3.

Assume that the matrix A has p distinct eigenvalues $\lambda_1, \lambda_2, \ldots, \lambda_p$ with multiplicities $\nu_1, \nu_2, \ldots, \nu_p$ and associated Weyr characteristics

$$\mathcal{W}_1, \mathcal{W}_2, \ldots, \mathcal{W}_p; \quad \mathcal{W}_i = \{w_{i1}, w_{i2}, \ldots, w_{i,h_i}\},$$

respectively, and let this matrix be reduced to the block-diagonal form according to (7.2). In this step, the matrix B is reduced into

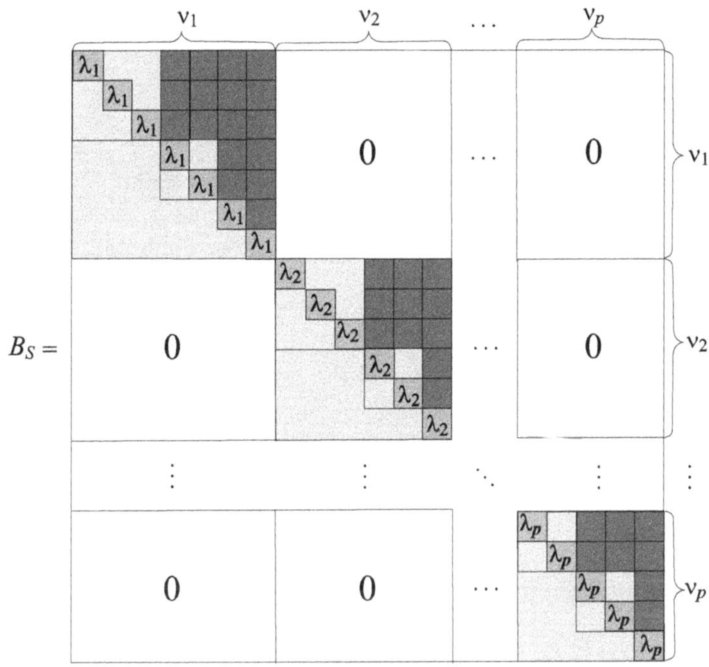

Fig. 7.4. The block-diagonal staircase form.

the *block-diagonal staircase form* (see Figure 7.4)

$$B_S = V^H B V = \begin{bmatrix} S(\mathcal{W}_1, \lambda_1) & & & \\ & S(\mathcal{W}_2, \lambda_2) & & \\ & & \ddots & \\ & & & S(\mathcal{W}_p, \lambda_p) \end{bmatrix} \tag{7.10}$$

using only unitary similarity transformations. The block $S(\mathcal{W}_i, \lambda_i)$, $i = 1, 2, \ldots, p$, is in the staircase form

$$S(\mathcal{W}_i, \lambda_i) = \begin{bmatrix} \lambda_i I_{w_{i1}} & N_{12}^{(i)} & \cdots & N_{1,h_i-1}^{(i)} & N_{1,h_i}^{(i)} \\ & \lambda_i I_{w_{i2}} & \cdots & N_{2,h_i-1}^{(i)} & N_{2,h_i}^{(i)} \\ & & \ddots & \vdots & \vdots \\ & & & \lambda_i I_{w_{i,h_i-1}} & N_{h_i-1,h_i}^{(i)} \\ & & & & \lambda_i I_{w_{i,h_i}} \end{bmatrix} \in \mathbb{C}^{\nu_i \times \nu_i}. \tag{7.11}$$

Algorithm 7.3. Reduction into block-diagonal staircase form
Reduction of a block-diagonal matrix into block-diagonal staircase form and determining the Weyr characteristics of the blocks.
function $[B_S, X_S, \mathcal{W}] = \text{block\_staircase}(B, X_B)$
Input: Block-diagonal matrix $B = \text{diag}(T_{11},\ T_{22}, \ldots,\ T_{pp}) \in \mathbb{C}^{n\times n}$,
whose p diagonal blocks $T_{ii} \in \mathbb{C}^{\nu_i\times\nu_i}$ correspond
to distinct numerically multiple eigenvalues $\lambda_i,\ i = 1, \ldots, p$.
The matrix X_B of the transformation of the initial matrix A
into block-diagonal form.
Output: The block-diagonal staircase form
$B_S = \text{diag}\left(S(\mathcal{W}_1, \lambda_1),\ S(\mathcal{W}_2, \lambda_2), \ldots,\ S(\mathcal{W}_p, \lambda_p)\right)$;
the matrix X_S of the similarity transformation of A into
block-diagonal staircase form; the vector $\mathcal{W}$ containing
the corresponding Weyr characteristics $\mathcal{W}_1,\ \mathcal{W}_2, \ldots,\ \mathcal{W}_p$.
Set $X_S = X_B$
for $i = 1,\ 2, \ldots, p$
  Make the diagonal block nilpotent
  $T_{ii} \leftarrow T_{ii} - \lambda_i I_{\nu_i}$
  Reduce by Algorithm 6.3 the nilpotent block T_{ii} into staircase form
  $S_i = Q_i^H T_{ii} Q_i$,
  where S_i is given by (6.13)
  $T_{ii} \leftarrow S_i + \lambda_i I_{\nu_i}$
  Accumulate the transformations
  $Y_i = [I_{\nu_1+\cdots+\nu_{i-1}}, Q_i, I_{\nu_{i+1}+\cdots+\nu_p}]$
  $X_S \leftarrow X_S Y_i$
end

At this step, the Weyr characteristic and consequently the corresponding Segre characteristic of each diagonal block are determined so that the complete Jordan structure becomes known. Note that due to the use of unitary similarity transformations at this step, the 2-condition number of the transformation matrix of A into block-diagonal staircase form B_S is the same as the condition number of the transformation matrix into block-diagonal form B.

The reduction of a block-diagonal matrix into a block-diagonal staircase form is done by Algorithm 7.3. The reduction of the diagonal blocks into staircase form requires

$$2\left(\frac{1}{3}\frac{n^3h}{p} + \frac{2n^3h}{p^2}\right)\text{flops},$$

where h is the average index of nilpotence of the diagonal blocks. This volume of work is slightly less than the volume of computations in the reduction into the Jordan–Schur form due to the absence of super-diagonal blocks that have to be transformed. Since the algorithm uses only unitary similarity transformation, it is backward stable. However, in case of ill-conditioned eigenvalue problems, the determination of the Weyr characteristics can be done erroneously, which corresponds to a transition to another bundle of Jordan structures.

Example 7.3. Consider the 8×8 matrix

$$A = \begin{bmatrix} 8 & 3 & -15 & 10 & -16 & -1 & 16 & -12 \\ -7 & -3 & 1 & 0 & 13 & 12 & -13 & -17 \\ -2 & -6 & 22 & -16 & 0 & -18 & 6 & 60 \\ 3 & 5 & -17 & 12 & -5 & 12 & 2 & -38 \\ -5 & 0 & -6 & 2 & 13 & 15 & -15 & -33 \\ 1 & 5 & -11 & 12 & -7 & 8 & 3 & -25 \\ -9 & -4 & 16 & -12 & 16 & 0 & -16 & 17 \\ 2 & 5 & -17 & 14 & -4 & 13 & 0 & -42 \end{bmatrix}$$

with well-conditioned eigenvalue problem. The matrix has a 5-tuple eigenvalue at -2.0 participating in two Jordan blocks of sizes 2 and 3 and a 3-tuple eigenvalue at 4.0 participating in two Jordan blocks of sizes 2 and 1. The associated Segre characteristics are $\mathcal{S}(A,-2) = \{3,2\}$, $\mathcal{S}(A,4) = \{2,1\}$ and the associated Weyr characteristics are $\mathcal{W}(A,-2) = \{2,2,1\}$, $\mathcal{W}(A,4) = \{2,1\}$, respectively. Hence the matrix is derogatory and defective.

Using tolerances $ein = 1 \cdot 10^{-12}$ and $tol = 1 \cdot 10^{-12}$, Algorithm 7.3 produces the block-diagonal staircase form

$$B_S = X_S^{-1} A X_S = \operatorname{diag}\left(\left[\begin{array}{cc|c} 4 & 0 & \times \\ & 4 & \times \\ \hline & & 4 \end{array}\right], \left[\begin{array}{cc|cc|c} -2 & 0 & \times & \times & \times \\ & -2 & \times & \times & \times \\ \hline & & -2 & 0 & \times \\ & & & -2 & \times \\ \hline & & & & -2 \end{array}\right] \right),$$

where the position of the horizontal and vertical lines corresponds to the values of the elements of the Weyr characteristics. The condition number of the transformation matrix X_S is equal to 44.80. The block-diagonal staircase form is found with a relative backward error

$$\|X_S B_S X_S^{-1} - A\|_F / \|A\|_F = 1.18 \cdot 10^{-13}.$$

7.4 Reduction into Numerical Weyr Form

The next step to the reduction into Jordan form is the reduction into the Weyr form. The *numerical Weyr form* of A is obtained as the direct sum

$$W = X_W^{-1} A X_W = \mathrm{diag}(W(\mathcal{W}_1, \lambda_1), \quad W(\mathcal{W}_2, \lambda_2), \ldots, W(\mathcal{W}_p, \lambda_p))$$

of Weyr blocks in the form

$$W(\mathcal{W}_i, \lambda_i) = \begin{bmatrix} \lambda_i I_{w_{i1}} & G_{w_{i1}, w_{i2}} & & \\ & \lambda_i I_{w_{i2}} & \ddots & \\ & & \ddots & G_{w_{i,h_i-1}, w_{i,h_i}} \\ & & & \lambda_i I_{w_{i,h_i}} \end{bmatrix}, \quad i = 1, 2, \ldots, p \tag{7.12}$$

each of them corresponding to a distinct eigenvalue λ_i with associated Weyr characteristic

$$\mathcal{W}_i = \{w_{i1}, w_{i2}, \ldots, w_{i,h_i}\}, \quad h_i = s_{1i}.$$

These blocks differ from the definition of Weyr block (2.43) in two ways. First, the superdiagonal blocks are in the un-normalized form

$$G_{w_{ij}, w_{i,j+1}} = \begin{bmatrix} 0_{(w_{ij} - w_{i,j+1}) \times w_{i,j+1}} \\ R_{w_{i,j+1}} \end{bmatrix} \in \mathbb{C}^{w_{ij} \times w_{i,j+1}}, \quad 1 \leq j \leq h_i - 1, \tag{7.13}$$

where

$$R_{w_{i,j+1}} = \mathrm{diag}\left(r_{1,j+1}^{(i,j+1)}, r_{2,j+1}^{(i,j+1)}, \ldots, r_{w_{i,j+1},j+1}^{(i,j+1)}\right),$$

$$\mathrm{rank}(R_{w_{i,j+1}}) = w_{i,j+1}, \quad 1 \leq j \leq h_i - 1$$

is a matrix, whose diagonal elements are different from 1 (Figure 7.5; the superscripts of the diagonal elements are omitted). These elements are the *coupling elements* in the blocks of the Jordan form

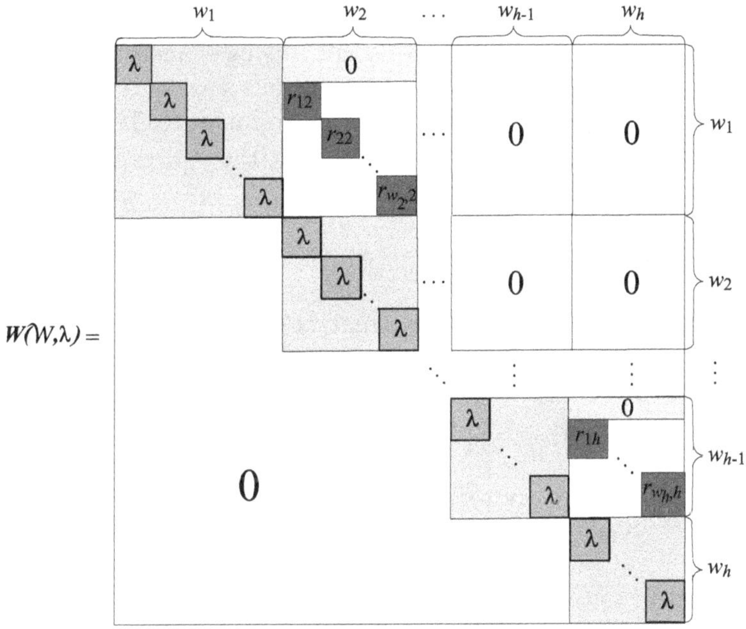

Fig. 7.5. Un-normalized Weyr block.

and they can be made equal to 1 with the aid of additional diagonal transformations. Such transformations, however, increase the condition number of the transformation matrix into Weyr and Jordan forms and that is why their usage is avoided.

The second peculiarity of the numerical Weyr form (7.12) is that the superdiagonal blocks are reduced into the form

$$\begin{bmatrix} 0_{(w_{ij}-w_{i,j+1})\times w_{i,j+1}} \\ R_{w_{i,j+1}} \end{bmatrix},$$

instead of into the form

$$\begin{bmatrix} R_{w_{i,j+1}} \\ 0_{(w_{ij}-w_{i,j+1})\times w_{i,j+1}} \end{bmatrix},$$

as it follows from the definition of the Weyr block given in Section 2.2.5. Since these blocks have full column rank, the matrix W is again in Weyr form, but this peculiarity of the reduction leads to a specific ordering of the blocks in the final Jordan form which is considered in more detail in the following section.

The determination of the coupling elements is the last step in the reduction into Jordan form. It is done by using elementary triangular matrices and not by unitary transformations so that it may increase the condition number of the transformation matrix. The determination of the coupling elements implies that the staircase form of the matrix is reduced to Weyr form.

7.4.1 *A Simple Example*

Let us for definiteness consider a matrix of fifth-order with multiple eigenvalue λ and associated Weyr characteristic $\mathcal{W} = \{3, 2\}$, which is reduced to the staircase form

$$S(\mathcal{W}, \lambda) = \left[\begin{array}{ccc|cc} \lambda & 0 & 0 & b_{14} & b_{15} \\ 0 & \lambda & 0 & b_{24} & b_{25} \\ 0 & 0 & \lambda & b_{34} & b_{35} \\ \hline 0 & 0 & 0 & \lambda & 0 \\ 0 & 0 & 0 & 0 & \lambda \end{array}\right].$$

Further on for convenience we shall use the nilpotent matrix

$$S := S(\mathcal{W}, 0) = S(\mathcal{W}, \lambda) - \lambda I_5$$

$$= \left[\begin{array}{ccc|cc} 0 & 0 & 0 & b_{14} & b_{15} \\ 0 & 0 & 0 & b_{24} & b_{25} \\ 0 & 0 & 0 & b_{34} & b_{35} \\ \hline 0 & 0 & 0 & 0 & 0 \\ 0 & 0 & 0 & 0 & 0 \end{array}\right].$$

In the general case, the elements b_{ij}, $i = 1, 2, 3$, $j = 4, 5$, are different from zero. Our aim is, by using elementary similarity transformations, to reduce the matrix S into the form

$$W(\mathcal{W}, 0) = \left[\begin{array}{ccc|cc} 0 & 0 & 0 & 0 & 0 \\ 0 & 0 & 0 & r_{11} & 0 \\ 0 & 0 & 0 & 0 & r_{21} \\ \hline 0 & 0 & 0 & 0 & 0 \\ 0 & 0 & 0 & 0 & 0 \end{array}\right],$$

from which we can find the coupling elements r_{11} and r_{21}. The matrix $W(\mathcal{W}, 0)$ is in an un-normalized form in the sense that the elements r_{11} and r_{21} are different from 1. To achieve this aim, two passes of transformations are performed. In the first pass of the reduction, one zeroes the entries which are located above the coupling elements and in the second pass are zeroed the entries which are located below the coupling elements. In the first pass, one starts from the last column of the staircase form going from right to left, and in the second pass, one starts from the first column of the superdiagonal block going from left to right.

In the first pass, one constructs the elementary upper triangular matrix

$$E_5 = I_5 - m_5 e_3^T,$$

where

$$m_5 = \begin{bmatrix} m_{15} \\ m_{25} \\ 0 \\ 0 \\ 0 \\ 0 \end{bmatrix},$$

e_3 is the third column of the unit fifth-order matrix and elements m_{15}, m_{25} are chosen as

$$m_{15} = b_{15}/b_{35},$$
$$m_{25} = b_{25}/b_{35}$$

(it is assumed that the element b_{35} is nonzero). In this way,

$$E_5 = \begin{bmatrix} 1 & 0 & -m_{15} & 0 & 0 \\ 0 & 1 & -m_{25} & 0 & 0 \\ 0 & 0 & 1 & 0 & 0 \\ 0 & 0 & 0 & 1 & 0 \\ 0 & 0 & 0 & 0 & 1 \end{bmatrix}.$$

Since

$$\begin{bmatrix} b_{15}^{(1)} \\ b_{25}^{(1)} \end{bmatrix} = \begin{bmatrix} 1 & 0 & -m_{15} \\ 0 & 1 & -m_{25} \end{bmatrix} \begin{bmatrix} b_{15} \\ b_{25} \\ b_{35} \end{bmatrix} = \begin{bmatrix} b_{15} \\ b_{25} \end{bmatrix} - \begin{bmatrix} m_{15} \\ m_{25} \end{bmatrix} b_{35} = \begin{bmatrix} 0 \\ 0 \end{bmatrix},$$

the matrix E_5S has the form

$$\left[\begin{array}{ccc|cc} 0 & 0 & 0 & b_{14}^{(1)} & 0 \\ 0 & 0 & 0 & b_{24}^{(1)} & 0 \\ 0 & 0 & 0 & b_{34} & b_{35} \\ \hline 0 & 0 & 0 & 0 & 0 \\ 0 & 0 & 0 & 0 & 0 \end{array}\right],$$

where

$$\begin{bmatrix} b_{14}^{(1)} \\ b_{24}^{(1)} \end{bmatrix} = \begin{bmatrix} b_{14} \\ b_{24} \end{bmatrix} - \begin{bmatrix} m_{15} \\ m_{25} \end{bmatrix} b_{34}.$$

Since $E_5^{-1} = I_5 + m_5 e_3^T$, the multiplication of E_5S on the right by E_5^{-1} does not lead to changes. In this way, the matrix $S_1 = X_1^{-1}SX_1 = E_5SE_5^{-1}$, $X_1 = E_5^{-1}$, has the same form as E_5S.

To avoid the situation in which the element b_{35} is zero or is very small, before the construction of the matrix E_5 the elements in the fifth column of S are exchanged, such that the maximum in module element is located in the position $(3,5)$. This corresponds to the case of partial pivoting in the Gaussian elimination and decreases the condition number of the transformation matrix and hence decreases the rounding errors during the transformation. The exchange of the rows of S is equivalent to its transformation by elementary permutation matrix P_1 on the left and by $P_1^{-1} = P_1^T$ on the right, $S_1 = E_5P_1SP_1^TE_5^{-1} = X_1^{-1}SX_1$, $X_1 = P_1^TE_5^{-1}$. This transformation changes the initial values of the elements b_{ij}, $i = 1,2,3$, $j = 4,5$, and does not affect the form of the transformations shown earlier.

At the second step of the first pass, one constructs an elementary upper triangular matrix

$$E_4 = I_5 - m_4 e_2^T = \begin{bmatrix} 1 & -m_{14} & 0 & 0 & 0 \\ 0 & 1 & 0 & 0 & 0 \\ 0 & 0 & 1 & 0 & 0 \\ 0 & 0 & 0 & 1 & 0 \\ 0 & 0 & 0 & 0 & 1 \end{bmatrix},$$

where $m_{14} = b_{14}^{(1)}/b_{24}^{(1)}$, with which one zeroes the element $b_{14}^{(1)}$. Before the construction of the matrix E_4, the elements in the fourth column of S are exchanged, such that the maximum in module element is located in the position $(2,4)$. This is done by using a permutation matrix P_2. As a result of the second step, we obtain

$$S_2 = X_2^{-1} S X_2 = \left[\begin{array}{ccc|cc} 0 & 0 & 0 & 0 & 0 \\ 0 & 0 & 0 & b_{24}^{(2)} & 0 \\ 0 & 0 & 0 & b_{34} & b_{35} \\ \hline 0 & 0 & 0 & 0 & 0 \\ 0 & 0 & 0 & 0 & 0 \end{array}\right],$$

where $X_2 = X_1 P_2 E_4 = P_1^T E_5^{-1} P_2^T E_4^{-1}$.

With this, the first pass of the transformation ends.

In the given case, the second pass of the transformation consists of only one step in which we zero the element b_{34}. This is done by using an elementary lower triangular matrix

$$E_3 = I_5 - m_2 e_3^T = \begin{bmatrix} 1 & 0 & 0 & 0 & 0 \\ 0 & 1 & 0 & 0 & 0 \\ 0 & -m_{23} & 1 & 0 & 0 \\ 0 & 0 & 0 & 1 & 0 \\ 0 & 0 & 0 & 0 & 1 \end{bmatrix},$$

where the element m_{23} is determined from

$$m_{23} = \frac{b_{34}}{b_{24}^{(2)}}.$$

As a result, we obtain that

$$S_3 = E_3 S_2 E_3^{-1} = X_3^{-1} S X_3 = \left[\begin{array}{ccc|cc} 0 & 0 & 0 & 0 & 0 \\ 0 & 0 & 0 & b_{24}^{(3)} & 0 \\ 0 & 0 & 0 & 0 & b_{35} \\ \hline 0 & 0 & 0 & 0 & 0 \\ 0 & 0 & 0 & 0 & 0 \end{array}\right],$$

where $X_3 = X_2 E_3^{-1} = P_1^T E_5^{-1} P_2^T E_4^{-1} E_3^{-1}$. Note that the multiplication by the matrix E_3 on the left leads to changes in the fourth column but does not introduce changes in the fifth column of S_2.

After zeroing the elements in the fourth column, it is possible to perform an additional diagonal transformation with which to make the element $b_{24}^{(3)}$ real and positive. For this aim, we use the matrix D_2 which differs from the unit matrix only in position $(2, 2)$ so that $d_{22} = \widehat{b}_2$,

$$\widehat{b}_2 = \frac{\overline{b}_{24}^{(3)}}{|b_{24}^{(3)}|},$$

where $\overline{b}_{24}^{(3)}$ is the conjugated of $b_{24}^{(3)}$. As a result, we obtain

$$S_4 = D_2 S_3 D_2^{-1} = X_4^{-1} S X_4 = \left[\begin{array}{ccc|cc} 0 & 0 & 0 & 0 & 0 \\ 0 & 0 & 0 & r_{12} & 0 \\ 0 & 0 & 0 & 0 & b_{35} \\ \hline 0 & 0 & 0 & 0 & 0 \\ 0 & 0 & 0 & 0 & 0 \end{array}\right],$$

where

$$r_{12} = \widehat{b}_2 b_{24}^{(3)} = |b_{24}^{(3)}|$$

and

$$X_4 = X_3 D_2^{-1} = P_1^T E_5^{-1} P_2^T E_4^{-1} E_3^{-1} D_2^{-1}.$$

For the given example, at the second step of the second pass, one performs only a transformation by the diagonal matrix D_3, $d_{33} = \widehat{b}_3$,

$$\widehat{b}_3 = \frac{\overline{b}_{35}}{|b_{35}|},$$

which makes the element b_{35} real and positive.

Finally, we obtain that

$$W(\mathcal{W},0) = D_3S_4D_3^{-1} = X^{-1}SX = \left[\begin{array}{ccc|cc} 0 & 0 & 0 & 0 & 0 \\ 0 & 0 & 0 & r_{12} & 0 \\ 0 & 0 & 0 & 0 & r_{22} \\ \hline 0 & 0 & 0 & 0 & 0 \\ 0 & 0 & 0 & 0 & 0 \end{array}\right],$$

where

$$X = X_4D_3^{-1} = P_1^TE_5^{-1}P_2^TE_4^{-1}E_3^{-1}D_2^{-1}D_3^{-1}, \quad X := [x_1, x_2, x_3, x_4, x_5].$$

We have that

$$X^{-1}S(\mathcal{W},\lambda)X = W(\mathcal{W},\lambda),$$

where the matrix $W(\mathcal{W},\lambda) = W(\mathcal{W},0) + \lambda I_5$ is an un-normalized Weyr block in the sense that the elements r_{11} and r_{22} are different from 1. The matrix equation

$$S(\mathcal{W},\lambda)X = XW(\mathcal{W},\lambda)$$

can be written as the following five equations:

$$\begin{aligned}
S(\mathcal{W},\lambda)x_1 &= x_1\lambda, \\
S(\mathcal{W},\lambda)x_2 &= x_2\lambda, \\
S(\mathcal{W},\lambda)x_3 &= x_3\lambda, \\
S(\mathcal{W},\lambda)x_4 &= x_4\lambda + r_{12}z_2, \\
S(\mathcal{W},\lambda)x_5 &= x_5\lambda + r_{22}z_3.
\end{aligned}$$

It follows from these equation that the vectors x_1, x_2 and x_3 are eigenvectors of the matrix $S(\mathcal{W},\lambda)$, and the vectors x_4 and x_5 are principal vectors of second grade. The equations can be written in the permuted form

$$\begin{aligned}
S(\mathcal{W},\lambda)x_1 &= x_1\lambda, \\
S(\mathcal{W},\lambda)x_2 &= x_2\lambda,
\end{aligned}$$

$$S(\mathcal{W},\lambda)x_4 = x_4\lambda + r_{12}x_2,$$
$$S(\mathcal{W},\lambda)x_3 = x_3\lambda,$$
$$S(\mathcal{W},\lambda)x_5 = x_5\lambda + r_{22}x_3,$$

i.e.,

$$S(\mathcal{W},\lambda)XP_J = XP_J J,$$

where

$$XP_J = [x_1,\ x_2,\ x_4,\ x_3,\ x_5],$$
$$P_J = \begin{bmatrix} 1 & 0 & 0 & 0 & 0 \\ 0 & 1 & 0 & 0 & 0 \\ 0 & 0 & 0 & 1 & 0 \\ 0 & 0 & 1 & 0 & 0 \\ 0 & 0 & 0 & 0 & 1 \end{bmatrix}$$

and

$$J = \left[\begin{array}{c|cc|cc} \lambda & 0 & 0 & 0 & 0 \\ \hline 0 & \lambda & r_{12} & 0 & 0 \\ 0 & 0 & \lambda & 0 & 0 \\ \hline 0 & 0 & 0 & \lambda & r_{22} \\ 0 & 0 & 0 & 0 & \lambda \end{array}\right].$$

We see that the transformation with the matrix XP_J reduces $S(\mathcal{W},\lambda)$ into the Jordan form J, in which the 5-tuple eigenvalue λ participates in three Jordan blocks with dimensions $s_1 = 1$, $s_2 = 2$ and $s_3 = 2$. Hence, the Weyr form $W(\mathcal{W},\lambda)$ is permutational similar to the Jordan form J:

$$J = P_J^T W(\mathcal{W},\lambda) P_J.$$

7.4.2 *The Case of n-Tuple Eigenvalue*

Consider now the case when the matrix A has an n-tuple eigenvalue λ with associate Weyr characteristic $\mathcal{W} = \{w_1, w_2, \ldots, w_h\}$ and Segre characteristic $\mathcal{S} = \{s_1, s_2, \ldots, s_q\}$, $h = s_1$, $q = w_1$. According to (7.11), the staircase form of $A - \lambda I_n$ can be written in the form

$$S(\mathcal{W}, 0) = \begin{bmatrix} 0_{w_1 \times w_1} & B_{12} & \cdots & B_{1,h-1} & B_{1h} \\ 0 & 0_{w_2 \times w_2} & \cdots & B_{2,h-1} & B_{2h} \\ \vdots & \vdots & \ddots & \vdots & \vdots \\ 0 & 0 & \cdots & 0_{w_{h-1} \times w_{h-1}} & B_{h-1,h} \\ 0 & 0 & \cdots & 0 & 0_{w_h \times w_h} \end{bmatrix},$$

where $B_{ij} \in \mathbb{C}^{w_i \times w_j}$.

The determination of the coupling elements starts with the transformation of the blocks in position h (most right), combining these blocks in the matrix

$$B_h = \begin{bmatrix} B_{1h} \\ B_{2h} \\ \vdots \\ B_{h-1,h} \end{bmatrix} \in \mathbb{C}^{(w_1 + \cdots + w_{h-1}) \times w_h}.$$

The transformation of B_h is done as described earlier with the exception that the partial pivoting is done only in the columns of the block $B_{h-1,h}$. This restriction is imposed by the requirement to avoid the changes in the structure of the rest blocks of S due to the row exchanges. The eliminations are carried out in rows working down and the columns from right to the left (see Figure 7.6). Thus the elements above b_{ik} in the kth column are eliminated by using the elementary transformation

$$E_j = I_n - m e_j^T, \quad m = b_{ij}/b_{ik},$$

where b_{ik} is maximized inside the kth column of $B_{h-1,h}$

The second pass eliminates the elements below the coupling elements. Now the order of elimination is columnwise from the left to the right inside the block $B_{h-1,h}$ (see Figure 7.7). In this pass, pivoting

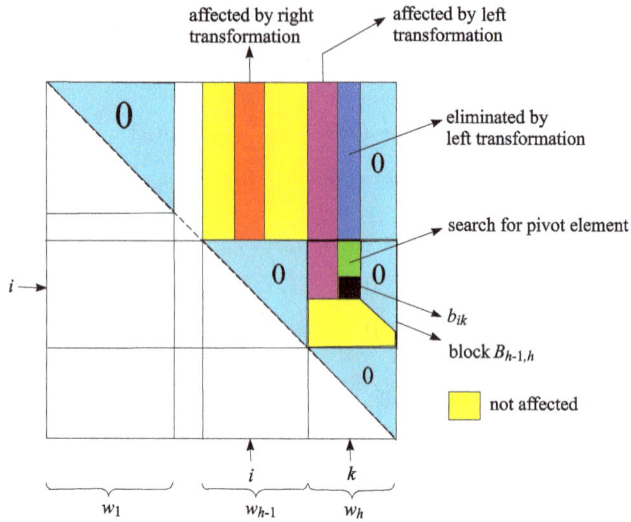

Fig. 7.6. The elimination process to find the coupling elements: Pass 1.

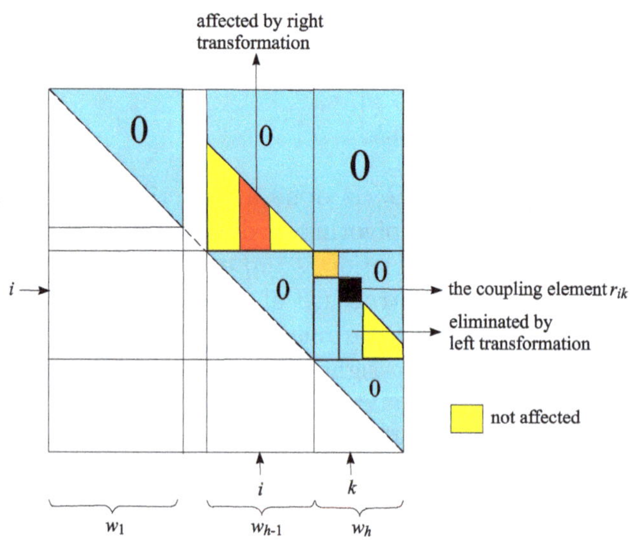

Fig. 7.7. The elimination process to find the coupling elements: Pass 2.

is not possible. As a result, we obtain the matrix

$$\widetilde{B}_h = \begin{bmatrix} 0 & 0 & \dots & 0 \\ 0 & 0 & \dots & 0 \\ \vdots & \vdots & \vdots & \vdots \\ r_{1h} & 0 & \dots & 0 \\ 0 & r_{2h} & \dots & 0 \\ \vdots & \vdots & \ddots & 0 \\ 0 & 0 & \dots & r_{w_h,h} \end{bmatrix}.$$

The transformations continue from right to left, for

$$B_{h-1} = \begin{bmatrix} B_{1,h-1} \\ B_{2,h-1} \\ \vdots \\ B_{h-2,h-1} \end{bmatrix} \in \mathbb{C}^{(w_1+\cdots+w_{h-2})\times w_{h-1}},$$

the pivoting being done in the framework of the transformed block $\widetilde{B}_{h-2,h-1}$, for B_{h-2} — in the block $\widetilde{B}_{h-3,h-2}$ and so on. The multiplication by elementary triangular matrices on the right leads to changes in the blocks located on the left of the current block.

Finally, we obtain the matrix

$$W(\mathcal{W}, 0) = X^{-1}S(\mathcal{W}, 0)X$$

in which all coupling elements are determined (Figure 7.8). Up to the addition of the diagonal elements (the eigenvalues), this matrix represents the un-normalized Weyr form of $S(\mathcal{W}, \lambda)$ represented in Figure 7.5.

As shown in Kågström and Ruhe (1980b), it is fulfilled that

$$\begin{aligned} |r_{1,j+1}r_{2,j+1}\dots r_{w_{j+1},j+1}| &> \sigma_{n_{j+1}}(N^j), \quad N = A - \lambda I_n \\ |r_{i,j+1}| &> \sigma_{n_{j+1}}(N^j)/\sigma_1(N^{j-1}), \\ i &= 1, 2, \dots, w_{j+1}, \\ j &= 1, 2, \dots, h-1, \end{aligned} \tag{7.14}$$

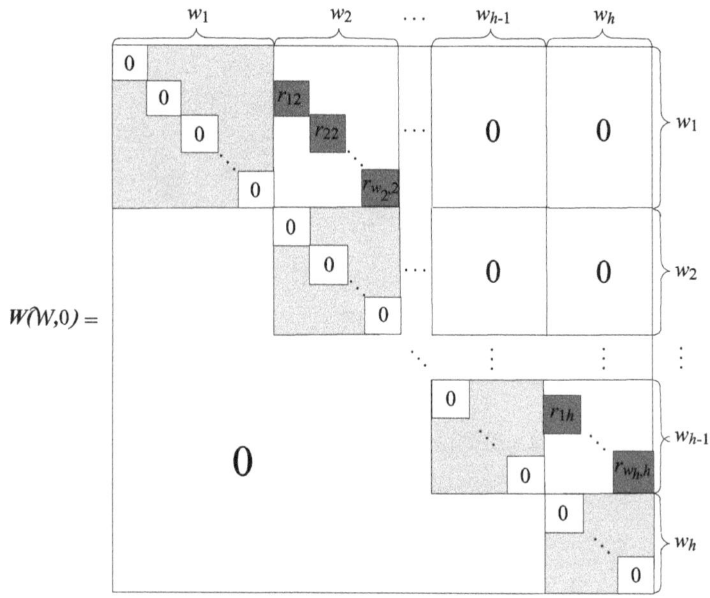

Fig. 7.8. The nilpotent block $W(\mathcal{W}, 0)$.

where $r_{i,j+1}$ are the coupling elements shown in Figure 7.8. Hence the element $r_{i,j+1}$ can be considered as an upper bound estimate of the singular value

$$\sigma_{n_{j+1}}(N^j),$$

i.e., of the first singular value of N^j interpreted as nonzero. On the other hand, the backward error (the distance to the nilpotent matrix) $\|E\|_F$ (6.14) estimates $\sigma_{n_j}(N^j)$ (the singular values interpreted as zeros). From the definition (5.23) of numerically multiple eigenvalue, we see that all elements $r_{i,j+1}$ should be significantly larger than $\|E\|_F$. If some $r_{i,j+1}$ is very small, this is an indication that the chain under consideration is weakly coupled and eventually can be divided into several shorter chains, for instance, by increasing the geometric multiplicity w_1. This can be achieved by increasing the tolerance *tol* and can lead to the increase in $\|E\|_F$ (6.14).

As a result of the reduction of the matrix $S(\mathcal{W}, 0)$ into the form $W(\mathcal{W}, 0)$, we obtain $w_1 = q$ eigenvectors (the first w_1 columns of X), w_2 principal vectors of grade 2 (the columns of X from $w_1 + 1$ to $w_1 + w_2$) and so on and w_h principal vectors of grade h (the columns

Algorithm 7.4. Reduction into Numerical Weyr form
Reduces the staircase form of a matrix block into Weyr form using nonsingular similarity transformations and computes the coupling elements of the Weyr and Jordan form
function $[W, X, nxt, supd] = \text{weyr}(S, \mathcal{W})$
Input: Staircase matrix $S \in \mathbb{C}^{n\times n}$ with zero multiple eigenvalue
and Weyr characteristic $\mathcal{W} = \{w_1, w_2, \ldots, w_h\}$.
Output: The Weyr form W; matrix $X \in \mathbb{C}^{n\times n}$ of the nonsingular
transformation which reduces the staircase form S into W;
vector nxt containing the column indices of the vector $supd$,
$nxt(i) = j$ implies that the vector with a coupling element
$supd(i)$ is placed in the column j of X;
vector $supd$ containing the super diagonal coupling elements
of the Weyr form.
Set $X = I_n$, $nxt = 0_{1\times n}$, $supd = 0_{1\times n}$
for $j = h, h-1, \ldots, 2$
$\nu_j = w_1 + \cdots + w_{j-1}$

Set $B = \begin{bmatrix} B_{1j} \\ \vdots \\ B_{j-1,j} \end{bmatrix} \in \mathbb{C}^{\nu_j \times w_j}$

Pass 1. Annihilate the elements above the coupling elements
for $k = w_j, w_{j-1}, \ldots, 1$
Finding the row in which is located the coupling element
$i_p = \nu_j + k - w_j$, $i_{p1} = i_p - 1$
Determine the row i_m, in which is located the maximum element
in the kth column of the block $B_{j-1,j}$
Exchange the rows i_p and i_m of S and columns i_p and i_m of S and X
Compute an elementary upper triangular matrix that annihilates
the elements in the rows from 1 to i_p in the kth columns of B

$$m_k = \begin{bmatrix} b_{1k}/b_{i_p,k} \\ \vdots \\ b_{i_{p1},k}/b_{i_p,k} \\ 0 \\ \vdots \\ 0 \end{bmatrix} \in \mathbb{C}^n$$

$E_k = I_n - m_k e_{i_p}^T$
Transform S and X
$S \leftarrow E_k S E_k^{-1}$, $X \leftarrow X E_k^{-1}$
end

Pass 2. Annihilate the elements below the coupling elements
for $k = 1, 2, \ldots, w_j$
$i_p = \nu_j + k - w_j$
$i_{p1} = i_p + 1$
Compute an elementary lower triangular matrix that annihilates the entries in rows from i_{p1} to ν_j in the kth column of B

$$m_k = \begin{bmatrix} 0 \\ b_{i_{p1},k}/b_{i_p,k} \\ \vdots \\ b_{\nu_j,k}/b_{i_p,k} \\ 0 \end{bmatrix} \in \mathbb{C}^n$$

$E_k = I_n - m_k e_{i_p}^T$
Transform S and X
$S \leftarrow E_k S E_k^{-1}$, $X \leftarrow X E_k^{-1}$
Make the element $b_{i_p,k}$ real
$\widehat{b}_{i_p} = \bar{b}_{i_p,k}/|b_{i_p,k}|$
$D = I_n$, $d_{i_p,i_p} = \widehat{b}_{i_p}$
$S \leftarrow DSD^{-1}$
$X \leftarrow XD^{-1}$
$nxt(i_p) = k$
$supd(i_p) = b_{i_p,k}$
end
end
Determine the Weyr form
Set $W = 0_{n\times n}$
for $i = 1, 2, \ldots, n$
if $nxt(i) > 0$
$W(i, nxt(i)) = supd(i)$
end

of X from $w_1 + \cdots + w_{h-1} + 1$ to $w_1 + \cdots + w_h$). Note that the columns of X constitute nonunitary bases of the corresponding null spaces. From these vectors, one can construct the transformation which reduces the staircase form $W(\mathcal{W}, \lambda)$ into the Jordan form J. These transformations should be done for all blocks with numbers from 1 to p corresponding to distinct eigenvalues.

The transformations considered are summarized in Algorithm 7.4. This algorithm implements transformations only with elementary triangular and permutation matrices. As mentioned previously, the algorithm is conditionally stable due to the use of nonunitary transformations.

To determine the Weyr form of the initial matrix, Algorithm 7.4 should be applied separately to each of the staircase forms corresponding to a block in the block-diagonal form of the matrix. The determination of the coupling elements of all p blocks requires about

$$\frac{5}{3}\frac{n^3}{p^2} \text{ flops.}$$

Example 7.4. Given is the matrix

$$A = \begin{bmatrix} -18 & 8 & -29 & 50 & -96 \\ 7 & -5 & 13 & -22 & 42 \\ -11 & 4 & -18 & 30 & -58 \\ 4 & -2 & 7 & -13 & 23 \\ 9 & -4 & 15 & -26 & 49 \end{bmatrix},$$

which has a single eigenvalue $\lambda = -1$, participating in three Jordan blocks of sizes $2, 2$ and 1. By the aid of an unitary similarity transformation with a matrix (with accuracy up to 5 decimal digits)

$$X_S = \begin{bmatrix} 0.72098 & -0.20297 & -0.19745 & 0.61414 & -0.15110 \\ -0.29693 & 0.35092 & 0.58087 & 0.66824 & 0.06880 \\ 0.46643 & -0.05885 & 0.75073 & -0.38822 & -0.25430 \\ -0.16965 & -0.84044 & 0.24445 & 0.10820 & 0.43980 \\ -0.38168 & -0.35476 & -0.01610 & 0.11771 & -0.84519 \end{bmatrix},$$

the matrix $A - \lambda I_n$ is reduced to the staircase form

$$S = X_S^H(A - \lambda I_n)X_S = \left[\begin{array}{ccc|cc} 0 & 0 & 0 & -0.05165 & 155.84411 \\ 0 & 0 & 0 & -0.17203 & -0.69730 \\ 0 & 0 & 0 & -1.39865 & -5.66923 \\ \hline 0 & 0 & 0 & 0 & 0 \\ 0 & 0 & 0 & 0 & 0 \end{array}\right]$$

with Weyr characteristic $\mathcal{W} = \{3, 2\}$. This implies that the coupling elements are in positions $(3, 5)$ and $(2, 4)$.

At the beginning of the first pass, we exchange the first and third rows of the matrix S so that the maximum element in the fifth column to go into the 3rd row. This is done by using the permutation matrix (note that in the given case this matrix is symmetric)

$$P_1 = \begin{bmatrix} 0 & 0 & 1 & 0 & 0 \\ 0 & 1 & 0 & 0 & 0 \\ 1 & 0 & 0 & 0 & 0 \\ 0 & 0 & 0 & 1 & 0 \\ 0 & 0 & 0 & 0 & 1 \end{bmatrix}.$$

As a result of the nonsingular transformation with this matrix, we obtain

$$P_1 S P_1^T = \left[\begin{array}{ccc|cc} 0 & 0 & 0 & -1.39865 & -5.66923 \\ 0 & 0 & 0 & -0.17203 & -0.69730 \\ 0 & 0 & 0 & -0.05165 & 155.84411 \\ \hline 0 & 0 & 0 & 0 & 0 \\ 0 & 0 & 0 & 0 & 0 \end{array}\right].$$

The annihilation of the elements of $P_1 S P_1^T$ in the 5th column that are located above the coupling element in the position $(3, 5)$ is done by transformation with upper triangular matrix

$$\begin{aligned} E_5 &= I_5 - m_5 e_3^T \\ &= \begin{bmatrix} 1 & 0 & -m_{15} & 0 & 0 \\ 0 & 1 & -m_{25} & 0 & 0 \\ 0 & 0 & 1 & 0 & 0 \\ 0 & 0 & 0 & 1 & 0 \\ 0 & 0 & 0 & 0 & 1 \end{bmatrix} = \begin{bmatrix} 1 & 0 & 0.03638 & 0 & 0 \\ 0 & 1 & 0.00447 & 0 & 0 \\ 0 & 0 & 1 & 0 & 0 \\ 0 & 0 & 0 & 1 & 0 \\ 0 & 0 & 0 & 0 & 1 \end{bmatrix}, \end{aligned}$$

$$m_{15} = b_{15}/b_{35} = -5.66923/155.84411 = -0.03638,$$

$$m_{25} = b_{25}/b_{35} = -0.69730/155.84411 = -0.00447.$$

As a result, we obtain

$$S_1 = E_5 P_1 S P_1^T E_5^{-1} = \left[\begin{array}{ccc|cc} 0 & 0 & 0 & -1.40053 & 0 \\ 0 & 0 & 0 & -0.17226 & 0 \\ 0 & 0 & 0 & -0.05165 & 155.84411 \\ \hline 0 & 0 & 0 & 0 & 0 \\ 0 & 0 & 0 & 0 & 0 \end{array}\right].$$

In the second step of the first pass, we exchange the first and second row of the matrix S_1 so that the maximum element in the fourth column goes into the third row. This is done by using the permutation matrix

$$P_2 = \begin{bmatrix} 0 & 1 & 0 & 0 & 0 \\ 1 & 0 & 0 & 0 & 0 \\ 0 & 0 & 1 & 0 & 0 \\ 0 & 0 & 0 & 1 & 0 \\ 0 & 0 & 0 & 0 & 1 \end{bmatrix}$$

which yields

$$S_2 = P_2 E_5 P_1 S P_1^T E_5^{-1} P_2^T = \left[\begin{array}{ccc|cc} 0 & 0 & 0 & -0.17226 & 0 \\ 0 & 0 & 0 & -1.40053 & 0 \\ 0 & 0 & 0 & -0.05165 & 155.84411 \\ \hline 0 & 0 & 0 & 0 & 0 \\ 0 & 0 & 0 & 0 & 0 \end{array}\right].$$

The annihilation of the $(1,4)$-element of the matrix S_2, which is located above the coupling element in position $(2,4)$, is done by a transformation with the upper triangular matrix

$$E_4 = I_5 - m_4 e_2^T$$

$$= \begin{bmatrix} 1 & -m_{14} & 0 & 0 & 0 \\ 0 & 1 & 0 & 0 & 0 \\ 0 & 0 & 1 & 0 & 0 \\ 0 & 0 & 0 & 1 & 0 \\ 0 & 0 & 0 & 0 & 1 \end{bmatrix} = \begin{bmatrix} 1 & -0.12300 & 0 & 0 & 0 \\ 0 & 1 & 0 & 0 & 0 \\ 0 & 0 & 1 & 0 & 0 \\ 0 & 0 & 0 & 1 & 0 \\ 0 & 0 & 0 & 0 & 1 \end{bmatrix},$$

$$m_{14} = b_{14}^{(1)} / b_{24}^{(1)} = -0.17226/(-1.40053) = 0.12300.$$

This transformation produces

$$S_3 = E_4P_2E_5P_1SP_1^TE_5^{-1}P_2^TE_4^{-1} = \left[\begin{array}{ccc|cc} 0 & 0 & 0 & 0 & 0 \\ 0 & 0 & 0 & -1.40053 & 0 \\ 0 & 0 & 0 & -0.05165 & 155.84411 \\ \hline 0 & 0 & 0 & 0 & 0 \\ 0 & 0 & 0 & 0 & 0 \end{array}\right].$$

With this, the transformations in the first pass finish.

In the second pass, we perform a single transformation with the lower triangular matrix

$$\begin{aligned} E_3 &= I_5 - m_2e_3^T \\ &= \begin{bmatrix} 1 & 0 & 0 & 0 & 0 \\ 0 & 1 & 0 & 0 & 0 \\ 0 & -m_{23} & 1 & 0 & 0 \\ 0 & 0 & 0 & 1 & 0 \\ 0 & 0 & 0 & 0 & 1 \end{bmatrix} = \begin{bmatrix} 1 & 0 & 0 & 0 & 0 \\ 0 & 1 & 0 & 0 & 0 \\ 0 & -0.03688 & 1 & 0 & 0 \\ 0 & 0 & 0 & 1 & 0 \\ 0 & 0 & 0 & 0 & 1 \end{bmatrix}, \\ m_{23} &= \frac{b_{34}}{b_{24}^{(2)}} = -0.05165/(-1.40053) = 0.03688, \end{aligned}$$

which annihilates the element in position $(3, 4)$. Besides this, we perform also a transformation with the diagonal matrix

$$D_2 = \text{diag}(1,\ -1,\ 1,\ 1,\ 1),$$

which makes the coupling element in position $(2, 4)$ positive. In the given case, D_3 is equal to the unit matrix, since the element b_{35} is positive. In this way, we finally obtain

$$W(\mathcal{W}, 0) = X^{-1}SX = \left[\begin{array}{ccc|cc} 0 & 0 & 0 & 0 & 0 \\ 0 & 0 & 0 & 1.40053 & 0 \\ 0 & 0 & 0 & 0 & 155.84411 \\ \hline 0 & 0 & 0 & 0 & 0 \\ 0 & 0 & 0 & 0 & 0 \end{array}\right],$$

where

$$X = P_1^T E_5^{-1} P_2^T E_4^{-1} E_3^{-1} D_2^{-1}$$
$$= \begin{bmatrix} 0 & -0.03688 & 1 & 0 & 0 \\ 1 & -0.12283 & -0.00447 & 0 & 0 \\ 0 & -0.99866 & -0.03638 & 0 & 0 \\ 0 & 0 & 0 & 1 & 0 \\ 0 & 0 & 0 & 0 & 1 \end{bmatrix}.$$

The matrix X has condition number $\mathrm{cond}_2(X) = 1.13$.

Taking into account the initial orthogonal transformation of A to staircase form, we obtain the Weyr form

$$W(\mathcal{W}, \lambda) = X_W^{-1} A X_W = \left[\begin{array}{ccc|cc} -1 & 0 & 0 & 0 & 0 \\ 0 & -1 & 0 & 1.40053 & 0 \\ 0 & 0 & -1 & 0 & 155.84411 \\ \hline 0 & 0 & 0 & -1 & 0 \\ 0 & 0 & 0 & 0 & -1 \end{array}\right],$$

where the similarity transformation matrix

$$X_W = \begin{bmatrix} -0.20297 & 0.19553 & 0.72907 & 0.61414 & -0.15110 \\ 0.35092 & -0.61224 & -0.31963 & 0.66824 & 0.06880 \\ -0.05885 & -0.75969 & 0.43938 & -0.38822 & -0.25430 \\ -0.84044 & -0.13463 & -0.17478 & 0.10820 & 0.43980 \\ -0.35476 & 0.07373 & -0.37951 & 0.11771 & -0.84519 \end{bmatrix}$$

has the same condition number as X.

7.5 Reduction into Numerical Jordan Form

7.5.1 *Transforming the Weyr Form into Jordan Form*

The construction of the Jordan form can be done based on the information obtained in the determination of the coupling elements. As it was shown in the previous section for a fifth-order example, for this aim it is necessary to reorder appropriately the equations obtained from the matrix equation

$$S(\mathcal{W}, \lambda)X = XW(\mathcal{W}, \lambda), \tag{7.15}$$

where

$$X = [x_1, \ x_2, \ x_4, \ x_3, \ x_5].$$

The reordering can be done exploiting the structure of the matrix $W(\mathcal{W}, \lambda)$. In fact, let us consider again the fifth-order matrix with associated Weyr characteristic $\mathcal{W} = \{3, \ 2\}$. We have that

$$W(\mathcal{W}, \lambda) = \left[\begin{array}{ccc|cc} \lambda & 0 & 0 & 0 & 0 \\ 0 & \lambda & 0 & r_{12} & 0 \\ 0 & 0 & \lambda & 0 & r_{22} \\ \hline 0 & 0 & 0 & \lambda & 0 \\ 0 & 0 & 0 & 0 & \lambda \end{array}\right].$$

It follows from the first equation of (7.15) for the column x_1 that this equation corresponds to a scalar Jordan block since the first row of the matrix contains only zeros in the columns from 2 to 5, i.e., the vector x_1 does not participate in a chain with a length larger than 1. As we see from the second row of the matrix $W(\mathcal{W}, \lambda)$, the element in the fourth column (r_{12}) is different from zero, i.e., the column x_2 participates in a chain together with the column x_4. Actually, we have that

$$S(\mathcal{W}, \lambda)x_4 = x_4\lambda + r_{12}x_2.$$

Going to the fourth row, we see that the column x_4 does not participate in another equation for a principal eigenvector, since all elements in this row, except the fourth one, are equal to zero. This implies that the columns x_2 and x_4 participate in a chain with length 2 i.e., they correspond to a Jordan block of second order. In a similar way, going to the third row, we observe that the element in the fifth column (r_{13}) is nonzero, i.e., the column x_3 participates together with the vector x_5 in a chain of a length 2. In this way, we obtain three Jordan blocks with sizes $s_1 = 1, \ s_2 = 2$ and $s_3 = 2$.

The construction of chains of eigenvectors and principal vectors is convenient to be done using the information from the vector nxt, obtained in the execution of Algorithm 7.4 (the function `weyr`). For a given eigenvector with a subscript $k_1 = j$, the element $k = nxt_{k_1}$ shows the subscript of the principal vector x_k, which participates in a chain with x_{k_1} (if $k > 0$). The next principal vector participating in this chain has a subscript nxt_{k_1} for $k_1 = k$ and so on. The chain is broken at the moment when $k = nxt_{k_1} = 0$. In this way, one also determines the length of the chain, i.e., the size of the corresponding Jordan block.

To find the Jordan form, it is necessary to reorder appropriately the equations obtained from the matrix equation

$$S(\mathcal{W}_i, \lambda_i)X_i = X_i W(\mathcal{W}_i, \lambda_i), \tag{7.16}$$

where $S(\mathcal{W}_i, \lambda_i)$ is a staircase block and $W(\mathcal{W}_i, \lambda_i)$ is the un-normalized Weyr block (7.12) corresponding to λ_i, $i = 1, 2, \ldots, p$. It can be shown that the form of the superdiagonal blocks $G_{w_{ij}, w_{i,j+1}}$ of $W(\mathcal{W}_i, \lambda_i)$ determines a specific ordering of the Jordan blocks, namely, these blocks appear in nondecreasing size. As shown in (7.13), the blocks are determined as

$$\begin{bmatrix} 0_{(w_{ij} - w_{i,j+1}) \times w_{i,j+1}} \\ R_{w_{i,j+1}} \end{bmatrix},$$

where $R_{w_{i,j+1}}$ is a diagonal $w_{i,j+1} \times w_{i,j+1}$ matrix. Further on to simplify the notation, we will assume without loss of generality that $R_{w_{i,j+1}} = I_{w_{i,j+1}}$. According to the structure of the Weyr block (Figure 7.5), each block of the transformation matrix into Weyr form,

$$X = [X_1, X_2, \ldots, X_p], \quad X_i \in \mathbb{C}^{n \times \nu_i},$$

can be represented as

$$X_i = [X_i^{(1)}, X_i^{(2)}, \ldots, X_i^{(h_i)}],$$

$$X_i^{(j)} \in \mathbb{C}^{n \times w_{ij}}, \quad j = 1, 2, \ldots, h_i.$$

The block $X_i^{(1)}$ contains the eigenvectors, corresponding to the eigenvalue λ_i, the block $X_i^{(2)}$ contains the principal vectors of grade 2 and so on. From equation (7.16), it follows that

$$\begin{aligned}
S(\mathcal{W}_i, \lambda_i)X_1 &= \lambda X_1, \\
S(\mathcal{W}_i, \lambda_i)X_2 &= \lambda X_2 + X_1 \begin{bmatrix} 0_{(w_{i1}-w_{i2})\times w_{i2}} \\ I_{w_{i2}} \end{bmatrix}, \\
\vdots \quad & \quad \vdots \\
S(\mathcal{W}_i, \lambda_i)X_{h_i} &= \lambda X_{h_i} + X_{h_i-1} \begin{bmatrix} 0_{(w_{i,h_i-1}-w_{i,h_i})\times w_{i,h_i}} \\ I_{w_{i,h_i}} \end{bmatrix}.
\end{aligned} \tag{7.17}$$

Let us partition each of the matrices $X_1,\ X_2, \ldots, X_{h_i}$ as

$$\begin{aligned}
X_1 &= [X_1^{(1)}, X_1^{(2)}, \ldots, X_1^{(h_i)}], \\
X_2 &= [X_2^{(2)}, \ldots, X_2^{(h_i)}], \\
\vdots &= \vdots \\
X_{h_i} &= [X_{h_i}^{(h_i)}],
\end{aligned}$$

where $X_i^{(j)} \in \mathbb{C}^{n\times(w_{ij}-w_{i,j+1})},\ j = 1, 2, \ldots, h_i - 1,\ X_i^{h_i} \in \mathbb{C}^{n\times w_{i,h_i}}$.
The equations (7.17) can be rewritten as

$$\begin{aligned}
S(\mathcal{W}_i, \lambda_i)\left[X_1^{(1)}, X_1^{(2)}, \ldots, X_1^{(h_i)}\right] &= \lambda\left[X_1^{(1)}, X_1^{(2)}, \ldots, X_1^{(h_i)}\right], \\
S(\mathcal{W}_i, \lambda_i)\left[X_2^{(2)}, \ldots, X_2^{(h_i)}\right] &= \lambda\left[X_2^{(2)}, \ldots, X_2^{(h_i)}\right] \\
&\quad + \left[X_1^{(2)}, \ldots, X_1^{(h_i)}\right], \\
\vdots \quad & \quad \vdots \\
S(\mathcal{W}_i, \lambda_i)X_{h_i}^{(h_i)} &= \lambda X_{h_i}^{(h_i)} + X_{h_i-1}^{(h_i)}.
\end{aligned} \tag{7.18}$$

The relationships (7.18) can be represented as the following chain relations:

$$
\begin{aligned}
S(\mathcal{W}_i,\lambda_i)X_1^{(1)} = \lambda X_1^{(1)},\ S(\mathcal{W}_i,\lambda_i)X_1^{(2)} &= \lambda X_1^{(2)},\ldots,\\
S(\mathcal{W}_i,\lambda_i)X_1^{(h_i)} &= \lambda X_1^{(h_i)},\\
S(\mathcal{W}_i,\lambda_i)X_2^{(2)} &= \lambda X_2^{(2)} + X_1^{(2)},\ldots,\\
S(\mathcal{W}_i,\lambda_i)X_2^{(h_i)} &= \lambda X_2^{(h_i)} + X_1^{(h_i)},\\
&\ \ \vdots\\
S(\mathcal{W}_i,\lambda_i)X_{h_i}^{(h_i)} &= \lambda X_{h_i}^{(h_i)} + X_{h_i-1}^{(h_i)}.
\end{aligned}
\tag{7.19}
$$

It is seen from (7.19) that the scalar Jordan blocks (i.e., the Jordan blocks of smallest size) associated with λ_i correspond to the eigenvectors contained in the first block $X_1^{(1)}$ of the matrix $X_i^{(1)}$. In turn, the longest chains and consequently the largest Jordan blocks correspond to the last w_{i,h_i} columns of the matrix $X_i^{(1)}$. These Jordan blocks have sizes equal to h_i. Thus, to have a correspondence between the sizes of the Jordan blocks and the elements s_{ji} of the Segre characteristics which appear in decreasing order, it is necessary to take the eigenvectors contained in the block $X_i^{(1)}$ in reverse order, i.e., first is determined the Jordan block for the last column of X_1, then the block for the column before the last column of X_1 and so on. In this way, the largest Jordan block will appear first and the smallest blocks will be the last.

These considerations are summarized in Algorithm 7.5 which determines the Jordan blocks and the corresponding chains of eigenvectors and principal vectors for each block with number i, $i = 1, 2, \ldots, p$, corresponding to distinct eigenvalue λ_i with multiplicity ν_i. To execute the algorithm, it is necessary to know the Weyr characteristic $\mathcal{W}_i = \{w_{i1}, w_{i2}, \ldots, w_{i,h_i}\}$, associated with λ_i. This characteristic is determined by Algorithm 6.3 (the function `staircase`). The first element w_{i1} of the Weyr characteristic is equal to the number of the eigenvectors corresponding to λ_i, i.e., to the number of the Jordan blocks that contain this eigenvalue.

Algorithm 7.5. Reduction into Numerical Jordan form
Obtains the Jordan form of a matrix from the Weyr form
function $[J, Z, \mathcal{S}] = \text{jord}(X_W, \mathcal{W}, \nu, nxt, supd)$
Input: The matrix $X_W \in \mathbb{C}^{n\times n}$ of the transformation of the initial matrix A into Weyr form; the Weyr characteristics $\mathcal{W} = [\mathcal{W}_1, \mathcal{W}_2, \dots \mathcal{W}_p]$, $\mathcal{W}_i = \{w_{i1}, w_{i2}, \dots, w_{i,h_i}\}$ associated with the distinct eigenvalues; vector $\nu = [\nu_1, \dots, \nu_p]$ containing the multiplicities of the distinct eigenvalues $\lambda_1, \dots, \lambda_p$; the vector nxt containing the column indices of the vector *supd*; the vector *supd* containing the super diagonal coupling elements of the Jordan form.
Output: The un-normalized Jordan form J; the nonsingular matrix $Z \in \mathbb{C}^{n\times n}$ which reduces the matrix A into Jordan form; The Segre characteristics $\mathcal{S} = [\mathcal{S}_1,\ \mathcal{S}_2, \dots,\ \mathcal{S}_p]$, $\mathcal{S}_i = \{s_{1i},\ s_{2i}, \dots,\ s_{q_i,i}\}$ associated with the distinct eigenvalues.

```
Set J = 0_{n×n}, Z = 0_{n×n}
ℓ = 0
for i = 1, 2, ..., p
    τ_i = ν_1 + ν_2 + ··· + ν_{i−1}
    Take the eigenvectors in reverse order
    for j = τ_i + w_{i1}, τ_i + w_{i1} − 1, ..., τ_i + 1
        ℓ = ℓ + 1, J_{ℓ,ℓ} = λ_i
        Eigenvector (A − λ_i I_n) x_ℓ = 0
        Z_{1:n,ℓ} = X_{W 1:n,j}
        Scalar Jordan block njb = 1
        k_1 = j
        while nxt(k_1) > 0,
            k = nxt(k_1)
            Jordan block of size larger than 1, njb = njb + 1
            ℓ = ℓ + 1, J_{ℓ,ℓ} = λ_i
            J_{ℓ−1,ℓ} = supd(k_1)
            Principal eigenvector (A − λ_i I_n) z_k = supd(k_1) x_{k_1}
            Z_{1:n,ℓ} = X_{W 1:n,k}
            k_1 = k
        end
        Determine an element of the Segre characteristic S_i
        s_{ji} = njb
    end
end
```

This step is free of rounding errors.

Example 7.5. Given is the matrix

$$A = \begin{bmatrix} -18 & 8 & -29 & 50 & -96 \\ 7 & -5 & 13 & -22 & 42 \\ -11 & 4 & -18 & 30 & -58 \\ 4 & -2 & 7 & -13 & 23 \\ 9 & -4 & 15 & -26 & 49 \end{bmatrix}$$

of Example 7.4 with a 5-tuple eigenvalue $\lambda = -1$ and a Weyr characteristic $\mathcal{W} = \{3, 2\}$. By using a nonsingular transformation with the matrix

$$X_W = \begin{bmatrix} -0.20297 & 0.19553 & 0.72907 & 0.61414 & -0.15110 \\ 0.35092 & -0.61224 & -0.31963 & 0.66824 & 0.06880 \\ -0.05885 & -0.75969 & 0.43938 & -0.38822 & -0.25430 \\ -0.84044 & -0.13463 & -0.17478 & 0.10820 & 0.43980 \\ -0.35476 & 0.07373 & -0.37951 & 0.11771 & -0.84519 \end{bmatrix},$$

the matrix A is reduced into the Weyr form

$$W(\mathcal{W}, \lambda) = X_S^{-1} A X_S = \left[\begin{array}{ccc|cc} -1 & 0 & 0 & 0 & 0 \\ 0 & -1 & 0 & 1.40053 & 0 \\ 0 & 0 & -1 & 0 & 155.84411 \\ \hline 0 & 0 & 0 & -1 & 0 \\ 0 & 0 & 0 & 0 & -1 \end{array}\right].$$

The location of the superdiagonal elements is given by the vector

$$nxt = [0,\ 4,\ 5,\ 0,\ 0]$$

and the values of these elements are stored in the vector

$$supd = [0,\ 1.40053,\ 155.84411,\ 0,\ 0].$$

Denote the columns of X_W as x_1, x_2, x_3, x_4 and x_5. From the Weyr characteristics of λ, it follows that the first three columns are eigenvectors of A and the next two are principal vectors of grade 2.

The determination of the Jordan form starts with obtaining the Jordan block corresponding to the eigenvector x_3. For $k_1 = j = 3$, we have $k = nxt_3 = 5$, i.e., there is a chain in which participate the eigenvector x_3 and the principal vector x_5. This chain has a length equal to 2, since $nxt_5 = 0$. The superdiagonal element, participating in this chain is $J_{12} = supd_3 = 155.84411$. For $k_1 = j = 2$, we have that $k = nxt_2 = 4$, i.e., there is a chain in which participate x_2 and x_4. This chain also has a length equal to 2, since $nxt_4 = 0$. The superdiagonal element in this case is $J_{34} = supd_2 = 1.40053$. For $j = 1$, it is fulfilled that $nxt_1 = 0$, which implies that we have a scalar Jordan block with corresponding eigenvector x_1. In this way, permuting the columns of the matrix X_W so that

$$\begin{aligned} Z = X_W P_J &= [x_3,\ x_5,\ x_2,\ x_4,\ x_1] \\ &= \begin{bmatrix} 0.72907 & -0.15110 & 0.19553 & 0.61414 & -0.20297 \\ -0.31963 & 0.06880 & -0.61224 & 0.66824 & 0.35092 \\ 0.43938 & -0.25430 & -0.75969 & -0.38822 & -0.05885 \\ -0.17478 & 0.43980 & -0.13463 & 0.10820 & -0.84044 \\ -0.37951 & -0.84519 & 0.07373 & 0.11771 & -0.35476 \end{bmatrix} \end{aligned}$$

where

$$P_J = \begin{bmatrix} 0 & 0 & 0 & 0 & 1 \\ 0 & 0 & 1 & 0 & 0 \\ 1 & 0 & 0 & 0 & 0 \\ 0 & 0 & 0 & 1 & 0 \\ 0 & 1 & 0 & 0 & 0 \end{bmatrix},$$

we obtain the Jordan form

$$J = P_J^T W(\mathcal{W}, \lambda) P_J = Z^{-1} A Z = \left[\begin{array}{cc|cc|c} -1 & 155.84411 & 0 & 0 & 0 \\ 0 & -1 & 0 & 0 & 0 \\ \hline 0 & 0 & -1 & 1.40053 & 0 \\ 0 & 0 & 0 & -1 & 0 \\ \hline 0 & 0 & 0 & 0 & -1 \end{array}\right].$$

Thus, the Segre characteristic of $\lambda = -1$ is

$$\mathcal{S} = \{2,\ 2,\ 1\}.$$

7.5.2 *The Full Algorithm*

The full reduction of a complex $n \times n$ matrix A into numerical Jordan form is done by the following sequence of algorithms presented in Chapters 6 and 7 and constituting the algorithm of Kågström and Ruhe (the preliminary reduction into Schur form is done by the corresponding LAPACK or MATLAB® programs):

- Algorithm 6.1 — reordering of the Schur form,
- Algorithm 6.2 — clustering of multiple eigenvalues,
- Algorithm 7.1 — reduction into block-diagonal form,
- Algorithm 7.2 — orthonormalization of the vectors corresponding to multiple eigenvalues,
- Algorithm 7.3 — reduction into block-diagonal staircase form,
- Algorithm 7.4 — determining the coupling elements and reduction into Weyr form,
- Algorithm 7.5 — obtaining the Jordan form from the Weyr form.

The reduction is controlled by two parameters. The parameter *ein* is used in the clustering of the numerically multiple eigenvalues, while the parameter *tol* sets the tolerance in finding the Weyr characteristics and constructing the nilpotent matrices.

In case of large Jordan blocks, the volume of the computational work required by the algorithm for full reduction approaches $O(n^4)$ flops which restricts its use to small- and medium-sized problems. According to the user's desire, the algorithm can be used to perform only some of the eight steps described at the beginning of this chapter. This modular structure of the algorithm allows obtaining at the user desire the following intermediate forms:

- Schur form,
- ordered Schur form,
- block-diagonal form,
- block-diagonal staircase form or
- Weyr form.

Apart from this, the separate application of Algorithms 6.1–6.4, considered in Chapter 6, produces the Jordan–Schur form of A. In the reduction into the staircase, Weyr and Jordan forms, the algorithm gives information about the size of the neglected elements during the reduction to staircase form and determination of the Weyr characteristics. For instance, it is possible to find only the Jordan structure avoiding potentially unstable step 7 for finding the coupling elements.

7.5.3 *Numerical Examples*

In this section, we represent the numerical results of the reduction into Jordan form of nine matrices taken from Kågström (1981). These test examples allow obtaining impression about the numerical behavior of the algorithm considered, especially the role of the tolerances.

Matrix A: Well-conditioned eigenvalue problem with multiple and defect eigenvalues,

$$A = \begin{bmatrix} 1 & 1 & 1 & -2 & 1 & -1 & 2 & -2 & 4 & -3 \\ -1 & 2 & 3 & -4 & 2 & -2 & 4 & -4 & 8 & -6 \\ -1 & 0 & 5 & -5 & 3 & -3 & 6 & -6 & 12 & -9 \\ -1 & 0 & 3 & -4 & 4 & -4 & 8 & -8 & 16 & -12 \\ -1 & 0 & 3 & -6 & 5 & -4 & 10 & -10 & 20 & -15 \\ -1 & 0 & 3 & -6 & 2 & -2 & 12 & -12 & 24 & -18 \\ -1 & 0 & 3 & -6 & 2 & -5 & 15 & -13 & 28 & -21 \\ -1 & 0 & 3 & -6 & 2 & -5 & 12 & -11 & 32 & -24 \\ -1 & 0 & 3 & -6 & 2 & -5 & 12 & -14 & 37 & -26 \\ -1 & 0 & 3 & -6 & 2 & -5 & 12 & -14 & 36 & -25 \end{bmatrix}.$$

The matrix has one simple eigenvalue at 1.0, a 5-tuple eigenvalue at 2.0 with one chain of length 2 and one chain with length 3, and a 4-tuple eigenvalue at 3.0 with two chains with lengths equal to 2. The tolerances used are $ein = 0.5 \cdot 10^{-8}$ and $tol = 0.5 \cdot 10^{-8}$.

Matrix B: Ill-conditioned eigenvalue problem with groups of close but distinct eigenvalues,

$$B = \begin{bmatrix} 0 & I_4 \\ -B_2 & -B_1 \end{bmatrix},$$

$$B_1 = \begin{bmatrix} 3\alpha & -(1+\alpha^2+2\beta^2) & \alpha(1+2\beta^2) & -\beta^2(\alpha^2+\beta^2) \\ 2 & 0 & 0 & 0 \\ 0 & 2 & 0 & 0 \\ 0 & 0 & 2 & 0 \end{bmatrix},$$

$$B_2 = \begin{bmatrix} -1+2\alpha^2 & \alpha-\alpha(\alpha^2+2\beta^2) & 2\alpha^2\beta^2 & -\alpha\beta^2(\alpha^2+\beta^2) \\ 2\alpha & -(\alpha^2+2\beta^2) & 2\alpha\beta^2 & -\beta^2(\alpha^2+\beta^2) \\ 1 & 0 & 0 & 0 \\ 0 & 1 & 0 & 0 \end{bmatrix}$$

and $\beta = 1+\alpha$. The used values are $\alpha = 10^{-7}$ and $\alpha = 10^{-8}$. Depending on the tolerance ein, the algorithm performs different clustering of the close eigenvalues. For tolerances $ein = 1.0 \cdot 10^{-14}$ and $tol = 1.0 \cdot 10^{-10}$, one achieves relatively small norms of the spectral projectors, i.e., good separation of the eigenvalue groups at the price of a slight increase in the residual norms.

Matrices C and D: Ill-conditioned eigenvalue problem with a single multiple and defect eigenvalue. The matrices

$$C = (H_2 \Sigma H_1) \cdot J_1 \cdot (H_1 \Sigma^{-1} H_2)$$

and

$$D = (H_2 \Sigma H_1) \cdot J_2 \cdot (H_1 \Sigma^{-1} H_2),$$

where

$$J_1 = \left[\begin{array}{cccc|cccccc} 1 & 1 & 0 & 0 & & & & & & \\ 0 & 1 & 1 & 0 & & & 0 & & & \\ 0 & 0 & 1 & 1 & & & & & & \\ 0 & 0 & 0 & 1 & & & & & & \\ \hline & & & & 0.8 & & & & & \\ & & & & & 0.7 & & & 0 & \\ & & & & & & 0.6 & & & \\ & 0 & & & & & & 0.5 & & \\ & & & & & 0 & & & 0.4 & \\ & & & & & & & & & 0.3 \end{array}\right],$$

$$
J_2 = \left[\begin{array}{cccc|cccccc}
1 & 1 & 0 & 0 & & & & & & \\
0 & 1 & 1 & 0 & & & & 0 & & \\
0 & 0 & 1 & 0 & & & & & & \\
0 & 0 & 0 & 1 & & & & & & \\
\hline
 & & & & 0.8 & & & & & \\
 & & & & & 0.7 & & & 0 & \\
 & & & & & & 0.6 & & & \\
 & & 0 & & & & & 0.5 & & \\
 & & & & & 0 & & & 0.4 & \\
 & & & & & & & & & 0.3
\end{array}\right]
$$

are constructed as shown in Example 4.4 with

$$
\begin{aligned}
H_1 &= I - \frac{2}{n} ee^T, \quad e = [1,\ 1, \ldots,\ 1]^T, \quad n = 10, \\
H_2 &= I - \frac{2}{n} ff^T, \quad f = [1,\ -1, \ldots,\ (-1)^{n-1}]^T, \\
\Sigma &= \mathrm{diag}(1, \sigma, \sigma^2, \ldots, \sigma^{n-1}).
\end{aligned}
$$

The matrix $H_2 \Sigma H_1$ can be made arbitrary ill conditioned by increasing the parameter $\sigma \geq 1$, since $\mathrm{cond}(H_2 \Sigma H_1) = \sigma^{n-1}$. For both matrices C and D, the values $\sigma = 1, 2$ and 3 are used. The difference between the two matrices is the defect of the eigenvalue λ with multiplicity equal to 4. The matrix C has only one eigenvector, corresponding to $\lambda = 1$, so that to the quadruple eigenvalue corresponds only one Jordan block. The matrix D has two eigenvectors, corresponding to $\lambda = 1$, so it is a derogatory matrix. The tolerances used are shown in Table 7.1

In Table 7.2, we show the results of the reduction of the nine matrices in block-diagonal and Jordan form. In the second and third columns of the table, we show the spectral condition numbers $\mathrm{cond}(M) = \|M\|_2 \|M^{-1}\|_2$ and $\mathrm{cond}(Z) = \|Z\|_2 \|Z^{-1}\|_2$ of the transformation matrices into block-diagonal and Jordan form. In the fourth column, we give the size $\|AZ - ZJ\|_F$ of the residual.

Table 7.1. Tolerances used in computing the Jordan forms of the matrices C and D.

Matrix	σ	*ein*	*tol*
C	1	$1.0 \cdot 10^{-10}$	$1.0 \cdot 10^{-10}$
	2	$1.0 \cdot 10^{-15}$	$1.0 \cdot 10^{-15}$
	3	$1.0 \cdot 10^{-16}$	$1.0 \cdot 10^{-16}$
D	1	$1.0 \cdot 10^{-9}$	$1.0 \cdot 10^{-9}$
	2	$1.0 \cdot 10^{-9}$	$1.0 \cdot 10^{-9}$
	3	$1.0 \cdot 10^{-15}$	$1.0 \cdot 10^{-15}$

In the sixth, seventh and eighth columns, we show the multiplicities of the eigenvalues, the norms of the projectors P_k which separate the kth group of eigenvalues from the rest and the mean value of the eigenvalues of the kth diagonal block, respectively. For the matrices C and D, information is given only for the most interesting eigenvalues 1.0 and 0.8.

It is seen from the results that the algorithm determines reliably the Jordan structure of the matrices as far as this is allowed by the conditioning of the respective problem.

7.6 Numerical Issues

7.6.1 *Operation Count*

As already mentioned in Section 6.7.1, the reduction of the diagonal blocks to staircase form needs a large volume of operations if these blocks have high indices of nilpotence, i.e., if their Jordan forms involve high-order blocks.

As in Chapter 6, in the approximate counting of the number of computational operations, we will assume that the eigenvalue multiplicities are equal ($\nu = n/p$). The first three steps of the full reduction algorithm (reduction to Schur form, grouping of close eigenvalues and clustering the multiple eigenvalues) require the same numbers of operations as in the reduction into Jordan–Schur form (Table 6.3).

Table 7.2. Numerical results for the nine test matrices.

Matrix	cond(M)	cond(Z)	$\|AZ - ZJ\|_F / \|A\|_F$	k	n_k	$\|P_k\|_2$	λ_0
A	$7.42 \cdot 10^1$	$5.83 \cdot 10^2$	$1.23 \cdot 10^{-15}$	1	1	$2.77 \cdot 10^1$	1.0
				2	5	$2.06 \cdot 10^1$	2.0
				3	4	$3.44 \cdot 10^1$	3.0
B, $\alpha = 10^{-7}$	$1.87 \cdot 10^1$	$1.87 \cdot 10^1$	$2.07 \cdot 10^{-7}$	1	3	$6.26 \cdot 10^0$	$\lambda_1^{(1)}$
				2	3	$6.26 \cdot 10^0$	$\lambda_2^{(1)}$
				3	2	$9.17 \cdot 10^0$	$\lambda_3^{(1)}$
B, $\alpha = 10^{-8}$	$1.87 \cdot 10^1$	$1.87 \cdot 10^1$	$5.66 \cdot 10^{-9}$	1	3	$6.26 \cdot 10^0$	$\lambda_1^{(2)}$
				2	3	$6.26 \cdot 10^0$	$\lambda_2^{(2)}$
				3	2	$9.17 \cdot 10^0$	$\lambda_3^{(2)}$
C, $\sigma = 1$	1.0	1.0	$1.90 \cdot 10^{-15}$	1	4	1.0	1.0
				2	1	1.0	0.8
				7	1	1.0	0.3
C, $\sigma = 2$	$2.82 \cdot 10^2$	$7.53 \cdot 10^3$	$2.33 \cdot 10^{-14}$	1	4	$2.82 \cdot 10^1$	1.0
				2	1	$2.81 \cdot 10^1$	0.8
C, $\sigma = 3$	$1.0 \cdot 10^4$	$6.37 \cdot 10^6$	$3.86 \cdot 10^{-13}$	1	4	$8.87 \cdot 10^2$	1.0
				2	1	$8.87 \cdot 10^2$	0.8
D, $\sigma = 1$	1.0	1.30	$1.61 \cdot 10^{-15}$	1	4	1.0	1.0
				2	1	1.0	0.8
D, $\sigma = 2$	$2.82 \cdot 10^2$	$5.45 \cdot 10^3$	$3.23 \cdot 10^{-15}$	1	4	$2.82 \cdot 10^1$	1.0
				2	1	$2.81 \cdot 10^1$	0.8
D, $\sigma = 3$	$1.0 \cdot 10^4$	$5.11 \cdot 10^6$	$2.44 \cdot 10^{-13}$	1	4	$8.87 \cdot 10^2$	1.0
				2	1	$8.87 \cdot 10^2$	0.8

$\lambda_1^{(1)} = -0.000000033333333 + j_o 1.000000066666667$
$\lambda_2^{(1)} = -0.000000033333333 - j_o 1.000000066666667$
$\lambda_3^{(1)} = -0.000000050000000$
$\lambda_1^{(2)} = -0.000000003333333 + j_o 1.000000006666666$
$\lambda_2^{(2)} = -0.000000003333333 - j_o 1.000000006666666$
$\lambda_3^{(2)} = -0.000000005000000$

The total volume of work is given in Table 7.3 and depends on the Jordan structure of the matrix. For $p = 1,\ h = n$ (nonderogatory but defective matrix with a single Jordan block), this maximum volume of work is of order

$$4\frac{2}{3}n^4 + 46\frac{1}{3}n^3 \text{ flops}$$

Table 7.3. Operation count for the reduction into Weyr and Jordan forms.

Step	Operation	Average	Maximum	Maximum for
1	Reduction into Schur form	$12kn^3 + \frac{20}{3}n^3$	$30\frac{2}{3}n^3$	2 iterations per eigenvalue
2	Grouping of close eigenvalues	$O(n^2)$	$6n^3$	Totally unordered Schur form
3	Clustering of multiple eigenvalues	$2n^3$	$6n^3$	No clustering possible
4	Block-diagonalization	$\frac{(p-1)n^3}{p^2}$	—	$p = 1,\ h = n$
		$+2\frac{(p-1)n^3}{p}$	$2n^3$	$p = n,\ h = 1$
5	Orthonormalization	$2\frac{n^3}{p}$	$2n^3$	$p = 1,\ h = n$
			$2n^2$	$p = n,\ h = 1$
6	Reduction into staircase form	$2(\frac{1}{3}\frac{n^3h}{p} + \frac{2n^3h}{p^2})$	$4\frac{2}{3}n^4$	$p = 1,\ h = n$
			$\frac{2}{3}n^3$	$p = n,\ h = 1$
7	Reduction into Weyr form	$\frac{5}{3}\frac{n^3}{p^2}$	$\frac{5}{3}n^3$	$p = 1,\ h = n$
			$\frac{5}{3}n$	$p = n,\ h = 1$
8	Obtaining the Jordan form	$O(n)$		
	Total	$34\frac{2}{3}n^3 + \frac{(p-1)n^3}{p^2}$	$4\frac{2}{3}n^4 + 46\frac{1}{3}n^3$	$p = 1,\ h = n$
		$+\frac{2}{3}\frac{n^3h}{p} + \frac{4n^3h}{p^2} + \frac{5}{3}\frac{n^3}{p^2}$	$47\frac{1}{3}n^3$	$p = n,\ h = 1$

and for $p = n,\ h = 1$ (derogatory but nondefective matrix with n scalar Jordan blocks), it is approximately

$$47\frac{1}{3}n^3 \text{ flops.}$$

Interestingly, the maximum volume of work for reduction into Jordan and Weyr forms is related to the matrix residing in a most generic bundle (single Jordan block with $\text{codim}(\mathcal{B}) = n - 1$). As in the case of reducing a matrix to the Jordan–Schur form, for matrices with large Jordan blocks the reduction into Weyr and Jordan forms requires a volume of work of order n^4. This number of operations does not include the accumulation of the similarity transformation matrix which requires additional $O(n^3)$ operations.

7.6.2 *Well-Conditioned Numerical Structures*

For λ, which is well conditioned in accordance with the definition (5.23) for a numerically multiple eigenvalue, the residual

$\|AZ - ZJ\|_F$ is of size $\mathbf{u}\|A\|_F\|Z\|_F$, and the condition number $\text{cond}(Z) = \|Z\|_2\|Z^{-1}\|_2$ is of size $O(\|Z\|_F)$. Note that with respect to the Jordan form of the initial matrix A, in the best case we can expect a transformation matrix Z, whose condition number is $O(\|M\|_2\|M^{-1}\|_2)$, where M is the transformation matrix from the Schur form T into block-diagonal form. That is why the number $\text{cond}_2(M) = \|M\|_2\|M^{-1}\|_2$ is an appropriate measure of the conditioning of the eigenstructure, inherent to A.

Example 7.6. Consider the defective and derogatory eighth-order matrix

$$A = \begin{bmatrix} 6 & 0 & 3 & -6 & -5 & -11 & 14 & 33 \\ -2 & -10 & 0 & -3 & 9 & 10 & -2 & 4 \\ 5 & 7 & 4 & -2 & -10 & -16 & 13 & 25 \\ 1 & -6 & 0 & 0 & 2 & 1 & 4 & 11 \\ -1 & -3 & -1 & 1 & 3 & 4 & -2 & -4 \\ 1 & -6 & 0 & -3 & 3 & 3 & 3 & 11 \\ 0 & 6 & -2 & 4 & -5 & -2 & -2 & -13 \\ -1 & -7 & 0 & -2 & 6 & 6 & -1 & 4 \end{bmatrix}$$

used in Example 6.6 to illustrate the reduction into Jordan–Schur form of a matrix with a well-conditioned numerical structure. This matrix has an eigenvalue λ equal to 1 with multiplicity eight, which participate in two Jordan blocks — one of size 6 and the other with size 2 so that the Segre characteristic associated with λ is $\mathcal{S}(A, \lambda) = \{6, 2\}$. The Weyr characteristic of this matrix is $\mathcal{W}(A, \lambda) = \{2, 2, 1, 1, 1, 1\}$ and the index of nilpotence is 6. The presence of Jordan block of size 6 makes the associate eigenvalue $\lambda = 1$ very sensitive to perturbations. On the other hand, as shown previously, the matrix has a very well-conditioned numerical structure. The eigenvector matrix is ill conditioned and its condition number is equal to $2.11 \cdot 10^{13}$. This is because the corresponding eigenvalue routine tries to find eight linearly independent eigenvectors while there are only two such vectors.

In the given case, the full reduction algorithm finds the following Jordan form (the superdiagonal elements are shown up to seven decimal digits)

$$J = \left[\begin{array}{cccccc|cc} 1 & 2.6412325 & 0 & 0 & 0 & 0 & 0 & 0 \\ 0 & 1 & 2.2508068 & 0 & 0 & 0 & 0 & 0 \\ 0 & 0 & 1 & 1.9193867 & 0 & 0 & 0 & 0 \\ 0 & 0 & 0 & 1 & 2.0779531 & 0 & 0 & 0 \\ 0 & 0 & 0 & 0 & 1 & 1.9490826 & 0 & 0 \\ 0 & 0 & 0 & 0 & 0 & 1 & 0 & 0 \\ \hline 0 & 0 & 0 & 0 & 0 & 0 & 1 & 5.5229416 \\ 0 & 0 & 0 & 0 & 0 & 0 & 0 & 1 \end{array}\right].$$

The sizes of the Jordan blocks are correctly determined, the diagonal elements (the eigenvalues) being accurate up to 15 digits. The transformation matrix into Jordan form is relatively well conditioned with a condition number equal to $7.62 \cdot 10^2$. The relative backward error in computing the Jordan form is

$$\|ZJZ^{-1} - A\|_F / \|A\|_F = 4.78 \cdot 10^{-15},$$

which is a very good result.

The un-normalized Weyr form of the matrix A is

$$W = \left[\begin{array}{cc|cc|c|c|c|c} 1 & 0 & 5.5229416 & 0 & 0 & 0 & 0 & 0 \\ 0 & 1 & 0 & 2.6412325 & 0 & 0 & 0 & 0 \\ \hline 0 & 0 & 1 & 0 & 0 & 0 & 0 & 0 \\ 0 & 0 & 0 & 1 & 2.2508068 & 0 & 0 & 0 \\ \hline 0 & 0 & 0 & 0 & 1 & 1.9193867 & 0 & 0 \\ \hline 0 & 0 & 0 & 0 & 0 & 1 & 2.0779531 & 0 \\ \hline 0 & 0 & 0 & 0 & 0 & 0 & 1 & 1.9490826 \\ \hline 0 & 0 & 0 & 0 & 0 & 0 & 0 & 1 \end{array}\right].$$

The following example illustrates the relationship between the coupling elements of the Weyr and Jordan forms and the singular values of the powers of the nilpotent matrix N.

Table 7.4. Parameters, characterizing the numerical structure of A.

k	n_k	w_k	$\sigma_{n_k+1}(N^k)$	$gap \cdot tol\|N\|_2^k$	$tol\|N\|_2^k$	$\sigma_{n_k}(N^k)$
1	1	1	$1.595 \cdot 10^{-1}$	$1.429 \cdot 10^{-8}$	$1.429 \cdot 10^{-11}$	$1.230 \cdot 10^{-18}$
2	2	1	$2.930 \cdot 10^{-1}$	$2.041 \cdot 10^{-7}$	$2.041 \cdot 10^{-10}$	$6.281 \cdot 10^{-16}$
3	3	1	$5.686 \cdot 10^{-1}$	$2.916 \cdot 10^{-6}$	$2.916 \cdot 10^{-9}$	$5.255 \cdot 10^{-16}$
4	4	1	$1.158 \cdot 10^{0}$	$4.165 \cdot 10^{-5}$	$4.165 \cdot 10^{-8}$	$1.658 \cdot 10^{-16}$
5	5	1	$1.166 \cdot 10^{1}$	$5.951 \cdot 10^{-4}$	$5.951 \cdot 10^{-7}$	$3.643 \cdot 10^{-16}$
k	n_k	w_k			$tol\|N\|_2^k$	$\sigma_{n_k}(N^k)$
6	6	1			$8.501 \cdot 10^{-6}$	0

Example 7.7. Given is the matrix

$$A = \begin{bmatrix} 2 & 1 & -2 & 2 & 4 & -4 \\ 2 & 3 & 3 & 0 & 0 & 4 \\ 1 & 0 & -3 & 1 & 4 & -6 \\ 2 & 2 & 2 & 1 & 1 & 2 \\ -3 & -2 & 0 & -2 & -3 & 1 \\ -2 & -1 & 3 & -2 & -5 & 6 \end{bmatrix},$$

which is similar to a unique Jordan block with eigenvalue $\lambda = 1$ and size equal to 6.

In Table 7.4, we give the parameters (up to 7 digits), characterizing the numerical Jordan structure determined according to Definition 5.23. These parameters are obtained by singular value decompositions of the matrices $N^k, N = A - \lambda I_8,\ k = 1, 2, \ldots, 6$. In computing the parameters, we used the tolerances $gap = 1000$ and $tol = 10^{-12}$. The parameters show that the numerical structure of the matrix is well conditioned.

The numerical Jordan form of A is computed by Algorithm 7.5 using tolerances $ein = tol = 10^{-12}$. The form is determined with relative backward error $\|E\|_F/\|A\|_F = 3.128 \cdot 10^{-15}$. In Table 7.5, we give the products of the coupling elements and the elements themselves along with the corresponding bounds (7.14). The lower bound $\sigma_{n_k+1}(N^k)/\sigma_1(N^{k-1})$ on the coupling elements is sufficiently large for all k so that $r_{1,k+1}$ are large and the Jordan chain is strongly coupled.

Table 7.5. Relationship between the coupling elements and the singular values of N^k.

k	n_k	$\vert r_{12}r_{13}\dots r_{1,k+1}\vert$	$\sigma_{n_k+1}(N^k)$	$\vert r_{1,k+1}\vert$	$\sigma_{n_k+1}(N^k)/\sigma_1(N^{k-1})$
1	1	2.065591	0.159546	2.065591	0.159546
2	2	1.800901	0.292997	0.871857	0.020509
3	3	4.442222	0.568572	2.466667	0.061237
4	4	2.656845	1.158455	0.598089	0.068517
5	5	11.66190	11.66190	4.389381	0.900653

7.6.3 *Ill-Conditioned Numerical Structures*

Example 7.8. Consider the performance of the algorithm in case of ill-conditioned Jordan structure. The matrix

$$A = \begin{bmatrix} 1153 & 4009 & 9132 & -14860 & -36532 & 109632 \\ -1308 & -4567 & -10409 & 16966 & 41686 & -125094 \\ 722 & 2522 & 5749 & -9372 & -23028 & 69102 \\ -617 & -2147 & -4891 & 7961 & 19570 & -58728 \\ -408 & -1416 & -3226 & 5248 & 12911 & -38739 \\ -244 & -848 & -1932 & 3144 & 7732 & -23201 \end{bmatrix},$$

used in Example 6.7, has a 6-tuple eigenvalue equal to 1 participating in two Jordan blocks of sizes 4 and 2. Hence, A lays in a bundle with codimension equal to 9.

Using tolerances $ein = 10^{-12}$ and $tol = 10^{-12}$, the full algorithm obtains accurately the Jordan form as

$$J = \left[\begin{array}{cccccc} 1 & 225.05317 & 0 & 0 & 0 & 0 \\ 0 & 1 & 11.08562 & 0 & 0 & 0 \\ 0 & 0 & 1 & 105.12667 & 0 & 0 \\ 0 & 0 & 0 & 1 & 0 & 0 \\ \hline 0 & 0 & 0 & 0 & 1 & 0.0438458 \\ 0 & 0 & 0 & 0 & 0 & 1 \end{array}\right].$$

The eigenvalues are computed exactly up to 12 digits and the condition number of the similarity transformation matrix Z is $3.16 \cdot 10^6$. The relative backward error in computing the Jordan form is

$$\|ZJZ^{-1} - A\|_F / \|A\|_F = 2.56 \cdot 10^{-14}.$$

Let us increase now the tolerance *tol* to 10^{-6}. As a result, we obtain

$$J = \left[\begin{array}{cc|c|c|c|c} 1 & 197961.68 & 0 & 0 & 0 & 0 \\ 0 & 1 & 0 & 0 & 0 & 0 \\ \hline 0 & 0 & 1 & 0 & 0 & 0 \\ \hline 0 & 0 & 0 & 1 & 0 & 0 \\ \hline 0 & 0 & 0 & 0 & 1 & 0 \\ \hline 0 & 0 & 0 & 0 & 0 & 1 \end{array}\right]$$

and the condition number of the similarity transformation is 1.37. The Jordan form is computed with relative backward error $7.18 \cdot 10^{-5}$. The Segre characteristic of A associated with $\lambda = 1$ is $\mathcal{S} = \{2, 1, 1, 1, 1\}$ so that A resides in a bundle with codimension equal to 25. In this case, the problem is regularized to a Jordan form with five blocks but with larger backward error. As a result, the Jordan form goes to a bundle with higher codimension.

Note that the Jordan form of the given matrix can be computed exactly by using the MATLAB® function `jordan` which implements symbolic computations. The application of this function, however, is restricted to matrices whose elements are integers or ratios of small integers and thus free of rounding errors.

Example 7.9. Consider the mutual disposition of the invariant subspaces of a matrix whose eigenstructure conditioning increases gradually depending on a parameter. Given is a family of 3×3 matrices $A = QJQ^{-1}$ constructed as shown in Example 4.4, where

$$J = \left[\begin{array}{cc|c} 1 & 1 & 0 \\ 0 & 1 & 0 \\ \hline 0 & 0 & 0.9 \end{array}\right]$$

is the exact Jordan form:

$$\begin{aligned} Q &= H_2 \Sigma H_1, \\ H_1 &= I_n - \frac{2}{3} e e^T, \quad e = [1,\ 1,\ 1]^T, \\ H_2 &= I_n - \frac{2}{3} f f^T, \quad f = [1,\ -1,\ 1]^T, \\ \Sigma &= \mathrm{diag}(1,\ \sigma,\ \sigma^2), \end{aligned}$$

and the parameter $\sigma > 0$ determines the conditioning of the similarity transformation matrix Q. With the increase in the parameter σ, the condition number of the transformation matrix Q increases which leads to the worsening of eigenstructure conditioning and increase in eigenvalue sensitivity.

In the given case, we are interested in the minimum angle $\Theta_{\min}(\mathcal{X}_1, \mathcal{X}_2)$ between the two-dimensional invariant subspace $\mathcal{X}_1$ corresponding to the eigenvalue $\lambda_1 = 1$ and the one-dimensional invariant subspace $\mathcal{X}_2$ corresponding to the eigenvalue $\lambda_2 = 0.9$. (The basis vectors of these subspaces are the corresponding columns of the matrix Z of the similarity transformation into numerical Jordan form.) If this angle is very small, this means that the basis vector of $\mathcal{X}_2$ is almost parallel to the plane spanned by the basis vectors of $\mathcal{X}_1$ and the condition number of the matrix Z is high. Hence, the value of $\Theta_{\min}(\mathcal{X}_1, \mathcal{X}_2)$ is an indication of the conditioning of the Jordan structure.

The Jordan form of A and the similarity transformation matrix Z are computed by the algorithms described in Section 7.5.2 for $ein = tol = 10^{-12}$ and the minimum angle $\Theta_{\min}(\mathcal{X}_1, \mathcal{X}_2)$ between the invariant subspaces $\mathcal{X}_1$ and $\mathcal{X}_2$ is determined by the expression (4.23). In Table 7.6,we give the values of $\Theta_{\min}(\mathcal{X}_1, \mathcal{X}_2)$, the condition number $\mathrm{cond}(Z)$ of the similarity transformation matrix Z into the Jordan form and the relative backward error

$$berr = \|ZJZ^{-1} - A\|_F / \|A\|_F$$

in computing the Jordan form, as functions of the parameter σ. With the increase in σ, the minimum angle between the invariant subspaces decreases which leads to increase in the conditioning of the transformation into Jordan form.

Table 7.6. The minimum angle between the invariant subspaces of the matrix A of Example 7.9.

σ	cond(Z)	$\Theta_{\min}(\mathcal{X}_1, \mathcal{X}_2)$ deg	*berr*
1.0	$1.00 \cdot 10^0$	90.00	$4.86 \cdot 10^{-16}$
2.0	$2.62 \cdot 10^0$	41.81	$1.91 \cdot 10^{-16}$
3.0	$4.96 \cdot 10^0$	22.80	$7.86 \cdot 10^{-16}$
4.0	$8.13 \cdot 10^0$	14.03	$6.02 \cdot 10^{-16}$
5.0	$1.22 \cdot 10^1$	9.40	$1.29 \cdot 10^{-15}$
6.0	$1.71 \cdot 10^1$	6.71	$8.56 \cdot 10^{-16}$
7.0	$2.29 \cdot 10^1$	5.01	$6.01 \cdot 10^{-16}$
8.0	$2.95 \cdot 10^1$	3.88	$7.35 \cdot 10^{-16}$

In Figure 7.9, we show the interposition of $\mathcal{X}_1$ and $\mathcal{X}_2$ for $\sigma = 1$, 3 and 8. For $\sigma = 8$, the angle between $\mathcal{X}_1$ and $\mathcal{X}_2$ is less than 4 degrees.

Example 7.10. Given is the family of 5×5 matrices $A = QJQ^{-1}$, described in Example 6.8, where

$$J = \left[\begin{array}{cccc|c} 1 & 1 & 0 & 0 & 0 \\ 0 & 1 & 1 & 0 & 0 \\ 0 & 0 & 1 & 1 & 0 \\ 0 & 0 & 0 & 1 & 0 \\ \hline 0 & 0 & 0 & 0 & 0.8 \end{array}\right]$$

is the exact Jordan form and the matrix Q is constructed as in the previous example. There are two invariant subspaces $\mathcal{X}_1$ and $\mathcal{X}_2$ with dimensions 4 and 1, corresponding to the eigenvalues 1 and 0.8, respectively.

In Table 7.7, we present the results of the reduction into Jordan form as functions of the parameter σ. The tolerances used in the reduction are same for all σ and are equal to $ein = tol = 10^{-12}$. With the increase in the condition number of the matrix Z, the conditioning of the Jordan structure gets worse, the accuracy of the computed eigenvalues λ_1 and λ_2 deteriorates and the angle between the subspaces $\Theta_{\min}(\mathcal{X}_1, \mathcal{X}_2)$ decreases. For $\sigma = 12$, instead of in the bundle associated with the Segre characteristics $\mathcal{S}(\lambda_1) = \{4\}$ and $\mathcal{S}(\lambda_2) = \{1\}$, the program finds a Jordan structure in the bundle

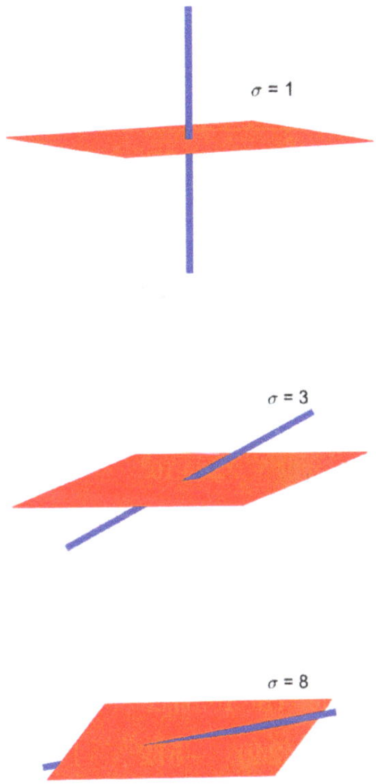

Fig. 7.9. Interposition of the invariant subspaces for $\sigma = 1, 3$ and 8.

with a single Jordan block with eigenvalue $\lambda = 0.959999999996148$ and Segre characteristic $\mathcal{S}(\lambda)$ equal to $\{5\}$. At this moment, the angle $\Theta_{\min}(\mathcal{X}_1, \mathcal{X}_2)$ becomes equal to 0. The error in the determination of the sizes of the Jordan blocks is due to a failure of the clustering algorithm as a result of the ill conditioning of the numerical structure.

7.6.4 *Stratification of Jordan Forms*

As demonstrated in Section 6.7.4, the tolerance parameters *ein* and *tol* can be used to find different Jordan structures in the vicinity of the given matrix A. Here we show the Jordan forms of the same matrix as in Example 6.10, obtained for several values of *ein* and *tol*.

Table 7.7. Results from the reduction into Jordan form for the matrix A of Example 7.10.

σ	λ_1, λ_2	$\mathcal{S}(\lambda)$	cond(Z)	$\Theta_{\min}(\mathcal{X}_1, \mathcal{X}_2)$ deg	*berr*
1	1.000000000000000	{4}	$1.00 \cdot 10^0$	$90.00 \cdot 10^0$	$1.94 \cdot 10^{-15}$
	0.800000000000001	{1}			
2	0.999999999999999	{4}	$2.72 \cdot 10^1$	$12.14 \cdot 10^0$	$3.27 \cdot 10^{-15}$
	0.800000000000001	{1}			
3	0.999999999999977	{4}	$3.17 \cdot 10^2$	$2.71 \cdot 10^0$	$4.37 \cdot 10^{-14}$
	0.800000000000072	{1}			
4	0.999999999999947	{4}	$2.05 \cdot 10^3$	$8.90 \cdot 10^{-1}$	$4.06 \cdot 10^{-12}$
	0.800000000000164	{1}			
6	0.999999999998177	{4}	$3.12 \cdot 10^4$	$1.80 \cdot 10^{-1}$	$1.93 \cdot 10^{-12}$
	0.800000000007105	{1}			
8	1.000000000018368	{4}	$2.24 \cdot 10^5$	$5.76 \cdot 10^{-2}$	$1.93 \cdot 10^{-11}$
	0.799999999926848	{1}			
10	0.999999999674030	{4}	$1.04 \cdot 10^6$	$2.37 \cdot 10^{-2}$	$2.80 \cdot 10^{-10}$
	0.800000001301106	{1}			
12	0.959999999996148	{5}	$1.22 \cdot 10^7$	0	$9.45 \cdot 10^{-9}$

Example 7.11. Consider the matrix

$$A = \begin{bmatrix} -111 & -240 & -572 & 1752 & -4272 \\ -149 & -335 & -778 & 2340 & -5688 \\ 132 & 288 & 657 & -1980 & 4744 \\ -38 & -84 & -201 & 613 & -1504 \\ -22 & -48 & -112 & 340 & -823 \end{bmatrix}$$

which is obtained as $A = QJQ^{-1}$, where

$$J = \operatorname{diag}\left(\begin{bmatrix} 1 & 1 \\ & 1 \end{bmatrix}, 1, \begin{bmatrix} -1 & 1 \\ & -1 \end{bmatrix} \right)$$

and

$$Q = \begin{bmatrix} 56 & 187 & 64 & -8 & 6 \\ 109 & 261 & 90 & -16 & 6 \\ -100 & -234 & -80 & 14 & -5 \\ 20 & 64 & 22 & -3 & 2 \\ 14 & 38 & 13 & -2 & 1 \end{bmatrix}.$$

The matrix A is in a bundle with codimension equal to 5.

Table 7.8. Distances of the computed 5×5 Jordan forms from A.

k	ein, tol	Matrix	Segre characteristic	codim($\mathcal{B}$)	cond(Z)	dist($\widehat{A}_k - A$)
1	10^{-9}	$\widehat{A}_1$	{2},{2,1}	5	$9.25 \cdot 10^3$	$6.16 \cdot 10^{-13}$
2	10^{-8}	$\widehat{A}_2$	{4,1}	6	$1.25 \cdot 10^6$	$1.80 \cdot 10^{-5}$
3	10^{-7}	$\widehat{A}_3$	{3,2}	8	$4.76 \cdot 10^2$	$1.32 \cdot 10^{-4}$
4	10^{-6}	$\widehat{A}_4$	{3,1,1}	10	$1.18 \cdot 10^4$	$2.39 \cdot 10^{-3}$
5	10^{-5}	$\widehat{A}_5$	{2,1,1,1}	16	$1.02 \cdot 10^0$	$3.45 \cdot 10^{-3}$
6	1	$\widehat{A}_6$	{1,1,1,1,1}	24	$1.00 \cdot 10^0$	$1.00 \cdot 10^0$

In Table 7.8, we give the Segre characteristics and bundle codimensions of the six matrices $\widehat{A}_k = Z\widehat{J_S}Z^{-1}$ corresponding to the Jordan forms $\widehat{J}_k$, $k = 1, 2, \ldots, 6$, obtained for different tolerances *ein* and *tol*. In the fifth column is shown the condition number of the respective similarity transformation matrices z and in the last column is shown the relative distance

$$\text{dist}(\widehat{A}_k, A) = \frac{\|\widehat{A}_k - A\|_F}{\|A\|_F}, \quad k = 1, 2, \ldots, 6$$

between the matrix $\widehat{A}_k$ and the original matrix A. The matrix $\widehat{A}_1$ is in the same bundle as the matrix A and its distance to A is very small which justifies accepting the corresponding Jordan form

$$\widehat{J}_1 = \left[\begin{array}{cc|ccc} -1.0000000 & 7.7376900 & 0 & 0 & 0 \\ 0 & -1.0000000 & 0 & 0 & 0 \\ \hline 0 & 0 & 1.0000000 & 73.4907379 & 0 \\ 0 & 0 & 0 & 1.0000000 & 0 \\ 0 & 0 & 0 & 0 & 1.0000000 \end{array}\right]$$

as the numerical Jordan form of A. Like in the case of computing the Jordan–Schur form, with the increase in the tolerances, the codimension of the corresponding bundle increases. It is interesting that at the same time the relative distance to the matrix A also increases and for the most degenerate bundle ($\text{cod}(\mathcal{B}) = 24$) it becomes equal to 1. This behavior of the distance is different from the case of the Jordan–Schur form obtained by unitary similarity transformations (compare with Example 6.10).

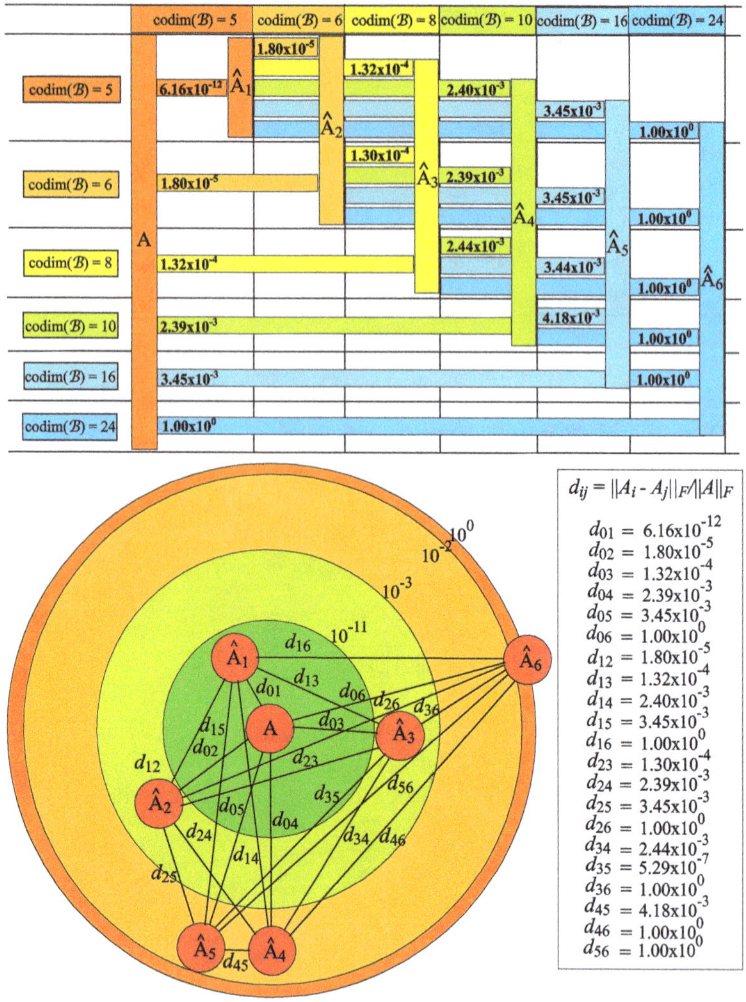

Fig. 7.10. Distances between the computed matrices $\widehat{A}_k$ and the matrix A.

In Figure 7.10, we show the distances between all computed matrices $\widehat{A}_k$ and between them and A, as well as a graph displaying the matrix position.

7.6.5 *Choice of Appropriate Tolerances*

The choice of appropriate tolerances *ein* and *tol* in the reduction into Jordan form depends on the same factors as in the reduction into Jordan–Schur form (see Section 6.7.5). In addition to this, in the transformation to Jordan form it is possible to exploit the information obtained from the estimates of the norms of the spectral projectors P_k, $k = 1, 2, \ldots, p$, determined after the block-diagonalization. If these estimates have relatively low values (say, less than 100), this means that the clusters of numerically multiple eigenvalues are determined correctly. If, however, some of the projectors have very large norms in comparison with the others, this implies that there are eigenvalues in different clusters which are very close and cannot be separated. This may require increasing the value of *ein*.

7.7 Notes and References

The presentation in this chapter is based on the papers of Kågström and Ruhe (1980a, 1980b) and Kågström (1981). The reduction into block-diagonal form can be done in different ways. In the original algorithm, Kågström and Ruhe implemented sequences of elementary nonsingular transformations that annihilate successively the elements outside the block-diagonal. In this chapter, we implement Algorithm 7.6.3 described in Golub and Van Loan (2013), Chapter 7, which is based on solving the Sylvester equation. The experiments show that both methods for block-diagonalization have nearly the same numerical properties. The considerations about the ordering of the eigenvectors and principal eigenvectors in obtaining the numerical Jordan form are taken from Petkov and Konstantinov (2022). In the same paper are presented several misconceptions about the use of the Jordan form. Test matrices with different Jordan forms can be found in Gregory and Karney (1969).

Chapter 8

Case Study 1: Eigenvalue Sensitivity Analysis

In this case study, we present the application of the block-diagonal and the numerical Jordan form of a matrix to the computing of eigenvalue sensitivity estimates. In order to obtain tighter bounds, the estimates obtained by using the Henrichi theorem are optimized. It is shown that the implementation of the Jordan form leads to minimum sensitivity bounds as a result of using diagonal blocks of minimum size.

8.1 Block-Diagonal Form Method

A straightforward approach to analyze the eigenvalue sensitivity of a matrix $A \in \mathbb{C}^{n \times n}$ with multiple eigenvalues is to apply the Henrichi theorem (Section 3.5) to each block of the block-diagonal form of A:

$$T = Z^{-1}AZ = \text{diag}(T_1,\ T_2, \dots,\ T_p),$$

where the block T_i corresponds to a distinct multiple eigenvalue λ_i. Unfortunately, this approach may produce pessimistic sensitivity estimates, as shown by the following example.

Example 8.1. Consider a tenth-order matrix

$$A = QJQ^{-1},\ \ Q = H_2 \Sigma H_1,\ \ Q^{-1} = H_1 \Sigma^{-1} H_2,$$

obtained as described in Example 4.5. The Jordan form of A is set equal to

$$J = \left(\begin{bmatrix} -1 & 1 & 0 \\ 0 & -1 & 1 \\ 0 & 0 & -1 \end{bmatrix}, -1, \begin{bmatrix} -2 & 1 & 0 & 0 \\ 0 & -2 & 1 & 0 \\ 0 & 0 & -2 & 1 \\ 0 & 0 & 0 & -2 \end{bmatrix}, \begin{bmatrix} -2 & 1 \\ 0 & -2 \end{bmatrix} \right)$$

and the matrices H_1 and H_2 are determined as

$$H_1 = I - \frac{2}{10} ee^T, \quad e = [1,\ 1, \ldots,\ 1]^T,$$

$$H_2 = I - \frac{2}{10} ff^T, \quad f = [1,\ -1, \ldots,\ (-1)^9]^T.$$

The diagonal matrix

$$\Sigma = \text{diag}(1, \sigma, \sigma^2, \ldots, \sigma^9),$$

depending on the parameter σ, is used to control the condition number of Q and consequently the sensitivity of the eigenvalues of A.

It is assumed that the matrix A is subject to the perturbation

$$\Delta_A = 10^{-8} \cdot \begin{bmatrix} 8 & 4 & -9 & 1 & -1 & -3 & 9 & -2 & -4 & 4 \\ -6 & -3 & -5 & -4 & -6 & 2 & 5 & -3 & 8 & -6 \\ -7 & 5 & 6 & -3 & 8 & -5 & -2 & -8 & 6 & 2 \\ 3 & -8 & -3 & -8 & -2 & 7 & 3 & -8 & -2 & -6 \\ 6 & -4 & 8 & -6 & 7 & 5 & 1 & 1 & 3 & 6 \\ 4 & 2 & 6 & 9 & -3 & 8 & 5 & 4 & -2 & -3 \\ -9 & -3 & -5 & 4 & -8 & -2 & 3 & -9 & 5 & 7 \\ -3 & 1 & -7 & 6 & -8 & -2 & -1 & 5 & 3 & 2 \\ 7 & 3 & 2 & -4 & -6 & 2 & 7 & -3 & -1 & -5 \\ -3 & -7 & 5 & 3 & 2 & -4 & 6 & -9 & -7 & 6 \end{bmatrix}.$$

The aim of the analysis is to estimate the changes in the eigenvalues of $\tilde{A} = A + \Delta A$ with respect to the perturbation ΔA for different values of σ. In the given case, the block-diagonal form has two blocks T_1 and T_2 of order 4 and 6 corresponding to the eigenvalues $\lambda_1 = -1$

and $\lambda_2 = -2$, respectively. According to the Henrichi theorem, the spectral deviation of $\tilde{T}_i$ with respect to T_i is given by

$$\mathrm{sv}_{T_i}(\tilde{T}_i) \leq \mathrm{dev}(T_i)\omega(x), \tag{8.1}$$

where $\mathrm{dev}(T_i)$ is the deviation from normality of T_i and $\omega(x)$ is the unique nonnegative solution of the equation

$$\psi^{\nu_i} - x\psi^{\nu_i - 1} - x\psi^{\nu_i - 2} - \cdots - x\psi - x = 0,$$

where $x = \mathrm{cond}(Z)\|\Delta A\|/\mathrm{dev}(T_i)$. In the given case, better results are obtained if the value of the spectral deviation is determined as the 2-norm of the strictly upper triangular part of T_i:

$$\mathrm{dev}_2(T_i) = \|T_i - \lambda_i I_{\nu_i}\|_2.$$

The using of (8.1) for $\sigma = 1,\ 1.5,\ 2,\ 2.5$ and 3 produces the results given in Table 8.1. In the third and fifth columns of the table, we give the estimates $\Delta\lambda_1^{\mathrm{est}}$ and $\Delta\lambda_2^{\mathrm{est}}$ obtained for the first and second diagonal blocks, respectively, and in the fourth and sixth column, we give the spectral deviations $\mathrm{sv}_{T_1}(\tilde{T}_1)$ and $\mathrm{sv}_{T_2}(\tilde{T}_2)$, respectively. Obviously, the sensitivity estimates obtained for large values of σ exceed significantly the corresponding actual changes in the eigenvalues.

The detailed investigation of the obtained numerical results reveals that the main factor influencing the sensitivity estimate produced by the Henrichi theorem is the size of the deviation from normality. This leads to the idea of using diagonal scaling transformations in the form $D_i^{-1}T_iD_i,\ i = 1, 2, \ldots, p$, where

$$D_i = \mathrm{diag}(\rho_1,\ \rho_2, \ldots,\ \rho_{\nu_i}), \quad \rho_i \neq 0.$$

These transformations preserve the eigenvalues of T_i but may decrease the value of $\mathrm{dev}_2(D_i^{-1}T_iD_i)$ if the scaling parameters $\rho_1, \rho_2, \ldots, \rho_{\nu_i}$ are appropriately chosen. More specifically, the

Table 8.1. Estimates of the eigenvalue sensitivity, obtained using the block-diagonal form and the Henrichi theorem.

σ	$\mathrm{cond}_2(Q)$	$\Delta\lambda_1^{\mathrm{est}}$	$\mathrm{sv}_{T_1}(\tilde{T}_1)$	$\Delta\lambda_2^{\mathrm{est}}$	$\mathrm{sv}_{T_2}(\tilde{T}_2)$
1.0	$1.00 \cdot 10^0$	$2.94 \cdot 10^{-2}$	$4.30 \cdot 10^{-3}$	$1.44 \cdot 10^{-1}$	$1.41 \cdot 10^{-2}$
1.5	$3.84 \cdot 10^1$	$1.52 \cdot 10^{-1}$	$1.96 \cdot 10^{-3}$	$4.72 \cdot 10^{-1}$	$1.62 \cdot 10^{-2}$
2.0	$5.12 \cdot 10^2$	$1.30 \cdot 10^0$	$4.72 \cdot 10^{-3}$	$2.28 \cdot 10^0$	$1.15 \cdot 10^{-2}$
2.5	$3.81 \cdot 10^3$	$7.67 \cdot 10^0$	$1.32 \cdot 10^{-2}$	$8.02 \cdot 10^0$	$3.58 \cdot 10^{-2}$
3.0	$1.97 \cdot 10^4$	$3.34 \cdot 10^1$	$2.48 \cdot 10^{-2}$	$2.25 \cdot 10^1$	$5.88 \cdot 10^{-2}$

decreasing of the deviation from normality and consequently the decreasing of the sensitivity estimate can be realized by choosing some of the scaling parameters different from one. This, however, will increase the condition number of the updated similarity transformation matrix ZD, increasing in its turn the value of the sensitivity estimate. Hence, the opposite effects of parameter change (the decreasing of the deviation from the normality and increasing of the condition number) show that there should be a minimum of the eigenvalue sensitivity estimate as a function of the scaling parameters. These qualitative considerations are supported by the following example.

Example 8.2. Consider again the matrix A presented in Example 8.1. The elements of the scaling matrices of the blocks T_1 and T_2 are determined to minimize the eigenvalue sensitivity bound by using the function `fminsearch` of MATLAB®. This function implements the multidimensional unconstrained nonlinear minimization method of Nelder–Mead and does not require the use of the gradient of the objective function.

The processes of minimizing the eigenvalue sensitivity bound obtained by the Henrichi theorem in the case of $\sigma = 3$ for the first and second diagonal blocks are illustrated in Figures 8.1 and 8.2, respectively. For the first block, the local minimum is found for 219 iterations using 400 objective function evaluations, and for the second block, the determining of the minimum is done for 746 iterations using 1195 objective function evaluations. The diagonal scaling matrices found for the two blocks are

$$\begin{aligned} D_1 = \text{diag}(&3.156615633397823, 0.000312189050724, \\ &0.002602726123477, 0.000117356134342, \\ &1, 1, 1, 1, 1, 1) \end{aligned}$$

and

$$\begin{aligned} D_2 = \text{diag}(&1, 1, 1, 1, \\ &10.779257278719435, 0.132185937168270, \\ &0.440269867180786, 0.0010490837632701, \\ &0.000468311419819, 0.000272038362370), \end{aligned}$$

respectively. In both the cases, we see that the most important factor in minimizing the eigenvalue sensitivity estimate is the size of the

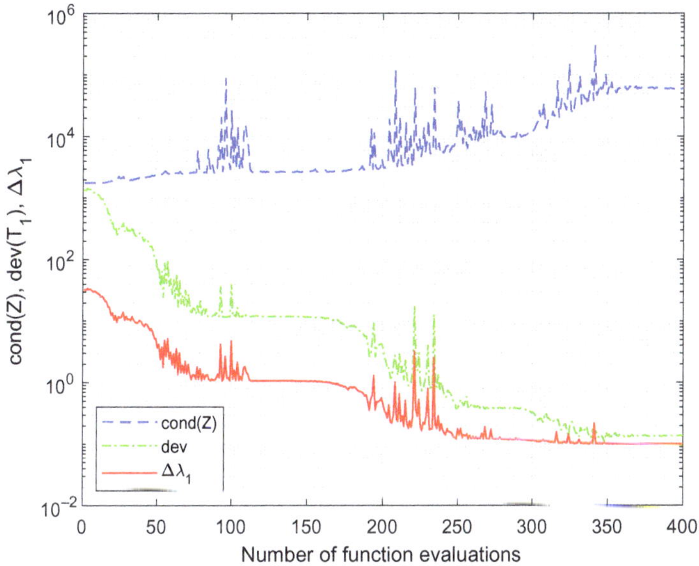

Fig. 8.1. Minimization of the eigenvalue sensitivity estimate for the first diagonal block.

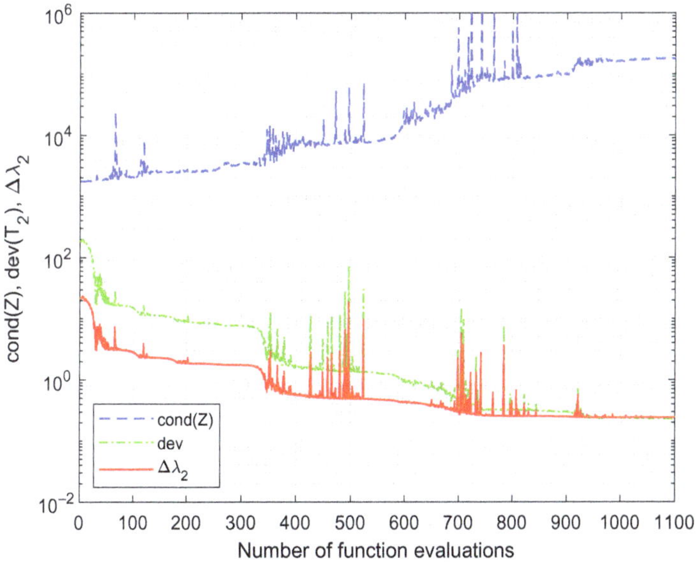

Fig. 8.2. Minimization of the eigenvalue sensitivity estimate for the second diagonal block.

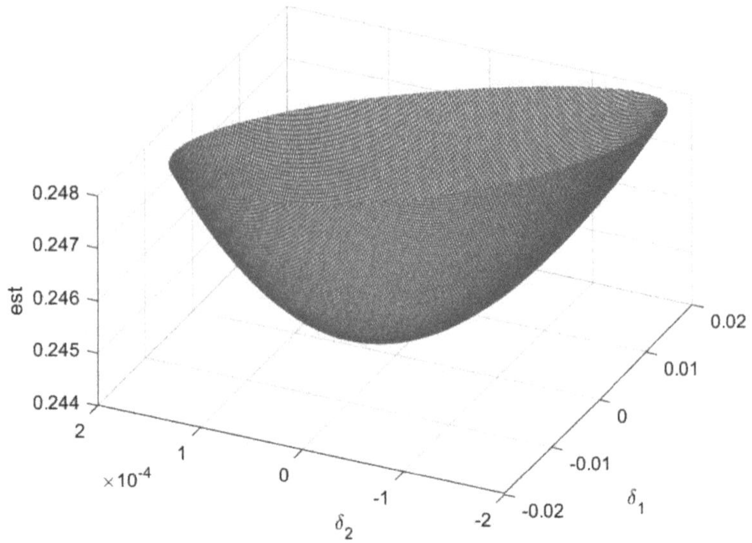

Fig. 8.3. Eigenvalue estimate around the minimum for the second block.

deviation from normality. It can be observed that the minimum value of the estimate is obtained at the price of increasing the condition number of the transformation into the block-diagonal form since this is a factor which is not so influential as the decreasing of the deviation from normality.

The values of the eigenvalue sensitivity bound for $\sigma = 3$ around the minimum for the second diagonal block for small perturbations δ_1 and δ_2 of the elements ρ_6 and ρ_8 of D_2 are shown in Figure 8.3.

The determining of the optimized eigenvalue sensitivity estimates by using the block-diagonal form of the matrix and the Henrichi theorem is done by Algorithm 8.1.

Example 8.3. In Table 8.2, we give the minimized eigenvalue sensitivity bounds for the matrix A of Example 8.2 found by Algorithm 8.1. The results show considerable improvement in the sensitivity estimates for both the blocks. Note that these estimates can be improved by using more sophisticated nonlinear optimization methods.

Algorithm 8.1. Determine optimized eigenvalue sensitivity using the block-diagonal form

Determines bounds on the eigenvalue sensitivity of the matrix $A \in \mathbb{C}^{n\times n}$ to perturbation ΔA of a given size using the block-diagonal form of A

function $[est, ev] = \text{sens\_block\_diag}(A, \delta)$

Input: The matrix $A \in \mathbb{C}^{n\times n}$; the 2-norm δ of the perturbation ΔA.

Output: $est \in \mathbb{C}^p$ contains the sensitivity bound for each distinct eigenvalue of A; $ev \in \mathbb{C}^p$ contains the distinct eigenvalues of A.

Reduce the matrix A into the block-diagonal form
$T = Z^{-1}AZ = \text{diag}(T_1, T_2, \ldots, T_p),\ T_i \in \mathbb{C}^{\nu_i\times\nu_i}$
by Algorithm 6.1 and orthonormalize the bases of the invariant subspaces corresponding to the blocks with distinct eigenvalues using Algorithm 6.2
for $i = 1, 2, \ldots, p$
  $nb = \nu_i$
  $ev_i = \lambda_i$
  Set the initial value $D = I_n$ of the diagonal scaling matrix
  $sum_i = \sum_{q=1}^{i} \nu_q$
  for $iter = 1, 2, \ldots$
    Find the values of the parameters $\rho_1, \rho_2, \ldots, \rho_{nb}$ which minimize the estimate $est_i = \text{fun}(T, D, \rho_1, \rho_2, \ldots, \rho_{nb})$
  end
end
function $est = \text{fun}(T, D, \rho_1, \rho_2, \ldots, \rho_{nb})$
Determines the eigenvalue sensitivity bound of a scaled diagonal block
for $q = 1, 2, \ldots nb$
  $d_{sum_i+q,sum_i+q} = |\rho_q|$
end
$T \leftarrow D^{-1}TD$
$T_i = T_{sum_i+1:sum_i+\nu_i,sum_i+1:sum_i+\nu_i}$
Compute the deviation from normality of T_i, $\text{dev}_2(T_i) = \|T_i - \lambda_i I_{nb}\|_2$
$Z \leftarrow ZD$
$poly_1 = 1$
for $q = 2, 3, \ldots, nb + 1$
  $poly_q = -\text{cond}_2(Z)\delta/\text{dev}_2(T_i)$
end
Compute the roots $r_1, r_2, \ldots, r_{nb}$ of $poly$
for $q = 1, 2, \ldots, nb$
  if $\text{real}(r_q) > 0$ **and** $\text{imag}(r_q) = 0,\ \omega = r_q$
end
$est = \text{dev}_2(T_i) \cdot \omega$

Table 8.2. Optimized eigenvalue sensitivity bounds, obtained using the block-diagonal form and the Henrichi theorem.

σ	$\Delta\lambda_1^{\text{est}}$	$\text{sv}_{T_1}(\tilde{T}_1)$	$\Delta\lambda_2^{\text{est}}$	$\text{sv}_{T_2}(\tilde{T}_2)$
1.0	$9.77 \cdot 10^{-3}$	$4.29 \cdot 10^{-3}$	$7.22 \cdot 10^{-2}$	$1.41 \cdot 10^{-2}$
1.5	$1.65 \cdot 10^{-2}$	$1.96 \cdot 10^{-3}$	$6.58 \cdot 10^{-2}$	$1.62 \cdot 10^{-2}$
2.0	$3.32 \cdot 10^{-2}$	$4.72 \cdot 10^{-3}$	$8.84 \cdot 10^{-2}$	$1.15 \cdot 10^{-2}$
2.5	$6.06 \cdot 10^{-2}$	$1.32 \cdot 10^{-2}$	$2.53 \cdot 10^{-1}$	$3.58 \cdot 10^{-2}$
3.0	$1.02 \cdot 10^{-1}$	$2.48 \cdot 10^{-2}$	$2.44 \cdot 10^{-1}$	$5.88 \cdot 10^{-2}$

8.2 Numerical Jordan Form Method

Better eigenvalue sensitivity estimates can be obtained by using the numerical Jordan form of the matrix due to the fact that the Jordan blocks associated with the eigenvalues are of the minimum possible size.

Assume that the numerical Jordan form of the matrix A is determined by implementing the algorithm of Kågström and Ruhe. Consider the determination of the sensitivity of the eigenvalues λ_i, $i = 1, 2, \ldots, p$, where p is the number of distinct eigenvalues of the matrix A. The part of the Jordan form of A,

$$\begin{bmatrix} J_{s_{1i}}(\lambda_i) & & & \\ & J_{s_{2i}}(\lambda_i) & & \\ & & \ddots & \\ & & & J_{s_{q_i,i}}(\lambda_i) \end{bmatrix},$$

corresponding to the multiple eigenvalue λ_i, consists of q_i Jordan blocks $J_{s_{ji}}(\lambda_i) \in \mathbb{C}^{s_{ji} \times s_{ji}}$, $j = 1, 2, \ldots, q_i$, where $\nu_i = s_{1i} + s_{2i} + \cdots + s_{q_i,i}$ is the algebraic multiplicity of λ_i and $\mathcal{S}_i = \{s_{1i}, s_{2i}, \ldots, s_{q_i,i}\}$ is the Segre characteristic of A associated with λ_i.

Since the Jordan blocks corresponding to λ_i are independent of each other, the sensitivity of this eigenvalue can be determined as the maximum sensitivity of the blocks $J_{s_{1j}}(\lambda_i)$, $j = 1, 2, \ldots, q_i$. That is why, if the matrix has several Jordan blocks containing the same eigenvalue, then it is necessary to compute the estimate corresponding to each block and take the maximum one.

In estimating the eigenvalue sensitivity by using the Jordan form, it is convenient to apply the modified Henrichi Theorem 3.13. According to it, if $J = Z^{-1}AZ$ is the Jordan form of A and nb is the order of the Jordan block J_k, then the spectral deviation of $\tilde{J}_k$ with respect to J_k is

$$\mathrm{sv}_{J_k}(\tilde{J}_k) \leq \omega(x)\mathrm{dev}_2(J_k), \tag{8.2}$$

where $\omega(x)$ is the unique positive root of

$$\psi^{nb} - x\psi^{nb-1} - x\psi^{nb-2} - \cdots - x\psi - x = 0,$$
$$x = \mathrm{cond}_2(Z)\|\Delta A\|_2/\mathrm{dev}_2(J_k)$$

and $\mathrm{dev}_2(J_k)$ is the deviation from normality of the block J_k. Note that if all superdiagonal elements of J_k are equal to 1,

$$\mathrm{dev}_2(J_k) = \|J_k - \lambda_i I_{nb}\|_2 = 1.$$

Using this value of the deviation from normality, the quantity $x = \mathrm{cond}_2(Z)\|\Delta A\|_2$ is the same for all Jordan blocks and the value of the estimate will be determined by the size nb of the block. Since for a given λ_i the first Jordan block $J_{s_{1i}}(\lambda_i)$ is of the largest size, then in accordance with Theorem 3.13, it suffices to determine the sensitivity estimate only for this block.

Example 8.4. If (8.2) is applied to the matrix A of the previous three examples, we may obtain the sensitivity estimates of $\lambda_1 = -1$ and $\lambda_2 = -2$ using only the first and third Jordan blocks since they have the highest order. The numerical Jordan form is preliminary normalized by using a diagonal scaling transformation so that for each block $\mathrm{dev}_2(J_k) = 1$. In Table 8.3, we show the eigenvalue sensitivity estimates for the first and third blocks which give the sensitivity estimates of λ_1 and λ_2, respectively. The estimates are much smaller than in the case of using the block-diagonal form and are nearly the same as the optimized estimates obtained for the block-diagonal form.

The eigenvalue sensitivity estimates obtained by using the Jordan form can be minimized similar to the case of using the block-diagonal form. For this aim, the elements of the diagonal scaling matrix D, corresponding to a given Jordan block J_k, are optimized iteratively

Table 8.3. Eigenvalue sensitivity estimates, obtained for the blocks J_1 and J_3 using the normalized Jordan form and Theorem 3.13.

σ	$\Delta\lambda_1^{\text{est}}$	$\text{sv}_{T_1}(\tilde{T}_1)$	$\Delta\lambda_2^{\text{est}}$	$\text{sv}_{T_2}(\tilde{T}_2)$
1.0	$6.85 \cdot 10^{-3}$	$4.30 \cdot 10^{-3}$	$2.39 \cdot 10^{-2}$	$1.41 \cdot 10^{-2}$
1.5	$1.85 \cdot 10^{-2}$	$1.96 \cdot 10^{-3}$	$5.07 \cdot 10^{-2}$	$1.62 \cdot 10^{-2}$
2.0	$4.60 \cdot 10^{-2}$	$4.72 \cdot 10^{-3}$	$1.01 \cdot 10^{-1}$	$1.15 \cdot 10^{-2}$
2.5	$9.18 \cdot 10^{-2}$	$1.32 \cdot 10^{-2}$	$1.71 \cdot 10^{-1}$	$3.58 \cdot 10^{-2}$
3.0	$1.59 \cdot 10^{-1}$	$2.48 \cdot 10^{-2}$	$2.69 \cdot 10^{-1}$	$5.88 \cdot 10^{-2}$

in order to find a local minimum of the sensitivity estimate. The numerical experiments show that for an initial value of the scaling matrix, it is appropriate to use the matrix D which reduces the numerical Jordan form into the Jordan normal form.

Example 8.5. Consider the minimization of the sensitivity estimates of the eigenvalues $\lambda_1 = -1$ and $\lambda_2 = -2$ for the matrix A considered in Examples 8.1–8.4. The sensitivity of λ_1 is determined by using the 3×3 block J_1 and the sensitivity of λ_2 is determined by using the 4×4 block J_3. The variation of the parameters determining the estimates of the two eigenvalues during the optimization by the function `fminsearch`, is shown for $\sigma = 3$ in Figures 8.4 and 8.5, respectively. The optimum estimates obtained are $\Delta\lambda_1^{\text{est}} = 7.75 \cdot 10^{-2}$ (after 192 iterations and 337 objective function evaluations) and $\Delta\lambda_2^{\text{est}} = 1.57 \cdot 10^{-1}$ (after 291 iterations and 485 function evaluations). The minimum values of the estimates are obtained for the diagonal scaling matrices:

$$D_1 = \text{diag}(10.364776156654361, 2.195467992767757,\\ 0.465044264771622, 1, 1, 1, 1, 1, 1, 1)$$

and

$$D_2 = \text{diag}(1, 1, 1, 1, 5.162969790889805, 1.611486724588428,\\ 0.502983664002355, 0.156993267399606, 1, 1),$$

respectively. The optimally scaled first and third Jordan blocks are

$$J_1 = \begin{bmatrix} -1.000000000038791 & 0.211820106829633 & 0 \\ 0 & -1.000000000038791 & 0.211820106830779 \\ 0 & 0 & -1.000000000038791 \end{bmatrix}$$

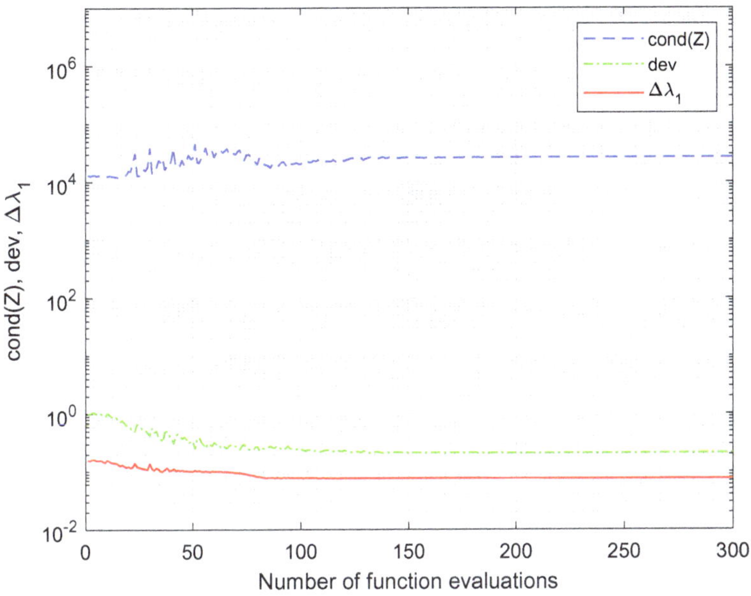

Fig. 8.4. Minimization of the eigenvalue sensitivity estimate for the first Jordan block.

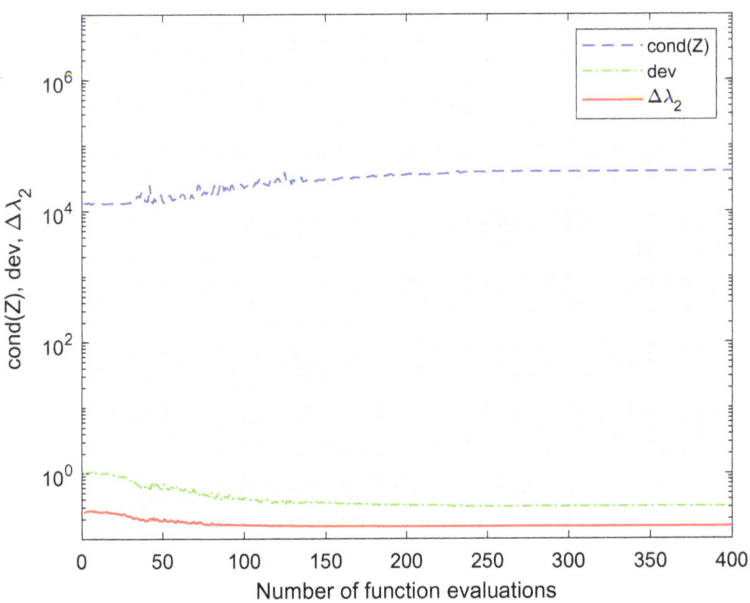

Fig. 8.5. Minimization of the eigenvalue sensitivity estimate for the third Jordan block.

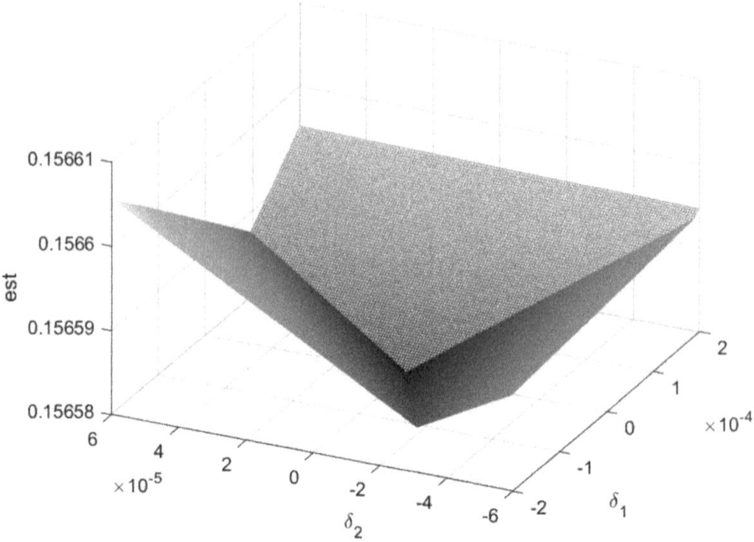

Fig. 8.6. Eigenvalue estimate around the minimum for the third Jordan block.

and

$$J_3 = \begin{bmatrix} -1.999999999974181 & 0.312123988684175 \\ 0 & -1.999999999974181 \\ 0 & 0 \\ 0 & 0 \end{bmatrix} \cdots \rightarrow$$

$$\leftarrow \cdots \begin{bmatrix} 0 & 0 \\ 0.312123988567648 & 0 \\ -1.999999999974181 & 0.312123988581210 \\ 0 & -1.999999999974181 \end{bmatrix}.$$

In Figure 8.6, we show the values of the eigenvalue sensitivity estimate $\Delta\lambda_2^{\text{est}}$ for small perturbations δ_1 and δ_2 of the (6, 6) and (7, 7) elements of the matrix D_2.

The optimized eigenvalue sensitivity estimates using the numerical Jordan form are obtained by Algorithm 8.2.

Example 8.6. In Table 8.4, we show the optimized sensitivity estimates of the eigenvalues of the matrix A considered in Examples 8.1–8.4 for different values of σ. The values obtained are minimal

Algorithm 8.2. Determine optimized eigenvalue sensitivity estimates using the numerical Jordan form

Determines bounds on the sensitivity of the eigenvalues of the matrix $A \in \mathbb{C}^{n\times n}$ to perturbation ΔA of a given size using the numerical Jordan form of A

function $[est, ev] = \text{sens\_jord}(A, \delta)$

Input: The matrix $A \in \mathbb{C}^{n\times n}$; the 2-norm δ of the perturbation ΔA.

Output: $est \in \mathbb{C}^p$ contains the sensitivity bound for each distinct eigenvalue of A; $ev \in \mathbb{C}^p$ contains the eigenvalues of A.

Determine the Jordan form $J = Z^{-1}AZ$ of A using Algorithm 7.5, the Weyr characteristics $\mathcal{W} = [\{w_{11}, \ldots\}\{w_{21}, \ldots\} \ldots \{w_{p1}, \ldots\}]$ and the Segre characteristics $\mathcal{S} = [\{s_{11}, \ldots\}\{s_{12}, \ldots\} \ldots \{s_{1p}, \ldots\}]$ associated with $\lambda_1, \lambda_2, \ldots, \lambda_p$, where p is the number of distinct eigenvalues of A.

Set $D = I_n$
for $i = 2, 3, \ldots, n$
    $scale = d_{i-1,i-1} j_{i-1,i}$
    if $scale \neq 0$
        $d_{ii} = scale$
    end
end
$J \leftarrow DJD^{-1}, \ Z \leftarrow ZD^{-1}$
$j0 = 1, \ k = 0$
for $i = 1, 2, \ldots, p$
    Determine the number $njb = w_{i1}$ of the Jordan blocks corresponding to λ_i
    Set $v_j = 0, \ j = 1, 2, \ldots, njb$
    $ev_i = \lambda_{j0}$
    for $j = 1, 2, \ldots, njb$
        $k = k + 1$
        Find the order $nb = s_{ki}$ of the kth Jordan block
        Set the initial value $D = I_n$ of the diagonal scaling matrix
        for $iter = 1, 2, \ldots$
            Find the the parameters $\rho_1, \rho_2, \ldots, \rho_{nb}$ which minimize the estimate $v_j = \text{fun}(J, D, j0, \rho_1, \rho_2, \ldots, \rho_{nb})$
        end
        Determine the position $j0 = j0 + nb$ of the next Jordan block
    end
    $est_i = \max(v_1, v_2, \ldots, v_{njb})$
end

function $est = \text{fun}(J, D, j0, \rho_1, \rho_2, \ldots, \rho_{nb})$
Determines the eigenvalue sensitivity estimate of a scaled Jordan block by the Henrichi theorem
for $q = 1, 2, \ldots nb$
  $d_{j0+q-1,j0+q-1} = |\rho_q|$
end
$J \leftarrow D^{-1}JD,\ Z \leftarrow ZD$
$J_k = J_{j0:j0+nb-1,j0:j0+nb-1}$
Compute the deviation from normality of J_k, $\text{dev}_2(J_k) = \|J_k - \lambda_i I_{nb}\|_2$
$poly_1 = 1$
for $q = 2, 3, \ldots, nb + 1$
  $poly_q = -\text{cond}_2(Z)\delta/\text{dev}_2(J_k)$
end
Compute the roots $r_1, r_2, \ldots, r_{nb}$ of $poly$
for $q = 1, 2, \ldots, nb$
  if $\text{real}(r_q) > 0$ **and** $\text{imag}(r_q) = 0$
    $\omega = r_q$
  end
end
$est = \text{dev}_2(J_k) \cdot \omega$

Table 8.4. Optimized eigenvalue sensitivity estimates, obtained using the Jordan form and the Henrichi theorem.

σ	$\Delta\lambda_1^{\text{est}}$	$\text{sv}_{T_1}(\tilde{T}_1)$	$\Delta\lambda_2^{\text{est}}$	$\text{sv}_{T_2}(\tilde{T}_2)$
1.0	$6.40 \cdot 10^{-3}$	$4.30 \cdot 10^{-3}$	$2.28 \cdot 10^{-2}$	$1.41 \cdot 10^{-2}$
1.5	$1.12 \cdot 10^{-2}$	$1.96 \cdot 10^{-3}$	$3.16 \cdot 10^{-2}$	$1.62 \cdot 10^{-2}$
2.0	$2.26 \cdot 10^{-2}$	$4.72 \cdot 10^{-3}$	$5.76 \cdot 10^{-2}$	$1.15 \cdot 10^{-2}$
2.5	$4.29 \cdot 10^{-2}$	$1.32 \cdot 10^{-2}$	$9.83 \cdot 10^{-2}$	$3.58 \cdot 10^{-2}$
3.0	$7.75 \cdot 10^{-2}$	$2.48 \cdot 10^{-2}$	$1.57 \cdot 10^{-1}$	$5.88 \cdot 10^{-2}$

among the respective values of all estimates considered up to the moment (compared with the estimates given in Tables 8.1–8.3).

8.3 Notes and References

The theoretical and computational issues in estimating eigenvalues, sensitivity are considered in many sources, see, for instance, Bhatia (2007), Stewart and Sun (1990), and Bai *et al.* (1993). In case of simple eigenvalues, the sensitivity analysis reduces to finding the eigenvalue condition numbers which can be done by the LAPACK routines `xGEESX` or MATLAB® function `condeig`. Algorithms to determine the eigenvalue condition numbers without using the eigenvectors are proposed in Chen *et al.* (1977) and Petkov (2021). Residual perturbation bounds of the eigenvalues are discussed in Kahan *et al.* (1982).

Chapter 9

Case Study 2: Computing the Matrix Exponential

The computation of the matrix exponential e^A of a real or complex $n \times n$ matrix A is an important task in the solution of differential equations, obtaining the time response and discretization of control systems, computing the integral of the exponential and several other problems arising in applied mathematics. In this case study, we consider some methods for numerical computation of the matrix exponential putting the emphasis on the using of Jordan and Jordan–Schur forms. After an introduction into the properties of the exponential, we present a brief review of some numerical methods intended for computing of e^A. We present then in detail four methods for evaluation of the exponential based on the numerical Jordan form, block-diagonal staircase form, Schur–Parlett and Schur–Fréchet methods.

9.1 Definition and Basic Properties

Consider the system of homogeneous linear differential equations

$$\dot{x}(t) = Ax(t), \quad x(0) = x_0, \tag{9.1}$$

where $x(t)$ is an n-vector and A is a constant $n \times n$ matrix. The solution of (9.1) is given by

$$x(t) = e^{At}x_0,$$

where e^{At} is the *matrix exponential* defined by the convergent matrix series

$$e^{At} \equiv \sum_{k=0}^{\infty}(At)^k/k! = I_n + At/1! + (At)^2/2! + \cdots \qquad (9.2)$$

The matrix exponential has the properties

(1) $e^{A(t+s)} = e^{At}e^{As}$,
(2) e^{At} is nonsingular for each t,
(3) $(e^{At})^{-1} = e^{-At}$,
(4) $\frac{d}{dt}(e^{At}) = Ae^{At} = e^{At}A$.

The following property of the exponential can be particularly useful. Assume that

$$A = \begin{bmatrix} A_{11} & A_{12} \\ 0 & A_{22} \end{bmatrix}. \qquad (9.3)$$

Then

$$e^A = \begin{bmatrix} e^{A_{11}} & \int_0^1 e^{A_{11}(1-s)}A_{12}e^{A_{22}s}ds \\ 0 & e^{A_{22}} \end{bmatrix}. \qquad (9.4)$$

Consider next a system of nonhomogeneous linear differential equations

$$\dot{x}(t) = Ax(t) + f(t), \quad x(0) = x_0, \qquad (9.5)$$

where $f(t)$ is an n-vector. The solution of (9.5) is

$$x(t) = e^{At}x_0 + \int_0^t e^{A(t-s}f(s)ds$$

Thus, the computation of the matrix exponential is a basic problem in the solution of linear differential equations.

9.2 Sensitivity of the Matrix Exponential

In assessing the accuracy of the various methods for computing the matrix exponential, it is necessary to know the sensitivity of e^A to perturbations ΔA of the matrix A. In practice, the elements of A are known with some error which leads to an error in the computed exponential e^A. The methods attempting to compute e^A also produce an approximation to the exact exponential due to the properties of the method used and the rounding errors. Hence from backward error point of view, instead of e^A we obtain the exact exponential of a perturbed matrix $e^{A+\Delta A}$. For a given perturbation ΔA, the sensitivity of the exponential is measured by the absolute change

$$\|e^{A+\Delta A} - e^A\|$$

or by the relative change

$$\|e^{A+\Delta A} - e^A\|/\|e^A\|.$$

To characterize the sensitivity of e^A, we need to introduce the derivative of the exponential in respect to the perturbation of A. The linear operator L_A is the *Fréchet derivative* of e^A, if

$$\lim_{\|\Delta A\|\to 0}\frac{\|e^{A+\Delta A} - e^A - L_A\Delta A\|}{\|\Delta A\|} = 0.$$

Equivalently, this expression is written as

$$e^{A+\Delta A} - e^A = L_A\Delta A + o(\|\Delta A\|),$$

where the term $h = o(\delta)$ means that $\|h\|/\delta \to 0$ as $\delta \to 0$.

In Figure 9.1, we give an illustrative interpretation of the Fréchet derivative L_A at the point e^A. The vector $L_A\Delta A$ which approximates the difference $e^{A+\Delta A} - e^A$ belongs to the range $\mathcal{R}(L_A)$ of the Fréchet derivative. The range of L_A is shown as a plane which is tangent to the surface of the exponents at the point e^A.

It is possible to prove that (Van Loan, 1977)

$$L_A\Delta A = \int_0^1 e^{A(1-s)}\Delta A e^{As} ds.$$

According to this expression, and taking into account (9.3) and (9.4), the Fréchet derivative of e^A can be found by computing the exponential

$$\exp\left(\begin{bmatrix} A & \Delta A \\ 0 & A \end{bmatrix}\right) = \begin{bmatrix} e^A & L_A \\ 0 & e^A \end{bmatrix}.$$

There is a close connection between the Fréchet derivative and the conditioning of e^A. The sensitivity of e^A to small perturbations

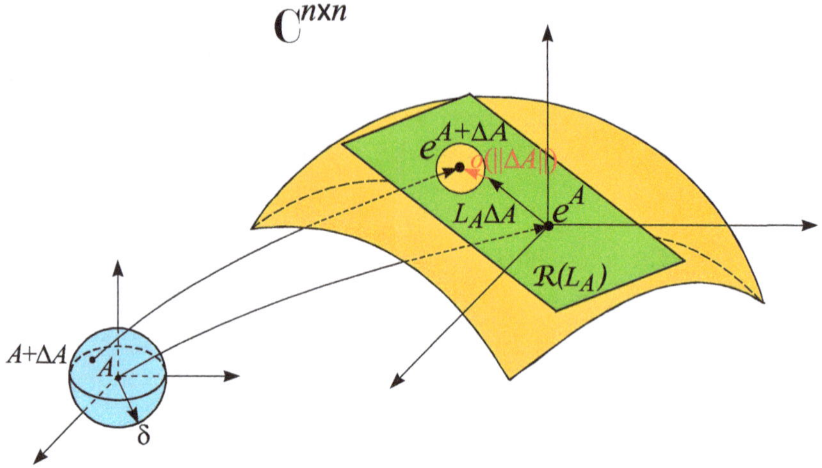

Fig. 9.1. Illustrative interpretation of the Fréchet derivative.

ΔA of A is characterized by the (relative) *condition number of the matrix exponential* which, according to (1.36), is defined as

$$\text{cond}_E := \lim_{\varepsilon \to 0} \max_{\|\Delta A\| \leq \delta \|A\|} \frac{\|e^{A+\Delta A} - e^A\|}{\delta \|e^A\|}.$$

Hence, we have that

$$\frac{\|e^{A+\Delta A} - e^A\|}{e^A} \leq \text{cond}_E \frac{\|\Delta A\|}{\|A\|} + o(\|\Delta A\|).$$

The condition number of the matrix exponential is determined by using the Fréchet derivative as (Van Loan, 1977, Theorem 5)

$$\text{cond}_E = \frac{\|L\| \|A\|}{\|e^A\|},$$

where

$$\|L\| = \max_{\|\Delta A\| \leq 1} \|L_A \Delta A\| = \max_{\|\Delta A\| \leq 1} \int_0^1 e^{A(1-s)} \Delta A e^{As} ds.$$

Note that if A is normal ($A^H A = A A^H$), then $\text{cond}_E = \|A\|$.

Efficient algorithms for computing $\|L\|$ are presented in Higham (2008), Chapter 10 (Algorithm 10.27), Mathias (1992) and Al-Mohy and Higham (2009a).

The function `expm_cond` from the Matrix Function Toolbox (Higham (2008), Appendix D) determines an approximation of cond_E in Frobenius form.

Example 9.1. The exact matrix exponential of

$$A = \begin{bmatrix} 48 & -49 & 50 & 49 \\ 0 & -2 & 100 & 0 \\ 0 & -1 & -2 & 1 \\ -50 & 50 & 50 & -52 \end{bmatrix}$$

is (up to six decimal digits)

$$e^A = \begin{bmatrix} 10.2855 & -10.0148 & 10.1501 & 10.0148 \\ -112.779 & 106.148 & -99.2459 & -106.013 \\ -3.38338 & 3.24805 & -3.24805 & -3.24805 \\ -119.546 & 112.779 & -106.013 & -112.644 \end{bmatrix}.$$

If the $(1, 2)$-element of A is changed from -49 to -48.996, we obtain

$$e^{A+\Delta A} = \begin{bmatrix} 8.86648 & -8.72082 & 9.30856 & 8.73846 \\ -110.101 & 103.733 & -98.8081 & -103.709 \\ -3.19616 & 3.07883 & -3.17227 & -3.08323 \\ -115.747 & 109.342 & -104.895 & -109.331 \end{bmatrix}$$

so that a relative perturbation

$$\|\Delta A\|_F / \|A\|_F = 2.316 \cdot 10^{-5}$$

leads to a relative change of the result equal to

$$\|e^{A+\Delta A} - e^A\|_F / \|e^A\|_F = 2.558 \cdot 10^{-2}.$$

In the given case, the matrix A is similar to a single Jordan block with an eigenvalue equal to -2 and the matrix exponential is ill conditioned. The condition number of the exponential, computed by the function `expm_cond`, is equal to

$$\text{cond}_E = 3.303 \cdot 10^3,$$

and an upper bound on the change is given by

$$\text{cond}_E \|\Delta A\|_F = 7.652 \cdot 10^{-2}.$$

We see that the actual change of the result is close to the bound.

9.3 Series Methods

The series methods for computing the matrix exponential e^A are derived by the direct application to matrices of standard approximation techniques e^a. The commonly used series methods involve truncated Taylor series or Padé approximations.

9.3.1 *Taylor Series*

By definition, the matrix exponential of A is expressed as

$$e^{At} = I_n + A/1! + (A)^2/2! + \cdots = \sum_{k=0}^{\infty} A^k/k! \tag{9.6}$$

The terms of the series (9.6) may be computed recursively, the kth term being A/k times the $(k-1)$th term. The summation of the series in floating-point arithmetic is done up to the moment when the addition of a new term does not alter the sum. This method may produce unsatisfactory results even in the scalar case because of rounding errors. The theoretical analysis and numerical experiments show that the truncation and rounding errors associated with the Taylor series approach increase with the norm of the matrix A. That is why it is appropriate to implement a technique which ensures that the matrix norm is bounded by unity. Such a technique is the *scaling and squaring* which employs the fact that if A is divided by a nonzero scalar q, then e^A/q should be raised to the qth power to obtain e^A, i.e.,

$$e^A = \left(e^{A/q}\right)^q.$$

For this aim, one determines an integer $m > 0$ such that

$$2^{m-1} \leq \|A\| \leq 2^m.$$

The matrix A is then scaled by the factor 2^{-m} resulting in a matrix with the desired property. Then we have

$$A_s = 2^{-m}A$$

and

$$e^A = \left(e^{A_s}\right)^{2^m}.$$

The computation of the matrix $\left(e^{A_s}\right)^{2^m}$ can now be carried out by squaring e^{A_s} m times.

9.3.2 *Padé Approximations*

Instead of Taylor series, it is possible to use Padé approximations, which are more efficient from computational point of view. Let

$$f(x) = s_0 + s_1 x_1 + s_2 x^2 + \cdots$$

be a power series with $s_0 \neq 0$ and let $n(x)$ and $d(x)$ be polynomials of degree p, q, respectively,

$$n(x) = \sum_{k=0}^{p} a_k x^k, \quad d(x) = \sum_{k=0}^{q} b_k x^k.$$

The ratio $n(x)/d(x)$ is referred to as *Padé approximation* to $f(x)$ if the terms $x^0, x^1, \ldots, x^{p+q}$ are canceled out in $f(x)d(x) - n(x)$. The sum $p + q$ is the *order* of the Padé approximation. It may be proved that the coefficients $a_0, a_1, \ldots, a_p; \quad b_0, b_1, \ldots, b_q$ always exist and the ratio $n(x)/d(x)$ is unique.

The Padé approximation to the matrix exponential e^A is defined by

$$R_{pq}(A) = [D_{pq}(A)]^{-1} N_{pq}(A),$$

where

$$N_{pq}(A) = \sum_{k=0}^{p} a_k A^k, \quad a_k = (p+q-k)!p!/[(p+q)!k!(p-k)!],$$

$$D_{pq}(A) = \sum_{k=0}^{q} b_k (-A)^k, \quad b_k = (p+q-k)!q!/[(p+q)!k!(q-k)!].$$

The nonsingularity of the matrix $D_{pq}(A)$ is assured if the parameters p and q are sufficiently large or if the eigenvalues of A have negative real parts (i.e., A is a *stability matrix*). The choice of p and q is done so that to achieve the prescribed accuracy of the approximation. From efficiency considerations for a given accuracy, it is preferable to use *diagonal Padé approximations* for which $p = q$.

The Padé approximations can be used when the norm of A is not very large. In case of large $\|A\|$, it is necessary to use scaling and squaring.

9.4 Decomposition Methods

If the matrix A is decomposed as

$$A = SBS^{-1}, \tag{9.7}$$

then the matrix exponential e^A satisfies

$$e^A = Se^B S^{-1}.$$

All methods for computing e^A which employ a decomposition of the form (9.7) involve two conflicting objectives:

(1) Make the matrix B close to a diagonal matrix so that e^B is easy to compute.
(2) Make the matrix S well conditioned so that the errors in computing $Se^B S^{-1}$ are small.

The use of the Jordan form of A puts emphasis on the first objective, while the implementation of Schur form puts emphasis on the second one. A compromise between these conflicting objective makes the method of block-diagonalization, considered in the following.

9.4.1 *Diagonalization and Block-Diagonalization*

If the matrix A is nondefective, the exponential e^A may be found from

$$e^A = X\text{diag}(e^{\lambda_1}, e^{\lambda_2}, \ldots, e^{\lambda_n})X^{-1},$$

where $\lambda_1, \lambda_2, \ldots, \lambda_n$ are the eigenvalues of A and X is the matrix of the corresponding eigenvectors. Unfortunately, this algorithm produces large errors when A is a nearly defective matrix, i.e., when the eigenvectors of A are almost linearly dependent. This situation is indicated by the large value of cond(X).

When the matrix A has multiple eigenvalues, it is possible to use its block-diagonal form

$$B = X_B^{-1} A X_B = \mathrm{diag}\,(T_{11}, T_{22}, \ldots, T_{pp}),$$

where each block $T_{ii} \in \mathbb{C}^{\nu_i \times \nu_i}$ corresponds to a distinct eigenvalues λ_i with multiplicity ν_i. According to the block-diagonalization method, the matrix exponential of A can be determined then from

$$e^A = X_B \mathrm{diag}\left(e^{T_{11}}, e^{T_{22}}, \ldots, e^{T_{pp}}\right) X_B^{-1},$$

where the exponentials of the blocks should be obtained by some other method.

9.4.2 *Jordan Decomposition*

In case of a defective matrix A with a unique multiple eigenvalue, we may use its Jordan form

$$J = Z^{-1} A Z = \mathrm{diag}(J_{s_1}, J_{s_2}, \ldots, J_{s_q}),$$

where

$$J_{s_k}(\lambda_k) = \begin{bmatrix} \lambda_k & 1 & 0 & \ldots & 0 & 0 \\ 0 & \lambda_k & 1 & \ldots & 0 & 0 \\ 0 & 0 & \lambda_k & \ldots & 0 & 0 \\ \vdots & \vdots & \vdots & \ddots & \vdots & \vdots \\ 0 & 0 & 0 & \ldots & \lambda_k & 1 \\ 0 & 0 & 0 & \ldots & 0 & \lambda_k \end{bmatrix}, \quad k = 1, 2, \ldots, q,$$

is a Jordan block of dimension s_k.

In this case,

$$e^A = Z e^J Z^{-1} = Z \mathrm{diag}(e^{J_{s_1}}, e^{J_{s_2}}, \ldots, e^{J_{s_q}}) Z^{-1},$$

where

$$e^{J_{s_k}} = e^{\lambda_k} \begin{bmatrix} 1 & 1/1! & 1/2! & \ldots & 1/(s_k-2)! & 1/(s_k-1)! \\ 0 & 1 & 1/1! & \ldots & 1/(s_k-3)! & 1/(s_k-2)! \\ 0 & 0 & 1 & \ldots & 1/(s_k-4)! & 1/(s_k-3)! \\ \vdots & \vdots & \vdots & \ddots & \vdots & \vdots \\ 0 & 0 & 0 & \ldots & 1 & 1/1! \\ 0 & 0 & 0 & \ldots & 0 & 1 \end{bmatrix}.$$

9.4.3 *Schur–Parlett Method*

Let the Schur decomposition of A be

$$A = UTU^H,$$

where U is unitary and T is upper triangular. Then

$$e^A = Ue^T U^H$$

and the problem for determining e^A reduces to the computation of the exponential of a triangular matrix. For this purpose, it is possible to use the recursive algorithm from Parlett (1976) which produces the upper triangular matrix $F = f(T)$ for a given analytic function f. This algorithm is based on the commutativity relation

$$FT = TF$$

which implies

$$\sum_{k=1}^{j} f_{ik}t_{kj} = \sum_{k=i}^{j} t_{ik}f_{kj}, \quad j > i.$$

Provided that $t_{ii} \neq t_{jj}$, it follows from the earlier identities that

$$f_{ij} = \left[t_{ij}(f_{ii} - f_{jj}) + \sum_{k=i+1}^{j-1} (f_{ik}t_{kj} - t_{ik}f_{kj}) \right] \Big/ (t_{ii} - t_{jj}). \quad (9.8)$$

In this way, it is possible to compute one superdiagonal of F at a time beginning with the diagonal $f(t_{11}), \ldots, f(t_{nn})$. In applying (9.8) to the determination of e^A, we obtain the following algorithm:

Parlett recurrence

Compute the Schur decomposition $A = UTU^H$ of A
for $i = 1, 2, \ldots, n$
  Set $f_{ii} = e^{t_{ii}}$
end
for $\ell = 1, 2, \ldots, n-1$
  for $i = 1, 2, \ldots, n - \ell$
    Set $j = i + \ell$
    Compute f_{ij} by (9.8)
  end
end
Compute $e^A = UFU^H$

A disadvantage of the Parlett recurrence is that it breaks down in the presence of equal diagonal elements (eigenvalues) t_{ii} and t_{jj} of T and can produce large errors in case of almost confluent eigenvalues.

Example 9.2. If

$$A = \begin{bmatrix} 1.0000001 & 0 \\ 0 & 0.9999999 \end{bmatrix}$$

and the computations are done in the single precision IEEE arithmetic ($\mathbf{u} = 2^{-24} \approx 5.96 \cdot 10^{-8}$), then we obtain (to eight decimal digits)

$$\hat{f}_{11} = 2.7182820, \quad \hat{f}_{22} = 2.7182815$$

and

$$\hat{f}_{12} = \text{fl}(\text{fl}(\hat{f}_{11} - \hat{f}_{22})/\text{fl}(t_{11} - t_{22})) = 2.3841858$$

while the exact values are

$$f_{11} = 2.7182821, \quad f_{22} = 2.7182816, \quad f_{12} = 2.7182818.$$

Note that the elements of A are represented exactly in the single precision arithmetic. The large error in $\hat{f}_{12}$ is due to the cancellation that takes place in computing $\hat{f}_{11} - \hat{f}_{22}$ and the subsequent division by $2 \cdot 10^{-7}$.

In Parlett (1974) gives the idea to use the block-diagonal form of

$$T = \begin{bmatrix} T_{11} & T_{12} & \cdots & T_{1p} \\ & T_{22} & \cdots & T_{2p} \\ & & \ddots & \vdots \\ & & & T_{pp} \end{bmatrix},$$

where the diagonal blocks T_{ii} have distinct eigenvalues. In such case, the off-diagonal blocks of F can be computed by using the following block-recurrence:

Schur–Parlett method

$$T_{ii}F_{ij} - F_{ij}T_{jj} = F_{ii}T_{ij} - T_{ij}F_{jj} + \sum_{d=i+1}^{j-1} (F_{id}T_{dj} - T_{id}F_{dj}), \quad (9.9)$$
$$j = 2, 3, \ldots, p, \quad i = j-1, \quad j-2, \ldots, 1.$$

This recurrence represents a Sylvester matrix equation with respect to F_{ij}. In order to have a well-conditioned Sylvester equation (9.9), the eigenvalues of the blocks T_{ii} and T_{jj} should be well separated for each i, j. Similar to the scalar case, it is necessary to know in advance the diagonal blocks $F_{ii} = f(T_{ii})$.

The Schur–Parlett method and the block-diagonalization method have much in common, since in both methods it is necessary to solve the matrix Sylvester equation. In the block-diagonalization method, this is done to zero a superdiagonal block of the Schur form (see Theorem 2.12), while in the Schur–Parlett method, the Sylvester equation is solved to find a superdiagonal block of the matrix exponential.

9.4.4 *Schur–Fréchet Method*

Another opportunity to compute the off-diagonal blocks of the matrix exponential is provided by the Schur–Fréchet algorithm developed and analyzed in Najfeld and Havel (1995), Kenney and Laub (1998) and Higham (2008), Chapter 9.

Consider a block-triangular matrix of the form

$$T = \begin{bmatrix} T_{11} & T_{12} \\ & T_{22} \end{bmatrix}$$

and assume that the blocks $T_{11} \in \mathbb{C}^{n_1 \times n_1}$ and $T_{22} \in Cs^{n_2 \times n_2}$ are scaled so that their norms are less than 1. The exponential of T has the form

$$e^T = \begin{bmatrix} E_{11} & E_{12} \\ & E_{22} \end{bmatrix}.$$

Suppose that one has already calculated $E_{11} = e^{T_{11}}$ and $E_{22} = e^{T_{22}}$. Then the block E_{12} can be computed by the following algorithm, implementing an eighth-order Padé approximants to the function

$x\coth(x)$. (See Higham, 2009 for a detailed exposition of the corresponding method.)

Schur–Fréchet method (Kenney and Laub, 1998)

(1) Let $\alpha_1, \alpha_2, \ldots, \alpha_8$ be the zeros of the polynomial $p(x) = x^8 + 990x^6 + 135135x^4 + 4729725x^2 + 34459425$.
(2) Let $\beta_1, \beta_2, \ldots, \beta_8$ be the zeros of the polynomial $q(x) = 45x^8 + 13860x^6 + 945945x^4 + 16216200x^2 + 34459425$.
(3) Set $D_{12} = T_{12},\ M_{11} = T_{11},\ M_{22} = T_{22}$.
(4) Evaluate the "Sylvester cascade."

for $i = 1, 2, \ldots, 8$
  $R = (I_{n1} + M_{11}/\alpha_i)D_{12} + D_{12}(I_{n_2} - M_{22}/\alpha_i)$
  Solve for D_{12} in $(I_{n1} + M_{11}/\beta_i)D_{12} + D_{12}(I_{n_2} - M_{22}/\alpha_i) = R$
end

(5) $D_{12} = (E_{11}D_{12} + D_{12}E_{22})/2$.
(6) $E_{12} = D_{12}$.

In exact arithmetic, this algorithm produces an approximation $\tilde{E}_{12}$ satisfying $\|\tilde{E}_{12} - E_{12}\|_F < 10^{-16}$ (Kenney and Laub, 1998). Note that the zeros α_i and β_i are *purely imaginary zeros*. The algorithm requires the solution of a "cascade" of eight Sylvester equations.

The Schur–Fréchet algorithm resembles the Schur–Parlett algorithm but as we see later on, the numerical properties of the two algorithms differ very much.

In the following sections, we consider in detail the implementation and properties of the matrix decomposition methods.

9.5 Numerical Jordan Form Method

Let the matrix A have p distinct eigenvalues $\lambda_1, \lambda_2, \ldots, \lambda_p$, each of them with numerical multiplicity ν_i and the set of Segre characteristics associated with the corresponding eigenvalues (Figure 9.2). Assume that the matrix A is reduced into the un-normalized numerical Jordan form

$$J = Z^{-1}AZ \tag{9.10}$$

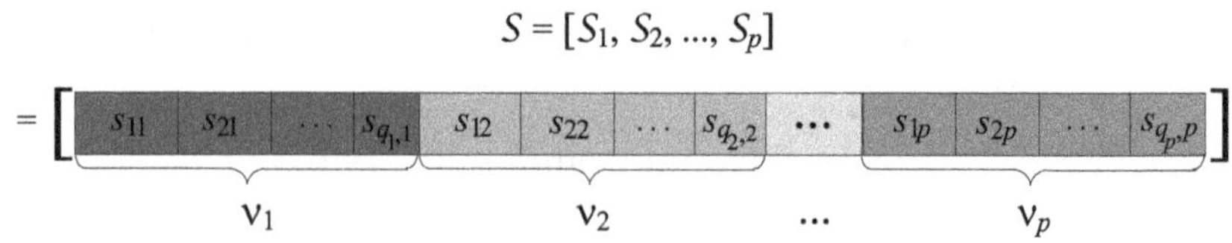

Fig. 9.2. Segre characteristics.

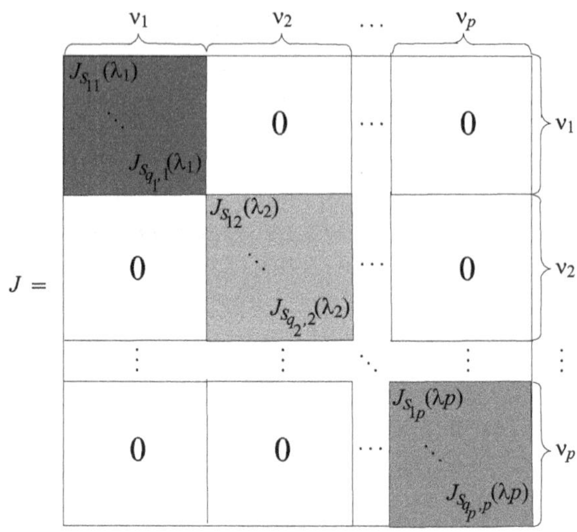

Fig. 9.3. Numerical Jordan form.

shown in Figure 9.3, where each Jordan block $J_{s_{ji}}(\lambda_i)$, $i = 1, 2, \ldots, p$; $j = 1, 2, \ldots, q_i$, is written in a simplified notation as

$$J_s(\lambda) = \begin{bmatrix} \lambda & c_1 & 0 & \ldots & 0 & 0 \\ 0 & \lambda & c_2 & \ldots & 0 & 0 \\ 0 & 0 & \lambda & \ldots & 0 & 0 \\ \vdots & \vdots & \vdots & \ddots & \vdots & \vdots \\ 0 & 0 & 0 & \ldots & \lambda & c_{s-1} \\ 0 & 0 & 0 & \ldots & 0 & \lambda \end{bmatrix}. \tag{9.11}$$

Note that in (9.10) we have $q_i = w_{i1}$.

According to (9.2) and (9.10), the matrix exponential of A is given by

$$e^A = Ze^J Z^{-1},$$

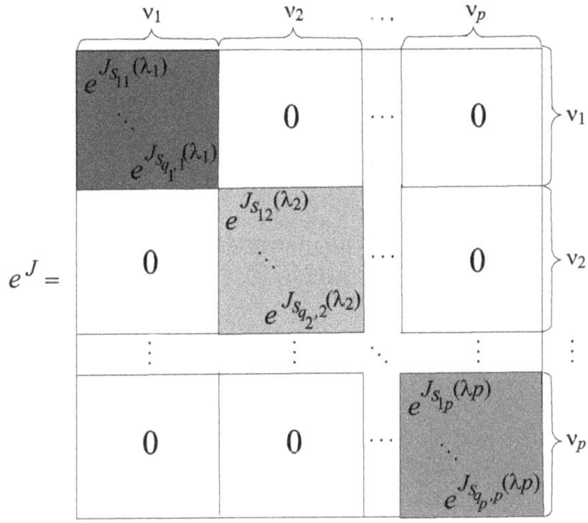

Fig. 9.4. Exponential of the Jordan form.

where e^J is shown in Figure 9.4, and

$$e^{J_s(\lambda)} = e^{\lambda} \begin{bmatrix} 1 & c_1 & c_1c_2/2 & \dots & c_1c_2\dots c_{s-2}/(s-2)! & c_1c_2\dots c_{s-1}/(s-1)! \\ 0 & 1 & c_2 & \dots & c_2c_3\dots c_{s-2}/(s-3)! & c_2c_3\dots c_{s-1}/(s-2)! \\ 0 & 0 & 1 & \dots & c_3c_4\dots c_{s-2}/(s-4)! & c_3c_4\dots c_{s-1}/(s-3)! \\ \vdots & \vdots & \vdots & \ddots & \vdots & \vdots \\ 0 & 0 & 0 & \dots & 1 & c_{s-1} \\ 0 & 0 & 0 & \dots & 0 & 1 \end{bmatrix}. \tag{9.12}$$

The computation of the matrix exponential using the numerical Jordan form is done by Algorithm 9.1.

In Table 9.1, we give estimates of the volume of computations associated with Algorithm 9.1. The estimates of the computation cost are obtained under the assumptions $\nu = n/p,\ s_j = h$.

Example 9.3. This example demonstrates the properties of Algorithm 9.1 for a family of problems with defective matrices and with increasing conditioning of the exponential. In the given case, the tenth-order matrix A is constructed as

$$A = QJQ^{-1},$$

Algorithm 9.1. Computing the matrix exponential using the numerical Jordan form

Determines the exponential e^A *of the matrix* $A \in \mathbb{C}^{n\times n}$ *using the numerical Jordan form of* A

function $[exp\_A, ev] = \mathrm{jord\_exp}(A, ein, tol)$

Input: The matrix $A \in \mathbb{C}^{n\times n}$; ein and tol are tolerance parameters, used in the reduction into Jordan form.

Output: $exp\_A \in \mathbb{C}^{n\times n}$ contains the matrix exponential of A;
$ev \in \mathbb{C}^n$ contains the eigenvalues of A.

```
if ||A||_F = 0
    exp_A = I_n, ev = 0_{n×1}
else
    Use Algorithm 7.5 to determine the numerical Jordan form
    J = Z^{-1}AZ of A and the Segre characteristics
    S = [{s_11,...,s_{q1,1}}, {s_12,... s_{q2,2}},...,{s_1p,...,s_{qp,p}}],
    associated with the distinct eigenvalues λ_1, λ_2,..., λ_p of A
    with multiplicities ν_1, ν_2,..., ν_p.
    Set exp_A = 0_{n×n}
    Set the position of the first Jordan block
    q = 1
    for i = 1,2,...,p
        ev_i = λ_i
        for j = 1,2,...,q_i
            Determine the order of the Jordan block J_{s_ji}(λ_i)
            jblock = s_ji
            if jblock = 1
                exp_A_{q,q} = e^{λ_i}
            else
                exp_A_{q:q+jblock-1,q:q+jblock-1} = I_jblock
                for ii = 1,2,...,jblock - 1
                    f = 0, c = 0
                    for jj = ii + 1, ii + 2,...,jblock - 1
                        f = f + 1
                        c = c · J_{q+jj-2,q+jj-1}/f
                        exp_A_{q+ii-1,q+jj-1} = c
                    end
                end
                exp_A_{q:q+jblock-1,q:q+jblock-1}
                        = e^{λ_i} exp_A_{q:q+jblock-1,q:q+jblock-1}
            end
            Determine the position of the next Jordan block
            q = q + jblock
        end
    end
    exp_A ← Z · exp_A · Z^{-1}
end
```

Table 9.1. Number of computational operations for reduction into Jordan form and computing the diagonal blocks of the matrix exponential.

Operation	Number of flops
Reduction of A into Jordan form J	$34\frac{2}{3}n^3 + \frac{(p-1)n^3}{p^2} + \frac{2}{3}\frac{n^3 h}{p} + \frac{4n^3 h}{p^2} + \frac{5}{3}\frac{n^3}{p^2}$
Case of $p = 1, h = n, \nu = n$	$4\frac{2}{3}n^4 + 46\frac{1}{3}n^3 47\frac{1}{3}n^3$
Case of $p = n, h = 1, \nu = 1$	
Computing the exponentials of the diagonal blocks	$\frac{2}{3}(h^3 - h)$
Case of $p = 1, h = n, \nu = n$	$\frac{2}{3}(n^3 - n)$
Case of $p = n, h = 1, \nu = 1$	—
Total: Reduction into Jordan form and computing the exponentials of the diagonal blocks	$34\frac{2}{3}n^3 + \frac{(p-1)n^3}{p^2} + \frac{2}{3}\frac{n^3 h}{p} + \frac{4n^3 h}{p^2} + \frac{5}{3}\frac{n^3}{p^2} + \frac{2}{3}(h^3 - h)$
Case of $p = 1, h = n$	$4\frac{2}{3}n^4 + 47n^3 47\frac{1}{3}n^3$
Case of $p = n, h = 1$	

where the desired Jordan form

$$J = \operatorname{diag}\left(\begin{bmatrix} \lambda_1 & 1 & 0 & 0 & 0 \\ 0 & \lambda_1 & 1 & 0 & 0 \\ 0 & 0 & \lambda_1 & 1 & 0 \\ 0 & 0 & 0 & \lambda_1 & 1 \\ 0 & 0 & 0 & 0 & \lambda_1 \end{bmatrix}, \begin{bmatrix} \lambda_2 & 1 & 0 & 0 & 0 \\ 0 & \lambda_2 & 1 & 0 & 0 \\ 0 & 0 & \lambda_2 & 1 & 0 \\ 0 & 0 & 0 & \lambda_2 & 1 \\ 0 & 0 & 0 & 0 & \lambda_2 \end{bmatrix} \right)$$

consists of two blocks with distinct eigenvalues. The nonsingular transformation matrix Q is constructed as described in Example 4.4:

$$\begin{aligned} &Q = H_2 \Sigma H_1, \quad Q^{-1} = H_1 \Sigma^{-1} H_2, \\ &H_1 = I_n - \frac{2}{10} u u^T, \quad H_2 = I_n - \frac{2}{10} v v^T, \\ &u = [1,\ 1,\ 1, \dots,\ 1]^T, \quad v = [1,\ -1,\ 1, \dots,\ (-1)^9]^T, \\ &\Sigma = \operatorname{diag}(1, \sigma,\ \sigma^2, \dots,\ \sigma^9), \end{aligned}$$

where H_1, H_2 are elementary reflections and the condition number $\text{cond}_2(Q)$ of Q with respect to the inversion is equal to σ^9. An accurate approximation of the exact exponential of A is determined from $e^A = Qe^JQ^{-1}$, where e^J is evaluated using its analytical expression (9.12). The Segre characteristics of A are $\mathcal{S}(A, \lambda_1) = \{5\}$, $\mathcal{S}(A, \lambda_2) = \{5\}$ and the corresponding Weyr characteristics are $\mathcal{W}(A, \lambda_1) = \{1,\ 1,\ 1,\ 1,\ 1\}$, $\mathcal{W}(A, \lambda_2) = \{1,\ 1,\ 1,\ 1,\ 1\}$. In the given case, the parameter σ controlling the conditioning of Q is chosen to vary between 1 and 6 with a step equal to 0.2. Note that for $\sigma = 6$ one has that $\text{cond}(Q) = 1.01 \cdot 10^7$, i.e., for this value of σ, the Jordan decomposition is ill conditioned. In computing the numerical Jordan form, we take $ein = tol = 10^{-14}$.

The relative forward error

$$\|\hat{e^A} - e^A\|_F / \|e^A\|_F$$

in the computed e^A as a function of σ evaluated by Algorithm 9.1 for $\lambda_1 = 5$, $\lambda_2 = 1$ and $ein = tol = 10^{-14}$ is shown in Figure 9.5 along with the errors of the exponential evaluated by the MATLAB®functions **expm** and **funm**. The relative error is less than

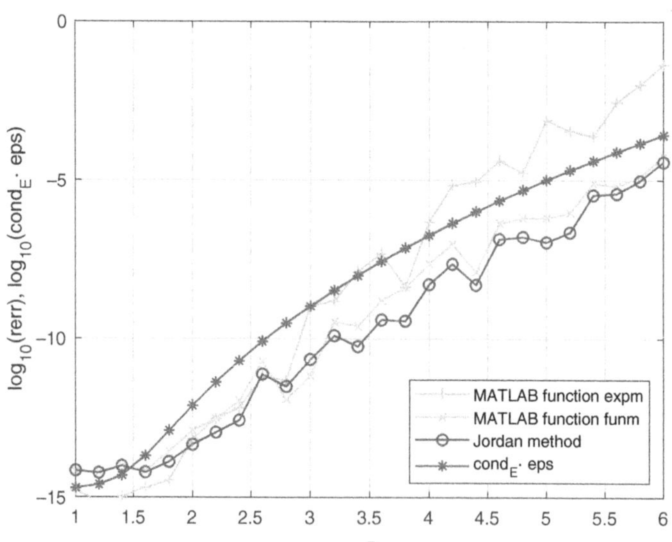

Fig. 9.5. Relative errors in computing the matrix exponential ($\lambda_1 = 5$, $\lambda_2 = 1$).

the bound $\text{cond}_E \cdot eps$, where eps is the MATLAB®function **eps**, $eps = 2^{-52} = 2\mathbf{u}$. For this example, Algorithm 9.1 produces the best results.

In case of close eigenvalues of the Jordan blocks and ill-conditioned matrix Q, it is possible that the numerical Jordan form is computed erroneously which may introduce large errors in the computed exponential.

In Figure 9.6, we show the errors computed for the same example but in case of $\lambda_1 = 1$ and $\lambda = 0.1$. The large errors in e^A computed by Algorithm 9.1 for $\sigma \geq 5.0$ are due to the errors in computing the dimensions of the Jordan blocks. For $s = 5.0$, instead of two Jordan blocks of order 5, Algorithm 6.5 determines one Jordan block of dimension 10 with associated eigenvalue equal to 0.55 (the mean value of 1.0 and 0.1). This error is due to a failure of the clustering algorithm to determine correctly the eigenvalue multiplicity. Note that for this value of σ the condition number of Q is $1.01 \cdot 10^7$ which makes the eigensystem of A ill conditioned. If we change the values of the tolerances to $ein = 10^{-16}$, $tol = 10^{-16}$, then the Jordan form is computed correctly for values of σ up to 5.4.

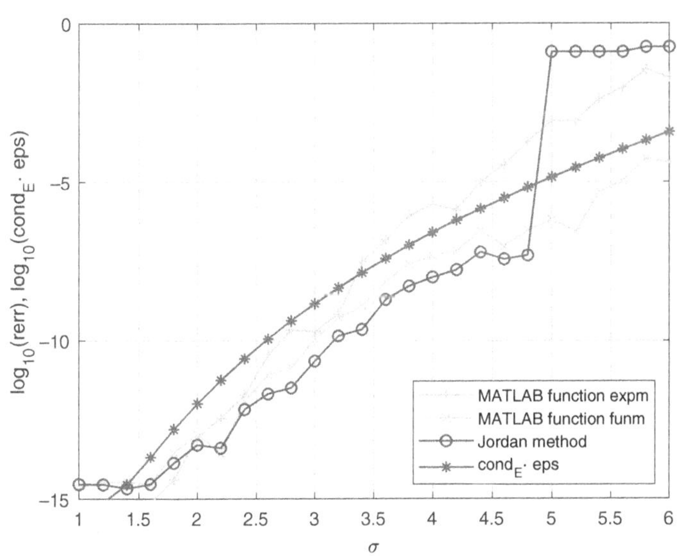

Fig. 9.6. Relative errors in computing the matrix exponential ($\lambda_1 = 1$, $\lambda_2 = 0.1$).

9.6 Block-Diagonal Staircase Form Method

The implementation of the Jordan form in computing the matrix exponential has the disadvantage that in some cases the corresponding similarity transformation matrix can be ill conditioned. This leads to the idea instead of the Jordan form to use some of the associated block-diagonal forms which are better conditioned than the Jordan form. For this aim, it is appropriate to use the block-diagonal staircase form (see Algorithm 7.3 in Section 7.3).

Assume that the matrix A has p distinct eigenvalues $\lambda_1, \lambda_2, \ldots, \lambda_p$ with multiplicities $\nu_1, \nu_2, \ldots, \nu_p$, respectively, and associated Weyr characteristics $\mathcal{W}_1, \mathcal{W}_2, \ldots, \mathcal{W}_p$. Then the matrix A can be reduced into the block-diagonal staircase form

$$B_S = X_S^{-1} A X_S = \begin{bmatrix} S(\mathcal{W}_1, \lambda_1) & & & \\ & S(\mathcal{W}_2, \lambda_2) & & \\ & & \ddots & \\ & & & S(\mathcal{W}_p, \lambda_p) \end{bmatrix} \tag{9.13}$$

where the diagonal blocks are in the staircase form

$$S(\mathcal{W}_i, \lambda_i) = \begin{bmatrix} \lambda_i I_{w_{i1}} & N_{12}^{(i)} & \cdots & N_{1,h_i-1}^{(i)} & N_{1,h_i}^{(i)} \\ & \lambda_i I_{w_{i2}} & \cdots & N_{2,h_i-1}^{(i)} & N_{2,h_i}^{(i)} \\ & & \ddots & \vdots & \vdots \\ & & & \lambda_i I_{w_{i,h_i-1}} & N_{h_i-1,h_i}^{(i)} \\ & & & & \lambda_i I_{w_{i,h_i}} \end{bmatrix} \in \mathbb{C}^{\nu_i \times \nu_i}. \tag{9.14}$$

To simplify the notation, further on the diagonal blocks will be written as

$$S(\mathcal{W}, \lambda) = \begin{bmatrix} \lambda I_{w_1} & N_{12} & \cdots & N_{1,h-1} & N_{1,h} \\ & \lambda I_{w_2} & \cdots & N_{2,h-1} & N_{2,h} \\ & & \ddots & \vdots & \vdots \\ & & & \lambda I_{w_{h-1}} & N_{h-1,h} \\ & & & & \lambda I_{w_h} \end{bmatrix} \in \mathbb{C}^{\nu \times \nu},$$

$$N_{ij} \in \mathbb{C}^{w_i \times w_j}, \tag{9.15}$$

where the Weyr characteristic associated with λ is denoted by

$$\mathcal{W} = \mathcal{W}(S, \lambda) = \{w_1, \ w_2, \ldots, w_h\},$$

the corresponding Segre characteristic by

$$\mathcal{S} = \mathcal{S}(S, \lambda) = \{s_1, \ s_2, \ldots, s_q\}, \ \ \nu = \sum_{k=1}^{h} w_k = \sum_{\ell=1}^{q} s_\ell$$

and $h = s_1$, $q = w_1$. Due to the staircase structure of the block $S(\mathcal{W}, \lambda)$, its exponential can be found by finite Taylor series accessing only the nonzero blocks $N_{k\ell}$.

Denote by $N = S(\mathcal{W}, \lambda) - \lambda I_\nu \in \mathbb{C}^{\nu \times \nu}$ the nilpotent part of a diagonal block, which has an index of nilpotence equal to

$$1 \leq h = s_1 \leq \nu.$$

To determine the exponentials of the diagonal blocks, we exploit the relationship

$$e^{S(\mathcal{W},\lambda)} = e^{\lambda} e^{N}.$$

Thus the problem for computing $e^{S(\mathcal{W},\lambda)}$ reduces to the evaluation of the exponential of a nilpotent matrix N with a staircase structure. This gives several advantages from numerical point of view.

The Taylor series expansion

$$e^N = I_\nu + N/1! + N^2/2! + \cdots + N^{h-1}/(h-1)!,$$

used to determine the exponential contains only the first $h - 1$ powers of N (if one doesn't count $N^0 - I_\nu$). Note that $h = \nu$ only in the extreme case of one Jordan block associated with λ of order ν when $\mathcal{W}(S, \lambda) = \{1, \ 1, \ldots, 1\}$ and $\mathcal{S}(S, \lambda) = \{\nu\}$ (nonderogatory but defective matrix). In the other extreme case of $\mathcal{S}(S, \lambda) = \{1, 1, \ldots, 1\}$, $\mathcal{W}(S, \lambda) = \{\nu\}$ (derogatory but nondefective matrix), the index of nilpotence h is equal to 1 and the exponential is found directly from

$$e^{S(\mathcal{W},\lambda)} = e^{\lambda} I_\nu.$$

For $h = 2$, one has that $\exp(N) = I_\nu + N$ so that the raising of N to power is not necessary. In the generic case, it is fulfilled that $1 < h < \nu$ which allows obtaining the exponential with a few terms

Algorithm 9.2. Matrix exponential of a staircase block
Given a $\nu \times \nu$ block S in the staircase form (9.15), *this algorithm computes its exponential using finite term Taylor series combined with scaling and squaring.*

```
function exp_S = exp_staircase(S, W)
Input: A matrix S = λI_ν + N ∈ C^{ν×ν} with an eigenvalue λ and strictly
        upper triangular part N in staircase form; the Weyr characteristic
        W = {w_1, w_2, ..., w_h} of λ
Output: The matrix exponential exp_S = exp(S)
λ = S(1,1),  S = S − λI_ν
Scale S so that ||S||_∞ ≤ 1,
if ||S||_∞ ≤ 1
    s = 0
else
    while ||S||_∞ > 1 do
        S = S/2
        s = s + 1
    end
end
Compute the exponential using h − 1 terms Taylor series
if h = 1
    E = I_ν
else
    E = I_ν,  F = S,  E = E + F
    if h > 2
        for m = 2 : h − 1
            G = [0_{w_i×w_j}]_{i=1:h, j=1:h}
            for k = 1 : h − m
                for ℓ = m + 1 : h
                    if k + 1 ≤ ℓ − m + 1
                        for d = k + 1 : ℓ − m + 1
                            G_{kℓ} = G_{kℓ} + S_{kd} F_{dℓ}
                        end
                    end
                end
            end
            F = G/m,  E = E + F
        end
    end
end
```

Square s times
if $s \geq 1$
  for $m = 1 : s$
    $E \leftarrow E^2$
  end
end
$exp\_S = e^{\lambda} E$

of the Taylor series expansion. Apart from the increasing efficiency, this allows avoiding some numerical difficulties related to the decision about the termination of the expansion. Since the nilpotent matrix is in staircase form, it is possible to achieve further reduction of the number of computational operations in evaluating the powers of this matrix. Denoting by F^m the mth power of N, one has that

$$F^m = NF^{m-1} = \begin{bmatrix} 0 & N_{12} & \dots & N_{1,h-1} & N_{1h} \\ & 0 & \dots & N_{2,h-1} & N_{2h} \\ & & \ddots & \vdots & \vdots \\ & & & 0 & N_{h-1,h} \\ & & & & 0 \end{bmatrix}$$

$$\cdot \begin{bmatrix} 0 \ (F^{m-1})_{1m} & \dots & (F^{m-1})_{1,h-1} & (F^{m-1})_{1h} \\ 0 & \dots & (F^{m-1})_{2,h-1} & (F^{m-1})_{2h} \\ & \ddots & \vdots & \vdots \\ & & 0 & (F^{m-1})_{h-m+1,h} \\ & & & 0 \end{bmatrix},$$

$$m = 2, 3, \dots, h-1, \quad F^1 = N.$$

Hence, the power F^m, $m = 2, 3, \dots, h-1$, of the nilpotent matrix N with staircase structure can be evaluated efficiently from F^{m-1} for $h > 2$. In computing the powers of the nilpotent matrix, one can use the scaling and squaring technique which makes the norm of N less than one. The scaling is necessary in cases of large norms of N and small values of the index of nilpotence h, when the norm of the power N^m decreases quickly with m which is related to cancellations in the floating point arithmetic and introduction of large rounding errors (see Example 1.10). The scaling and squaring exploits the relationship $e^N = (e^{N/q})^q$, where $q = 2^s$ and s is chosen so that $\|N/q\|_\infty < 1$. Hence

$$e^N = (e^{N/2^s})^{2^s}.$$

The computation of the exponential $e^{S(\mathcal{W},\lambda)}$ of a diagonal block in staircase form is done by Algorithm 9.2. Note that the squaring of the matrix E can be done taking into account that the matrix E is upper triangular.

The matrix exponential of a matrix A can be found by implementing Algorithm 9.2 to each diagonal block of the matrix B_S. This is done by Algorithm 9.3.

Algorithm 9.3. Computing the matrix exponential using the block-diagonal staircase form

Computes the exponential e^A of the matrix $A \in \mathbb{C}^{n\times n}$ using the block-diagonal staircase form of A

function $[exp\_A, ev]$ = exp_diag_staircase(A)

Input: The matrix $A \in \mathbb{C}^{n\times n}$; *ein* and *tol* are tolerance parameters, used in the reduction into block-diagonal staircase form.

Output: $exp\_A \in \mathbb{C}^{n\times n}$ contains the matrix exponential of A; $ev \in \mathbb{C}^n$ contains the eigenvalues of A.

if $\|A\|_F = 0$
    $exp\_A = I_n$
    $ev = 0_{n\times 1}$
else
    Using Algorithm 7.3, determine the block-diagonal staircase form
    $B_S = X_S^{-1}AX_S = \text{diag}\,(S(\mathcal{W}_1,\lambda_1), S(\mathcal{W}_2,\lambda_2), \ldots, S(\mathcal{W}_p,\lambda_p))$
    of A, where $S(\mathcal{W}_i,\lambda_i) \in \mathbb{C}^{\nu_i\times\nu}$
    Set $exp\_S = 0_{n\times n}$
    $q = 1$
    for $i = 1, 2, \ldots, p$
        $ev(i) = \lambda_i$
        Compute $e^{S(\mathcal{W}_i,\lambda_i)}$ using Algorithm 9.2
        $exp\_S_{q:q+\nu_i-1,q:q+\nu_i-1} = e^{S(\mathcal{W}_i,\lambda_i)}$
        $q = q + \nu_i$
    end
    $exp\_A = X_S \cdot exp\_S \cdot X_S^{-1}$
end

Table 9.2. Number of computational operations for reduction into block-diagonal staircase form and computing the diagonal blocks of the matrix exponential.

Operation	Number of flops
Reduction of A into block-diagonal staircase form B_S	$34\frac{2}{3}n^3 + \frac{(p-1)n^3}{p^2} + \frac{2}{3}\frac{n^3h}{p} + \frac{4n^3h}{p^2}$
Case of $p = 1, h = n, \nu = n$	$4\frac{2}{3}n^4 + 34\frac{1}{3}n^3$
Case of $p = n, h = 1, \nu = 1$	$34\frac{2}{3}n^3$
Computing the exponentials of the diagonal blocks	$(h^4 + 3\frac{1}{3}h^3 - 29h^2 + 48\frac{2}{3}h - 24)\frac{n^3}{12h^3}$
Case of $p = 1, h = n, \nu = n$	$\frac{1}{12}n^4 + \frac{5}{18}n^3$
Case of $p = n, h = 1, \nu = 1$	-
Total: Reduction into block-diagonal staircase form and computing the exponentials of the diagonal blocks	$34\frac{2}{3}n^3 + \frac{(2h+3)p + 12h - 3}{3p^2}n^3 + (3h^4+10h^3-87h^2+146h-72)\frac{n^3}{36h^3}$
Case of $p = 1, h = n, \nu = n$	$4\frac{3}{4}n^4 + 34\frac{11}{18}n^3$
Case of $p = n, h = 1, \nu = 1$	$34\frac{2}{3}n^3$

Consider the volume of the computational work associated with determining the exponentials of the diagonal blocks done by Algorithm 9.3. In Table 9.2, we give an approximate number of the operations necessary for reduction into block-diagonal staircase form and computing the diagonal blocks of the exponential. The volume of work is determined under the assumptions for a matrix with equal eigenvalue multiplicities ($\nu = n/p$) and equal elements of the Weyr characteristic ($w_j = \nu/h$).The number of operations is given for the two extreme cases:

- derogatory but nondefective matrix with scalar Jordan blocks ($h = 1$); and
- nonderogatory but defective matrix with a single Jordan block ($h = n$).

As seen from Table 9.2, the number of operations in these cases varies between $34\frac{2}{3}n^3$ and $4\frac{3}{4}n^4 + 34\frac{17}{18}n^3$.

Example 9.4. This example demonstrates the advantage of using the block-diagonal staircase form in cases when the matrix A has one eigenvalue of multiplicity $\nu = n$. The matrix A is of even order and is taken as

$$A = QJQ^{-1},$$

where the Jordan form

$$J = \operatorname{diag}\left(\begin{bmatrix} \lambda & 1 \\ & \lambda \end{bmatrix}, \begin{bmatrix} \lambda & 1 \\ & \lambda \end{bmatrix}, \dots, \begin{bmatrix} \lambda & 1 \\ & \lambda \end{bmatrix}\right)$$

consists of 2×2 blocks with equal eigenvalues and the nonsingular transformation matrix Q is constructed as in Example 9.3. In the given case, we take $\lambda = 0.1$ and the parameter σ controlling the conditioning of Q is chosen equal to 1.25. The Segre characteristic of A associated with λ is equal to $\mathcal{S}(A, \lambda) = \{2,\ 2,\ 2, \dots,\ 2\}$ and the corresponding Weyr characteristic is $\mathcal{W}(A, \lambda) = \{n/2,\ n/2\}$. The block-diagonal staircase form is determined for $ein = 10^{-14}$, $tol = 10^{-14}$. By increasing n, the condition number cond_E of the matrix exponential increases quickly, but in the same, the 2-norm condition

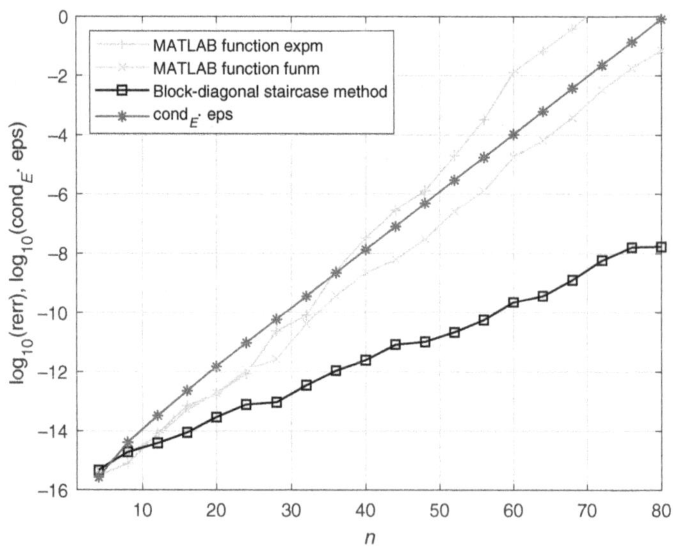

Fig. 9.7. Relative errors in computing the matrix exponential using the block-diagonal staircase form.

number of the transformation matrix into block-diagonal staircase form remains equal to 1. This is due to the fact that the eigenvalue λ participates in only one block ($\nu_1 = n$) and the reduction of this block into staircase form is done by using unitary similarity transformations which are perfectly conditioned. As shown in Figure 9.7, this allows obtaining the matrix exponential with high accuracy, as opposite to the MATLAB®functions `expm` and `funm` which do not "recognize" the special structure of the matrix A.

9.7 Jordan–Schur Methods

The two methods presented in this section are modifications of the Schur–Parlett and Schur–Fréchet methods for computing the matrix exponential exploiting the Jordan–Schur form instead of the Schur form of the given matrix. This form is obtained by the version of the algorithm of Kågström and Ruhe present in Chapter 6. The usage of the Jordan–Schur form

$$J_S = U_S^H A U_S = \begin{bmatrix} S(\mathcal{W}_1, \lambda_1) & T_{12} & \dots & T_{1p} \\ & S(\mathcal{W}_2, \lambda_2) & \dots & T_{2p} \\ & & \ddots & \vdots \\ & & & S(\mathcal{W}_p, \lambda_p) \end{bmatrix} \in \mathbb{C}^{n \times n} \tag{9.16}$$

in which the diagonal blocks $S(\mathcal{W}_i, \lambda_i)$ have distinct eigenvalues λ_i, $i = 1, 2, \dots, p$, simplifies the corresponding computations and increases the accuracy. More specifically, the benefit of using the Jordan–Schur form is twofold.

First, the diagonal blocks are in staircase form so that their exponentials can be found by finite Taylor series accessing only the nonzero blocks $N_{k\ell}$ (Section 9.6). Second, the Jordan–Schur form is in upper block-triangular form with the numerical multiple eigenvalues already clustered into blocks. This facilitates significantly the usage of Parlett block-recurrence (9.9). Also, due to the special structure of the matrices T_{ii} and T_{jj}, the solution of the Sylvester equation in the Schur–Parlett method and the "Sylvester cascade" in the Schur–Fréchet method can be done more efficiently in comparison with other methods.

It should be pointed out that if the matrix A has distinct eigenvalues, its Jordan–Schur form coincides with the Schur form. In such cases, the Jordan–Schur algorithms for computing the exponential do not have some advantage over the Schur algorithms.

9.7.1 *Schur–Parlett Method*

Using the Schur–Parlett method, the exponential of the Jordan–Schur form J_S is determined as

$$e^{J_S} = \begin{bmatrix} E_{11} & E_{12} & \cdots & E_{1p} \\ & E_{22} & \cdots & E_{2p} \\ & & \ddots & \vdots \\ & & & E_{pp} \end{bmatrix},$$

where the diagonal blocks $E_{ii} = e^{S(\mathcal{W}_i, \lambda_i)}$, $i = 1, \ldots, p$, are already evaluated by using Algorithm 9.2.

The computation of the off-diagonal blocks E_{ij} of the matrix exponential implementing the Parlett block-recurrence involves the solution of the Sylvester equations

$$S_{ii}E_{ij} - E_{ij}S_{jj} = E_{ii}T_{ij} - T_{ij}E_{jj} + Z,$$
$$j = 2, 3, \ldots, p, \quad i = j-1, \quad j-2, \ldots, 1, \tag{9.17}$$

where the blocks S_{ii} and S_{jj} are of the form (9.15), T_{ij} are the superdiagonal blocks of the Jordan–Schur form (9.16) and

$$R = \sum_{d=i+1}^{j-1} (E_{id}T_{dj} - T_{id}E_{dj}).$$

The block R appears in (9.17) only if $i + 1 \leq j - 1$. Note that the off-diagonal blocks are computed in a specific order shown in the following for $p = 5$:

$$\begin{bmatrix} \square & 1 & 3 & 6 & 10 \\ & \square & 2 & 5 & 9 \\ & & \square & 4 & 8 \\ & & & \square & 7 \\ & & & & \square \end{bmatrix}.$$

Due to the specific structure of the blocks S_{ii} and S_{jj}, the equation (9.17) is called *staircase Sylvester equation.* It can be solved efficiently taking into account the structure of the diagonal blocks S_{ii} and S_{jj}.

Consider the staircase Sylvester equation

$$S_A X + X S_B = C, \tag{9.18}$$

where for simplicity we use the following notation:

$$S_A = \begin{bmatrix} \lambda I_{wa_1} & N_{12} & \dots & N_{1,ha} \\ & \lambda I_{wa_2} & \dots & N_{2,ha} \\ & & \ddots & \vdots \\ & & & \lambda I_{wa_{ha}} \end{bmatrix} \in \mathbb{C}^{\kappa\times\kappa},$$

$$S_B = \begin{bmatrix} \mu I_{wb_1} & R_{12} & \dots & R_{1,hb} \\ & \mu I_{wb_2} & \dots & R_{2,hb} \\ & & \ddots & \vdots \\ & & & \mu I_{wb_{hb}} \end{bmatrix} \in \mathbb{C}^{\nu\times\nu}.$$

It is assumed that S_A and S_B have eigenvalues λ and μ with the associated Weyr characteristics

$$\mathcal{W}(S_A, \lambda) = \{wa_1,\ wa_2, \dots, wa_{ha}\},$$
$$\mathcal{W}(S_B, \mu) = \{wb_1,\ wb_2, \dots, wb_{hb}\},$$

respectively.

Partitioning the right-hand side $C \in \mathbb{C}^{\kappa\times\nu}$ and the unknown matrix $X \in \mathbb{C}^{\kappa\times\nu}$ as

$$C = \begin{bmatrix} C_{11} & C_{12} & \dots & C_{1,hb} \\ C_{21} & C_{22} & \dots & C_{2,hb} \\ \vdots & \vdots & \ddots & \vdots \\ C_{ha,1} & C_{ha,2} & \dots & C_{ha,hb} \end{bmatrix},$$

$$X = \begin{bmatrix} X_{11} & X_{12} & \dots & X_{1,hb} \\ X_{21} & X_{22} & \dots & X_{2,hb} \\ \vdots & \vdots & \ddots & \vdots \\ X_{ha,1} & X_{ha,2} & \dots & X_{ha,hb} \end{bmatrix},$$

where $C_{k\ell}, X_{k\ell} \in \mathbb{C}^{wa_k \times wb_\ell}$, one obtains

$$X_{k\ell} = \frac{1}{\lambda + \mu}(C_{k\ell} - \sum_{f=k+1}^{ha} N_{kf} X_{f\ell} + \sum_{g=1}^{\ell-1} X_{kg} R_{g\ell}), \tag{9.19}$$

$$\ell = 1, 2, \ldots, hb, \quad k = ha, \quad ha - 1, \ldots, 1.$$

Using (9.19), the staircase Sylvester equation (9.18) is solved by Algorithm 9.4.

Algorithm 9.4. Solution of a staircase Sylvester equation
Solution of the Sylvester equation $S_A X + X S_B = C$, where the strictly upper triangular parts of the matrices S_A and S_B are in staircase form

function $X = sylv\_stairs(S_A, S_B, C)$
Input: Matrices $S_A = \lambda I + N \in \mathbb{C}^{\kappa \times \kappa}$, $S_B = \mu I + R \in \mathbb{C}^{\nu \times \nu}$
with strictly upper triangular parts N and R in staircase form with Weyr characteristics
$\mathcal{W}(S_A, \lambda) = \{wa_1, \; wa_2, \ldots, wa_{ha}\}$ and
$\mathcal{W}(S_B, \mu) = \{wb_1, \; wb_2, \ldots, wb_{hb}\}$, respectively;
matrix $C \in \mathbb{C}^{\kappa \times \nu}$
Output: Solution $X \in \mathbb{C}^{\kappa \times \nu}$

```
for ℓ = 1 : hb
    for k = ha : −1 : 1
        Set W = 0_{wa_k, wb_ℓ}
        if k + 1 ≤ ha
            for f = i + 1 : ha
                W = W + N_{kf} Y_{fℓ}
            end
        end
        Set Z = 0_{wa_k, wb_ℓ}
        if ℓ > 1
            for g = 1 : ℓ − 1
                Z = Z + X_{kg} R_{gℓ}
            end
        end
        if λ + μ ≠ 0
            X_{kℓ} = (C_{kℓ} − W − Z)/(λ + μ)
        else
            Error: The equation has no solution
        end
    end
end
```

Algorithm 9.4 resembles the Bartels–Stewart algorithm (Algorithm 4.1 in Section 4.2.3) for solving the Sylvester equation with one significant difference. If the elements wa_k and wb_ℓ of the Weyr characteristics are large, then Algorithm 9.4 finds at once a large block $X_{k\ell}$ instead of several blocks of size 1×1. Hence, in cases of multiple eigenvalues and large elements of the Weyr characteristics, Algorithm 9.4 is more efficient than the Bartels–Stewart algorithm. In fact, the solution of the staircase Sylvester equation (9.18) in which the matrix S_A is $\kappa \times \kappa$ and the matrix S_B is $\nu \times \nu$ by Algorithm 9.4 requires

$$(ha-1)\frac{\kappa^2\nu}{ha} + (hb-1)\frac{\kappa\nu^2}{hb} \text{ flops.} \tag{9.20}$$

For large indices of nilpotence ($ha = \kappa, hb = \nu$), all elements of the Weyr characteristics are equal to 1 and Algorithm 9.4 reduces to the Bartels–Stewart algorithm which requires

$$\kappa(\kappa-1)\nu + \kappa\nu(\nu-1) \text{ flops.}$$

By decreasing the indices of nilpotence and increasing the elements of the Weyr characteristics, the computational work associated with Algorithm 9.4 decreases.

Example 9.5. Consider the solution of a staircase equation $S_A X + X S_B = C$, whose matrices are given as follows:

The eighth-order matrix

$$S_A = \left[\begin{array}{ccccc|ccc}
\lambda & & & & & \times & \times & \times \\
 & \lambda & & & & \times & \times & \times \\
 & & \lambda & & & \times & \times & \times \\
 & & & \lambda & & \times & \times & \times \\
 & & & & \lambda & \times & \times & \times \\
\hline
 & & & & & \lambda & & \\
 & & & & & & \lambda & \\
 & & & & & & & \lambda
\end{array}\right] \in \mathbb{C}^{\kappa\times\kappa},$$

where ($\times$) denotes a nonzero element, has an eigenvalue λ with multiplicity $\kappa = 8$ and associated Weyr characteristic $\mathcal{W}(S_A, \lambda) = \{5,\ 3\}$.

The fifth-order matrix

$$S_B = \left[\begin{array}{ccc|cc} \mu & & & \times & \times \\ & \mu & & \times & \times \\ & & \mu & \times & \times \\ \hline & & & \mu & \\ & & & & \mu \end{array}\right] \in \mathbb{C}^{\nu\times\nu}$$

has an eigenvalue μ with $\nu = 5$ and a Weyr characteristic $\mathcal{W}(S_B, \mu) = \{3,\ 2\}$. Hence the solution matrix X has the form (the order of computing is shown with numbers in bold face)

$$\begin{array}{rl} & \quad\; wb_1 = 3 \quad wb_2 = 2 \\ X = \begin{array}{c} wa_1 = 5 \\ \hline wa_2 = 3 \end{array} & \left[\begin{array}{c|c} \mathbf{2} & \mathbf{4} \\ \hline \mathbf{1} & \mathbf{3} \end{array}\right], \end{array}$$

where the straight lines correspond to the partition of blocks with sizes determined by the elements of the respective Weyr characteristics. Note that there are four blocks with dimensions 5×3, 5×2, 3×3 and 3×2, respectively. Thus, instead of solving $8 \cdot 5 = 40$ scalar equations as in the Bartels–Stewart algorithm, it is necessary to solve 4 block matrix equations.

The evaluation of the off-diagonal blocks of the exponential of a matrix in Jordan–Schur form by the Schur–Parlett method is done by Algorithm 9.5. After computing e^{J_S}, the exponential of A is determined by

$$e^A = U_S e^{J_S} U_S^H,$$

where U_S is the matrix of the unitary transformation of A into J_S.

In Table 9.3, we give estimates of the volume of computations associated with Algorithm 9.5. The estimates of the computation cost are obtained under the assumptions $\nu = n/p$, $w_j = \nu/h$.

Algorithm 9.5. Computing the matrix exponential using the Schur–Parlett method

Evaluating the matrix exponential of a matrix in Jordan–Schur form using the Parlett block-recurrence method.

function $E = jsf\_parlett(J_S)$

Input: A matrix in Jordan–Schur form

$$J_S = \begin{bmatrix} S(\mathcal{W}_1, \lambda_1) & T_{12} & \dots & T_{1p} \\ & S(\mathcal{W}_2, \lambda_2) & \dots & T_{2p} \\ & & \ddots & \vdots \\ & & & S(\mathcal{W}_p, \lambda_p) \end{bmatrix} \in \mathbb{C}^{n \times n},$$

with diagonal blocks $S(\mathcal{W}_i, \lambda_i) \in \mathbb{C}^{\nu_i \times \nu_i}$ that correspond to distinct multiple eigenvalues λ_i and have the form $S(\mathcal{W}_i, \lambda_i) = \lambda_i I_{\nu_i} + N_i$, where N_i is a staircase matrix, $i = 1, 2, \dots, p$.

Output: $E = e^{J_S}$

if $\|J_S\|_\infty = 0$
    $E = I_n$
else
    for $i = 1, 2, \dots, p$
        Compute $E_{ii} = e^{S(w_i, \lambda_i)}$ using Algorithm 9.2
    end
    if $p > 1$
        for $j = 2, 3, \dots, p$
            for $i = j - 1 : -1 : 1$
                Set $R = 0_{\nu_i \times \nu_j}$
                if $i + 1 \le j - 1$
                    for $d = i + 1 : j - 1$
                        $R = R + E_{id}T_{dj} - T_{id}E_{dj}$
                    end
                end
                Solve the staircase Sylvester equation
                    $S_{ii}E_{ij} - E_{ij}S_{jj} = E_{ii}T_{ij} - T_{ij}E_{jj} + R$
                for E_{ij} using Algorithm 9.4
            end
        end
    end
end

Table 9.3. Number of computational operations required by the Schur–Parlett algorithm.

Operation	Number of flops
Reduction into Jordan–Schur form and computing the exponentials of the diagonal blocks	$32\frac{2}{3}n^3 + \frac{2}{3}\frac{(h+3)p+6h-3}{p^2}n^3 + (3h^4+10h^3-87h^2+146h-72)\frac{n^3}{36h^3}$
Case of $p=1, h=n, \nu=n$	$4\frac{3}{4}n^4 + 32\frac{17}{18}n^3$
Case of $p=n, h=1, \nu=1$	$42\frac{2}{3}n^3$
Computing the off-diagonal blocks $1 \le h \le n,\ 1 \le p \le n$ $\nu = n/p,\ w_j = \nu/h$	$\frac{2(p^3-p)\nu^3}{3h^3}$
Case of $p=1, h=n, \nu=n$	—
Case of $p=n, h=1, \nu=1$	$\frac{2(n^3-n)}{3}$
Total: Reduction into Jordan–Schur form and computing its exponential $1 \le h \le n,\ 1 \le p \le n$ $\nu = n/p,\ w_j = \nu/h$	$32\frac{2}{3}n^3 + \frac{2}{3}\frac{(h+3)p+6h-3}{p^2}n^3 + (3h^4+10h^3-87h^2+146h-72)\frac{n^3}{36h^3} + \frac{2(p^3-p)\nu^3}{3h^3}$
Case of $p=1, h=n, \nu=n$	$4\frac{3}{4}n^4 + 32\frac{17}{18}n^3$
Case of $p=n, h=1, \nu=1$	$43\frac{1}{3}n^3$

Example 9.6. Consider the matrix

$$A = \begin{bmatrix} 1 & -5 & 7 & -10 & 9 \\ 0 & -1 & -1 & 2 & -2 \\ -3 & 6 & -10 & 12 & -11 \\ 0 & -2 & 3 & -6 & 5 \\ 2 & -5 & 7 & -10 & 8 \end{bmatrix}.$$

This matrix has the Jordan form

$$J = \operatorname{diag}\left(\begin{bmatrix} -1 & 1 \\ & -1 \end{bmatrix}, \begin{bmatrix} -2 & 1 & \\ & -2 & \\ & & -2 \end{bmatrix}\right).$$

The eigenvalue $\lambda_1 = -1$ has a Weyr characteristic equal to $\{1,\ 1\}$ and Segre characteristic equal to $\{2\}$. The eigenvalue $\lambda_2 = -2$ has a Weyr characteristic equal to $\{2,\ 1\}$ and the same Segre characteristic. The Jordan–Schur form of A has a 2×2 diagonal block (to eight decimal digits)

$$S_{11} = \begin{bmatrix} -1 & 4.53557368 \\ & -1 \end{bmatrix},$$

associated with $\lambda_1 = -1$, a 3×3 diagonal block

$$S_{22} = \begin{bmatrix} -2 & 0 & 3.06799138 \\ & -2 & -1.85595575 \\ & & -2 \end{bmatrix},$$

associated with $\lambda_2 = -2$, and a 2×3 superdiagonal block

$$S_{12} = \begin{bmatrix} -6.29342726 & -0.44310809 & 30.05203820 \\ -0.84090739 & -0.95805781 & -0.13363062 \end{bmatrix}.$$

The exponential of the first diagonal block is computed by Algorithm 9.2 as

$$e^{S_{11}} = \begin{bmatrix} 0.36787944 & 1.66854431 \\ 0 & 0.36787944 \end{bmatrix}$$

and the exponential of the second diagonal block is

$$e^{S_{22}} = \begin{bmatrix} 0.13533528 & 0 & 0.41520748 \\ & 0.13533528 & -0.25117630 \\ & & 0.13533528 \end{bmatrix}.$$

The solution of the staircase Sylvester equation (9.10), obtained by Algorithm 9.4, is

$$E_{12} = \begin{bmatrix} -1.97966816 & -0.69112026 & 4.97076879 \\ -0.19554810 & -0.22279075 & -0.10901551 \end{bmatrix}.$$

The computed matrix exponential is

$$\widehat{e^A} = \begin{bmatrix} 1.23897361 & -0.96830304 & 1.23897361 & -1.93660608 & 1.10363832 \\ -0.13533528 & 0.36787944 & -0.23254416 & 0.46508832 & -0.32975303 \\ -1.23897361 & 1.20084720 & -1.33618248 & 2.40169440 & -1.43339136 \\ 0.13533528 & -0.36787944 & 0.50321472 & -0.60042360 & 0.73575888 \\ 0.87109417 & -0.96830304 & 1.23897361 & -1.93660608 & 1.47151776 \end{bmatrix}.$$

The exponential is computed with a relative error

$$\|\widehat{e^A} - e^A\|_F / \|e^A\|_F = 4.54 \times 10^{-15}.$$

Consider the numerical properties of Algorithm 9.5.

In the proposed version of the Schur–Parlett method, the matrix A is transformed into Jordan–Schur form and, similar to the original Schur–Parlett method, the computation of its exponential is reduced to the determination of the exponentials of the diagonal blocks followed by the computation of the off-diagonal blocks using the Parlett block-recurrence.

As mentioned in Section 6.7, the reduction into Jordan–Schur form is numerically stable due to the implementation of unitary transformations. However, as previously noted, the numerical stability doesn't guarantee that the clustering of the numerically multiple eigenvalues and the determination of the sizes of the Jordan blocks are always done in the precise way. The extensive experiments with this algorithm show that it works reliably except in case of very ill-conditioned system of eigenvectors and principal vectors when any clustering method will have difficulties in grouping the eigenvalues.

The exponentials of the diagonal blocks are found by using finite term Taylor series combined with scaling and squaring. As concluded by Higham (Higham, 2008), Chapter 10, no clear general conclusions can be drawn about the stability of the scaling and squaring algorithm. Nevertheless, this algorithm is generally considered stable and is implemented, for instance, in the MATLAB®function `expm`.

The extensive experiments confirm that the exponentials of the diagonal blocks are usually found with high relative accuracy.

The next step of the algorithm is the Parlett block-recursion which requires solution of a Sylvester equation of the form (9.17) with respect to the unknown off-diagonal block E_{ij}. In case of close eigenvalues of the blocks T_{ii} and T_{jj}, this equation can be ill conditioned. More specifically, let the Sylvester equation be written as

$$S_A X - X S_B = C, \tag{9.21}$$

where $S_A \in \mathbb{C}^{\kappa\times\kappa}, S_B \in \mathbb{C}^{\nu\times\nu}, C \in \mathbb{C}^{\kappa\times\nu}$, and let $\Omega = I_\nu \otimes S_A - S_B^T \otimes I_\kappa$. Then the relative change $\|\Delta X\|_F/\|X\|_F$ in the solution of this equation due to small relative perturbations $\Delta S_A/\|S_A\|_F$, $\Delta S_B/\|S_B\|_F$ and $\Delta C/\|C\|_F$ is given by (see Section 4.2.5)

$$\|\Delta X\|_F/\|X\|_F \leq \sqrt{3}\mathrm{cond}_{\mathrm{sylv}}\eta, \tag{9.22}$$

where

$$\begin{aligned}\mathrm{cond}_{\mathrm{sylv}} &= \|\Omega^{-1}\|_F\,[\|S_A\|_F(X^T \otimes I_\kappa),\\ &\quad -\|S_B\|_F(I_\nu \otimes X), -\|C\|_F I_{\kappa\cdot\nu}]\,\|_2/\|X\|_F,\\ \Omega &= I_\nu \otimes S_A - S_B^T \otimes I_\kappa \in \mathbb{C}^{\kappa\nu\times\kappa\nu},\end{aligned}$$

is the relative condition number of the Sylvester equation and

$$\eta = \max\{\|\Delta S_A\|_F/\|S_A\|_F, \|\Delta S_B\|_F/\|S_B\|_F, \|\Delta C\|_F/\|C\|_F\}$$

is the maximum relative perturbation in the data. Now, if the separation

$$\mathrm{sep}(S_A, S_B) = \min_{X\neq 0}\frac{\|S_A X - X S_B\|_F}{\|X\|_F} = \sigma_{\min}(\Omega)$$

between the matrices S_A and S_B is small, then the condition number of the Sylvester equation is large and the sensitivity of this equation can be large even in the case of well-conditioned matrix exponential. This follows from the fact that $\|\Omega^{-1}\|_2 = 1/\mathrm{sep}(S_A, S_B)$.

Example 9.7. This example reveals the reasons for poor accuracy of the Schur–Parlett algorithm in case of a small separation between the diagonal blocks of the Jordan–Schur form.

The matrix A in this example is constructed as

$$A = QJQ^{-1}, \quad Q = V_Q \cdot V_Q^{-T},$$

where the matrix V_Q is taken as

$$V_Q = \begin{bmatrix} 1 & -1 & 0 & 0 & 1 & 1 & -1 & 1 \\ & 1 & 1 & -1 & 0 & 0 & 1 & 1 \\ & & 1 & 0 & 0 & 1 & -1 & -1 \\ & & & 1 & 1 & 0 & 0 & 1 \\ & & & & 1 & -1 & 1 & 0 \\ & & & & & 1 & 0 & 1 \\ & & & & & & 1 & -1 \\ & & & & & & & 1 \end{bmatrix}.$$

Note that the matrices Q and $Q^{-1} = V_Q^T \cdot V_Q^{-1}$ have integer entries. The condition number with respect to the inversion of the similarity transformation matrix Q is $\text{cond}_F(Q) = 333.40$. The Jordan form is taken as

$$J = \text{diag}\left(\begin{bmatrix} \lambda_1 & 1 & & & \\ & \lambda_1 & 1 & & \\ & & \lambda_1 & 1 & \\ & & & \lambda_1 & 1 \\ & & & & \lambda_1 \end{bmatrix}, \begin{bmatrix} \lambda_2 & 1 & \\ & \lambda_2 & 1 \\ & & \lambda \end{bmatrix}\right),$$

where the eigenvalue λ_1 varies between 3 and 3.9 and eigenvalue λ_2 is constant and equal to 4. As in Example 9.6, the exact exponential is obtained by using the analytical expression of the exponential e^J. The Jordan–Schur form of A is computed for $ein = 10^{-15}$, $tol = 10^{-15}$. With the increase in λ_1, the separation between the two blocks T_{11} and T_{22} decreases quickly, which leads to large errors in the solution of Sylvester equation (9.17) appearing in the block-recurrence. The relative error in computing the exponential along with the error bound $\text{cond}_E \cdot eps$ where eps is the MATLAB® function `eps`, $eps = 2^{-52}$, as a function of $\tau = \lambda_1 - \lambda_2$, is shown in Figure 9.8.

The example presented shows the potential instability of the Schur–Parlett method and its Jordan–Schur version in the presence

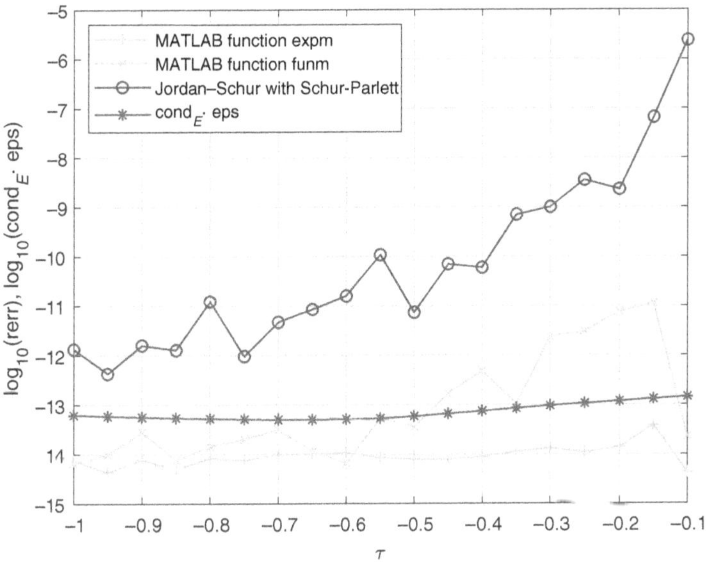

Fig. 9.8. Relative errors of the computed e^A in Example 9.7.

of blocks with close eigenvalues. Note that this instability is common for all methods implementing the Parlett block-recurrence and is not specific to the use of the Jordan–Schur form itself. The function `funm` also produces large errors, as a result of implementing the Schur–Parlett method. For such matrices, it is appropriate to use the Schur–Fréchet algorithm, the Padé approximation combined with scaling and squaring, or the alternative methods that are based on the Taylor series (Caliari and Zivcovich, 2019).

9.7.2 *Schur–Fréchet Method*

In the following, we present a modification of the Schur–Fréchet method for computing e^A which takes advantage of the reduction into Jordan–Schur form of the original matrix.

The Schur–Fréchet method determines recursively an approximation of the off-diagonal block E_{12} of the exponential of

$$T = \begin{bmatrix} T_{11} & T_{12} \\ & T_{22} \end{bmatrix},$$

where the matrices T_{11} and T_{22} are diagonal blocks of the Jordan–Schur form of the given matrix A. The recursion involves the solution of the Sylvester equation

$$(I_{n_1} + M_{11}/\beta_i)D_{12} + D_{12}(I_{n_2} - M_{22}/\beta_i) = Q, \tag{9.23}$$

with matrices $M_{11} = T_{11}$, $M_{22} = T_{22}$ and initial approximation $D_{12} = T_{12}$. Note that T_{11} is a single diagonal block of order ν_i (the multiplicity of the eigenvalue λ_i) but the matrix T_{22} may have on its diagonal several blocks associated with distinct eigenvalues μ_j of multiplicities ν_j.

The key observation here is that the matrices M_{11}, M_{22} and $I_{n_1} + M_{11}/\beta_i$, $I_{n_2} - M_{22}/\beta_i$, respectively, are in Jordan–Schur form. This allows solving equation (9.23) efficiently taking into account the structure of these matrices. More specifically, in simplified notation, one has to solve the block Sylvester equation

$$FX + X\begin{bmatrix} S(\mathcal{WG}_1, \mu_1) & G_{12} & \cdots & G_{1p} \\ & S(\mathcal{WG}_2, \mu_2) & \cdots & G_{2p} \\ & & \ddots & \vdots \\ & & & S(\mathcal{WG}_p, \mu_p) \end{bmatrix} = H, \tag{9.24}$$

where $F = \lambda I_{nf} + M \in \mathbb{C}^{nf\times nf}$ and M is a staircase matrix with Weyr characteristic $\mathcal{WF}$ associated with λ, the diagonal block $S(\mathcal{WG}_j, \mu_j) = \mu_j I_{\nu_j} + N_j \in \mathbb{C}^{\nu_j\times\nu_j}$ corresponds to a different numerically multiple eigenvalue μ_j and N_j is a staircase matrix with Weyr characteristic $\mathcal{WG}_j$ associated with μ_j, $j = 1, \ldots, p$.

Representing the matrices C and X as

$$H = [H_1, H_2, \ldots, H_p], \quad H_j \in \mathbb{C}^{nf\times\nu_j},$$
$$X = [X_1, X_2, \ldots, X_p], \quad X_j \in \mathbb{C}^{nf\times\nu_j},$$

one obtains from (9.24) the system of p staircase Sylvester equations

$$FX_j + X_j S(\mathcal{WG}_j, \mu_j) = H_j - \sum_{d=1}^{j-1} X_d G_{dj}, \quad j = 1, 2, \ldots, p. \tag{9.25}$$

It is important that (9.25) is a staircase Sylvester equation of the same type as (9.18). Thus, it can be solved by Algorithm 9.4.

The system (9.25) of p staircase Sylvester equations is solved recursively by Algorithm 9.6.

Let us return to the problem of evaluating the exponential

$$e^T = \begin{bmatrix} E_{11} & E_{12} \\ 0 & E_{22} \end{bmatrix}$$

Algorithm 9.6. Solution of a block Sylvester equation
Solution of the Sylvester matrix equation $FX + XG = H$,
where F and G are the diagonal blocks of a certain Jordan–Schur form.
function $X = sylv\_frechet(F, G, H)$
Input: A matrix $F \in \mathbb{C}^{nf \times nf}$ in the form $F = \lambda I_{nf} + M$, where M is a staircase matrix with Weyr characteristic $\mathcal{WF}$ associated with λ; a matrix

$$G = \begin{bmatrix} S(\mathcal{WG}_1, \mu_1) & G_{12} & \cdots & G_{1p} \\ & S(\mathcal{WG}_2, \mu_2) & \cdots & G_{2p} \\ & & \ddots & \vdots \\ & & & S(\mathcal{WG}_p, \mu_p) \end{bmatrix} \in \mathbb{C}^{ng \times ng}$$

in Jordan–Schur form whose p diagonal blocks $S(\mathcal{WG}_j, \mu_j) \in \mathbb{C}^{\nu_j \times \nu_j}$ correspond to distinct numerically multiple eigenvalues μ_j and have the form $S(\mathcal{WG}_j, \mu_j) = \mu_j I_{\nu_j} + N_j$, where N_j is a staircase matrix with Weyr characteristic $\mathcal{WG}_j$ associated with μ_j, $j = 1, \ldots, p$, $H = [H_1, H_2, \ldots, H_p]$, $H_j \in \mathbb{C}^{nf \times \nu_j}$, $j = 1, \ldots, p$.
Output: $X \in \mathbb{C}^{nf \times ng}$
Set $S_F = F$
$X = [\ ]$
for $j = 1, 2, \ldots, p$
  Set $S_G = S(\mathcal{WG}_j, \mu_j)$
  Set $R = 0_{nf, \nu_j}$
  if $j > 1$
    for $d = 1, 2, \ldots, j - 1$
      $R = R + X_d G_{dj}$
    end
  end
  Solve the staircase Sylvester equation
    $S_F X_j + X_j S_G = H_j - R$
  for $X_j \in \mathbb{C}^{nf \times \nu_j}$ using Algorithm 9.4
  $X \leftarrow [X,\ X_j]$
end

of the block triangular matrix

$$T = \begin{bmatrix} T_{11} & T_{12} \\ 0 & T_{22} \end{bmatrix}.$$

Having an algorithm to solve (9.23), it is possible to find the off-diagonal block E_{12} of the matrix exponential of T according to the Schur–Fréchet method present in Section 9.4.4. This can be done by Algorithm 9.7.

Algorithm 9.7. Computing an off-diagonal block of the matrix exponential using the Schur–Fréchet method

Evaluating the off-diagonal block E_{12} of the exponential

$$\begin{bmatrix} E_{11} & E_{12} \\ & E_{22} \end{bmatrix} = \exp\left(\begin{bmatrix} T_{11} & T_{12} \\ & T_{22} \end{bmatrix}\right)$$

of a matrix in Jordan–Schur form, using the Schur–Fréchet method. The corresponding block Sylvester equation (9.24) is solved taking into account the staircase structure of the individual equations.

function $E_{12} = block\_frechet(T_{11}, E_{11}, T_{22}, E_{22}, T_{12})$

Input: A block $T_{11} \in \mathbb{C}^{n_1 \times n_1}$ in the form $T_{11} = \lambda_1 I_{n_1} + M_{11}$, where the matrix M_{11} is in staircase form with Weyr characteristic $\mathcal{W}_1$ associated with λ_1;
the matrix exponential E_{11} of the block T_{11};
a matrix

$$T_{22} = \begin{bmatrix} S(\mathcal{W}_{21}, \mu_1) & T_{12} & \dots & T_{1,p_2} \\ & S(\mathcal{W}_{22}, \mu_2) & \dots & T_{2,p_2} \\ & & \ddots & \vdots \\ & & & S(\mathcal{W}_{2,p_2}, \mu_{p_2}) \end{bmatrix} \in \mathbb{C}^{n_2 \times n_2}$$

in Jordan–Schur form with diagonal blocks $S(\mathcal{W}_{2j}, \mu_j) \in \mathbb{C}^{\nu_j \times \nu_j}$, corresponding to distinct numerically multiple eigenvalues μ_j which have the form $S(\mathcal{W}_{2j}, \mu_j) = \mu_j I_{\nu_j} + N_j$, where N_j is a staircase matrix with Weyr characteristic $\mathcal{W}_{2j}$ associated with μ_j, $j = 1, \dots, p_2$;
the matrix exponential E_{22} of the block T_{22};
the off-diagonal block $T_{12} \in \mathbb{C}^{n_1 \times n_2}$

Output: $E_{12} \in \mathbb{C}^{n_1 \times n_2}$ contains the off-diagonal block of the matrix exponential
Set $M_{11} = T_{11}$, $M_{22} = T_{22}$, $D_{12} = T_{12}$
Let $\alpha_1, \alpha_2, \ldots, \alpha_8$ be the zeros of the polynomial
$p(x) = x^8 + 990x^6 + 135135x^4 + 4729725x^2 + 34459425$
and $\beta_1, \beta_2, \ldots, \beta_8$ be the zeros of the polynomial
$q(x) = 45x^8 + 13860x^6 + 945945x^4 + 16216200x^2 + 34459425$
Set $m = 8$
Evaluate the hyperbolic tangent Sylvester cascade
for $i = 1, 2, \ldots, m$
  $R = (I_{n_1} + M_{11}/\alpha_i)D_{12} + D_{12}(I_{n_2} - M_{22}/\alpha_i)$
  Solve the system of staircase Sylvester equations
  $(I_{n1} + M_{11}/\beta_i)D_{12} + D_{12}(I_{n_2} - M_{22}/\beta_i) = R$
  using Algorithm 9.4.
end
$D_{12} = (E_{11}D_{12} + D_{12}E_{22})/2$
$E_{12} = D_{12}$

Consider now an $n \times n$ matrix

$$J_S = \begin{bmatrix} S(\mathcal{W}_1, \lambda_1) & T_{12} & \cdots & T_{1p} \\ & S(\mathcal{W}_2, \lambda_2) & \cdots & T_{2p} \\ & & \ddots & \vdots \\ & & & S(\mathcal{W}_p, \lambda_p) \end{bmatrix} \in \mathbb{C}^{n \times n}$$

which is in Jordan–Schur form and assume that the exponentials $e^{S(\mathcal{W}_i, \lambda_i)}$, $i = 1, 2, \ldots, p$, of the diagonal blocks are already found by using Algorithm 9.2.

Using the Schur–Fréchet method, the off-diagonal blocks of

$$e^{J_S} = \begin{bmatrix} E_{11} & \cdots & E_{1,p-2} & E_{1,p-1} & E_{1,p} \\ & \ddots & \vdots & \vdots & \vdots \\ & & E_{p-2,p-2} & E_{p-2,p-1} & E_{p-2,p} \\ & & & E_{p-1,p-1} & E_{p-1,p} \\ & & & & E_{pp} \end{bmatrix}$$

can be evaluated recursively by the following way. First, one can find the block $E_{p-1,p}$ using the blocks $S(\mathcal{W}_{p-1}, \lambda_{p-1})$ and $S(\mathcal{W}_p, \lambda_p)$.

Then, using the diagonal blocks $S(\mathcal{W}_{p-2}, \lambda_{p-2})$ and

$$\begin{bmatrix} S(\mathcal{W}_{p-1}, \lambda_{p-1}) & T_{p-1,p} \\ & S(\mathcal{W}_p, \lambda_p) \end{bmatrix},$$

it is possible to evaluate the off-diagonal blocks $E_{p-2,p-1}$, $E_{p-2,p}$ and so on, finding ultimately the off-diagonal blocks in the first row of e^{J_S}.

The matrix exponential of J_S is found by the Schur–Fréchet method implementing Algorithm 9.8.

The volume of the computational work necessary to determine of the off-diagonal blocks by Schur–Fréchet method is estimated as follows. Assuming as in the case of Schur–Parlett method equal eigenvalue multiplicities ($\kappa_i = n/p, \nu_i = n/p$), equal elements of the Weyr

Algorithm 9.8. Computing the matrix exponential using the Schur–Fréchet method

Evaluating the matrix exponential of a matrix in Jordan–Schur form using the Schur–Fréchet method. The diagonal blocks of the exponential are computed by finite series Taylor expansion with scaling and squaring and the off-diagonal blocks are obtained by solving systems of block staircase Sylvester equations.

function $E = jsf\_frechet(J_S)$

Input: A matrix

$$J_S = \begin{bmatrix} S(\mathcal{W}_1, \lambda_1) & T_{12} & \dots & T_{1p} \\ & S(\mathcal{W}_2, \lambda_2) & \dots & T_{2p} \\ & & \ddots & \vdots \\ & & & S(\mathcal{W}_p, \lambda_p) \end{bmatrix} \in \mathbb{C}^{n \times n},$$

in Jordan–Schur form with diagonal blocks $S(\mathcal{W}_i, \lambda_i) \in \mathbb{C}^{\nu_i \times \nu_i}$ which correspond to distinct numerically multiple eigenvalues μ_i and have the form $S(\mathcal{W}_i, \mu_i) = \mu_i I_{\nu_i} + N_i$, where N_i is a staircase matrix with Weyr characteristic $\mathcal{W}_i$ associated with λ_i, $i = 1, 2, \dots, p$.

Output: $E = e^{J_S}$

if $\|J_S\|_\infty = 0$
    $E = I_n$
else
    for $i = 1, 2, \dots, p$
        Evaluate $E_{ii} = e^{S(\mathcal{W}_i, \lambda_i)}$ using Algorithm 9.2
    end

$E = \mathrm{diag}(E_{11}, E_{22}, \ldots, E_{pp})$
Compute the off-diagonal blocks of the exponential
if $p > 1$
for $i = p-1, p-2, \ldots, 1$
$T_{11} = S(\mathcal{W}_i, \lambda_i)$,
$n_2 = \sum_{j=i+1}^{p} \nu_j$

$$T_{22} = \begin{bmatrix} S(\mathcal{W}_{i+1}, \lambda_{i+1}) & T_{i+1,2} & \cdots & T_{i+1,p} \\ & S(\mathcal{W}_{i+2}, \lambda_{i+2}) & \cdots & T_{i+2,p} \\ & & \ddots & \vdots \\ & & & S(\mathcal{W}_p, \lambda_p) \end{bmatrix}$$

$\in \mathbb{C}^{n_2 \times n_2}$,
$T_i = [T_{i,i+1}, T_{i,i+2}, \ldots, T_{ip}]$
Solve the Sylvester equation
$T_{11}E_i + E_i T_{22} = T_i$
for E_i using Algorithm 9.6

$$E \leftarrow \begin{bmatrix} E_{11} & & & & \\ & \ddots & & & \\ & & E_{ii} & E_i & \\ & & & \ddots & \vdots \\ & & & & E_{pp} \end{bmatrix}$$

end
end
end

characteristics ($w = \mathcal{W}\mathcal{F}_k = \mathcal{W}\mathcal{G}_\ell = \nu/h, h = hf = hg$), and using the estimate (9.20), the necessary computational work for solving eight times the Sylvester equation of the type (9.25) by Algorithm 9.7 is estimated as

$$\frac{8(p^3 - 3p^2)\nu^3}{3h^3} \text{ flops.}$$

The maximum of this volume of work is obtained for $h = 1$, $p = n$, $\nu = 1$ and for large n it is approximately equal to $\frac{8}{3}n^3$. Thus, for a matrix in Jordan–Schur form, the Schur–Fréchet method for

Table 9.4. Number of computational operations required by the Schur–Fréchet algorithm.

Assumptions	Number of flops
Reduction into Jordan–Schur form and computing the exponentials of the diagonal blocks	$32\frac{2}{3}n^3 + \frac{2}{3}\frac{(h+3)p+6h-3}{p^2}n^3 + (3h^4+10h^3-87h^2+146h-72)\frac{n^3}{36h^3}$
Case of $p=1, h=n, \nu=n$	$4\frac{3}{4}n^4 + 32\frac{17}{18}n^3$
Case of $p=n, h=1, \nu=1$	$42\frac{2}{3}n^3$
Computing the off-diagonal blocks $1 \le h \le n,\ 1 \le p \le n$ $\nu = n/p,\ w = \nu/h$	$\frac{8(p^3-3p^2)\nu^3}{3h^3}$
Case of $p=1, h=n, \nu=n$	—
Case of $p=n, h=1, \nu=1$	$\frac{8(n^3-3n^2)}{3}$
Total: Reduction into Jordan–Schur form and computing its exponential $1 \le h \le n,\ 1 \le p \le n$ $\nu = n/p,\ w = \nu/h$	$32\frac{2}{3}n^3 + \frac{2}{3}\frac{(h+3)p+6h-3}{p^2}n^3 + (3h^4+10h^3-87h^2+146h-72)\frac{n^3}{36h^3} + \frac{8(p^3-3p^2)\nu^3}{3h^3}$
Case $p=1, h=n, \nu=n$	$4\frac{3}{4}n^4 + 32\frac{17}{18}n^3$
Case $p=n, h=1, \nu=1$	$45\frac{1}{3}n^3$

computing the exponential requires the same work as the Schur–Parlett method. Note that this operation count does not involve the computations for scaling and squaring the diagonal blocks.

In Table 9.4, we give the number of operations for the Schur–Fréchet algorithm including the extreme cases $h = 1$ and $h = n$. The estimates of the computation cost are obtained under the assumptions $\nu = n/p$, $w = \nu/h$. Clearly, the work necessary for the reduction into Jordan–Schur form dominates the total operation count which is approximately the same for Schur–Parlett and

Schur–Fréchet methods. In the most difficult case ($h = n$), both methods require approximately $4\frac{3}{4}n^4$ flops.

Example 9.8. Consider again the matrix A used in Example 9.7. Using Algorithm 9.8, the matrix exponential e^A is computed with a relative forward error

$$\|\widehat{e^A} - e^A\|_F/\|e^A\|_F = 4.76 \times 10^{-15}.$$

The main computational task in the Schur–Fréchet method is the solution of the system of staircase Sylvester equations (9.25). Unlike the Sylvester equation (9.17) solved in the Schur–Parlett method, the sensitivity of (9.25) does not depend on the closeness of the eigenvalues of Jordan–Schur blocks. Moreover, the separation of matrices F and $-S(\mathcal{W}\mathcal{G}_j, \mu_j)$ in (9.25) is always of order 1 which leads to a low condition number of the Sylvester equation.

For this reason, the Schur–Fréchet Algorithm 9.8 almost always produces more accurate results than Algorithm 9.5 implementing the Schur–Parlett method. The Schur–Fréchet algorithm is especially appropriate in the cases when the blocks of the Jordan–Schur form have close eigenvalues.

Example 9.9. Let us apply the Jordan–Schur methods to the matrix A given in Example 9.4 which is obtained for a similarity transformation matrix Q constructed for $\sigma = 1.25$. The relative errors of the computed exponential for $n = 4, 8, 12, \ldots, 80$ with tolerances $ein = 10^{-14}$, $tol = 10^{-14}$ are shown in Figure 9.9 for the four methods used earlier. The function `expm` exhibits the worst performance, and for $n = 68$, the relative error of the computed result is equal to $4.50 \cdot 10^{-1}$. At the same time, the relative errors of the exponential computed by the two versions of the Jordan–Schur algorithm are identical and for $n = 68$ are approximately equal to $6.67 \cdot 10^{-4}$. Even for $n = 80$ the relative errors of these algorithms are smaller than 1. The goods results for the Jordan–Schur algorithms are due to the simple structure of the Jordan–Schur form

$$J_S = \begin{bmatrix} \lambda I_{n/2} & N_{12} \\ & \lambda I_{n/2} \end{bmatrix}$$

so that it is necessary to compute only one diagonal and one off-diagonal block. This facilities very much the computation of the

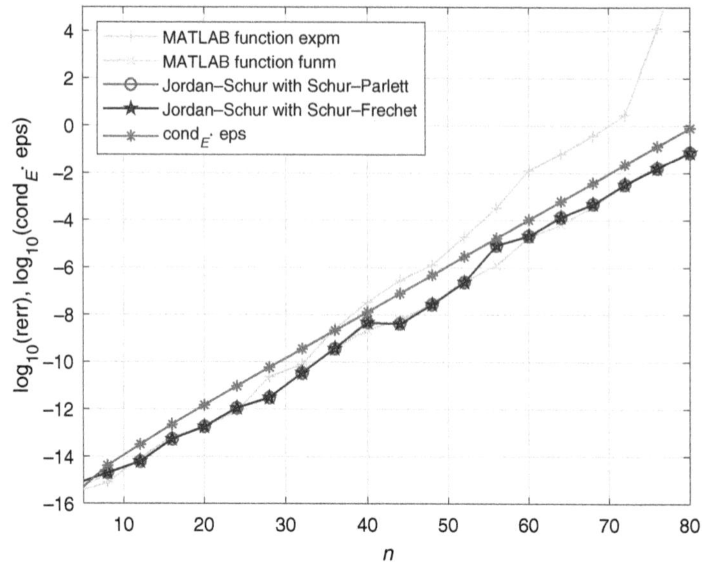

Fig. 9.9. Relative errors for Example 9.9.

exponential. The function `funm` also produces accurate results for this example.

9.7.3 *Numerical Properties of Jordan and Jordan–Schur Algorithms*

Let us compare the volume of computations and accuracy of the four Algorithms 9.1, 9.3, 9.5 and 9.8. In Table 9.5, we give an approximate operation count for these algorithms for two extreme cases. The four algorithms require nearly the same volume of work, but in some cases, the block-diagonal staircase and Jordan–Schur algorithms perform more reliably than the algorithm based on the Jordan form.

Compare now the accuracy of the Jordan and block-diagonal staircase algorithms. The following example shows that in some cases the transformation into Jordan form can be much more ill conditioned than the transformation into block-diagonal staircase form.

Example 9.10. Given is the matrix A in the form

$$A = QJQ^{-1},$$

Table 9.5. Comparison of the number of operations (flops) in computing the matrix exponential.

Method	Case of $p=1, h=n, \nu=n$ Defective, but non-derogatory matrix	Case of $p=n, h=1, \nu=n$ Non-defective, but derogatory matrix
Jordan	$4\frac{2}{3}n^4 + 47n^3$	$47\frac{1}{3}n^3$
Block-diagonal staircase	$4\frac{3}{4}n^4 + 34\frac{11}{18}n^3$	$34\frac{2}{3}n^3$
Jordan–Schur — Schur–Parlett	$4\frac{3}{4}n^4 + 32\frac{17}{18}n^3$	$43\frac{1}{3}n^3$
Jordan–Schur — Schur–Fréchet	$4\frac{3}{4}n^4 + 32\frac{17}{18}n^3$	$45\frac{1}{3}n^3$

where

$$J = \operatorname{diag}(J_{10}(\lambda_1), J_{10}(\lambda_2)),$$

$\lambda_1 = -2$ and $\lambda_2 = 1$. The matrix Q is constructed as in Example 9.4 for $\sigma = 1,\ 1.05,\ 1.10, \ldots, 2.50$. The Segre characteristics of A are $\mathcal{S}(A,\lambda_1) = \{10\}$, $\mathcal{S}(A,\lambda_2) = \{10\}$ and the corresponding Weyr characteristics are $\mathcal{W}(A,\lambda_1) = \{1,\ 1, \ldots, 1\}$, $\mathcal{W}(A,\lambda_2) = \{1,\ 1, \ldots, 1\}$. In the given case, we use small tolerances $ein = tol = 10^{-17}$, to avoid errors in determining the Jordan structure for the matrices with ill-conditioned eigensystems. Note the close relation between the condition number cond_E of the matrix exponential and the condition number of the transformation matrix to Jordan form both increasing quickly with the increase in σ, as shown in Table 9.6. At the same time, the conditioning of the transformation into block-diagonal staircase form remains moderate. The relative errors of the exponential computed by the Jordan and block-diagonal staircase form are given in the table for different values of σ and plotted in Figure 9.10. The two methods produce almost the same results for all σ but the conditioning of the Jordan form is much more worse than the conditioning of the block-diagonal form.

Consider now the properties of the block-diagonal staircase method and Jordan–Schur method.

Table 9.6. Conditioning of the transformation matrices and accuracy of Jordan and block-diagonal staircase algorithms.

σ	cond_E	$\text{cond}_2(Z)$	$\text{cond}_2(X_S)$	$rerr_J$	$rerr_{B_S}$
1.0	$4.64 \cdot 10^{0}$	$1.00 \cdot 10^{0}$	$1.00 \cdot 10^{0}$	$3.13 \cdot 10^{-15}$	$3.16 \cdot 10^{-15}$
1.2	$1.75 \cdot 10^{1}$	$3.96 \cdot 10^{1}$	$2.95 \cdot 10^{0}$	$4.05 \cdot 10^{-15}$	$3.80 \cdot 10^{-15}$
1.4	$2.09 \cdot 10^{3}$	$5.37 \cdot 10^{3}$	$8.31 \cdot 10^{0}$	$9.19 \cdot 10^{-15}$	$1.03 \cdot 10^{-14}$
1.6	$2.92 \cdot 10^{5}$	$5.46 \cdot 10^{5}$	$2.25 \cdot 10^{1}$	$6.14 \cdot 10^{-13}$	$6.20 \cdot 10^{-13}$
1.8	$2.41 \cdot 10^{7}$	$3.56 \cdot 10^{7}$	$5.74 \cdot 10^{1}$	$1.23 \cdot 10^{-10}$	$1.23 \cdot 10^{-10}$
2.0	$1.27 \cdot 10^{9}$	$1.57 \cdot 10^{9}$	$1.37 \cdot 10^{2}$	$7.86 \cdot 10^{-9}$	$7.86 \cdot 10^{-9}$
2.2	$4.64 \cdot 10^{10}$	$4.94 \cdot 10^{10}$	$3.04 \cdot 10^{2}$	$1.77 \cdot 10^{-8}$	$1.77 \cdot 10^{-8}$
2.4	$1.24 \cdot 10^{12}$	$1.16 \cdot 10^{12}$	$6.54 \cdot 10^{2}$	$2.94 \cdot 10^{-6}$	$2.95 \cdot 10^{-6}$

- cond_E is the condition number of the matrix exponential,
- $\text{cond}_2(Z)$ is the condition number of the transformation matrix into Jordan form,
- $\text{cond}_2(X_S)$ is the condition number of the transformation matrix into block-diagonal form,
- $rerr_J$ is the relative error in computing the exponential by the Jordan algorithm,
- $rerr_{B_S}$ is the relative error in computing the exponential by the block-diagonal staircase algorithm.

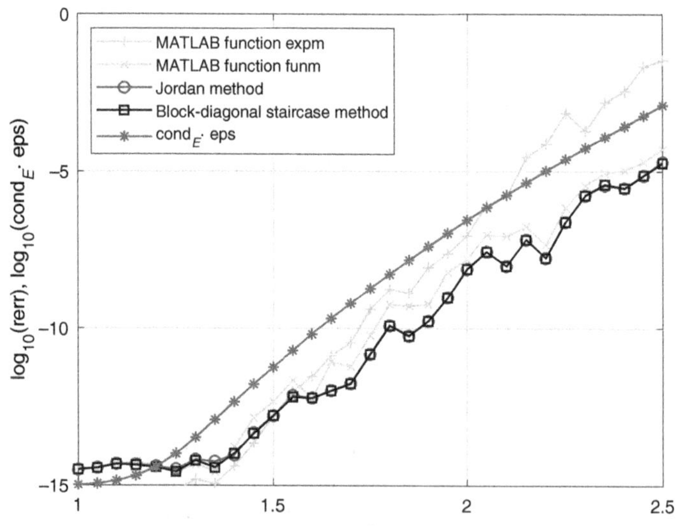

Fig. 9.10. Relative errors in computing the matrix exponential using the Jordan and block-diagonal staircase forms.

Example 9.11. In this example, we compare the accuracy of the block-diagonal staircase and Jordan–Schur algorithms using a matrix A, constructed as

$$A = QJQ^{-1},$$

where

$$J = \text{diag}\,(J_5(\lambda_1), J_5(\lambda_1), J_5(\lambda_2), J_5(\lambda_2),$$

with $\lambda_1 = -75$ and $\lambda_2 = -1.5$. The matrix Q is constructed as in Example 9.4 for different values of σ. The Segre characteristics of A are $\mathcal{S}(A, \lambda_1) = \{5,\ 5\}$, $\mathcal{S}(A, \lambda_2) = \{5,\ 5\}$ and the corresponding Weyr characteristics are $\mathcal{W}(A, \lambda_1) = \{2,\ 2,\ 2,\ 2,\ 2\}$, $\mathcal{W}(A, \lambda_2) = \{2,\ 2,\ 2,\ 2,\ 2\}$. The matrix A is a stability matrix with widely spread eigenvalues. Such matrices correspond to *stiff differential equations* having a slowly varying solution, determined by the small in module eigenvalues and a rapidly damping solution component determined by the large in module eigenvalues. For this matrix, the Jordan algorithm for e^A fails since the similarity transformation matrix Z is singular to working precision for all values of σ. In the given case, we use tolerances $ein = tol = 10^{-16}$.

In Figure 9.11, we show the errors in computing the exponential of A for $\sigma = 1,\ 1.05,\ 1.10, \ldots, 2.50$ by the block-diagonal staircase method, both Jordan–Schur algorithms and MATLAB®functions `expm` and `funm`. Both Jordan–Schur algorithms produce almost the same results which are close to the results obtained by the function `funm`. The best results are produced by the block-diagonal staircase method.

In conclusion, the Jordan–Schur methods compare favorably with the known methods for computing the matrix exponential provided the Jordan structure is determined correctly. These algorithms are reliable in the sense that the failure to determine accurately the Jordan structure can be discovered. The using of the Jordan form gives accurate results in cases when the similarity transformation matrix is well conditioned. The block-diagonal staircase algorithm produces accurate results, but it also requires correctly determined Jordan structure.

On the basis of this brief comparison, it is justified to suppose that in most cases the block-diagonal staircase method can be a

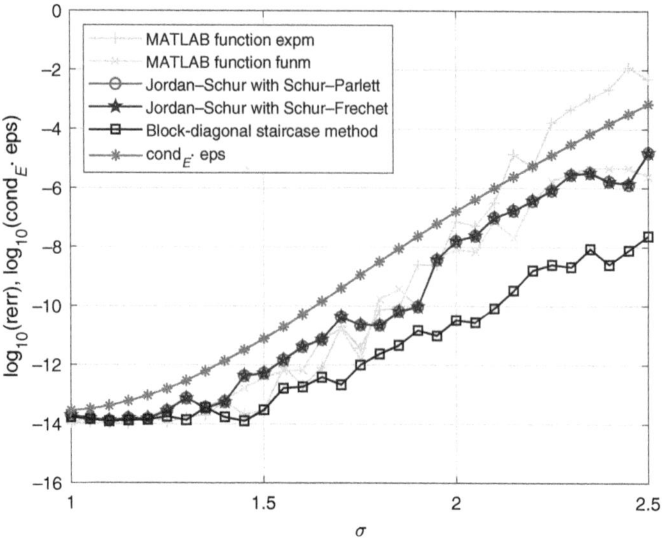

Fig. 9.11. Relative errors in computing the matrix exponential for five different methods.

good alternative to the known methods for computing the matrix exponential.

9.8 Notes and References

As pointed out in the classical paper Moler and Van Loan (1978) (reprinted in 2009; Moler and Van Loan, 2009), there exist at least 19 "dubious" methods for computing the matrix exponential and there is no a single "best" method for solving this important numerical problem. The new methods proposed in this chapter do not make an exception from this conclusion although they can over perform the known methods in some cases.

The sensitivity of the matrix exponential to perturbations is studied by several authors, see, for instance, Kågström (1977), Van Loan (1977), Kenney and Laub (1989) and Mathias (1995).

Among the good methods for computing the exponential that works in many cases is the scaling and squaring method proposed in Ward (1977) and improved in Higham (2009). An efficient method of this group combines the computation of the matrix exponential by using Padé approximations with scaling and squaring to improve the accuracy, see, for instance, Van Loan (1978), Golub and Van Loan (2013), Chapter 9 (Algorithm 9.3.1), Higham (2008), Chapter 10 (Algorithm 10.20), Al-Mohy and Higham (2009b), Dieci and Papini (2000) and Koikari (2007). An arbitrary precision scaling and squaring algorithm for the matrix exponential is proposed in Fasi and Higham (2003). Alternative methods of the same group are based on the Taylor series, see Caliari and Zivcovich (2019) and Sastre *et al.* (2015).

The methods for computing the matrix exponential of nonnormal matrices that are based on the Schur form of the matrix A (Golub and Van Loan, 2013, Chapter 9), Higham (2008), Chapter 10, are frequently preferred due to the numerical stability of the QR algorithm for computing the Schur form. Among the efficient methods for this aim is the Schur–Parlett algorithm (Davies and Higham, 2003), which has comparable performance with the scaling and squaring method. The numerical properties of this algorithm are investigated in Higham (2008), Chapter 9, see also Algorithm 10.23 in Higham (2008), Chapter 10. Another opportunity to compute the off-diagonal blocks of the matrix exponential by using the Schur form is provided by the Schur–Fréchet algorithm developed and analyzed in Najfeld and Havel (1995), Kenney and Laub (1998) and Higham (2008), Chapter 9. The Jordan–Schur algorithms present in this chapter are proposed in Petkov (2023).

Appendix

Review of Linear Algebra and Matrix Analysis

In this Appendix, we give the most important definitions and results about matrices, linear spaces and linear transformations which are used throughout this book. It is assumed that the reader has some knowledge about matrices and operations on matrices.

A.1 Vectors and Matrices

A *column vector* v with n elements or *n-vector* is written as

$$v = \begin{bmatrix} v_1 \\ v_2 \\ \vdots \\ v_n \end{bmatrix},$$

where v_i, $i = 1, \ldots, n$, are scalars. The scalars v_i are called the *components* of v. Formally, we write $v \in \mathbb{C}^n$, if the elements of v are complex numbers or $v \in \mathbb{R}^n$, if the elements of v are real. The *row vector* $v^T = [v_1, v_2, \ldots, v_n]$ denotes the *transposed* of the vector v.

The *zero vector* is a vector whose components are all zero and is denoted by 0.

Let v and w be n-vectors and α be a scalar. The *sum* of v and w is the vector

$$v + w = \begin{bmatrix} v_1 + w_1 \\ v_2 + w_2 \\ \vdots \\ v_n + w_n \end{bmatrix}.$$

The *scalar–vector product* αv is the vector

$$\alpha v = \begin{bmatrix} \alpha v_1 \\ \alpha v_2 \\ \vdots \\ \alpha v_n \end{bmatrix}.$$

Consider a complex $m \times n$ matrix A with elements $a_{ij} = \mathrm{Re}\, a_{ij} + j_o \mathrm{Im}\, a_{ij}$. Here m is the number of rows and n is the number of columns. Formally, we write $A \in \mathbb{C}^{m\times n}$ if A is a complex matrix or $A \in \mathbb{R}^{m\times n}$, if A is a real matrix. If a matrix A is square ($m = n$), we say that A is an nth order matrix.

Matrix with all elements equal to zero is called *zero matrix* and is denoted by 0 or by $0_{m\times n}$ when it is necessary to indicate its size. A matrix A with elements a_{ij} is *diagonal* if $a_{ij} = 0$ for $i \neq j$. The square diagonal matrix with diagonal elements equal to 1 is called *unit matrix* or *identity matrix* and is denoted by I (or by I_n when it is necessary to indicate its order). The columns of the $n \times n$ unit matrix are denoted by e_i, $i = 1, 2, \ldots, n$.

The *transposed* of the matrix A is denoted by A^T (with elements a_{ji} in position (i, j)), the *conjugate* of A – by $\bar{A}$ (with elements $\mathrm{Re}\, a_{ij} - j_o \mathrm{Im}\, a_{ij}$) and the *conjugate transpose* (Hermitian conjugate) of A – by $A^H := \bar{A}^T$ (with elements $\mathrm{Re} a_{ji} - j_o \mathrm{Im} a_{ji}$). The *trace* $\mathrm{trace}(A)$ of a matrix A is the sum of its diagonal elements and $\det A$ is the determinant of A.

A *block matrix* A is a representation of A in the form

$$A = \begin{bmatrix} A_{11} & A_{12} & \ldots & A_{1q} \\ A_{21} & A_{22} & \ldots & A_{2q} \\ \vdots & \vdots & \ddots & \vdots \\ A_{p1} & A_{p2} & \ldots & A_{pq} \end{bmatrix},$$

where $A_{ij} \in \mathbb{F}^{m_i \times n_j}$ are submatrices whose sizes satisfy $m_1 + m_2 + \cdots + m_p = m$ and $n_1 + n_2 + \cdots + n_q = n$.

The *Kronecker product* of the $m \times n$ matrix A and the matrix B is defined as

$$A \otimes B = \begin{bmatrix} a_{11}B & a_{12}B & \dots & a_{1n}B \\ a_{21}B & a_{22}B & \dots & a_{2n}B \\ \vdots & \vdots & \ddots & \vdots \\ a_{m1}B & a_{m2}B & \dots & a_{mn}B \end{bmatrix}.$$

A matrix $P \in \mathbb{R}^{n \times n}$ is a *permutation matrix* if exactly one entry in each row and column is equal to 1 and all other entries are 0. Multiplication by such a matrix affects a permutation of the rows or columns of the matrix multiplied. For example,

$$\begin{bmatrix} 0 & 1 & 0 & 0 \\ 0 & 0 & 0 & 1 \\ 1 & 0 & 0 & 0 \\ 0 & 0 & 1 & 0 \end{bmatrix} \begin{bmatrix} 1 \\ 2 \\ 3 \\ 4 \end{bmatrix} = \begin{bmatrix} 2 \\ 4 \\ 1 \\ 3 \end{bmatrix}$$

illustrates how the multiplication of a vector by a permutation matrix interchanges the vector elements. Left multiplication of a matrix $A \in \mathbb{C}^{m \times n}$ by an m-by-m permutation matrix P permutes the rows of A, while right multiplication of A by an n-by-n permutation matrix P permutes the columns of A.

Let $A \in \mathbb{C}^{m \times n}$ be partitioned columnwise as $A = [a_1, a_2, \dots, a_n]$. Then the *vec mapping* of A from $\mathbb{C}^{m \times n}$ to $\mathbb{C}^{mn}$ is defined as

$$\mathrm{vec}(A) = \begin{bmatrix} a_1 \\ a_2 \\ \vdots \\ a_n \end{bmatrix}.$$

We have that

$$\mathrm{vec}(AXB) = (B^T \otimes A)\mathrm{vec}(X).$$

The matrix Π_n determined by Henderson and Searle (1981) and Horn and Johnson (1991), Chapter 4,

$$\mathrm{vec}(A^T) = \Pi_n \mathrm{vec}(A),$$

is called the *vec-permutation matrix*. It is possible to show that (Horn and Johnson, 1991, Chapter 4)

$$\Pi_n = \sum_{i,j=1}^{n} (e_i e_j^T) \otimes (e_i e_j^T).$$

A square matrix $A \in \mathbb{C}^{n \times n}$ is *nonsingular*, if there is a matrix $A^{-1} \in \mathbb{C}^{n \times n}$ (the *inverse* of A) such that $AA^{-1} = I$. In such case, one also has that $A^{-1}A = I$. The matrix A^{-1} is unique whenever it exists. If A^{-1} does not exist, then A is a *singular* matrix. If $A \in \mathbb{C}^{n \times n}$ is nonsingular and $b \in \mathbb{C}^n$, then $x = A^{-1}b$ is the unique solution of the system of linear equations or simply *linear system* $Ax = b$.

A square matrix Q is *symmetric*, if $Q^T = Q$, and *Hermitian*, if $Q^H = Q$. A Hermitian matrix is *positive definite*, if $x^H Q x > 0$ for all nonzero vectors $x \in \mathbb{C}^n$. The positive definite matrices are nonsingular. A Hermitian matrix Q is *positive semidefinite*, if $x^H Q x \geq 0$ for all $x \in \mathbb{C}^n$. The matrix Q is called a *square root* of the positive semidefinite matrix P if $Q^2 = P$. This is denoted as $Q = P^{1/2}$.

Let A and B be $m \times n$ matrices and α be a scalar. The *sum* of A and B is the matrix

$$A + B = \begin{bmatrix} a_{11} + b_{11} & a_{12} + b_{12} & \dots & a_{1n} + b_{1n} \\ a_{21} + b_{21} & a_{22} + b_{22} & \dots & a_{2n} + b_{2n} \\ \vdots & \vdots & \ddots & \vdots \\ a_{m1} + b_{m1} & a_{m2} + b_{m2} & \dots & a_{mn} + b_{mn} \end{bmatrix}$$

and the *scalar-matrix product* of α and A is

$$\alpha A = \begin{bmatrix} \alpha a_{11} & \alpha a_{12} & \dots & \alpha a_{1n} \\ \alpha a_{21} & \alpha a_{22} & \dots & \alpha a_{2n} \\ \vdots & \vdots & \ddots & \vdots \\ \alpha a_{m1} & \alpha a_{m2} & \dots & \alpha a_{mn} \end{bmatrix}.$$

A.2 Linear Subspaces

Let $\mathbb{F}$ be the set $\mathbb{R}$ of real or the set $\mathbb{C}$ of complex numbers. A *linear space* over $\mathbb{F}$ is a nonempty set $\mathcal{X}$ of elements called *vectors* or *points*, satisfying several axioms (Halmosh, 1974, Chapter I), between them:

(1) (Addition of two vectors). To each two vectors x_1 and x_2 of $\mathcal{X}$ corresponds a vector $x_1 + x_2$ of $\mathcal{X}$ called *sum* of x_1 and x_2.
(2) (Multiplication of vector by a scalar). To each number $\alpha \in \mathbb{F}$ and to each vector $x \in \mathcal{X}$ corresponds a vector $\alpha x \in \mathcal{X}$ called *product* of α and x.

The linear space of real (respectively complex) n-vector columns is denoted by $\mathbb{R}^n$ (respectively, $\mathbb{C}^n$).

The nonempty subset $\mathcal{Y}$ of the linear space $\mathcal{X}$ is called *subspace*, if it itself represents a linear space with respect to the operations addition of two vectors and multiplication of a vector by a scalar. Otherwise speaking, $\mathcal{Y} \subset \mathcal{X}$ is a subspace if from $x_1, x_2 \in \mathcal{Y}$ it follows that $\alpha_1 x_1 + \alpha_2 x_2 \in \mathcal{Y}$ for arbitrary α_1 and α_2.

Trivial examples of subspaces are the zero vector 0 and the linear space $\mathbb{F}^n$ itself.

Let $\mathbb{F}$ be $\mathbb{R}$ or $\mathbb{C}$ and let $x_1, x_2, \ldots, x_k \in \mathbb{F}^n$. Then the vector

$$\alpha_1 x_1 + \alpha_2 x_2 + \cdots + \alpha_k x_k,$$

where $\alpha_1, \alpha_2, \ldots, \alpha_k \in \mathbb{F}$, is a *linear combination* over $\mathbb{F}$ of the vectors $x_1, x_2, \ldots, x_k$. The set of all linear combinations of $x_1, x_2, \ldots, x_k$ is a subspace which is called *span* of $x_1, x_2, \ldots, x_k$ and is denoted by

$$\text{span}(x_1, x_2, \ldots, x_k) = (\alpha_1 x_1 + \alpha_2 x_2 + \cdots + \alpha_k x_k : \alpha_1, \alpha_2, \ldots, \alpha_k \in \mathbb{F}).$$

The vectors $x_1, x_2, \ldots, x_k$ are called *linearly dependent* over $\mathbb{F}$ if there exist numbers $\alpha_1, \alpha_2, \ldots, \alpha_k \in \mathbb{F}$, which are not simultaneously zeros, such that

$$\alpha_1 x_1 + \alpha_2 x_2 + \cdots + \alpha_k x_k = 0$$

(we also say that the system $(x_1, x_2, \ldots, x_k)$ is linearly dependent). In the opposite case, these vectors are *linearly independent*. For instance, a system of one vector is linearly dependent exactly when this vector is zero. If the vectors $(x_1, x_2, \ldots, x_k)$ are linearly independent, then

$$\alpha_1 x_1 + \alpha_2 x_2 + \cdots + \alpha_k x_k = 0 \longrightarrow \alpha_1 = \alpha_2 = \cdots = \alpha_k = 0.$$

Note that in a system of vectors it is possible to have repeating elements.

Let $\mathcal{S}$ be a subspace of $\mathbb{F}^n$. The vector system $(x_1, x_2, \ldots, x_k) \in \mathcal{S}$ is called *basis* of $\mathcal{S}$ if $x_1, x_2, \ldots, x_k$ are linearly independent. The subspace spanned by the vectors $x_1, x_2, \ldots, x_k$ is written as

$$\mathcal{S} = \text{span}(x_1, x_2, \ldots, x_k).$$

Such basis of $\mathcal{S}$ is not unique, but all bases of $\mathcal{S}$ have the same number of elements. This number is called *dimension* of $\mathcal{S}$ and is denoted by $\dim(\mathcal{S})$. The dimension of $\text{span}(x_1, x_2, \ldots, x_k)$ is equal to the maximum number of linearly independent vectors in the system $x_1, x_2, \ldots, x_k$. It is not possible to have repeated elements between the basis vectors since such elements are linearly dependent. By convention, the dimension of the trivial subspace (0) is zero. Note that the set $\mathbb{C}$ of complex numbers can be considered as a one-dimensional vector space over itself with a basis vector (1). The same set can also be considered as a two-dimensional vector space over $\mathbb{R}$ with a basis $(1, \jmath_0)$. More generally, any scalar field $\mathbb{F}$ can be considered as an one-dimensional vector space over itself and has a basis of (1).

In n-dimensional space, an arbitrary system of n linearly independent vectors constitutes a basis and any system of $n+1$ vectors is linearly dependent.

Example A.1. Consider as examples a subspace of the space $\mathbb{R}^3$ of dimension 1 with a basis

$$x_1 = \begin{bmatrix} 1 \\ 0 \\ 0 \end{bmatrix}$$

and a subspace of dimension 2 with a basis $[x_1, x_2]$:

$$x_1 = \begin{bmatrix} 0.9 \\ -0.1 \\ -0.2 \end{bmatrix}, \quad x_2 = \begin{bmatrix} 0.33 \\ 2 \\ 1 \end{bmatrix}.$$

The geometric interpretation of a subspace of dimension 1 in $\mathbb{R}^3$ is a straight line and of dimension 2 is a plane (Figure A.1). Note the common vector (point) at 0.

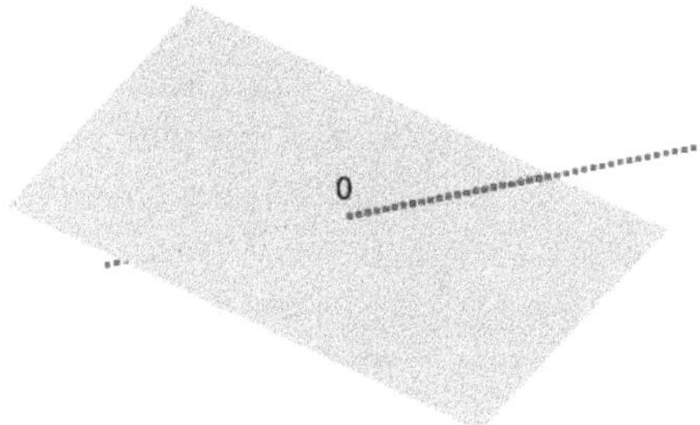

Fig. A.1. Subspaces in $\mathbb{R}^3$ of dimensions 1 and 2.

Any vector x of the linear space $\mathcal{X}$ over $\mathbb{F}$ can be represented as the linear combination

$$x = \alpha_1 x_1 + \alpha_2 x_2 + \cdots + \alpha_n x_n, \tag{A.1}$$

where $\alpha_1, \alpha_2, \ldots, \alpha_n \in \mathbb{F}$ are some scalars and $x_1, x_2, \ldots, x_n$ are basis of $\mathcal{X}$. The linear combination (A.1) is called *decomposition of the vector x along the basis* and the numbers $\alpha_1, \alpha_2, \ldots, \alpha_n$ are called *coordinates of the vector x with respect to this basis.*

In particular,

$$\mathbb{F}^n = \operatorname{span}(e_1, e_2, \ldots, e_n),$$

since for any $x \in \mathbb{F}^n$ we have that

$$x = \xi_1 e_1 + \xi_2 e_2 + \cdots + \xi_n e_n,$$

where $\xi_1, \xi_2, \ldots, \xi_n \in \mathbb{F}^n$ are the coordinates of x with respect to the basis $e_1, e_2, \ldots, e_n$. The basis $e_1, e_2, \ldots, e_n$ is called the *canonical basis* of $\mathbb{F}^n$.

Let $\mathcal{Y}$ be a subspace of the n-dimensional linear space $\mathcal{X}$. Then $\mathcal{Y}$ has dimension $k \leq n$. If $k = n$, then $\mathcal{Y}$ coincides with $\mathcal{X}$.

If the vectors $f_1, f_2, \ldots, f_k$ constitute a basis of the k-dimensional subspace $\mathcal{Y}$ of the n-dimensional space $\mathcal{X}$, then this basis can be expanded by the vectors $f_{k+1}, f_{k+2}, \ldots, f_n$ from the space $\mathcal{X}$, such that the collection of vectors $f_1, \ldots, f_k, f_{k+1}, \ldots, f_n$ constitute a basis of the whole space $\mathcal{X}$. In this basis, all vectors from $\mathcal{Y}$ and only such vectors will have coordinates $\alpha_{k+1} = \alpha_{k+2} = \cdots = \alpha_n = 0$.

Two linear spaces $\mathcal{X}$ and $\mathcal{X}'$ over the same scalar field $\mathbb{F}$ are called *isomorphic*, if between their vectors is possible to establish one-to-one correspondence. Such correspondence implies that if to the vectors x_1 and x_2 from the space $\mathcal{X}$ correspond respectively the vectors x_1' and x_2' from the space $\mathcal{X}'$, then to the vector $x_1 + x_2$ corresponds the vector $x_1' + x_2'$ and to the vector αx for arbitrary $\alpha \in \mathbb{F}$ corresponds the vector $\alpha x'$. The necessary and sufficient condition for isomorphic correspondence between two linear spaces is the equality of their dimensions.

Sum $\mathcal{Y} + \mathcal{Z}$ of two linear subspaces $\mathcal{Y}$ and $\mathcal{Z}$ is the set of all vectors of the form $y + z$, where $y \in \mathcal{Y}, z \in \mathcal{Z}$. *Intersection* $\mathcal{Y} \cap \mathcal{Z}$ of the linear subspaces $\mathcal{Y}$ and $\mathcal{Z}$ is the set of all vectors which belong simultaneously to $\mathcal{Y}$ and to $\mathcal{Z}$.

The sum and the intersection of subspaces are also subspaces. For each subspace $\mathcal{Y}$ of $\mathcal{X}$, it is fulfilled that

$$\mathcal{Y} + \{0\} = \mathcal{Y}, \quad \mathcal{Y} \cap \mathcal{X} = \mathcal{Y}.$$

The dimension of the sum of two arbitrary subspaces depends not only on the dimensions of these subspaces but also on the dimension of their common part (the intersection). In the general case, we have that

$$\dim(\mathcal{Y} + \mathcal{Z}) = \dim(\mathcal{Y}) + \dim(\mathcal{Z}) - \dim(\mathcal{Y} \cap \mathcal{Z}).$$

In Figure A.2, we show the intersection of two subspaces $\mathcal{Y}$ and $\mathcal{Z}$ whose dimension is equal to 2. The intersection $\mathcal{Y} \cap \mathcal{Z}$ has dimension equal to one.

If $\mathcal{Y}$ and $\mathcal{Z}$ are subspaces of $\mathcal{X}$ and $\mathcal{Y} \cap \mathcal{Z} = \{0\}$, then we say that the subspaces are *disjoint.* (Note that disjoint subspaces have a common element: the zero vector.) The sum of disjoint subspaces $\mathcal{Y}$ and $\mathcal{Z}$ is called the *direct sum* of $\mathcal{Y}$ and $\mathcal{Z}$ and is written as $\mathcal{Y} \oplus \mathcal{Z}$. Each vector x of the direct sum $\mathcal{Y} \oplus \mathcal{Z}$ is represented in an unique way as $x = y + z$, where $y \in Y, z \in Z$.

In Figure A.3, we show the direct sum $\mathcal{Y} \oplus \mathcal{Z}$ of two subspaces $\mathcal{Y}$ and $\mathcal{Z}$ whose dimension is equal to 1. The intersection $\mathcal{Y} \cap \mathcal{Z}$ has dimension equal to zero so that the sum $\mathcal{Y} + \mathcal{Z}$ has dimension equal to 2.

The sum of several subspaces is a direct sum, if each of them has a zero intersection with the rest. For a space $\mathcal{X}$ to be a direct sum of

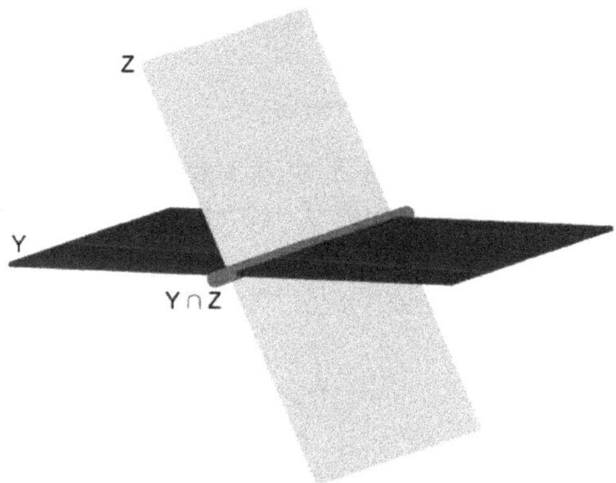

Fig. A.2. Intersection of two subspaces.

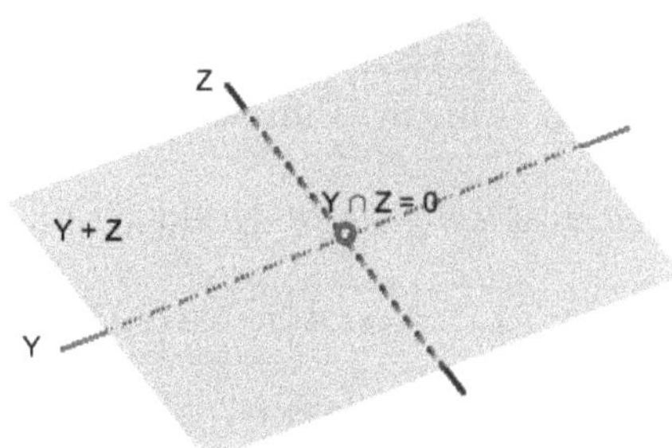

Fig. A.3. Direct sum of two subspaces.

its subspaces, it is necessary and sufficient for the union of the bases of these subspaces to be a basis of the whole space.

Let $\mathcal{Y}$ be a subspace of the linear space $\mathcal{X}$. The subspace $\mathcal{Y}'$ is called *complementary* to the subspace $\mathcal{Y}$, if

$$\mathcal{X} = \mathcal{Y} \oplus \mathcal{Y}'.$$

The complementary subspace always exists and

$$\dim(\mathcal{Y}') = \dim(\mathcal{X}) - \dim(\mathcal{Y}).$$

The dimension of the complementary subspace $\mathcal{Y}'$ is called the *codimension* of $\mathcal{Y}$. The codimension of $\mathcal{Y}$ is always nonnegative and is zero if and only if $\mathcal{Y} = \mathcal{X}$.

A.3 Linear Transformations and Matrices

A.3.1 *Linear Transformations*

Given are the two linear spaces $\mathcal{X}$ and $\mathcal{Y}$ over the same scalar field $\mathbb{F}$. Assume that there exists a rule f which to each vector of $\mathcal{X}$ puts in correspondence an uniquely defined vector of $\mathcal{Y}$. We say that f is a *mapping* or *transformation* of $\mathcal{X}$ into $\mathcal{Y}$ which is denoted by $f : \mathcal{X} \to \mathcal{Y}$ or $\mathcal{Y} = f(\mathcal{X})$.

The transformation $\mathcal{A} : \mathcal{X} \to \mathcal{Y}$ of the linear space $\mathcal{X}$ into the linear space $\mathcal{Y}$, where $\mathcal{X}$ and $\mathcal{Y}$ are over the same scalar field, is called *linear transformation* (or *linear operator*) if

(1) $\mathcal{A}(\alpha x) = \alpha\mathcal{A}(x)$ for all $x \in \mathcal{X}$ and all scalars α and
(2) $\mathcal{A}(x_1 + x_2) = \mathcal{A}(x_1) + \mathcal{A}(x_2)$ for all $x_1,\ x_2 \in \mathcal{X}$.

In the opposite case, this transformation is called *nonlinear*.

If the transformation is linear, the *image* $\mathcal{A}(x)$ is usually denoted as $\mathcal{A}x$.

A transformation $\mathcal{A}$ of $\mathbb{F}^n$ into $\mathbb{F}^m$ is linear if and only if

$$\mathcal{A}(\alpha_1 x_1 + \alpha_2 x_2 + \cdots + \alpha_k x_n) = \alpha_1\mathcal{A}(x_1) + \alpha_2\mathcal{A}(x_2) + \cdots + \alpha_n\mathcal{A}(x_n)$$

for all $x_1, x_2, \ldots, x_n \in \mathbb{F}^n$, all scalars $\alpha_1, \alpha_2, \ldots, \alpha_n$ and all finite n. This fact is called the *principle of superposition* and it plays an important role in the natural sciences.

If $\mathcal{A}$ is a linear transformation, then $\mathcal{A}(0) = 0$.

The *image* (or *range*) $\mathcal{R}(\mathcal{A})$ of the linear transformation $\mathcal{A}$ consists of all possible vectors of the form $\mathcal{A}x$, i.e.,

$$\mathcal{R}(\mathcal{A}) = \{\mathcal{A}x : x \in \mathbb{F}^n\} \subset \mathbb{F}^m.$$

The *rank* of $\mathcal{A}$ is the integer

$$\operatorname{rank}(\mathcal{A}) := \dim(\mathcal{R}(\mathcal{A})).$$

The *null space* (or *kernel*) $\mathcal{N}(\mathcal{A})$ of the linear transformation $\mathcal{A}$ is the set of all vectors x such that $\mathcal{A}x = 0$, i.e.,

$$\mathcal{N}(\mathcal{A}) = \{x \in \mathbb{F}^n : \mathcal{A}x = 0\} \subset \mathbb{F}^n.$$

The dimension of $\mathcal{N}(\mathcal{A})$ is called the *nullity* (or the *defect*) of $\mathcal{A}$ and is denoted by $\operatorname{null}(\mathcal{A})$.

Let $\mathcal{Y}$ be a subspace of $\mathbb{F}^n$ and $\mathcal{A} : \mathbb{F}^n \to \mathbb{F}^m$ be a linear transformation. Then the set of all vectors $\mathcal{A}y \in \mathbb{F}^m$, $y \in \mathcal{Y}$ is called the *image* of $\mathcal{Y}$ (under the action of $\mathcal{A}$) and is denoted by $\mathcal{A}\mathcal{Y}$. In particular, $\mathcal{A}\mathbb{F}^n = \mathcal{R}(\mathcal{A})$ and $\mathcal{A}\mathcal{N}(\mathcal{A}) = \{0\}$.

A.3.2 *Matrix of the Linear Transformation*

Let $\mathcal{X}$ and $\mathcal{Y}$ be linear spaces with $\dim(\mathcal{X}) = n$, $\dim(\mathcal{Y}) = m$ and let $\mathcal{A} : \mathcal{X} \to \mathcal{Y}$ be a linear transformation.

Assume that $B_1 = (x_1, x_2, \dots, x_n)$ and $B_2 = (y_1, y_2, \dots, y_m)$ are bases of $\mathcal{X}$ and $\mathcal{Y}$, respectively. Arbitrary vectors $x \in \mathcal{X}$ and $y \in \mathcal{Y}$ can be represented in the form

$$x = \alpha_1 x_1 + \alpha_2 x_2 + \cdots + \alpha_n x_n$$

and

$$y = \beta_1 y_1 + \beta_2 y_2 + \cdots + \beta_m y_m,$$

respectively. Hence for each $x \in \mathcal{X}$, $y = \mathcal{A}(x)$ can be represented in a unique way as

$$\mathcal{A}(x) = \beta_1 y_1 + \beta_2 y_2 + \cdots + \beta_m y_m$$

and from the linearity of $\mathcal{A}$ it follows that

$$\mathcal{A}(x) = \alpha_1 \mathcal{A}(x_1) + \alpha_2 (\mathcal{A}x_2) + \cdots + \alpha_n \mathcal{A}(x_n).$$

But $\mathcal{A}(x_1), (\mathcal{A}x_2), \dots, \mathcal{A}(x_n)$ are points in $\mathcal{Y}$ and can be represented in a unique way as

$$\begin{aligned}
\mathcal{A}(x_1) &= a_{11} y_1 + a_{21} y_2 + \cdots + a_{m1} y_m, \\
\mathcal{A}(x_2) &= a_{12} y_1 + a_{22} y_2 + \cdots + a_{m2} y_m, \\
\vdots \quad & \quad \vdots \\
\mathcal{A}(x_n) &= a_{1n} y_1 + a_{2n} y_2 + \cdots + a_{mn} y_m,
\end{aligned}$$

where a_{ij} are scalars. In this way, we have that

$$\begin{aligned}\mathcal{A}(x) &= \beta_1 y_1 + \beta_2 y_2 + \cdots + \beta_m y_m \\ &= \alpha_1(a_{11}y_1 + a_{21}y_2 + \cdots + a_{m1}y_m) \\ &\quad + \alpha_1(a_{12}y_1 + a_{22}y_2 + \cdots + a_{m2}y_m) + \cdots \\ &\quad \alpha_n(a_{1n}y_1 + a_{2n}y_2 + \cdots + a_{mn}y_m) \\ &= (a_{11}\alpha_1 + a_{12}\alpha_2 + \cdots + a_{1n}\alpha_n)y_1 \\ &\quad + (a_{21}\alpha_1 + a_{22}\alpha_2 + \cdots + a_{2n}\alpha_n)y_2 + \cdots \\ &\quad (a_{m1}\alpha_1 + a_{m2}\alpha_2 + \cdots + a_{mn}\alpha_n)y_m.\end{aligned}$$

Again, the representation of $\mathcal{A}(x)$ in the terms of the basis B_2 is unique. Hence

$$\begin{aligned}\beta_1 &= a_{11}\alpha_1 + a_{12}\alpha_2 + \cdots + a_{1n}\alpha_n, \\ \beta_2 &= a_{21}\alpha_1 + a_{22}\alpha_2 + \cdots + a_{2n}\alpha_n, \\ \vdots\ &\ \vdots \\ \beta_m &= a_{m1}\alpha_1 + a_{m2}\alpha_2 + \cdots + a_{mn}\alpha_n.\end{aligned}$$

These equations are written in a matrix form as

$$\begin{bmatrix}\beta_1 \\ \beta_2 \\ \vdots \\ \beta_m\end{bmatrix} = \begin{bmatrix} a_{11} & a_{12} & \cdots & a_{1n} \\ a_{21} & a_{22} & \cdots & a_{2n} \\ \vdots & \vdots & \ddots & \vdots \\ a_{m1} & a_{m2} & \cdots & a_{mn}\end{bmatrix}\begin{bmatrix}\alpha_1 \\ \alpha_2 \\ \vdots \\ \alpha_n\end{bmatrix}.$$

We say that the matrix

$$A = \begin{bmatrix} a_{11} & a_{12} & \cdots & a_{1n} \\ a_{21} & a_{22} & \cdots & a_{2n} \\ \vdots & \vdots & \ddots & \vdots \\ a_{m1} & a_{m2} & \cdots & a_{mn}\end{bmatrix}$$

represents the linear transformation $\mathcal{A}: \mathcal{X} \to \mathcal{Y}$. Note that $\mathcal{A}$ and A are not the same. $\mathcal{A}$ is the rule that is used to put in correspondence the vectors y with the vectors x, while A is an $m \times n$ array of scalars. The matrix A represents the linear transformation $\mathcal{A}$. More precisely, A represents $\mathcal{A}$ with respect to the bases B_1 and B_2 in a sense that A together with B_1 and B_2 can be used to solve the equation $y = Ax$.

Each $m \times n$ matrix A over $\mathbb{F}$ is a matrix of some linear transformation acting from the n-dimensional space $\mathbb{F}^n$ to the m-dimensional space $\mathbb{F}^m$ under fixed bases in these spaces. In this way, there exists a one-to-one correspondence between the linear transformations and the rectangular matrices over the same field under arbitrary fixed bases. This correspondence lies in the basis of the full analogy between the properties of the linear transformations and the properties of their corresponding matrices.

Further on we shall identify a linear transformation $\mathcal{A}$ with the matrix A which defines it and in both cases we shall use the symbol A.

Let $A \in \mathbb{F}^{m\times n}$. Then the null space of A and the image of A^H are subspaces for which

$$\mathcal{R}(A) \oplus \mathcal{N}(A^H) = \mathbb{F}^m, \quad \mathcal{R}(A^H) \oplus \mathcal{N}(A) = \mathbb{F}^n.$$

From these expressions, it follows that

$$\text{rank}(A) + \text{null}(A) = n.$$

Example A.2. The expression

$$X \to AX + XB, \tag{A.2}$$

where $A \in \mathbb{C}^{m\times m}$ and $B \in \mathbb{C}^{n\times n}$, defines a linear transformation

$$\mathcal{L}X = AX + XB$$

from $\mathbb{C}^{m\times n}$ to $\mathbb{C}^{m\times n}$. The transformation $\mathcal{L}X$ is called *Sylvester operator*. Expanding the matrix X columnwise as the vector $\text{vec}(X) \in \mathbb{C}^{m\cdot n}$, (A.2) is represented as the linear transformation

$$\text{vec}(X) \to (I_n \otimes A + B^T \otimes I_m)\text{vec}(X)$$

from $\mathbb{C}^{m\cdot n}$ to $\mathbb{C}^{m\cdot n}$. (Note that $\mathbb{C}^{m\cdot n}$ is isomorphic to $\mathbb{C}^{m\times n}$.) Hence

$$L = I_n \otimes A + B^T \otimes I_m \in \mathbb{C}^{mn\times mn}$$

is the matrix of the linear transformation $\mathcal{L}$ in $\mathbb{C}^{m\cdot n}$.

A.3.3 *Operations on Linear Transformations*

Let A and B be two linear transformations acting from $\mathbb{F}^n$ to $\mathbb{F}^m$. *Sum* is called the linear transformation S defined by the equality

$$Sx = Ax + Bx.$$

The transformation S is denoted by

$$S = A + B. \tag{A.3}$$

Equation (A.3) expresses the relationship between the matrices of the corresponding transformations in the same basis.

Product of the linear transformation A by the scalar α is called the linear transformation P determined by

$$Px = \alpha(Px).$$

The transformation P is denoted by

$$P = \alpha A$$

and this equation expresses the relationship between the matrices P and A in the same basis.

Zero transformation, denoted by the symbol 0, is called the transformation which maps all elements of the space $\mathbb{F}^n$ into the zero element of $\mathbb{F}^m$, i.e., $0x = 0$.

Let to each vector $x \in \mathbb{F}^n$ be put in correspondence the same vector x which defines a linear transformation I acting from $\mathbb{F}^n$ to $\mathbb{F}^n$. This transformation is called *identity* or *unit* transformation. By definition, $x = Ix$.

For each transformation A is defined the *reversing transformation* $-A$ by the equation

$$-A = (-1)A.$$

The set $\mathcal{M}_{\mathbb{F}^n\mathbb{F}^m}$ of all linear transformations $A : \mathbb{F}^n \to \mathbb{F}^m$ with the so-introduced operations addition and multiplication by a scalar represents a linear space which has a dimension equal to the product mn. The set M_{mn} of all $m \times n$ matrices whose elements are scalars from $\mathbb{F}$ is also a linear space which is isomorphic to the space $\mathcal{M}_{\mathbb{F}^n\mathbb{F}^m}$. The relation "$[T]$ represents A with respect to B_1 and B_2" is a mapping of $\mathcal{M}_{\mathbb{F}^n\mathbb{F}^m}$ into M_{mn}.

Latter on the linear space of all real (complex) matrices is denoted by $\mathbb{R}^{m\times n}(\mathbb{C}^{m\times n})$.

Since the matrices in $\mathbb{C}^{m\times n}$ can be identified with vectors in the space $\mathbb{C}^{m\cdot n}$, the notion of linear independence can be spread over matrices. The matrices $A_1,\ A_2,\dots,A_k \in \mathbb{C}^{m\times n}$ are called *linearly independent*, if

$$\alpha_1 A_1 + \alpha_2 A_2 + \cdots + \alpha_k A_k = 0_{m\times n} \longrightarrow \alpha_1 = \alpha_2 = \cdots = \alpha_k = 0.$$

The space $\mathbb{F}^{m\times n}$ regarded as a linear space has the basis

$$\left\{E_{ij} = e_i e_j^T : i = 1,2,\dots,m;\ j = 1,2,\dots,n\right\}.$$

Each matrix in $\mathbb{F}^{m\times n}$ can be represented with respect to this basis as

$$A = \sum_{i=1}^{m}\sum_{j=1}^{n} \alpha_{ij} E_{ij},$$

where $\alpha_{ij} \in \mathbb{F}$.

Given are the linear spaces $\mathcal{X}, \mathcal{Y}$ and $\mathcal{Z}$ over the same field and the transformations $A : \mathcal{X} \to \mathcal{Y},\ B : \mathcal{Y} \to \mathcal{Z}$. Then the transformation $P : \mathcal{X} \to \mathcal{Z}$ is called *product* of the transformations A and B if it is fulfilled that

$$Px = B(Ax)$$

for each vector $x \in \mathcal{X}$. We write that $P = BA$.

The following relationships hold:

$$\mathcal{R}(BA) = B\mathcal{R}(A),$$
$$\operatorname{rank}(BA) = \operatorname{rank}(A) - \dim(\mathcal{R}(A) \cap \mathcal{N}(B)),$$
$$\operatorname{null}(BA) = \operatorname{null}(A) + \dim(\mathcal{R}(A) \cap \mathcal{N}(B)).$$

In the general case, it is fulfilled that

$$\operatorname{rank}(BA) \le \min\{\operatorname{rank}(B), \operatorname{rank}(A)\}.$$

If
$\operatorname{rank}(A) = \dim(\mathcal{Y})$, then $\operatorname{rank}(BA) = \operatorname{rank}(B)$, and if $\operatorname{rank}(B) = \dim(\mathcal{Y})$, then $\operatorname{rank}(BA) = \operatorname{rank}(A)$.

The transformation $A : \mathbb{F}^n \to \mathbb{F}^n$ is called *nonsingular* if its kernel consists only of the zero vector. In the opposite case, the transformation is *singular.*

If A is nonsingular, then $\operatorname{rank}(A) = n$.

Let A be a nonsingular transformation. Then for each vector $b \in \mathbb{F}^n$ there exists an unique vector $x \in \mathbb{F}^n$ such that

$$b = Ax. \tag{A.4}$$

This correspondence determines a linear transformation which is called *inverse* to the transformation A and is denoted by A^{-1}. If (A.4) holds, then

$$x = A^{-1}b.$$

From the definition of the inverse transformation, it follows that

$$AA^{-1} = AA^{-1} = I.$$

Consider now the case of a rectangular system of linear equations

$$Ax = b, \tag{A.5}$$

where $A \in \mathbb{F}^{m\times n}$, $b \in \mathbb{F}^m$ and $x \in \mathbb{F}^n$. Clearly, (A.5) has a solution for x exactly when $b \in \mathcal{R}(A)$, which is equivalent to the condition (*Theorem of Kronecker–Capelli*)

$$\operatorname{rank}(A) = \operatorname{rank}([A, b]).$$

Under this condition, there exists at least one particular solution $x_0 \in \mathbb{F}^n$ of the equation. The general solution (i.e., the set of all solutions x) in this case is the *linear manifold*

$$x = x_0 + \mathcal{N}(A) = \{x_0 + z : Az = 0\} \subset \mathbb{F}^n.$$

From here it follows that if there exists solution, it is unique only when

$$\mathcal{N}(A) = 0.$$

For $m = n$, the null space $\mathcal{N}(A)$ of A is the zero subspace only when this matrix is nonsingular.

A.3.4 Change of Basis

Let $f_1, f_2, \ldots, f_m$ and $g_1, g_2, \ldots, g_m$ be two bases of the same m-dimensional space $\mathcal{X}$. The vectors $g_1, g_2, \ldots, g_m$ are uniquely determined by their decompositions

$$\begin{aligned} g_1 &= p_{11}f_1 + p_{21}f_2 + \cdots + p_{m1}f_m, \\ g_2 &= p_{12}f_1 + p_{22}f_2 + \cdots + p_{m2}f_m, \\ \vdots\ & \quad \vdots \\ g_m &= p_{1m}f_1 + p_{2m}f_2 + \cdots + p_{mm}f_m, \end{aligned}$$

along the vectors $f_1, f_2, \ldots, f_m$ so that

$$g_j = \sum_{i=1}^{m} p_{ij} f_i, \quad j = 1, 2, \ldots, m. \tag{A.6}$$

The coefficients p_{ij} determine the matrix

$$P = \begin{bmatrix} p_{11} & p_{12} & \cdots & p_{1m} \\ p_{21} & p_{22} & \cdots & p_{2m} \\ \vdots & \vdots & \ddots & \vdots \\ p_{m1} & p_{m2} & \cdots & p_{mm} \end{bmatrix}.$$

This matrix is called *coordinate transformation matrix* in the transition from the basis $f_1, f_2, \ldots, f_m$ to the basis $g_1, g_2, \ldots, g_m$.

Let us take an arbitrary vector $x \in \mathcal{X}$ and decompose it along the vectors of both bases. Let

$$x = \sum_{i=1}^{m} \xi_i f_i = \sum_{j=1}^{m} \eta_j g_j.$$

Using (A.6) it is possible to show that

$$\xi_i = \sum_{j=1}^{m} p_{ij} \eta_j \tag{A.7}$$

for $i = 1, 2, \ldots, m$. These formulas are called *formulas of coordinate transformation.* Denoting

$$x_e = \begin{bmatrix} \xi_1 \\ \xi_2 \\ \vdots \\ \xi_n \end{bmatrix}, \quad x_f = \begin{bmatrix} \eta_1 \\ \eta_2 \\ \vdots \\ \eta_n \end{bmatrix},$$

we obtain from (A.7) that

$$x_e = Px_f. \tag{A.8}$$

The matrix P should be nonsingular since in the opposite case there will exist a linear dependence between its columns and hence between the vectors $g_1, g_2, \ldots, g_m$. Multiplying (A.8) on the left by P^{-1}, we obtain

$$x_f = P^{-1}x_e.$$

Consider a linear transformation acting from the m-dimensional space $\mathcal{X}$ to the n-dimensional space $\mathcal{Y}$. Choose in the space $\mathcal{X}$ two bases

$$f_1, f_2, \ldots, f_m \quad \text{and} \quad g_1, g_2, \ldots, g_m$$

and in the space $\mathcal{Y}$ two bases

$$q_1, q_2, \ldots, q_n \quad \text{and} \quad t_1, t_2, \ldots, t_n.$$

To the same transformation A in the first pair of bases corresponds the coordinate equality

$$y_q = A_{qf}x_f \tag{A.9}$$

and in the second pair of bases

$$y_t = A_{tg}x_g. \tag{A.10}$$

Thus, for these pairs of bases we have two matrices A_{qf} and A_{tg} for the same transformation A.

Denote by P the matrix of coordinate transformation in the transition from the basis $f_1, f_2, \ldots, f_m$ to the basis $g_1, g_2, \ldots, g_m$ and by

Q the matrix of coordinate transformation in the transition from $q_1, q_2, \ldots, q_n$ to $t_1, t_2, \ldots, t_n$. We have that

$$x_f = Px_g, \quad y_q = Qy_t. \tag{A.11}$$

Substituting these expressions for x_f and x_g into (A.9), we find

$$Qy_t = A_{qf}Px_g$$

so that

$$y_t = (Q^{-1}A_{qf}P)x_g.$$

Comparing the obtained equality with (A.10), we conclude that

$$A_{tf} = Q^{-1}A_{qe}P. \tag{A.12}$$

Equation (A.12) relates the matrices of the same transformation in different bases.

Two rectangular matrices A and B with the same sizes are called *equivalent* if there exist two nonsingular matrices R and S such that

$$B = RAS.$$

It follows from (A.12) that two matrices corresponding to the same linear transformation for different choice of the bases in $\mathcal{X}$ and $\mathcal{Y}$ are always equivalent. The transformation $A \rightarrow RAS$ is called *equivalence transformation.*

Let the linear transformation A act from the space $\mathcal{X}$ to $\mathcal{X}$. In such case, the basis $q_1, q_2, \ldots, q_n$ can be chosen as $f_1, f_2, \ldots, f_m$ and the basis $a_1, a_2, \ldots, a_n$ can be chosen as $g_1, g_2, \ldots, g_m$ so that the matrices P and Q in (A.12) coincide. In this way, to each linear transformation acting in a linear space corresponds a class of matrices related by the equation

$$B = S^{-1}AS \tag{A.13}$$

for different nonsingular matrices S. Such matrices are called *similar* and the matrix S is called *similarity transformation matrix.*

The *similarity transformation* $A \rightarrow P^{-1}AP$ plays an important role in the analysis and solution of the eigenvalue problem.

A.4 Vector Norms

In lots of cases, it is useful to have a single number which characterizes the size of a vector or matrix. For this purpose, we use functions which are called norms.

A *norm* of the element v (vector or matrix) is a real number, denoted by $\|v\|$, which satisfies the following properties:

(1) *Nonnegativity*: $\|v\| \geq 0$, as $\|v\| = 0$ exactly when $v = 0$.
(2) *Homogeneity*: $\|\alpha v\| = |\alpha| \, \|v\|$ for all complex scalars α.
(3) *Triangle inequality*:

$$\|v + w\| \leq \|v\| + \|w\|. \tag{A.14}$$

Consider a column vector v with n real or complex elements v_i. We shall consider three norms which are special cases of the *vector p-norm*

$$\|v\|_p = \left(\sum |v_i|^p\right)^{1/p}, \tag{A.15}$$

where it is necessary to have $p \geq 1$ to satisfy the triangle inequality (property 3 of norm). Here $|v_i|$ is the module of the complex scalar v_i. To illustrate the different norms, we shall compute each of them for the vector

$$z = \begin{bmatrix} z_1 \\ z_2 \\ z_3 \end{bmatrix} = \begin{bmatrix} 1 \\ 2i_o \\ -3 \end{bmatrix}.$$

Vector 1-norm (or norm–sum). We have that

$$\|v\|_1 := \sum_i |v_i| \quad (\|z\|_1 = 1 + 2 + 3 = 6). \tag{A.16}$$

Vector 2-norm (Euclidean norm). This is the most frequently used vector norm which is interpreted as a vector length. We have that

$$\|v\|_2 := \sqrt{\sum_i^n |a_i|^2} \quad (\|z\|_2 = \sqrt{1 + 4 + 9} = 3.7417). \tag{A.17}$$

The 2-norm satisfies the property

$$v^H v = \|v\|_2^2, \tag{A.18}$$

where v^H is the Hermitian conjugate of the vector v.

Vector norm ∞. This is the maximum in module vector element. We use the notion $\|v\|_{\max}$, such that

$$\|v\|_{\max} \equiv \|v\|_\infty := \max_i |v_i| \quad (\|z\|_{\max} = |-3| = 3). \tag{A.19}$$

For all $p, q \geq 1$, there exists a constant c_{pq}, such that $\|a\|_p \leq c_{pq}\|a\|_q$ for each a. In this sense, the p-norms frequently are called *equivalent*. For instance, for a vector v with n elements, it is fulfilled that

$$\|v\|_{\max} \leq \|v\|_2 \leq \sqrt{n}\,\|v\|_{\max}, \tag{A.20}$$

$$\|v\|_2 \leq \|v\|_1 \leq \sqrt{n}\,\|v\|_2. \tag{A.21}$$

Figure A.4 illustrates the difference between the vector norms by plotting the line $\|v\|_p = 1$ in $\mathbb{R}^2$.

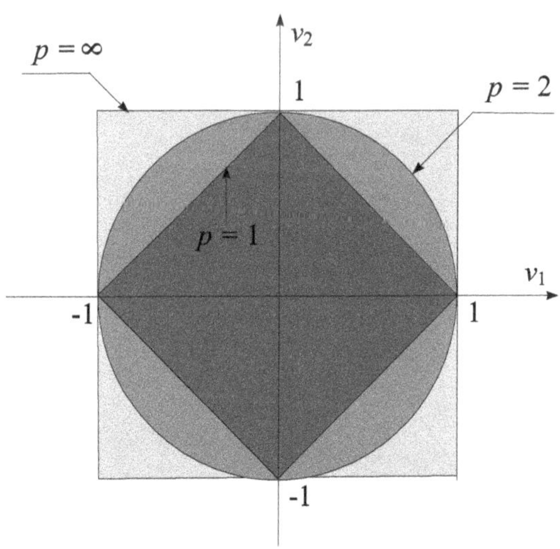

Fig. A.4. Contours for the vector p-norm, $\|v\|_p = 1$ for $p = 1,\ 2,\ \infty$.

A.5 Orthogonality

Let x and y be vectors in $\mathbb{R}^2$. The cosine of the angle between these vectors (see Figure A.5) is given by

$$\begin{aligned}\cos\Theta &= \cos(\alpha-\beta) = \cos\alpha\cos\beta + \sin\alpha\sin\beta \\ &= \frac{x_1y_1 + x_2y_2}{\sqrt{x_1^2+x_2^2}\sqrt{y_1^2+y_2^2}} = \frac{x^Ty}{\|x\|_2\|y\|_2}.\end{aligned}$$

By analogy, the acute angle between two vectors x and y in $\mathbb{C}^n$ is defined as

$$\Theta = \arccos\left(\frac{|y^Hx|}{\|x\|_2\|y\|_2}\right) \quad 0 \le \Theta \le \pi/2. \tag{A.22}$$

If $x^Hy = 0$, then $\Theta = \pi/2$ and we say that x and y are *orthogonal.* This is written as

$$x \perp y.$$

If $u_1, u_2, \ldots, u_k$ are vectors in $\mathbb{C}^n$ and $u_i^Hu_j = 0,\ i \neq j$, then $u_1, u_2, \ldots, u_k$ are said to be *pairwise orthogonal.* An orthogonal set of vectors is linearly independent.

Let $u_1, u_2, \ldots, u_k \in \mathbb{C}^n$. Then the vectors u_i are *orthonormal* if

$$u_i^Hu_j = \begin{cases} 0 & \text{if } i \neq j, \\ 1 & \text{if } i = j. \end{cases}$$

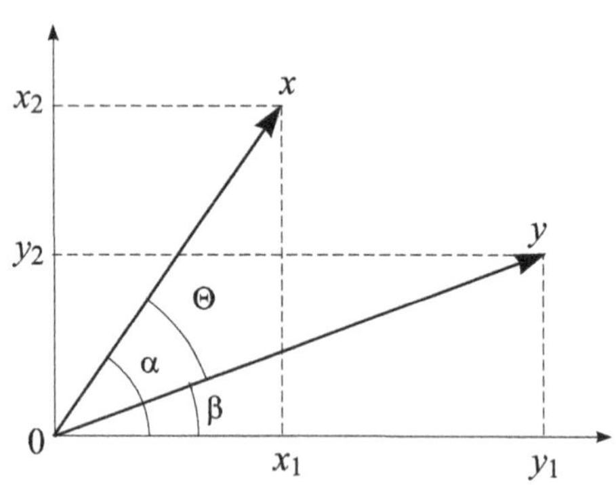

Fig. A.5. Angles between vectors.

Let the $n \times m$ matrix U have orthonormal columns. Then

$$U^H U = I_m.$$

Let $u_1, u_2, \ldots, u_n \in \mathbb{C}^n$ be orthonormal vectors. Then

$$U = [u_1, u_2, \ldots, u_n] \in \mathbb{C}^{n \times n}$$

is called a *unitary matrix*. By definition, $U^H U = I_n$ and hence U is nonsingular. From $U^{-1}U = I_n$, it follows that $U^{-1} = U^H$ and $UU^H = U^H U = I_n$. A square matrix $Q \in \mathbb{R}^{n \times n}$ which fulfills $Q^T Q = QQ^T = I_n$ is called *orthogonal*. If the matrix Q is orthogonal, then $Q^{-1} = Q^T$. Simple examples of orthogonal matrices are the unit and the permutation matrix. If the matrix U is unitary, then $U^H AU$ is referred to as *unitary similar* to A. The matrix B is called *permutation similar* to A if for some permutation matrix P it is fulfilled that $B = P^T AP$.

The unitary and orthogonal matrices are important tools used in the numerical matrix computations.

The set $\mathcal{S}^\perp$ of all vectors $y \in \mathbb{C}^n$, such that $y^H x = 0$ for each $x \in \mathcal{S}$, is said to be the *orthogonal complement* of $\mathcal{S} \subset \mathbb{C}^n$. The orthogonal complement $\mathcal{S}^\perp$ is a subspace of $\mathbb{C}^n$ (Figure A.6).

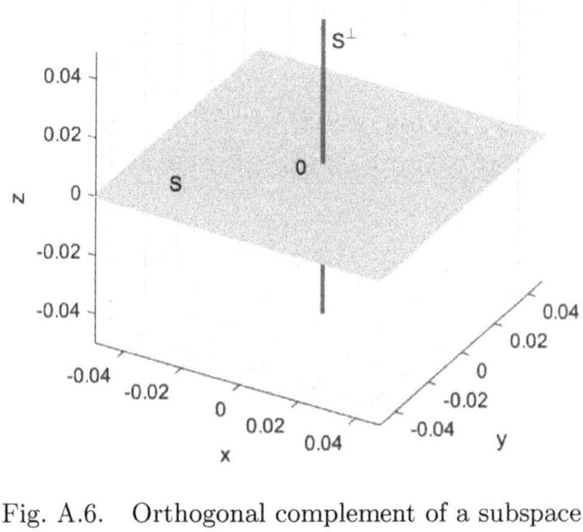

Fig. A.6. Orthogonal complement of a subspace.

Using the orthogonal complement, the space $\mathbb{C}^n$ is decomposed into the direct sum

$$\mathbb{C}^n = \mathcal{S} \oplus \mathcal{S}^\perp.$$

It may be shown that

$$\begin{aligned}(\mathcal{S}^\perp)^\perp &= \mathcal{S},\\ (\mathcal{S} + \mathcal{W})^\perp &= \mathcal{S}^\perp \cap \mathcal{W}^\perp,\\ (\mathcal{S} \cap \mathcal{W})^\perp &= \mathcal{S}^\perp + \mathcal{W}^\perp.\end{aligned} \tag{A.23}$$

A.6 Singular Value Decomposition

Each complex $m \times n$ matrix A can be represented by the *singular value decomposition* (SVD) as

$$A = U\Sigma V^H, \tag{A.24}$$

where the $m \times m$ matrix U and the $n \times n$ matrix V are unitary, and the $m \times n$ matrix Σ is diagonal:

$$\Sigma = \begin{bmatrix} \Sigma_1 \\ 0 \end{bmatrix}; \quad \text{if } m \geq n \tag{A.25}$$

or

$$\Sigma = \begin{bmatrix} \Sigma_1 & 0 \end{bmatrix}; \quad \text{if } m \leq n, \tag{A.26}$$

where

$$\Sigma_1 = \operatorname{diag}(\sigma_1, \sigma_2, \ldots, \sigma_k); \quad k = \min\{m, n\} \tag{A.27}$$

and

$$\sigma_{\max} := \sigma_1 \geq \sigma_2 \geq \cdots \geq \sigma_k := \sigma_{\min}. \tag{A.28}$$

The numbers $\sigma_i \geq 0$ are called *singular values* of the matrix A.

The columns of V, denoted by v_i, are called *right* singular vectors, and the columns of U, denoted by u_i, are called *left* singular vectors.

The decomposition (A.24) is not unique, since the matrices U and V are not unique. However, the singular values σ_i are unique.

The *rank*, rank(A), of an $m \times n$ matrix A is the number of its positive singular values. Let rank(A) $= r$. Then the matrix A is called *rank deficient* if $r < k = \min\{m, n\}$. In this case, $\sigma_i = 0$ for $i = r+1, \ldots, k$. A rank deficient square matrix is a singular matrix.

It follows from the SVD definition that

$$\sigma_{\max}(A^H) = \sigma_{\max}(A) \quad \text{and} \quad \sigma_{\max}(A^T) = \sigma_{\max}(A). \tag{A.29}$$

An important property of the singular values is that

$$\sigma_{\max}(AB) \leq \sigma_{\max}(A)\sigma_{\max}(B) \tag{A.30}$$

which in particular means that rank(A^H) $=$ rank(A).

For a nonsingular matrix A (or B), we have the inequalities

$$\sigma_{\min}(A)\sigma_{\max}(B) \leq \sigma_{\max}(AB) \quad \text{or} \quad \sigma_{\max}(A)\sigma_{\min}(B) \leq \sigma_{\max}(AB). \tag{A.31}$$

The following inequality also holds:

$$\sigma_{\min}(A)\sigma_{\min}(B) \leq \sigma_{\min}(AB). \tag{A.32}$$

For a nonsingular square matrix A, we have that

$$\sigma_{\max}(A^{-1}) = \frac{1}{\sigma_{\min}(A)} \quad \text{and} \quad \sigma_{\max}(A) = \frac{1}{\sigma_{\min}(A^{-1})}. \tag{A.33}$$

Other useful relationships for the singular values are (*Fan's inequalities*)

$$\sigma_{\min}(A) - \sigma_{\max}(B) \leq \sigma_{\min}(A+B) \leq \sigma_{\min}(A) + \sigma_{\max}(B)$$

and

$$\sigma_{\min}(A) - 1 \leq \frac{1}{\sigma_{\max}(I+A)^{-1}} \leq \sigma_{\min}(A) + 1,$$

provided that $I + A$ is nonsingular.

From geometric point of view, the real $m \times n$ matrix A maps the sphere with unit radius

$$B_{\mathbb{R}^n} = \left\{ v : \sqrt{v_1^2 + v_2^2 + \cdots + v_n^2} = 1 \right\}$$

in $\mathbb{R}^n$ into ellipsoid in $\mathbb{R}^m$ (Figure A.7). The singular values σ_i give the lengths of the principal axes of the ellipsoid. The singular vectors u_i determine the orthogonal directions of this ellipsoid and the

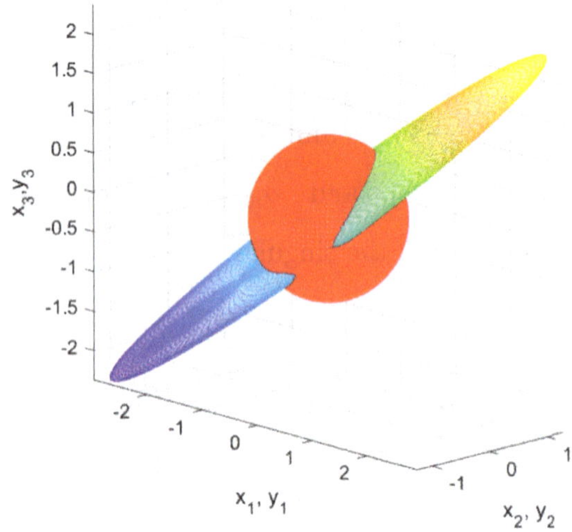

Fig. A.7. Image of the unit sphere under the action of a matrix.

singular vectors v_i determine the orthogonal directions of the vectors which are mapped into u_i. In the given case, the matrix A of the transformation which maps the unit sphere into ellipsoid is

$$A = \begin{bmatrix} -2.0 & -2.0 & 0.8 \\ -1.0 & 0.0 & 0.8 \\ -2.0 & -1.0 & 0.8 \end{bmatrix}$$

and the lengths of the principal axes of the ellipsoid are

$$\sigma_1 = 3.87, \quad \sigma_2 = 0.94, \quad \sigma_3 = 0.22.$$

A.7 Matrix Norms

A.7.1 *Definition and Properties*

Consider the matrices $A \in \mathbb{C}^{m\times n}$ and $B \in \mathbb{C}^{n\times p}$. The quantity $\|A\|$ is a *matrix norm*, if in addition to the three properties of norms, considered in Section A.4, it also satisfies the multiplicative property

$$\|AB\| \leq \|A\|\|B\|. \tag{A.34}$$

For a numerical illustration, we shall use the 2×2 matrix

$$M = \begin{bmatrix} -5 & 1 \\ 4 & -7 \end{bmatrix}.$$

One of the most frequently used matrix norms is the *Frobenius matrix norm*. It is defined as the square root of the sum of squares of the modules of matrix elements:

$$\|A\|_F = \sqrt{\sum_{i,j} |a_{i,j}|^2} = \sqrt{\mathrm{trace}(A^H A)}. \quad (\|M\|_F = \sqrt{91} = 9.5394). \tag{A.35}$$

The most important matrix norm is the *norm induced by the vector p-norm*, also referred to as the *operator norm associated with the vector p-norm*, which is defined as

$$\|A\|_p := \max_{w \neq 0} \frac{\|Aw\|_p}{\|w\|_p}, \tag{A.36}$$

where $\|w\|_p = (\sum_i |w_i|^p)^{1/p}$ denotes the vector p-norm. Other ways speaking, we look for a direction of the vector w, such that the ratio $\|z\|_p/\|w\|_p$ is maximized. In this way, the induced norm yields the maximum possible "amplification" of a matrix. The following equivalent definition is also used:

$$\|A\|_p = \max_{\|w\|_p \leq 1} \|Aw\|_p = \max_{\|w\|_p = 1} \|Aw\|_p. \tag{A.37}$$

The geometric interpretation of the induced norm is that it is the maximum norm of all vectors z, obtained after transforming by the matrix A of all vectors w, having unity norm. This is illustrated in Figure A.8 for vectors in $\mathbb{R}^3$ with 2-norm.

The induced 1, 2 and ∞ matrix norms can be determined from

$$\|A\|_1 = \max_j \left(\sum_i |a_{ij}| \right) \quad \text{'maximum column sum'}, \tag{A.38}$$

$$\|A\|_\infty = \max_i \left(\sum_j |a_{ij}| \right) \quad \text{'maximum row sum'}, \tag{A.39}$$

$$\|A\|_2 = \sigma_{\max}(A) = \sqrt{\rho(A^H A)} \quad \text{'spectral norm'}, \tag{A.40}$$

where the *spectral radius* $\rho(M) = \max_i |\lambda_i(A)|$ is the largest eigenvalue (for the definition of eigenvalue, see Chapter 2) of the matrix A.

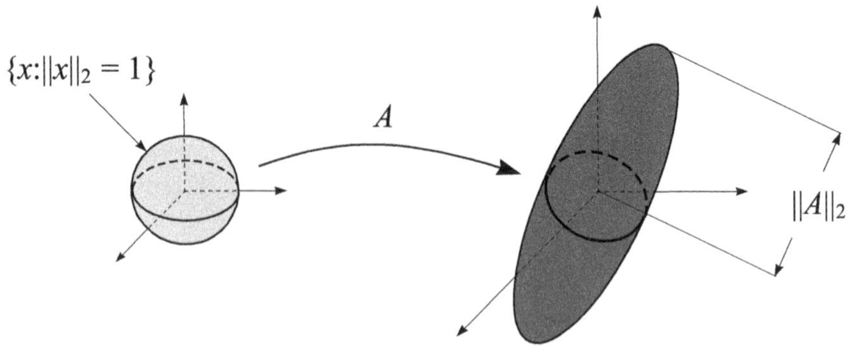

Fig. A.8. Geometric interpretation of the induced 2-norm.

The induced 2-norm of a matrix is equal to its maximum singular value and is called *spectral norm*. For the example matrix, we obtain

$$\|M\|_1 = 9; \quad \|M\|_\infty = 11; \quad \|M\|_2 = \sigma_{\max}(M) = 8.8772.$$

All induced norms $\|A\|_p$ are matrix norms and satisfy the multiplicative property

$$\|AB\|_p \leq \|A\|_p \|B\|_p. \tag{A.41}$$

From (A.41), it follows that

$$\|A^k\|_p \leq \|A\|_p^k, \quad k \geq 1.$$

A.7.2 *Unitary Equivalent Norms*

Let the $n \times p$ matrix U have orthonormal columns. Then

$$\|U\|_2 = 1, \quad \|U\|_F = \sqrt{p}.$$

A very important property of the 2-norm and Frobenius norm is that they are invariant to unitary transformations, i.e., for unitary matrices $U_i, i = 1, 2$, satisfying $U_i U_i^H = I$, we have that

$$\|U_1 A U_2\|_F = \|A\|_F, \tag{A.42}$$

$$\|U_1 A U_2\|_2 = \|A\|_2. \tag{A.43}$$

From the singular value decomposition of the matrix $A = U\Sigma V^H$ and (A.42), one obtains a relationship between the Frobenius norm and the singular values $\sigma_i(A)$:

$$\|A\|_F = \sqrt{\sum_i \sigma_i^2(A)}. \tag{A.44}$$

It follows from this equation that $\|A\|_F = \|A\|_2$ exactly when the rank r of A is equal to 1. Also it is seen that

$$\sqrt{r}\,\sigma_r \leq \|A\|_F \leq \sqrt{r}\,\sigma_1. \tag{A.45}$$

A.7.3 *Relationships between Matrix Norms*

Let $A \in \mathbb{C}^{m\times n}$, then

$$\|A\|_2 \leq \|A\|_F \leq \sqrt{\min\{m,n\}}\;\|A\|_2, \tag{A.46}$$

$$\|A\|_2 \leq \sqrt{\|A\|_1\|A\|_\infty}, \tag{A.47}$$

$$\frac{1}{\sqrt{n}}\|A\|_\infty \leq \|A\|_2 \leq \sqrt{m}\;\|A\|_\infty, \tag{A.48}$$

$$\frac{1}{\sqrt{m}}\|A\|_1 \leq \|A\|_2 \leq \sqrt{n}\;\|A\|_1. \tag{A.49}$$

All these norms satisfy (A.34).

A.8 Computations with Subspaces

Consider how to determine orthonormal bases of the subspaces related to a given matrix $A \in \mathbb{C}^{m\times n}$ using the SVD.

First, we are interested in the so-called *four fundamental subspaces*: the image $\mathcal{R}(A) \subset \mathbb{C}^m$ and its orthogonal complement $\mathcal{R}(A)^\perp \subset \mathbb{C}^m$ and the null space $\mathcal{N}(A) \subset \mathbb{C}^n$ and its orthogonal complement $\mathcal{N}(A)^\perp \subset \mathbb{C}^n$. We recall that $\dim(\mathcal{R}(A)) = r$ and $\dim(\mathcal{N}(A)) = n - r$, where $r = \operatorname{rank}(A)$. It may be shown that (Figure A.9)

$$\mathcal{R}(A)^\perp = \mathcal{N}(A^H),$$
$$\mathcal{N}(A)^\perp = \mathcal{R}(A^H)$$

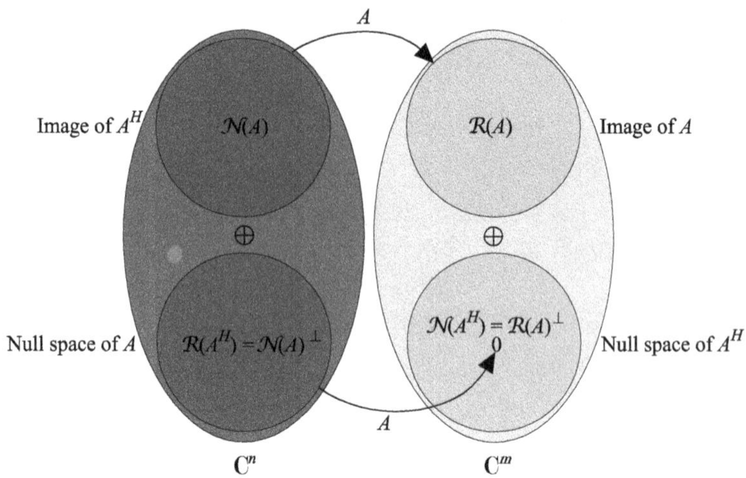

Fig. A.9. The four fundamental subspaces.

and

$$
\begin{aligned}
A\mathcal{R}(A^H) &= \mathcal{R}(A), \\
A\mathcal{N}(A) &= 0
\end{aligned}
$$

Orthonormal bases of the four fundamental subspaces can be found by performing only one singular value decomposition of A. Partitioning the SVD of A as

$$
A = [\underbrace{U_1}_{r}, \underbrace{U_2}_{m-r}] \begin{bmatrix} \Sigma & 0 \\ 0 & 0 \end{bmatrix} \begin{bmatrix} V_1^H \\ V_2^H \end{bmatrix} \begin{matrix} \} r \\ \} n-r \end{matrix}, \tag{A.50}
$$

$$
\Sigma = \mathrm{diag}(\sigma_1, \ldots, \sigma_r), \; \sigma_1 \geq \cdots \geq \sigma_r > 0,
$$

we have that

$$
\begin{aligned}
\mathcal{R}(A) &= \mathrm{span}(U_1), \\
\mathcal{R}(A)^\perp &= \mathrm{span}(U_2), \\
\mathcal{N}(A) &= \mathrm{span}(V_2), \\
\mathcal{N}(A)^\perp &= \mathrm{span}(V_1).
\end{aligned}
$$

Clearly, the accurate determination of orthonormal bases relies on the proper determination of the matrix rank.

Example A.3. Let

$$\begin{bmatrix} -7 & 2 & 11 \\ 4 & 9 & 14 \\ -6 & 1 & 8 \\ -8 & -3 & 2 \\ 4 & -5 & -14 \end{bmatrix}.$$

The SVD of A yields

$$\sigma_1 = 26.438, \quad \sigma_2 = 13.528, \quad \sigma_3 = 0$$

so that

$$\dim(\mathcal{R}(A)) = \text{rank}(A) = 2, \quad \dim(\text{null}(A)) = n - \text{rank}(A) = 1.$$

The bases of the four fundamental subspaces are given by (up to five decimal digits)

$$\mathcal{R}(A) = \mathcal{R}(U_1),$$

$$U_1 = \begin{bmatrix} -0.46695 & -0.34349 \\ -0.56211 & 0.62787 \\ -0.34186 & -0.32481 \\ -0.10167 & -0.61747 \\ 0.58205 & 0.03215 \end{bmatrix},$$

$$\mathcal{R}(A)^{\perp} = \mathcal{R}(U_2),$$

$$U_2 = \begin{bmatrix} -0.62007 & -0.40848 & 0.33561 \\ 0.01790 & 0.39105 & 0.36957 \\ -0.27434 & 0.59778 & -0.58739 \\ 0.47841 & 0.33434 & 0.51743 \\ -0.55772 & 0.45945 & 0.37153 \end{bmatrix},$$

$$\mathcal{N}(A) = \mathcal{R}(V_2),$$

$$V_2 = \begin{bmatrix} -0.40825 \\ 0.81650 \\ -0.40825 \end{bmatrix},$$

$$\mathcal{N}(A)^{\perp} = \mathcal{R}(V_1),$$

$$V_1 = \begin{bmatrix} 0.23500 & 0.88210 \\ -0.33814 & 0.46797 \\ -0.91128 & 0.05383 \end{bmatrix}.$$

Consider now how to compute orthonormal bases of the sum $\mathcal{Y} + \mathcal{Z}$ and the intersection $\mathcal{Y} \cap \mathcal{Z}$ of two subspaces $\mathcal{Y}$ and $\mathcal{Z}$, using the SVD. If $\mathcal{Y} = \mathcal{R}(A), \mathcal{Z} = \mathcal{R}(B), A \in \mathbb{C}^{m\times n}$ and $B \in \mathbb{C}^{m\times p}$, then

$$\mathcal{Y} + \mathcal{Z} = \mathcal{R}([A, B])$$

and an orthonormal basis of $\mathcal{Y} + \mathcal{Z}$ is found by using the SVD of $[A, B]$. If

$$[A, B] = [\underbrace{U_1}_{r}, \underbrace{U_2}_{n-r}] \begin{bmatrix} \Sigma & 0 \\ 0 & 0 \end{bmatrix} V^H,$$

$$\Sigma = \mathrm{diag}(\sigma_1, \ldots, \sigma_r), \quad \sigma_1 \geq \cdots \geq \sigma_r > 0,$$

we have

$$r = \mathrm{rank}([A, B]) = \dim(\mathcal{Y} + \mathcal{Z})$$

so that

$$\mathcal{Y} + \mathcal{Z} = \mathcal{R}([A, B]) = \mathcal{R}(U_1).$$

At the same time, the columns of U_2 constitute an orthonormal basis for the orthogonal complement

$$(\mathcal{Y} + \mathcal{Z})^{\perp} = \mathcal{Y}^{\perp} \cap \mathcal{Z}^{\perp}$$

of $\mathcal{Y} + \mathcal{Z}$.

Example A.4. Let

$$A = \begin{bmatrix} 4 & -1 & 8 \\ 3 & 6 & 1 \\ 7 & -9 & -2 \\ -5 & -4 & -6 \\ 3 & -7 & 0 \end{bmatrix} \quad \text{and} \quad B = \begin{bmatrix} -3 & 5 & 2 \\ -9 & -3 & 0 \\ 2 & 16 & -4 \\ 9 & -1 & 8 \\ 4 & 10 & -5 \end{bmatrix}.$$

In this case, $\dim(\mathcal{Y} + \mathcal{Z}) = \mathrm{rank}([A, B]) = 4$, and with the aid of SVD, we obtain

$$\mathcal{Y} + \mathcal{Z} = \mathcal{R}(A) + \mathcal{R}(B) = \mathcal{R}(U_1),$$

$$U_1 = \begin{bmatrix} -0.16263 & -0.37740 & -0.90706 & 0.07878 \\ 0.25299 & -0.48199 & 0.12304 & -0.66365 \\ -0.78274 & -0.09883 & 0.12118 & -0.47865 \\ -0.00997 & 0.78436 & -0.34894 & -0.43787 \\ -0.54476 & 0.01649 & 0.16021 & 0.36404 \end{bmatrix},$$

$$\mathcal{Y}^{\perp} \cap \mathcal{Z}^{\perp} = \mathcal{R}(A)^{\perp} \cap \mathcal{R}(B)^{\perp} = \mathcal{R}(U_2),$$

$$U_2 = \begin{bmatrix} -0.04642 & -0.49812 & 0.36574 & -0.26683 & -0.73808 \end{bmatrix}^T.$$

The computation of an orthonormal basis for the intersection $\mathcal{Y} \cap \mathcal{Z}$, where $\mathcal{Y} = \mathcal{R}(A)$ and $\mathcal{Z} = \mathcal{R}(B)$, can be done by using (A.23):

$$\mathcal{Y} \cap \mathcal{Z} = (\mathcal{Y}^{\perp} + \mathcal{Z}^{\perp})^{\perp}.$$

Let the orthonormal bases of $\mathcal{Y}^{\perp}$ and $\mathcal{Z}^{\perp}$, obtained by singular vale decomposition of A and B, be denoted by Y_{orth} and Z_{orth}, respectively. Then

$$\mathcal{Y} \cap \mathcal{Z} = \mathcal{R}(S_{\text{orth}}),$$

where S_{orth} is an orthonormal basis for the orthogonal complement of the subspace

$$\mathcal{Y}^{\perp} + \mathcal{Z}^{\perp} = \mathcal{R}([Y_{\text{orth}}, Z_{\text{orth}}]).$$

Another way to compute $\mathcal{Y} \cap \mathcal{Z}$ uses the SVD

$$[A, B] = U^H \begin{bmatrix} \Sigma & 0 \\ 0 & 0 \end{bmatrix} \begin{bmatrix} V_1^H \\ V_2^H \end{bmatrix} \begin{matrix} \}\, r \\ \}\, n + p - r \end{matrix},$$

$$\Sigma = \text{diag}(\sigma_1, \ldots, \sigma_r), \quad \sigma_1 \geq \cdots \geq \sigma_r > 0$$

of the matrix $[A, B]$. Partitioning the matrix V_2 as

$$V_2 = \begin{bmatrix} W_1 \\ W_2 \end{bmatrix} \begin{matrix} \}\, n \\ \}\, p \end{matrix}$$

and choosing

$$D = \begin{cases} AW_1, & \text{if } \text{rank}(A) < \text{rank}(B), \\ BW_2, & \text{if } \text{rank}(A) \geq \text{rank}(B), \end{cases}$$

then an orthonormal basis for the intersection of $\mathcal{Y}$ and $\mathcal{Z}$ is obtained as

$$\mathcal{Y} \cap \mathcal{Z} = \mathcal{R}(D).$$

Example A.5. If the subspaces $\mathcal{Y}$ and $\mathcal{Z}$ are spanned by the columns of the matrices A and B given in Example A.4, then

$$\mathcal{Y}^{\perp} = \mathcal{R}(A)^{\perp} = \mathcal{R}(Y_{\text{orth}}),$$

$$Y_{\text{orth}} = \begin{bmatrix} 0.46595 & -0.17607 \\ 0.53553 & 0.29268 \\ 0.02284 & -0.41825 \\ 0.70291 & -0.04656 \\ -0.03857 & 0.84038 \end{bmatrix},$$

$$\mathcal{Z}^{\perp} = \mathcal{R}(B)^{\perp} = \mathcal{R}(Z_{\text{orth}}),$$

$$Z_{\text{orth}} = \begin{bmatrix} 0.65457 & -0.13998 \\ -0.35066 & 0.61918 \\ -0.51934 & -0.23118 \\ -0.18589 & 0.33112 \\ 0.37988 & 0.65874 \end{bmatrix},$$

$$\begin{aligned} \mathcal{Y} \cap \mathcal{Z} &= (\mathcal{R}(A)^{\perp} + \mathcal{R}(B)^{\perp})^{\perp} \\ &= \mathcal{R}(Y_{\text{orth}} + Z_{\text{orth}})^{\perp} = \mathcal{R}(S_{\text{orth}}), \end{aligned}$$

$$S_{\text{orth}} = \begin{bmatrix} 0.33429 & 0.20233 \\ 0.60177 & -0.25049 \\ 0.20796 & 0.78610 \\ -0.69091 & 0.05994 \\ -0.07433 & 0.52418 \end{bmatrix}.$$

Using the alternative method, we obtain

$$W_2 = \begin{bmatrix} -0.08264 & -0.57140 \\ 0.57140 & -0.08264 \\ 0.00000 & 0.00000 \end{bmatrix},$$

$$D = \begin{bmatrix} 3.10495 & 1.30099 \\ -0.97042 & 5.39057 \\ 8.97719 & -2.46511 \\ -1.31520 & -5.06000 \\ 5.38347 & -3.11206 \end{bmatrix}.$$

The orthonormal basis of $\mathcal{Y} \cap \mathcal{Z}$ is found by SVD of D. As a result, we find

$$\mathcal{Y} \cap \mathcal{Z} = \mathcal{R}(D),$$
$$\text{span}\left(\begin{bmatrix} -0.18605 & -0.34362 \\ 0.27908 & -0.58906 \\ -0.77522 & -0.24544 \\ -0.09303 & 0.68723 \\ -0.52715 & 0.04909 \end{bmatrix}\right).$$

Note that the orthonormal bases of the intersection, determined by the two methods, are different.

A.9 Projections, Distances and Angles

A.9.1 *Projections*

A *projection* is an *idempotent* linear transformation defined by

$$P^2 = P.$$

If x is an arbitrary element of $\mathbb{C}^n$, we can write

$$x = Px + (x - Px),$$

where $Px \in \mathcal{R}(P)$ and $x - Px \in \mathcal{N}(P)$. Hence, to each projection there corresponds a decomposition of $\mathbb{C}^n$ into a direct sum

$$\mathbb{C}^n = \mathcal{S} \oplus \mathcal{W},$$

where $\mathcal{S} = \mathcal{R}(P)$ and $\mathcal{W} = \mathcal{N}(P)$. We say that P is a projection on $\mathcal{S}$ parallel to $\mathcal{W}$.

If $\dim(\mathcal{S}) = 1$, then P is called *elementary projection*.

When $\mathcal{W} = \mathcal{S}^\perp$, we obtain a projection with special properties.

Let $\mathcal{S} \subset \mathbb{C}^n$ be a subspace, then each vector $x \in \mathbb{C}^n$ is represented uniquely in the form $x = y + z$, where $y \in \mathcal{S}$ and $z \in \mathcal{S}^\perp$. The matrix $P \in \mathbb{C}^{n\times n}$ defined by the equation $Px = y$ is referred to as the *orthogonal projection* onto $\mathcal{S}$. The orthogonal projection is a linear transformation which maps each vector from $\mathbb{C}^n$ into a vector in $\mathcal{S}$. If $x \in \mathcal{S}$, then $Px = x$, and if $x \in \mathcal{S}^\perp$, then $Px = 0$.

The orthogonal projection satisfies the following conditions:

(1) $\mathcal{R}(P) = S$, $\mathcal{N}(P) = \mathcal{S}^\perp$.
(2) $P^2 = P$.
(3) $P^H = P$.

It may be shown that P is a positive definite matrix possessing the property $\|P\|_2 = 1$.

If P is the orthogonal projection onto $\mathcal{S}$, then $P^\perp =: I_n - P$ is the orthogonal projection onto $\mathcal{S}^\perp$ since the vectors $y = Px$ and $z = (I_n - P)x$ are orthogonal. Thus we have

$$x^H x = y^H y + z^H z,$$

i.e.,

$$\|x\|_2^2 = \|y\|_2^2 + \|z\|_2^2. \tag{A.51}$$

Equation (A.51) is a generalization of the Pythagorean theorem in $\mathbb{C}^n$ (Figure A.10). It follows from (A.51) that $\|Px\|_2 \leq \|x\|_2$ for all x.

Let the columns of $V = [v_1, v_2, \ldots, v_k]$ form an orthonormal basis for a subspace $\mathcal{S}$. It my be proved that

$$P = VV^H$$

is the unique orthogonal projection onto $\mathcal{S}$. Note that any other basis VZ, where Z is unitary, will produce the same P.

Consider how to determine the orthogonal projections on the four fundamental subspaces: the range $\mathcal{R}(A) \subset \mathbb{C}^m$ and its orthogonal complement $\mathcal{R}(A)^\perp \subset \mathbb{C}^m$ and the null space $\mathcal{N}(A) \subset \mathbb{C}^n$ and its orthogonal complement $\mathcal{N}(A)^\perp \subset \mathbb{C}^n$. The computation of these projections can be done by using the singular value decomposition

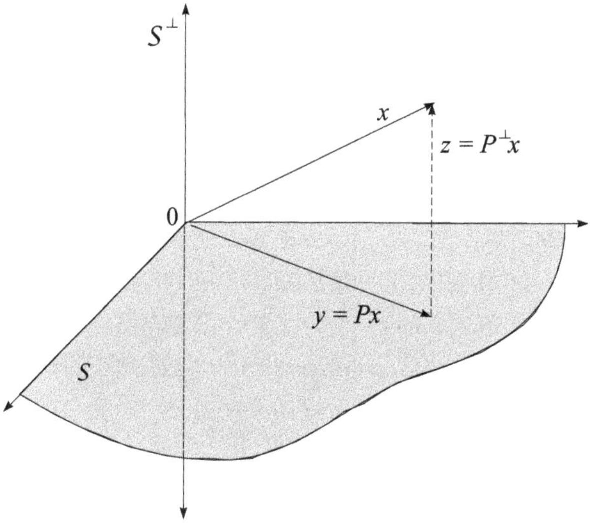

Fig. A.10. Orthogonal projection.

of A. Partitioning the SVD of A as

$$A = [\underbrace{U_1}_{r}, \underbrace{U_2}_{n-r}] \begin{bmatrix} \Sigma & 0 \\ 0 & 0 \end{bmatrix} \begin{bmatrix} V_1 \\ V_2 \end{bmatrix} \begin{matrix} \}\, r \\ \}\, n-r \end{matrix} \tag{A.52}$$

$$\Sigma = \mathrm{diag}(\sigma_1, \ldots, \sigma_r), \ \sigma_1 \geq \cdots \geq \sigma_r > 0,$$

we have that

$$\begin{aligned} P_{\mathcal{R}(A)} &= U_1 U_1^H, \\ P_{\mathcal{N}(A^H)} &= U_2 U_2^H, \\ P_{\mathcal{N}(A)} &= V_2 V_2^H, \\ P_{\mathcal{R}(A^H)} &= V_1 V_1^H. \end{aligned}$$

Example A.6. Let

$$A = \begin{bmatrix} -1 & 3 \\ 2 & -5 \\ -6 & -9 \\ 7 & 1 \end{bmatrix}.$$

Using the SVD of A we obtain

$$P_{\mathcal{R}(A)} = \begin{bmatrix} 0.14921 & -0.25934 & -0.16003 & -0.18461 \\ -0.25934 & 0.45157 & 0.25590 & 0.33899 \\ -0.16003 & 0.25590 & 0.77212 & -0.29130 \\ -0.18461 & 0.33899 & -0.29130 & 0.62709 \end{bmatrix},$$

$$P_{\mathcal{N}(A^H)} = \begin{bmatrix} 0.85079 & 0.25934 & 0.16003 & 0.18461 \\ 0.25934 & 0.54843 & -0.25590 & -0.33899 \\ 0.16003 & -0.25590 & 0.22788 & 0.29130 \\ 0.18461 & -0.33899 & 0.29130 & 0.37291 \end{bmatrix}.$$

In this case, $\text{rank}(A) = 2$ and

$$P_{\mathcal{R}(A^H)} = \begin{bmatrix} 1 & 0 \\ 0 & 1 \end{bmatrix}, \quad P_{\mathcal{N}(A)} = \begin{bmatrix} 0 & 0 \\ 0 & 0 \end{bmatrix}.$$

If $x = [-2, \ 8, \ 4, \ -6]^T$, then the projections of x onto the column space $\mathcal{R}(A)$ of A and onto its orthogonal complement are given by

$$y = P_{\mathcal{R}(A)} = \begin{bmatrix} -1.90560 \\ 3.12094 \\ 7.20354 \\ -1.84661 \end{bmatrix}, \quad z = P_{\mathcal{N}(A^H)} = \begin{bmatrix} -0.09440 \\ 4.87906 \\ -3.20354 \\ -4.15339 \end{bmatrix},$$

respectively, and

$$\|x\|_2^2 = \|y\|_2^2 + \|z\|_2 = 120.$$

A.9.2 *Distances between Vectors and Subspaces*

Let x and y be arbitrary vectors (points) in a linear space. By analogy with the notion of distance between points in $\mathbb{R}^n$, the distance $d(x, y)$ between x and y is defined by

$$d(x, y) = \|x - y\|_2. \tag{A.53}$$

The distance from x to the origin 0 is simply $\|x\|_2$ — the length of the vector x.

Let $\mathcal{S}$ be a subspace of $\mathbb{C}^n$. The *distance from the vector x to the subspace $\mathcal{S}$* is the quantity

$$d(x, \mathcal{S}) = \min_{y \in \mathcal{S}} \|x - y\|_2.$$

The vector y^* is referred to as the *best approximation* to the vector x by the vectors in $\mathcal{S}$ if

$$\|x - y^*\|_2 = d(x, \mathcal{S}).$$

Expanding the vector x as $x = x_{\mathcal{S}} + x_{\mathcal{S}^\perp}$, where $x_{\mathcal{S}}$ and $x_{\mathcal{S}^\perp}$ are the projections onto $\mathcal{S}$ and $\mathcal{S}^\perp$, respectively, we have that $x_{\mathcal{S}} - y$ and $x_{\mathcal{S}^\perp}$ are orthogonal (Figure A.11). Hence, by the Pythagorean theorem,

$$\|x - y\|_2^2 = \|x_{\mathcal{S}} - y\|_2^2 + \|x_{\mathcal{S}^\perp}\|_2^2. \tag{A.54}$$

Expression (A.54) implies that

$$\|x - y\|_2^2 \geq \|x_{\mathcal{S}^\perp}\|_2^2,$$

the equality holding only when $y = y^* = x_{\mathcal{S}}$. Thus

$$d(x, \mathcal{S}) = \|x - x_{\mathcal{S}}\|_2 = \|x_{\mathcal{S}^\perp}\|_2 = \|P_{\mathcal{S}^\perp} x\|_2 = \|(I_n - P_{\mathcal{S}})x\|_2.$$

In this way, the problem for finding the distance between a vector and a subspace reduces to the computation of the orthogonal projection

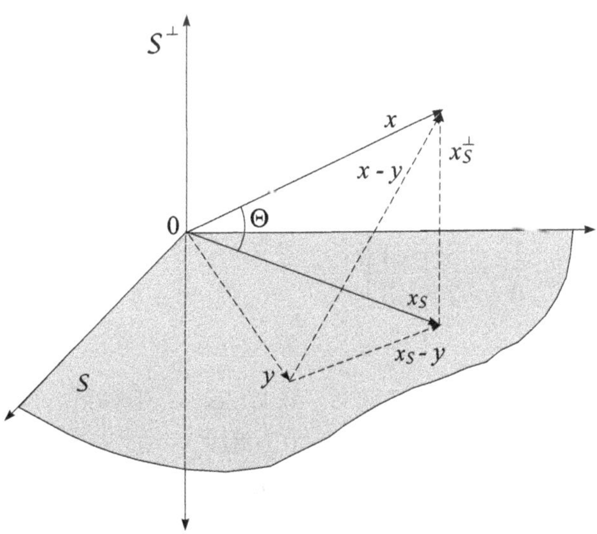

Fig. A.11. Distance from a vector to a subspace.

$P_{\mathcal{S}}$ or $P_{\mathcal{S}^\perp}$. Clearly, $x_{\mathcal{S}}$ is the best approximation of x by the vectors in $\mathcal{S}$.

Consider now a measure of the nearness of two subspaces bearing in mind that they have at least one common vector (the zero vector). For nontrivial subspaces $\mathcal{S}$ and $\mathcal{W}$, denote by $B_{\mathcal{S}}$ and $B_{\mathcal{W}}$ the unit spheres

$$B_{\mathcal{S}} = \{x \in \mathcal{S} : \|x\|_2 = 1\}, \quad B_{\mathcal{W}} = \{x \in \mathcal{W} : \|x\|_2 = 1\}.$$

If $\mathcal{S} = 0$, set $B_{\mathcal{S}} = 0$. Let $\max d(x, \mathcal{S}), x \in B_{\mathcal{W}}$, be the maximum distance between x and $\mathcal{S}$ as x varies on $B_{\mathcal{W}}$ and define analogically $\max d(x, \mathcal{W}), x \in B_{\mathcal{S}}$. The *gap* between subspaces $\mathcal{S}$ and $\mathcal{W}$ is the quantity

$$\gamma(\mathcal{S}, \mathcal{W}) = \max\{\max_{x \in B_{\mathcal{W}}} d(x, \mathcal{S}), \max_{x \in B_{\mathcal{S}}} d(x, \mathcal{W})\}.$$

The gap function satisfies the inequality

$$\gamma(\mathcal{S}, \mathcal{W}) \leq 1,$$

since $\|x\|_2 = 1$ implies that $d(x, \mathcal{S}), d(x, \mathcal{W}) \leq 1$.

The gap function satisfies

$$\gamma(\mathcal{S}, \mathcal{W}) = \|P_{\mathcal{W}} - P_{\mathcal{S}}\|_2, \tag{A.55}$$

where $P_{\mathcal{S}}$ and $P_{\mathcal{W}}$ are the projections onto $\mathcal{S}$ and $\mathcal{W}$, respectively.

Example A.7. If

$$A = \begin{bmatrix} 4 & -2 & 7 \\ -3 & -5 & 1 \\ -6 & 2 & -9 \\ -1 & 8 & -4 \\ 0 & -7 & 3 \end{bmatrix} \quad \text{and} \quad B = \begin{bmatrix} -6 & 1 & 5 \\ 4 & -3 & -2 \\ 11 & -7 & -4 \\ -3 & -6 & 9 \\ -8 & 6 & 2 \end{bmatrix},$$

then

$$P_{\mathcal{R}(A)} = \begin{bmatrix} 0.40530 & 0.13430 & -0.47029 & 0.04007 & -0.01459 \\ 0.13430 & 0.94423 & 0.10312 & 0.08578 & 0.12896 \\ -0.47029 & 0.10312 & 0.62771 & 0.04318 & 0.00370 \\ 0.04007 & 0.08578 & 0.043182 & 0.64384 & -0.46742 \\ -0.01459 & 0.12896 & 0.00370 & -0.46742 & 0.37891 \end{bmatrix},$$

$$P_{\mathcal{R}(B)} = \begin{bmatrix} 0.21359 & 0.00000 & -0.24021 & 0.29723 & 0.14806 \\ 0.00000 & 1.00000 & 0.00000 & 0.00000 & 0.00000 \\ -0.24021 & 0.00000 & 0.56438 & 0.03590 & -0.43228 \\ 0.29723 & 0.00000 & 0.03590 & 0.87934 & -0.12831 \\ 0.14806 & 0.00000 & -0.43228 & -0.12831 & 0.34269 \end{bmatrix}$$

and the gap between the ranges of A and B is

$$\gamma(\mathcal{R}(A), \mathcal{R}(B)) = \|P_{\mathcal{W}} - P_{\mathcal{S}}\|_2 = 0.66760.$$

A.9.3 *Angles between Vectors and Subspaces*

Let $\mathcal{S}$ be a subspace of $\mathbb{C}^n$ and x be a given vector. The smallest angle between x and $\mathcal{S}$ is defined by (Figure A.11)

$$\cos\Theta = \max_{y\in\mathcal{S}} \frac{x^H y}{\|x\|_2\|y\|_2}, \quad 0 \le \Theta \le \pi/2.$$

Obviously, the maximum is attended for $y = x_{\mathcal{S}} = P_{\mathcal{S}}x$. Thus,

$$\cos\Theta = \frac{x^H P_{\mathcal{S}}x}{\|x\|_2\|P_{\mathcal{S}}x\|_2} = \frac{\|P_{\mathcal{S}}x\|_2}{\|x\|_2}$$

and

$$\sin\Theta = \frac{\|P_{\mathcal{S}\perp}x\|_2}{\|x\|_2}.$$

Example A.8. The angle between the vector $x = [-2,\ 8,\ 4,\ -6]^T$ and the range of the matrix

$$A = \begin{bmatrix} -1 & 3 \\ 2 & -5 \\ -6 & -9 \\ 7 & 1 \end{bmatrix}$$

is given by

$$\Theta = \arccos\left(\frac{\|P_{\mathcal{R}(A)}x\|_2}{\|x\|_2}\right) = \arccos(0.756) = 0.713 \ \text{ rad} = 40.8 \ \text{ deg}.$$

Let $\mathcal{S}$ and $\mathcal{W}$ be subspaces of $\mathbb{C}^n$ with dimensions equal to p and let X and Y be orthonormal bases of these subspaces. Then the *canonical angles* between the subspaces are the numbers

$$\Theta_i = \arccos(\sigma_i(Y^H X)), \quad i = 1,\ 2,\ \ldots,\ p, \tag{A.56}$$

where $\sigma_i(Y^H X)$ are the singular values of the matrix $Y^H X$. Note that $\sigma_i(Y^H X) \leq \|X\|_2 \|Y\|_2 \leq 1$. The canonical angles are considered in decreasing order, and the ith canonical angle being written as $\Theta_i(\mathcal{S}, \mathcal{W})$. In practice, usually the most interesting are the maximum

$$\Theta_{\text{max}} = \arccos(\sigma_{\text{min}}(Y^H X)) \tag{A.57}$$

and the minimum

$$\Theta_{\text{min}} = \arccos(\sigma_{\text{max}}(Y^H X)), \tag{A.58}$$

angles between two subspaces, where $\sigma_{\text{min}}(Y^H X))$ and $\sigma_{\text{max}}(Y^H X))$ are the minimum and the maximum singular values of $Y^H X$, respectively.

Two subspace are identical if and only if the canonical angles between them are zeros.

The angles between subspaces are defined in the same way also in the case when $\dim(\mathcal{S}) < \dim(\mathcal{W})$.

Example A.9. If we compute the angles between the ranges of the matrices A and B from Example A.7, then we obtain

$$\begin{aligned}
\Theta_1 &= 0.731 \quad \text{rad} = 41.9 \ \text{deg}, \\
\Theta_2 &= 0.335 \quad \text{rad} = 19.2 \ \text{deg}, \\
\Theta_3 &= 0.0 \qquad \text{rad} = 0.0 \ \text{deg}.
\end{aligned}$$

The presence of a zero angle shows that $\mathcal{R}(A)$ and $\mathcal{R}(B)$ have a common subspace. In fact, the intersection between $\mathcal{R}(A) \cap \mathcal{R}(B)$ has a dimension equal to 1.

A.10 Notes and References

A rigorous and systematic presentation of the material in this Appendix (and much more) can be found in many works including the classical books of Bellman (1970), Halmosh (1974) and Strang (2006). A concise introduction to linear algebra from the vector spaces and matrices point of view can be found in Calvetti and Somersalo (2022) and a detailed presentation with many exercises is given in Jeffrey (2010). The book Naylor and Sell (1982) contains a clear presentation of the linear operator theory in a form suitable for engineers, scientists and applied mathematicians. A profound exposition of the theory of linear operators is given in the classical three-volume text of Dunford and Schwartz (1988a, 1988b, 1988c). A lot of facts about the matrices and transformations along with a reach bibliography can be found in Bernstein (2009).

Bibliography

Abdelfattah, A., Anzt, H., Boman, E., Carson, E., Cojean, T., Dongarra, J., Fox, A., Gates, M., Higham, N., Li, X. S., Liu, Y., Loe, J., Luszczek, P., Pranesh, S., Rajamanickam, S., Ribizel, T., Smith, B., Swirydowicz, K., Thomas, S., Tomov, S., Tzai, M., Yamazaki, I., and Yang, U. M. (2021). A survey of numerical linear algebra methods utilizing mixed-precision arithmetic, *Int. J. High Perform. Comput. Appl.* **35**, 344–369. doi:0.1177/109434202110033.

Aitken, A. (1933). On the canonical form of the singular matrix pencil, *Q. J. Math.* **os-4**, 241–257. doi:10.1093/qmath/os-4.1.241.

Al-Mohy, A. and Higham, N. (2009a). Computing the Fréchet derivative of the matrix exponential, with an application to condition number estimation, *SIAM J. Matrix Anal. Appl.* **30**, 1639–1657. doi:10.1137/080716426.

Al-Mohy, A. and Higham, N. (2009b). A new scaling and squaring algorithm for the matrix exponential, *SIAM J. Matrix Anal. Appl.* **31**, 970–989. doi:10.1137/09074721X.

Anderson, E., Bai, Z., Bischof, C., Blackford, S., Demmel, J., Dongarra, J., Du Croz, J., Greenbaum, A., Hammarling, S., McKenney, A., and Sorensen, D. (1999). *LAPACK Users' Guide*, 3rd edn. (SIAM, Philadelphia, PA).

Arioli, M., Demmel, J., and Duff, I. (1989). Solving sparse linear systems with sparse backward error, *SIAM J. Matrix Anal. Appl.* **10**, 165–190. doi:10.1137/0610013.

Arnold, V. (1971). On matrices depending on parameters, *Russ. Math. Surv.* **26**, 29–43. doi:10.1070/rm1971v026n02abeh003827.

Arnol'd, V. (1972). Lectures on bifurcations in versal families, *Russ. Math. Surv.* **27**, 54–123. doi:10.1070/rm1972v027n05abeh001385.

Arnold, V. (1988). *Geometrical Methods in the Theory of Ordinary Differential Equations*, 2nd edn. (Springer-Verlag, New York). doi:10.1007/978-1-4612-1037-5.

Arnold, V. (ed.) (1992). *Catastrophe Theory* (Springer-Verlag, Berlin). doi:10.1007/978-3-642-58124-3.

Arnol'd, V. (ed.) (1994). *Dynamical Systems V. Bifurcation Theory and Catastrophe Theory.* Encyclopaedia of Mathematical Sciences, Vol. 5 (Springer-Verlag, Berlin). doi:10.1007/978-3-642-57884-7.

Arnold, V., Goryunov, V., Lyashko, O., and Vasil'ev, V. (1993). *Dynamical Systems VIII. Singularity Theory II, Applications.* Encyclopaedia of Mathematical Sciences, Vol. 39 (Springer-Verlag, Berlin). doi:10.1007/978-3-662-06798-7.

Arnold, V., Goryunov, V., Lyashko, O., and Vasil'ev, V. (1998). *Singularity Theory I.* Encyclopaedia of Mathematical Sciences, Vol. 6 (Springer-Verlag, Berlin). doi:10.1007/978-3-642-58009-3.

Arrowsmith, D. and Place, C. (1990). *An Introduction to Dynamical Systems* (Cambridge University Press, Cambridge, UK).

Bai, Z. and Demmel, J. (1989). On a block implementation of Hessenberg multishift QR iteration, *Int. J. High Speed Comput.* **1**, 97–112. doi:10.1142/S0129053389000068.

Bai, Z. and Demmel, J. (1993). On swapping diagonal blocks in real Schur form, *Linear Algebra Appl.* **186**, 73–95. doi:10.1016/0024-3795(93)90286-W.

Bai, Z. and Demmel, J. (1998). Using the matrix sign function to compute invariant subspaces, *SIAM J. Matrix Anal. Appl.* **19**, 205–225. doi:10.1137/S0895479896297719.

Bai, Z.-Z. and Pan, J.-Y. (2021). *Matrix Analysis and Computations* (SIAM, Philadelphia, PA). doi:10.1137/1.9781611976632.

Bai, Z., Demmel, J., and Gu, M. (1997). An inverse free parallel spectral divide and conquer algorithm for nonsymmetric eigenproblems, *Numer. Math.* **76**, 279–308. doi:10.1007/s002110050264.

Bai, Z., Demmel, J., and Mckenney, A. (1993). On computing condition numbers for the nonsymmetric eigenproblem, *ACM Trans. Math. Softw.* **19**, 202–223. doi:10.1145/152613.152617.

Bai, Z., Demmel, J., Dongarra, J., Ruhe, A., and van der Vorst, H. (eds.) (2000). *Templates for the Solution of Algebraic Eigenvalue Problems: A Practical Guide* (SIAM, Philadelphia, PA). doi:10.1137/1.9780898719581.

Bartels, R. and Stewart, G. (1972). Algorithm 432: Solution of the matrix equation $AX+XB = C$, *Comm. ACM* **15**, 820–826. doi:10.1145/361573.361582.

Bauer, F. and Fike, C. (1960). Norms and exclusion theorems, *Numer. Math.* **2**, 137–141. doi:10.1007/BF01386217.

Baumgärtel, H. (1985). *Analytic Perturbation Theory for Matrices and Operators* (Birkhäuser Verlag, Basel, Switzerland).

Bavely, C. and Stewart, G. (1979). An algorithm for computing reducing subspaces by block diagonalization, *SIAM J. Numer. Anal.* **16**, 359–367. doi:10.1137/0716028.

Beattie, C. and Ipsen, I. (2003). Inclusion regions for matrix eigenvalues, *Linear Algebra Appl.* **358**, 281–291. doi:10.1016/S0024-3795(02)00279-3.

Beelen, T. and Van Dooren, P. (1988). An improved algorithm for the computation of Kronecker's canonical form of a singular pencil, *Linear Algebra Appl.* **105**, 9–65. doi:10.1016/0024-3795(88)90003-1.

Beelen, T. and Van Dooren, P. (1990). Computational aspects of the Jordan canonical form, in Cox, M. G. and Hammarling, S. (eds.), *Reliable Numerical Computation* (Oxford University Press, New York), pp. 57–72.

Bellman, R. (1959). Kronecker products and the second method of Lyapunov, *Math. Nachrichten* **20**, 17–19. doi:10.1002/mana.19590200103.

Bellman, R. (1970). *Introduction to Matrix Analysis*, 2nd edn. (SIAM, Philadelphia, PA).

Belyi, V., Khazanov, V., and Kublanovskaya, V. (1988). Spectral problems for matrix pencils. Methods and algorithms. III. *Sov. J. Numer. Anal. Math. Model.* **3**, 19–51. doi:10.1515/rnam.1989.4.1.19.

Benner, P., Mehrmann, V., and Xu, H. (1998). A numerically stable, structure preserving method for computing the eigenvalues of real Hamiltonian or symplectic pencils, *Numer. Math.* **78**, 329–358. doi:10.1007/s002110050315.

Benner, P., Byers, R., Mehrmann, V., and Xu, H. (2002). Numerical computation of deflating subspaces of skew Hamiltonian/Hamiltonian pencils, *SIAM J. Matrix Anal. Appl.* **24**, 165–190. doi:10.1137/S0895479800367439.

Berg, J. and Kwatny, H. (1995). A canonical parameterization of the Kronecker form of a matrix pencil, *Automatica* **31**, 669–680. doi:10.1016/0005-1098(94)00111-U.

Bernstein, D. (2009). *Matrix Mathematics: Theory, Facts and Formulas*, 2nd edn. (Princeton University Press, Princeton, NJ).

Bhatia, R. (2007). *Perturbation Bounds for Matrix Eigenvalues*. Classics in Applied Mathematics (SIAM, Philadelphia, PA). doi:10.1137/1.9780898719079.

Bhatia, R. and Rosenthall, P. (1997). How and why to solve the operator equation $ax - xb = c$, *Bull. Lond. Math. Soc.* **29**, 1–21. doi:10.1112/S0024609396001828.

Bickley, W. and McNamee, J. (1960). Matrix and other direct methods for the solution of systems of linear difference equations, *Philos. Trans. R. Soc. Lond. Ser. A* **252**, 69–131.

Bini, D., Iannazzo, B., and Meini, B. (2012). *Numerical Solution of Algebraic Riccati Equations* (SIAM, Philadelphia, PA).

Bischof, C. and Quintana-Ortí, G. (1998a). Algorithm 782: Codes for rank-revealing QR factorizations of dense matrices, *ACM Trans. Math. Softw.* **24**, 254–257. doi:10.1145/290200.287638.

Bischof, C. and Quintana-Ortí, G. (1998b). Computing rank-revealing QR factorizations of dense matrices, *ACM Trans. Math. Softw.* **24**, 226–253. doi:10.1145/290200.287637.

Björck, A. (2015). *Numerical Methods in Matrix Computations* (Springer, New York). doi:10.1007/978-3-319-05089-8.

Blackford, L., Choi, J., Cleary, A., D'Azevedo, E., Demmel, J., Dhillon, I., Dongarra, J., Hammarling, S., Henry, G., Petitet, A., Stanley, K., Walker, D., and Whaley, R. (1997). *ScaLAPACK Users' Guide* (SIAM, Philadelphia).

Bojanczyk, A. and Van Dooren, P. (1993). Reordering diagonal blocks in real Schur form, in Moonen, M., Golub, G., and Moor, B. D. (eds.), *Linear Algebra for Large Scale and Real-Time Applications* (Kluwer Academic Publishers, Dordrecht, The Netherlands), pp. 351–352. doi:10.1007/978-94-015-8196-7_23.

Boley, D. (1990). Estimating the sensitivity of the algebraic structure of pencils with simple eigenvalue estimates, *SIAM J. Matrix Anal. Appl.* **11**, 632–643. doi:10.1137/0611046.

Braman, K., Byers, R., and Mathias, R. (2002a). The multishift QR algorithm. Part I: Maintaining well-focused shifts and level 3 performance, *SIAM J. Matrix Anal. Appl.* **23**, 929–947. doi:10.1137/S0895479801384573.

Braman, K., Byers, R., and Mathias, R. (2002b). The multishift QR algorithm. Part II: Aggressive early deflation, *SIAM J. Matrix Anal. Appl.* **23**, 948–973. doi:10.1137/S0895479801384585.

Brandts, J. (2001). Matlab code for sorted real Schur forms, Preprint No. 1180, Mathematical Institute, Utrecht University.

Brechenmacher, F. (2006). Histoire du théorème de Jordan de la décomposition matricielle (1870–1930). Formes de représentation et méthodes de décomposition, Thèse de doctorat, École des Hautes Études en Sciences Sociales, Paris. https://tel.archives-ouvertes.fr/tel-00142786/document.

Brechenmacher, F. (2016). Algebraic generality versus arithmetic generality in the 1874 controversy between C. Jordan and L. Kronecker, in Chemla, K., Chorlay, R., and Rabouin, D. (eds.), *The Oxford Handbook of Generality in Mathematics and the Sciences* (Oxford University Press, Oxford, UK), pp. 433–467.

Brualdi, R. (1987). The Jordan canonical form: An old proof, *Am. Math. Mon.* **94**, 257–267. doi:10.2307/2323392.

Brualdi, R. and Mellendorf, S. (1994). Regions in the complex plane containing the eigenvalues of a matrix, *Am. Math. Mon.* **101**, 975–985. doi:10.1080/00029890.1994.12004577.

Bryan, K. and Leise, T. (2006). The $25,000,000,000 eigenvector: The linear algebra behind Google, *SIAM Rev.* **48**, 569–581. doi:10.1137/050623280.

Bunch, J. (1987). The weak and strong stability of algorithms in numerical linear algebra, *Linear Algebra Appl.* **88/89**, 49–66. doi:10.1016/0024-3795(87)90102-9.

Byers, R. (1984). A LINPACK-style condition estimator for the equation $AX - XB^T = C$, *IEEE Trans. Automat. Control* **AC-29**, 926–928. doi:10.1109/TAC.1984.1103389.

Byers, R., He, C., and Mehrmann, V. (1997). The matrix sign function method and the computation of invariant subspaces, *SIAM J. Matrix Anal. Appl.* **18**, 615–632. doi:10.1137/S0895479894277454.

Caliari, M. and Zivcovich, F. (2019). On-the-fly backward error estimate for matrix exponential approximation by Taylor algorithm, *J. Comput. Appl. Math.* **346**, 532–548. doi:10.1016/j.cam.2018.07.042.

Calvetti, D. and Somersalo, E. (2022). *The Less is More Linear Algebra of Vector Spaces and Matrices* (SIAM, Philadelphia, PA).

Campbell, S. and Meyer, C. (2009). *Generalized Inverses of Linear Transformations*. Classics in Applied Mathematics (SIAM, Philadelphia, PA). doi: 10.1137/1.9780898719048.

Cauchy, A. (1829). Sur l'équation à l'aide de laquelle on détermine les inégalités séculaires des mouvements des planètes, in *Exercices de Mathématiques*, Vol. IV (Bure Frères, Paris), pp. 140–160, Œuvres Complétes (IIe Série), t. IX (Gauthier-Villars, Paris), 1891, pp. 174–195.

Cauchy, A. (1840). Mémoire sur l'intégration des équationes linéares, in *Exercices d'Analyse et de Physique Mathématique*, Vol. I (Bachelier, Paris), pp. 53–100, Œuvres Complétes (Ie Série), t. IV (Gauthier-Villars, Paris), 1884, pp. 369–426.

Chaitin-Chatelin, F. and Frayssé, V. (1996). *Lectures on Finite Precision Computations* (SIAM, Philadelphia, PA).

Chaitin-Chatelin, F., Frayssé, V., and Braconnier, T. (1995). Computations in the neighbourhood of algebraic singularities, *Numer. Funct. Anal. Optim.* **16**, 287–302. doi:10.1080/01630569508816619.

Chan, T. (1987). Rank revealing QR factorizations, *Linear Algebra Appl.* **88–89**, 67–82. doi:10.1016/0024-3795(87)90103-0.

Chan, T. and Hansen, P. (1992). Some applications of the rank revealing QR factorization, *SIAM J. Sci. Statist. Comput.* **13**, 727–741. doi:10.1137/0913043.

Chandrasekaran, S. and Ipsen, I. (1994). On rank-revealing factorisations, *SIAM J. Matrix Anal. Appl.* **15**, 592–622. doi:10.1137/S0895479891223781.

Chang, X.-W., Paige, C., and Stewart, G. (1997). Perturbation analyses for the QR factorization, *SIAM J. Matrix Anal. Appl.* **18**, 1328–1340. doi:10.1137/S0895479896297720.

Chatelin, F. (1984). Simultaneous Newton's iteration for the eigenproblem, in Böhmer, K. and Stetter, H. (eds.), *Defect Correction Methods*. Computing Supplementum, Vol. 5 (Springer, Wien), pp. 67–74. doi:10.1007/978-3-7091-7023-6_4.

Chatelin, F. (2012). *Eigenvalues of Matrices*. Classics in Applied Mathematics (SIAM, Philadelphia, PA). doi:10.1137/1.9781611972467.

Chen, S., Feldman, R., and Parlett, B. (1977). Algorithm 517. A program for computing the condition numbers of matrix eigenvalues without computing eigenvectros, *ACM Trans. Math. Softw.* **3**, 186–203. doi:10.1145/355732.355741.

Chow, S.-N., Li, C., and Wang, D. (1994). *Normal Forms and Bifurcation of Planar Vector Fields* (Cambridge University Press, London, UK).

Coleman, T. and Sorensen, D. (1984). A note on the computation of an orthonormal basis for the null space of a matrix, *Math. Program.* **29**, 234–242. doi:10.1007/BF02592223.

Collatz, L. (1949). *Eigenwertaufgaben mit Technischen Anwendungen* (Geest & Portig K.-G., Leipzig).

Crawford, J. (1991). Introduction to bifurcation theory, *Rev. Mod. Phys.* **63**, 991–1037. doi:10.1103/RevModPhys.63.991.

Davies, P. and Higham, N. (2003). A Schur–Parlett algorithm for computing matrix functions, *SIAM J. Matrix Anal. Appl.* **25**, 464–485. doi:10.1137/S0895479802410815.

De Terán, F. and Dopico, F. (2023). On bundles of matrix pencils under strict equivalence, *Linear Algebra Appl.* **658**, 1–31. doi:10.1016/j.laa.2022.10.029.

Demmel, J. (1983a). The condition number of equivalence transformations that block diagonalize matrix pencils, *SIAM J. Numer. Anal.* **20**, 599–610. doi:10.1137/0720040.

Demmel, J. (1983b). A Numerical Analyst's Jordan Canonical Form, Ph.D. Thesis, Center for Pure and Applied Mathematics, University of California, Berkeley, CA.

Demmel, J. (1986). Computing stable eigendecompositions of matrices, *Linear Algebra Appl.* **79**, 163–193. doi:10.1016/0024-3795(86)90298-3.

Demmel, J. (1987a). On condition numbers and the distance to the nearest ill-posed problem, *Numer. Math.* **51**, 251–289. doi:10.1007/BF01400115.

Demmel, J. (1987b). Three methods for refining estimates of invariant subspaces, *Computing* **38**, 43–57. doi:10.1007/BF02253743.

Demmel, J. (1988). The probability that a numerical analysis problem is difficult, *Math. Comput.* **50**, 449–480. doi:10.2307/2008617.

Demmel, J. (1993). Steve Smale and the geometry of ill-conditioning, in Hirsch, M., Marsden, J., and Shub, M. (eds.), *From Topology to Computation: Proceedings of the Smalefest* (Springer-Verlag, New York), pp. 305–316. doi:10.1007/978-1-4612-2740-3.

Demmel, J. (1997). *Applied Numerical Linear Algebra* (SIAM, Philadelphia, PA).

Demmel, J. and Edelman, A. (1995). The dimension of matrices (matrix pencils) with given Jordan (Kronecker) canonical form, *Linear Algebra Appl.* **230**, 61–87. doi:10.1016/0024-3795(93)00362-4.

Demmel, J. and Kågström, B. (1986). Stably computing the Kronecker structure and reducing subspaces of singular pencils $A - \lambda B$ for uncertain data, in Cullum, J. and Willoughby, R. (eds.), *Large Scale Eigenvalue Problems* (Elsevier (North-Holland), Amsterdam, The Netherlands), pp. 283–323. doi:10.1016/S0304-0208(08)72650-1.

Demmel, J. and Kågström, B. (1987). Computing stable eigendecompositions of matrix pencils, *Linear Algebra Appl.* **88–89**, 139–186. doi:10.1016/0024-3795(87)90108-X.

Demmel, J. and Kågström, B. (1993a). The generalized Schur decomposition of an arbitrary pencil $A - \lambda B$: Robust software with error bounds and applications. Part I: Theory and algorithms, *ACM Trans. Math. Softw.* **19**, 160–174. doi:10.1145/152613.152615.

Demmel, J. and Kågström, B. (1993b). The generalized Schur decomposition of an arbitrary pencil $A - \lambda B$: Robust software with error bounds and applications. Part II: Software and applications, *ACM Trans. Math. Softw.* **19**, 175–201. doi:10.1145/152613.152616.

Dieci, L. and Papini, A. (2000). Padé approximation for the exponential of a block triangular matrix, *Linear Algebra Appl.* **308**, 183–202. doi:10.1016/S0024-3795(00)00042-2.

Dieci, L. and Russell, R. (1987). On the computation of invariant subspaces, Technical Report, LCCR TR 87-11, Laboratory for Computer and Communication Research, Simon Fraser University, Burnaby, BC, Canada.

Dmytryshyn, A., Johansson, S., and Kågström, B. (2017). Canonical structure transitions of system pencils, *SIAM J. Matrix Anal. Appl.* **38**, 1249–1267. doi:10.1137/16M1097857.

Dongarra, J., Grigori, L., and Higham, N. (2020). Numerical algorithms for high-performance computational science, *Philos. Trans. R. Soc. Lond. Ser. A* **378**, 1–18. doi:10.1098/rsta.2019.0066.

Dongarra, J., Hammarling, S., and Wilkinson, J. (1992). Numerical considerations in computing invariant subspaces, *SIAM J. Matrix Anal. Appl.* **13**, 145–161. doi:10.1137/0613013.

Dongarra, J., Moler, C., and Wilkinson, J. (1983). Improving the accuracy of computed eigenvalues and eigenvectors, *SIAM J. Numer. Anal.* **20**, 23–45. doi:10.1137/0720002.

Dongarra, J., Bunch, J., Moler, C., and Stewart, G. (1979). *LINPACK Users' Guide* (SIAM, Philadelphia, PA).

Drmač, Z. and Bujanović, Z. (2008). On the failure of rank-revealing QR factorization software — a case study, *ACM Trans. Math. Softw.* **35**, 1–28. doi:10.1145/1377612.1377616.

Dunford, N. and Schwartz, J. (1988a). *Linear Operators, Vol. 1: General Theory* (Wiley-Interscience, New York).

Dunford, N. and Schwartz, J. (1988b). *Linear Operators, Vol. 2: Spectral Theory, Self Adjoint Operators in Hilbert Space* (Wiley-Interscience, New York).

Dunford, N. and Schwartz, J. (1988c). *Linear Operators, Vol. 3: Spectral Operators* (Wiley-Interscience, New York).

Eckart, C. and Young, G. (1936). The approximation of one matrix by another of lower rank, *Psychometrika* **3**, 211–218. doi:10.1007/BF02288367.

Edelman, A. and Ma, Y. (2000). Staircase failures explained by orthogonal versal forms, *SIAM J. Matrix Anal. Appl.* **21**, 1004–1025. doi:10.1137/S089547989833574X.

Edelman, A., Elmroth, E., and Kågström, B. (1997). A geometric approach to perturbation theory of matrices and matrix pencils. Part I: Versal deformations, *SIAM J. Matrix Anal. Appl.* **18**, 653–692. doi:10.1137/S0895479895284634.

Edelman, A., Elmroth, E., and Kågström, B. (1999). A geometric approach to perturbation theory of matrices and matrix pencils. Part II: A stratification-enhanced staircase algorithm, *SIAM J. Matrix Anal. Appl.* **20**, 667–699. doi:10.1137/S0895479896310184.

Elmroth, E., Johansson, P., and Kågström, B. (2001). Computation and presentation of graphs displaying closure hierarchies of Jordan and Kronecker structures, *Numer. Linear Algebra Appl.* **8**, 381–399. doi:10.1002/nla.253.

Elmroth, E., Johansson, P., and Kågström, B. (2003). Bounds for the distance between nearby Jordan and Kronecker structures in a closure hierarchy, *J. Math. Sci. (N.Y.)* **114**, 1765–1779. doi:10.1023/A:1022498301583.

Elsner, L. (1982). On the variation of the spectra of matrices, *Linear Algebra Appl.* **47**, 127–138. doi:10.1016/0024-3795(82)90230-0.

Elsner, L. (1985). An optimal bound for the spectral variation of two matrices, *Linear Algebra Appl.* **71**, 77–80. doi:10.1016/0024-3795(85)90236-8.

Elsner, L. (1986). Perturbation theorems for the matrix eigenvalue problem, *Port. Math.* **43**, 69–76.

Elsner, L. and Lancaster, P. (1985). The spectral variation of pencils of matrices, *J. Comput. Math.* **3**, 262–274. doi:10.1023/A:1022498301583.

Elsner, L. and Sun, J. (1982). Perturbation thèorems for the generalized eigenvalue problem, *Linear Algebra Appl.* **48**, 341–357. doi:10.1016/0024-3795(82)90120-3.

Erxiong, J. (1994). Bounds for the smallest singular value of a Jordan block with an application to eigenvalue perturbation, *Linear Algebra Appl.* **167–198**, 691–707. doi:10.1016/0024-3795(94)90510-X.

Fasi, M. and Higham, N. J. (2003). An arbitrary precision scaling and squaring algorithm for the matrix exponential, *SIAM J. Matrix Anal. Appl.* **40**, 1233–1256. doi:10.1137/18M1228876.

Filippov, A. (1971). A short proof of the reduction to Jordan form, *Moskow Univ. Math. Bull.* **26**, 70–71.

Fletcher, R. and Sorensen, D. (1983). An algorithmic derivation of the Jordan canonical form, *Am. Math. Mon.* **90**, 12–16. doi:10.2307/2975686.

Forsythe, G. E. and Moler, C. B. (1967). *Computer Solution of Linear Algebraic Systems* (Prentice-Hall, Englewood Cliffs, NJ).

Francis, J. (1961). The QR transformation: A unitary analogue to the LR transformation — part 1, *Comput. J.* **4**, 265–271. doi:10.1093/comjnl/4.3.265.

Francis, J. (1962). The QR transformation — part 2, *Comput. J.* **4**, 332–345. doi:10.1093/comjnl/4.4.332.

Frobenius, G. (1879). Theorie der linearen Formen mit ganzen Coefficienten, *J. Reine Angew. Math.* **86**, 146–208. doi:10.3931/e-rara-18818.

Gantmacher, F. (1959a). *Theory of Matrices*, Vol. 1 (AMS Chelsea Publishing, New York), Reprinted by the AMS, 2000.

Gantmacher, F. (1959b). *Theory of Matrices*, Vol. 2 (AMS Chelsea Publishing, New York), Reprinted by the AMS, 2000.

Garbow, B., Boyle, J., Dongarra, J., and Moler, C. (1977). *Matrix Eigensystem Routines* — EISPACK *Guide Extension.* Lecture Notes in Computer Science, Vol. 51 (Springer-Verlag, New York).

Geršhgorin, S. (1931). Über die Abgrenzung der Eigenwerte einer Matrix, *Bulletin de l'Académie des Sciences de l'URSS. Classe des sciences mathématiques et naturelles* Issue 6, 749–754.

Geurts, A. (1982). A contribution to the theory of condition, *Numer. Math.* **39**, 85–96. doi:10.1007/BF01399313.

Ghavimi, A. and Laub, A. (1995). Backward error, sensitivity and refinement of computed solutions of algebraic Riccati equations, *Numer. Linear Algebra Appl.* **2**, 29–49. doi:10.1002/nla.1680020104.

Gilmore, R. (1993). *Catastrophe Theory for Scientists and Engineers* (Dover Publications, New York).

Gleich, D. and Constantine, P. (2015). Ranking web pages, in Higham, N. (ed.), *The Prinston Companion to Applied Mathematics* (Princeton University Press, Princeton, NJ), pp. 755–757.

Glendinning, P. (2015). Bifurcation theory, in Higham, N. (ed.), *The Prinston Companion to Applied Mathematics* (Princeton University Press, Princeton, NJ), pp. 393–402.

Gohberg, I. and Koltracht, I. (1993). Mixed, componentwise, and structured condition numbers, *SIAM J. Matrix Anal. Appl.* **14**, 688–704. doi:10.1137/0614049.

Gohberg, I., Lancaster, P., and Rodman, L. (2006). *Invariant Subspaces of Matrices with Applications* (SIAM, Philadelphia, PA). doi:10.1137/1.9780898719093.

Goldberg, D. (1991). What every computer scientist should know about floating-point arithmetic, *ACM Comput. Surv.* **23**, 5–48. doi:10.1145/103162.103163.

Golub, G. and Uhlig, F. (2009). The QR algorithm: 50 years later its genesis by John Francis and Vera Kublanovskaya and subsequent development, *IMA J. Numer. Anal.* **29**, 467–485. doi:10.1093/imanum/drp012.

Golub, G. and van der Vorst, H. (2000). Eigenvalue computation in the 20th century, *J. Comput. Appl. Math.* **123**, 35–65. doi:10.1016/S0377-0427(00)00413-1.

Golub, G. and Van Loan, C. (2013). *Matrix Computations*, 4th edn. (The John Hopkins University Press, Baltimore, MD).

Golub, G. and Wilkinson, J. (1976). Ill-conditioned eigensystems and the computation of the Jordan canonical form, *SIAM Rev.* **18**, 578–619. doi:10.1137/1018113.

Golub, G., Klema, V., and Stewart, G. (1976). Rank degeneracy and least squares problems, Technical Report TR-456, Department of Computer Science, University of Maryland, College Park, MD.

Golub, G., Nash, S., and Van Loan, C. (1979). A Hessenberg—Schur method for the problem $AX + XB = C$, *IEEE Trans. Automat. Control* **AC-24**, 909–913. doi:10.1109/TAC.1979.1102170.

Golubitsky, M. (1978). An introduction to catastrophe theory and its applications, *SIAM Rev.* **20**, 352–387. doi:10.1137/1020043.

Granat, R., Kågström, B., and Kressner, D. (2009). Parallel eigenvalue reordering in real Schur forms, *Concurrency and Computation: Practice and Experience* **21**, 1225–1250. doi:10.1002/cpe.1386.

Greenbaum, A., Li, R.-C., and Overton, M. (2020). First-order perturbation theory for eigenvalues and eigenvectors, *SIAM Rev.* **62**, 463–482. doi:10.1137/19M124784X.

Gregory, R. (1960). Defective and derogatory matrices, *SIAM Rev.* **2**, 134–139. doi:10.1137/1002025.

Gregory, R. and Karney, D. (1969). *A Collection of Matrices for Testing Computational Algorithms* (John Wiley and Sons, New York).

Gu, M. (1995). Finding well-conditioned similarities to block-diagonalize nonsymmetric matrices is NP-hard, *J. Complex.* **11**, 377–391. doi:10.1006/jcom.1995.1019.

Gu, M. and Eisenstat, S. (1996). Efficient algorithms for computing a strong rank-revealing QR factorization, *SIAM J. Sci. Statist. Comput.* **17**, 848–869. doi:10.1137/0917055.

Guckenheimer, J. (1978). The catastrophe controversy, *Math. Intell.* **1**, 15–20. doi:10.1007/BF03023037.

Guckenheimer, J. (2010). Bifurcations of dynamical systems, in Marchioro, C. (ed.), *Dynamical Systems*, Vol. 78 (Springer, Berlin), pp. 7–123. doi:10.1007/978-3-642-13929-1_1.

Guglielmi, N., Overton, M., and Stewart, G. (2015). An efficient algorithm for computing the generalized null space decomposition, *SIAM J. Matrix Anal. Appl.* **36**, 38–54. doi:10.1137/140956737.

Hadamard, J. (1923). *Lectures on Cauchy's Problem in Linear Partial Differential Equations* (Yale University Press, New Haven).

Hager, W. (1984). Condition estimates, *SIAM J. Sci. Statist. Comput.* **5**, 311–316. doi:10.1137/0905023.

Hahn, W. (1955). Eine bemerkung zur zweiten methode von Ljapunov, *Math. Nachrichten* **14**, 349–354. doi:10.1002/mana.19550140413.

Halmosh, P. (1974). *Finite-Dimensional Vector Spaces* (Springer-Verlag, New York).

Hansen, P. (1998). *Rank-Deficient and Discrete Ill-Posed Problems: Numerical Aspects of Linear Inversion* (SIAM, Philadelphia, PA).

Hawkins, T. (1975). Cauchy and the spectral theory of matrices, *Hist. Math.* **2**, 1–29. doi:10.1016/0315-0860(75)90032-4.

Hawkins, T. (1977). Weierstrass and the theory of matrices, *Arch. Hist. Exact Sci.* **17**, 119–163. doi:10.1007/BF02464978.

Henderson, H. and Searle, S. (1981). The vec-permutation matrix, the vec operator and Kronecker products: A review, *Linear Multilinear Algebra* **9**, 271–288. doi:10.1080/03081088108817379.

Henrici, P. (1962). Bounds for iterates, inverses, spectral variation and fields of values for non-normal matrices, *Numer. Math.* **4**, 24–40. doi:10.1007/BF01386294.

Hessenberg, K. (1940). Behandlung linearer Eigenwertaufgaben mit Hilfe der Hamilton-Cayleyschen Gleichung, Technical Report 1. Bericht der Reihe Numerische Verfahren, Institut für Praktische Mathematik (IPM), Technische Hochschule Darmstadt, Darmstadt, Germany.

Higham, N. (1988). FORTRAN codes for estimating the one-norm of a real or complex matrix, with applications to condition estimation, *ACM Trans. Math. Softw.* **14**, 381–396. doi:10.1145/50063.214386.

Higham, N. (1994). A survey of componentwise perturbation theory in numerical linear algebra, in Gautchi, W. (ed.), *Mathematics of Computation 1943–1993: A Half Century of Computational Mathematics, Proceedings of Symposia in Applied Mathematics*, Vol. 48 (American Mathematical Society, Providence, RI, USA), pp. 49–77.

Higham, N. (2002). *Accuracy and Stability of Numerical Algorithms*, 2nd edn. (SIAM, Philadelphia, PA).

Higham, N. (2008). *Functions of Matrices: Theory and Computation* (SIAM, Philadelphia, PA).

Higham, N. and Mary, T. (2022). Mixed precision algorithms in numerical linear algebra, *Acta Numer.* 347–414. doi:10.1017/S0962492922000022.

Higham, N. J. (1993). Perturbation theory and backward error for $AX - XB = C$, *BIT Numer. Math.* **33**, 124–136. doi:10.1007/BF01990348.

Higham, N. J. (2009). The scaling and squaring method for the matrix exponential revisited, *SIAM Rev.* **51**, 747–764. doi:10.1137/090768539.

Hilbert, D. (1904). Grundzüge einer allgemeinen Theorie der linearen Integralgleichungen, *Nachrichten von der Königl. Gesselschaft der Wissenschaften zu Göttingen. Mathematisch-physikalische Klasse*, 49–91, 213–259.

Hirsh, M. and Smale, S. (1974). *Differential Equations, Dynamical Systems, and Linear Algebra* (Academic Press, San Diego, CA).

Hong, Y. and Pan, C.-T. (1992). Rank-revealing factorizations and the singular value decomposition, *Math. Comput.* **58**, 213–232. doi:10.1090/S0025-5718-1992-1106970-4.

Horn, R. and Johnson, C. (1991). *Topics in Matrix Analysis* (Cambridge University Press, Cambridge, UK).

Horn, R. and Johnson, C. (2013). *Matrix Analysis*, 2nd edn. (Cambridge University Press, Cambridge, UK).

Horn, R. and Serra-Capizzano, S. (2006). A general setting for the parametric Google matrix, *Int. Math.* **3**, 385–411. doi:10.1080/15427951.2006.10129131.

Householder, A. (1958). Unitary triangularization of a nonsymmetric matrix, *J. ACM* **5**, 339–342. doi:10.1145/320941.320947.

Householder, A. (1964). *The Theory of Matrices in Numerical Analysis* (Blaisdell, New York).

Householder, A., Varga, R., and Wilkinson, J. (1970). A note on Gerschgorin's inclusion theorem for eigenvalues of matrices, *Numer. Math.* **16**, 141–144. doi:10.1007/BF02308866.

IEEE. (2019). 754-2019 Standard for Floating-Point Arithmetic (IEEE, New York). https://standards.ieee.org/standard/754-2019.html.

Ikramov, K. (1993). Matrix pencils: Theory, applications, and numerical methods, *J. Math. Sci.* **64**, 783–853. doi:10.1007/BF01098963.

Jeffrey, A. (2010). *Matrix Operations for Engineers and Scientists: An Essential Guide in Linear Algebra* (Springer, London). doi:10.1007/978-90-481-9274-8.

Johansson, P. (2003). StratiGraph User's Guide, Technical Report UMINF-03.21, Department of Computing Science, University of Umeå, Umeå, Sweden. https://archive.cs.umu.se/software/sgmcs/docs/Stratigraph2UsersGuide.pdf.

Johansson, P. (2006). Software Tools for Matrix Canonical Computations and Web-Based Software Library Environments, Ph.D. Thesis, Department of Computing Science, Umeå University, Umeå, Sweden.

Johansson, S. (2009). Tools for Control System Design, Ph.D. Thesis, Department of Computing Science, Umeå University, Umeå, Sweden.

Johnston, R. and Smith, B. (1970). Calculation of best isolated Gerschgorin discs, *Numer. Math.* **16**, 22–31. doi:10.1007/BF02162404.

Jordan, C. (1870). *Traité des Substitutions et des Équations Algébriques* (Gauthier-Villars, Paris).

Jordan, D. and Godbout, L. (1977). On computation of the canonical pencil of a linear system, *IEEE Trans. Automat. Control* **AC-22**, 126–128. doi: 10.1109/TAC.1977.1101436.

Kågström, B. (1977). Bounds and perturbation bounds for the matrix exponential, *BIT Numer. Math.* **17**, 39–57. doi:10.1007/BF01932398.

Kågström, B. (1981). How to compute the Jordan normal form — the choice between similarity transformations and methods using chain relations, Technical Report UMINF-91.81, Institute of Information Processing, University of Umeå, Umeå, Sweden.

Kågström, B. (1986). RGSVD — an algorithm for computing the Kronecker structure and reducing subspaces of singular $A - \lambda B$ pencils, *SIAM J. Sci. Statist. Comput.* **7**, 185–211. doi:10.1137/0907014.

Kågström, B. (1993). A direct method for reordering eigenvalues in the generalized real Schur form of a regular matrix pair (A, B), in Moonen, M., Golub, G., and Moor, B. D. (eds.), *Linear Algebra for Large Scale and Real-Time Applications* (Kluwer Academic Publishers, Dordrecht, The Netherlands), pp. 195–218. doi:10.1007/978-94-015-8196-7_11.

Kågström, B. (1994). A perturbation analysis of the generalized Sylvester equation $(AR - LB, DR - LE) = (C, F)$, *SIAM J. Matrix Anal. Appl.* **15**, 1045–1060. doi:10.1137/S0895479893246212.

Kågström, B. and Poromaa, P. (1992). Distributed and shared memory block algorithms for the triangular Sylvester equation with sep^{-1} estimators, *SIAM J. Matrix Anal. Appl.* **13**, 90–101. doi:10.1137/0613009.

Kågström, B. and Poromaa, P. (1996a). Computing eigenspaces with specified eigenvalues of a regular matrix pair (A, B) and condition estimation: Theory, algorithms and software, *Numer. Algorithms* **12**, 369–407. doi: 10.1007/BF02142813.

Kågström, B. and Poromaa, P. (1996b). LAPACK-style algorithms and software for solving the generalized Sylvester equation and estimating the separation between regular matrix pairs, *ACM Trans. Math. Softw.* **22**, 78–103. doi: 10.1145/225545.225552.

Kågström, B. and Ruhe, A. (1980a). Algorithm 560: JNF, an algorithm for numerical computation of the Jordan normal form of a complex matrix, *ACM Trans. Math. Softw.* **6**, 437–443. doi:10.1145/355900.355917.

Kågström, B. and Ruhe, A. (1980b). An algorithm for numerical computation of the Jordan normal form of a complex matrix, *ACM Trans. Math. Softw.* **6**, 398–419. doi:10.1145/355900.355912.

Kågström, B. and Westin, L. (1989). Generalized Schur methods with condition estimators for solving the generalized Sylvester equation, *IEEE Trans. Automat. Control* **34**, 745–751. doi:10.1109/9.29404.

Kågström, B. and Wiberg, P. (2000). Extracting partial canonical structure for large scale eigenvalue problems, *Numer. Algorithms* **24**, 195–237. doi:10.1023/A:1019153528915.

Kågström, B., Johansson, S., and Johansson, P. (2012). StratiGraph Tool: Matrix stratifications in control applications, in Biegler, L., Campbell, S., and Mehrmann, V. (eds.), *Control and Optimization with Differential-Algebraic Constraints* (SIAM, Philadelphia, PA), pp. 79–103. doi:10.1137/9781611972252.ch5.

Kahan, W. (1972). Conserving confluence curbs ill-condition, Technical Report AD-766 916, University of California, Berkeley, CA.

Kahan, W., Parlett, B., and Jiang, E. (1982). Residual bounds on approximate eigensystems of nonnormal matrices, *SIAM J. Numer. Anal.* **19**, 470–484. doi:10.1137/0719030.

Kalogeropoulos, G., Mitrouli, M., Pantelous, A., and Triantafyllou, D. (2009). The Weierstrass canonical form of a regular matrix pencil: Numerical issues and computational techniques, in Margenov, S., Vulkov, L., and Wasniewski, J. (eds.), *Numerical Analysis and Its Applications, NAA 2008*. Lecture Notes in Computer Science, Vol. 5434 (Springer, Berlin), pp. 322–329. doi:10.1007/978-3-642-00464-3_35.

Karkanias, N. (1987). On the characteristic, Weyr sequences, the Kronecker invariants and canonical form of a singular pencil, *IFAC Proc. Vol.* **20**, 109–114. doi:10.1016/S1474-6670(17)55018-1.

Karkanias, N. and Kalogeropoulos, G. (1986). On the Segré, Weyr characteristics of right (left) regular matrix pencils, *Int. J. Control* **44**, 991–1015. doi:10.1080/00207178608933647.

Karow, M., Kressner, D., and Tisseur, F. (2006). Structured eigenvalue condition numbers, *SIAM J. Matrix Anal. Appl.* **28**, 1052–1068. doi:10.1137/050628519.

Kato, T. (1995). *Perturbation Theory for Linear Operators*, 2nd edn. (Springer-Verlag, Berlin).

Kelvin, L. and Tait, P. (1912). *Treatise on Natural Philosophy*, Vol. I (Cambridge University Press, Cambridge, UK).

Kenney, C. S. and Laub, A. J. (1989). Condition estimates for matrix functions, *SIAM J. Matrix Anal. Appl.* **10**, 191–209. doi:10.1137/0610014.

Kenney, C. S. and Laub, A. J. (1998). A Schur–Fréchet algorithm for computing the logarithm and exponential of a matrix, *SIAM J. Matrix Anal. Appl.* **19**, 640–663. doi:10.1137/S0895479896300334.

Khazanov, V. and Kublanovskaya, V. (1988a). Spectral problems for matrix pencils. Methods and algorithms. I. *Sov. J. Numer. Anal. Math. Model.* **3**, 337–371. doi:10.1515/rnam.1988.3.5.337.

Khazanov, V. and Kublanovskaya, V. (1988b). Spectral problems for matrix pencils. Methods and algorithms. II. *Sov. J. Numer. Anal. Math. Model.* **3**, 467–484. doi:10.1515/rnam.1988.3.6.467.

Kline, M. (1972). *Mathematical Thought from Ancient to Modern Time*, Vol. 2 (Oxford University Press, New York).

Koikari, S. (2007). An error analysis of the modified scaling and squaring method, *Comput. Math. Appl.* **53**, 1293–1305. doi:10.1016/j.camwa.2006.04.032.

Konstantinov, M. and Petkov, P. (2020). *Perturbation Methods in Matrix Analysis and Control* (NOVA Science Publishers, New York). https://novapublishers.com/shop/perturbation-methods-in-matrix-analysis-and-control.

Konstantinov, M., Petkov, P., and Christov, N. (1994). Nonlocal perturbation analysis of the Schur system of a matrix, *SIAM J. Matrix Anal. Appl.* **15**, 383–392. doi:10.1137/S089547989120267X.

Koren, I. (2002). *Computer Arithmetic Algorithms*, 2nd edn. (A K Peters, Ltd., Natick, MA).

Korkmaz, E., Faverge, M., Pichon, G., and Ramet, P. (2022). Reaching the quality of SVD for low-rank compression through QR variants, Technical Report RR-9476, INRIA, Bordeaux — Sud Ouest.

Kressner, D. (2005). *Numerical Methods for General and Structured Eigenvalue Problems* (Springer, Berlin).

Kressner, D. (2006). Block algorithms for reordering standard and generalized Schur forms, *ACM Trans. Math. Soft.* **32**, 521–532. doi:10.1145/1186785.1186787.

Kronecker, L. (1890). Algebraische Reduction der Schaaren bilinearer Formen, *Sitzungsberihte der Königlich Preussischen Akademie der Wissenschaften zu Berlin vom Jahre 1890*, pp. S. 1225–1287. doi:10.3931/e-rara-17875; Werke, 3(2) (1931) (Teubner, Leipzig), pp. 139–155.

Kublanovskaya, V. (1961). Certain algorithms for the solution of the complete eigenvalue problem, *Sov. Math. Dokl.* **2**, 16–19.

Kublanovskaya, V. (1962). On some algorithms for the solution of the complete eigenvalue problem, *USSR Comput. Math. Math. Phys.* **1**, 637–657. doi:10.1016/0041-5553(63)90168-X.

Kublanovskaya, V. (1966). On a method of solving the complete eigenvalue problem for a degenerate matrix, *USSR Comput. Math. Math. Phys.* **6**, 1–14. doi:10.1016/0041-5553(66)90001-2.

Kublanovskaya, V. (1984). *AB*-algorithm and its modifications for the spectral problem of linear pencils of matrices, *Numer. Math.* **43**, 329–342. doi:10.1007/BF01390177.

Kublanovskaya, V. (1989). An algorithm for computing the spectral structure of a singular linear matrix pencil, *J. Math. Sci.* **47**, 2834–2842. doi:10.1007/BF01305211.

Kuznetsov, Y. (2004). *Elements of Applied Bifurcation Theory*, 4th edn. (Springer, New York). doi:10.1007/978-3-031-22007-4.

Lagrange, J.-L. (1762–1765). Solution de différents problèms de calcul intégral, *Miscellanea Taurinensia*, Œuvres de Lagrange, t. I (Gauthier-Villars, Paris), 1867, pp. 472–668.

Lagrange, J.-L. (1811). *Mécanique Analitique*, Vol. I, seconde edn. (Courcier, Paris).

Lagrange, J.-L. (1853). *Mécanique Analitique*, Vol. I, troisième edn. (Mallet-Bachelier, Paris), quatrième edn. in Œuvres de Lagrange, t. XI, Gauthier-Villars, Paris, 1888.

Lancaster, P. (1970). Explicit solutions of linear matrix equations, *SIAM Rev.* **12**, 544–566. doi:10.1137/1012104.

Lancaster, P. and Tismenetsky, M. (1985). *The Theory of Matrices*, 2nd edn. (Academic Press, Orlando, FL).

Laplace, P. (1799). *Traité de Mécanique Céleste*, Vol. I (Duprat, Paris).

Laplace, P. (1829). *Traité de Mécanique Céleste*, Vol. I, seconde edn. (Bachelier, Paris).

Laub, A. (2012). *Computational Matrix Analysis* (SIAM, Philadelphia, PA).

Lawson, C. and Hanson, R. (1995). *Solving Least Squares Problems*. Classics in Applied Mathematics (SIAM, Philadelphia, PA).

Le Verrier, U. (1840). Sur le variations séqulaires des elèménts elliptiques des sept planètes principales, *J. Math. Pures Appl.* **5**, 220–254.

Lefschetz, S. (1965). *Stability of Nonlinear Control Systems* (Academic Press, New York).

Lehouck, R. (1996). The computation of elementary unitary matrices, *ACM Trans. Math. Softw.* **22**, 393–400. doi:10.1145/235815.235817.

Levinger, B. and Varga, R. (1966). Minimal Gerschgorin sets. II, *Pac. J. Math.* **17**, 199–210. doi:10.2140/pjm.1966.17.199.

Li, H. and Wei, Y. (2015). Improved rigorous perturbation bounds for the LU and QR factorizations, *Numer. Linear Algebra Appl.* **252**, 1115–1130. doi: 10.1002/nla.1998.

Li, R. (2014). Matrix perturbation theory, in Hogben, L. (ed.), *Handbook of Linear Algebra*, 2nd edn. Discrete Mathematics and Applications (CRC Press, Boca Raton, FL), pp. (21–1)–(21–20).

Li, T. Y., Zhang, Z., and Wang, T. (1997). Determining the structure of the Jordan normal form of a matrix by symbolic computation, *Linear Algebra Appl.* **252**, 221–259. doi:10.1016/S0024-3795(97)00342-X.

Lidskii, V. (1966). Perturbation theory of non-conjugate operators, *USSR Comput. Math. Math. Phys.* **6**, 73–85. doi:10.1016/0041-5553(66)90033-4.

Lin, R., Wang, Z., and Lim, M. (1996). A practical algorithm for the efficient computation of eigenvector sensitivities, *Comput. Methods Appl. Mech. Eng.* **130**, 355–367. doi:10.1016/0045-7825(95)00929-9.

Lin, Y. and Wei, Y. (2007). Condition numbers of the generalized Sylvester equation, *IEEE Trans. Automat. Control* **52**, 2380–2385. doi:10.1109/TAC.2007.910727.

Ma, E.-C. (1966). A finite series solution of the matrix equation $AX - XB = C$, *SIAM J. Appl. Math.* **14**, 490–495. doi:10.1137/0114043.

Ma, Y. and Edelman, A. (1998). Nongeneric eigenvalue perturbations of Jordan blocks, *Linear Algebra Appl.* **273**, 45–63. doi:10.1016/S0024-3795(97)00342-X.

MacDuffee, C. (1946). *The Theory of Matrices* (Chelsea, New York).

Mailybaev, A. (2001). Transformation to versal deformations of matrices, *Linear Algebra Appl.* **337**, 87–108. doi:10.1016/S0024-3795(01)00346-9.

Mailybaev, A. (2006). Computation of multiple eigenvalues and generalized eigenvectors for matrices dependent on parameters, *Numer. Linear Algebra Appl.* **13**, 419–436. doi:10.1002/nla.471.

Malyshev, A. (1989). Computing invariant subspaces of a regular linear pencil of matrices, *Sib. Math. J.* **30**, 559–567. doi:10.1007/BF00971756.

Marcus, M. and Minc, H. (1964). *A Survey of Matrix Theory and Matrix Inequalities* (Allyn and Bacon, Boston).

Markus, A. and Parilis, E. (1983). The change of the Jordan structure of a matrix under small perturbations, *Linear Algebra Appl.* **54**, 139–152. doi:10.1016/0024-3795(83)90210-0.

Mastronardi, N. and Van Dooren, P. (2017). Computing the Jordan structure of an eigenvalue, *SIAM J. Matrix Anal. Appl.* **38**, 949–966. doi:10.1137/16M1083098.

Mastronardi, N. and Van Dooren, P. (2020). On QZ steps with perfect shifts and computing the index of a differential-algebraic equation, *IMA J. Numer. Anal.* **41**, 2516–2529. doi:10.1093/imanum/draa049.

Mathias, R. (1992). Evaluating the Frechet derivative of the matrix exponential, *Numer. Math.* **63**, 212–226. doi:10.1007/Bf01385857.

Mathias, R. (1995). Condition estimation for matrix functions via the Schur decomposition, *SIAM J. Matrix Anal. Appl.* **16**, 565–578. doi:10.1137/S0895479893244389.

MATLAB. (2020). MATLAB Version 9.9.0.1538559 (R2020b), The MathWorks, Inc., Natick, MA. http://www.mathworks.com.

McCartin, B. (2009). *Rayleigh-Schrödinger Perturbation Theory: Pseudoinverse Formulation* (Hikari Ltd., Rousse, Bulgaria).

MCST. (2019). Matrix Canonical Structure Toolbox, Department of Computing Science, University of Umeå, Umeå, Sweden. http://www.umu.se/en/stratigraph-mcs.

Medley, H. and Varga, R. (1968). On smallest isolated Gerschgorin disks for eigenvalues. III, *Numer. Math.* **11**, 361–369. doi:10.1007/BF01390177.

Mehrmann, V. and Voss, H. (2005). Nonlinear eigenvalue problems: A challenge for modern eigenvalue methods, *GAMM-Mitt.* **27**, 121–152. doi:10.1002/gamm.201490007.

Miller, P. (2015). Perturbation theory and asymptotics, in Higham, N. (ed.), *The Prinston Companion to Applied Mathematics* (Princeton University Press, Princeton, NJ), pp. 208–218.

Moler, C. (2002). The world's largest matrix computation. https://www.mathworks.com/company/newsletters/articles/the-world-s-largest-matrix-computation.html.

Moler, C. (2004). *Numerical Computing with MATLAB®* (SIAM, Philadelphia, PA). doi:10.1137/1.9780898717952.

Moler, C. and Stewart, G. (1973). An algorithm for generalized matrix eigenvalue problems, *SIAM J. Numer. Anal.* **10**, 241–256. doi:10.1137/0710024.

Moler, C. and Van Loan, C. (1978). Nineteen dubious ways to compute the exponential of a matrix, *SIAM Rev.* **20**, 801–836. doi:10.1137/1020098.

Moler, C. and Van Loan, C. (2009). Nineteen dubious ways to compute the exponential of a matrix, Twenty-five years later, *SIAM Rev.* **45**, 3–49. doi:10.1137/S00361445024180.

Moro, J. and Dopico, F. (1990). First order eigenvalue perturbation theory and the Newton diagram, in Drmač, Z., Hari, V., Sopta, L., Tutek, Z., and Veselić, K. (eds.), *Applied Mathematics and Scientific Computing* (Springer, New York), pp. 143–175. doi:10.1007/978-1-4757-4532-0.

Moro, J., Burke, J., and Overton, M. (1997). On the Lidskii-Vishik-Lyusternik perturbation theory for eigenvalues of matrices with arbitrary Jordan structure, *SIAM J. Matrix Anal. Appl.* **18**, 793–817. doi:10.1137/S0895479895294666.

Muller, J.-M., Brisebarre, N., de Dinechin, F., Jeannerod, C.-P., Lefèvre, V., Melquiond, G., Revol, N., Stehlé, D., and Torres, S. (2010). *Handbook of Floating-Point Arithmetic* (Birkhäuser, Boston). doi:10.1007/978-0-8176-4705-6.

Murnaghan, F. and Wintner, A. (1931). A canonical form for real matrices under orthogonal transformations, *Proc. Natl. Acad. Sci. USA* **17**, 417–420. doi:10.1073/pnas.17.7.417.

Najfeld, I. and Havel, T. (1995). Derivatives of the matrix exponential and their computation, *Adv. Appl. Math.* **16**, 321–375. doi:10.1006/aama.1995.1017.

Nakatsukasa, Y. (2011). Algorithms and Perturbation Theory for Matrix Eigenvalue Problems and the Singular Value Decomposition, Ph.D. Thesis, Office of Graduate Studies, University of California, Davis, CA.

Naylor, A. and Sell, G. (1982). *Linear Operator Theory in Engineering and Science* (Springer-Verlag, New York).

Obrechkoff, N. (2003). *Zeros of Polynomials* (Marin Drinov Academic Publishing House, Sofia).

O'Meara, K., Clark, J., and Vinsonhaler, C. (2011). *Advanced Topics in Linear Algebra: Weaving Matrix Problems through the Weyr Form* (Oxford University Press, Oxford).

Ortega, J. (1987). *Matrix Theory: A Second Course* (Plenum Press, New York).

Ortega, J. and Rheinboldt, W. (2000). *Iterative Solution of Nonlinear Equations in Several Variables*. Classics in Applied Mathematics (SIAM, Philadelphia, PA). doi:10.1137/1.9780898719468.

Ostrowski, A. (1957). Über die Stetigkeit von charakteristischen Wurzeln in Abhängigkeit von den Matrizenelementen, *Jahresbericht der Deutschen Mathematiker-Vereinigung* **60**, 40–42.

Overton, M. (2001). *Numerical Computing with IEEE Floating Point Arithmetic* (SIAM, Philadelphia).

Page, L., Brin, S., Motwani, R., and Winograd, T. (1999). The PageRank citation ranking: Bringing order to the web, Technical Report SIDL-WP-1999-0120, Stanford Digital Laboratoris.

Palis, J. and de Melo, W. (1982). *Geometric Theory of Dynamical Systems: An Introduction* (Springer-Verlag, New York).

Parlett, B. (1991). Symmetric matrix pencils, *J. Comput. Appl. Math.* **38**, 373–385. doi:10.1016/0377-0427(91)90183-K.

Parlett, B. (1998). The symmetric eigenvalue problem. *Classics in Applied Mathematics* (SIAM, Philadelphia, PA). doi:10.1137/1.9781611971163.

Parlett, B. N. (1974). Computation of functions of triangular matrices, Technical Report Memorandum ERL-M481, Electronics Research Laboratory, College of Engineering, University of California, Berkeley, CA.

Parlett, B. N. (1976). A recurrence among the elements of functions of triangular matrices, *Linear Algebra Appl.* **14**, 117–121. doi:10.1016/0024-3795(76)90018-5.

Parlett, B. N. (2009). A result complementary to Gersgorin's Circle Theorem, *Linear Algebra Appl.* **431**, 20–27. doi:10.1016/j.laa.2009.01.030.

Parodi, M. (1959). *La Localisation des Valeurs Caractéristiques des Matrices et ses Applications, Traité de Physique Théoretique et de Physique Mathématique*, Vol. XII (Gauthier-Villars, Paris).

Peters, G. and Wilkinson, J. (1970). $Ax = \lambda Bx$ and the generalized eigenproblem, *SIAM J. Numer. Anal.* **7**, 479–492. doi:10.1137/0707039.

Petkov, P. (2021). Componentwise perturbation analysis of the Schur decomposition of a matrix, *SIAM J. Matrix Anal. Appl.* **42**, 108–133. doi:10.1137/20M1330774.

Petkov, P. (2022). Componentwise perturbation analysis of the QR decomposition of a matrix, *Mathematics* **10**, 4687. doi:10.3390/math10244687.

Petkov, P. (2023). Jordan–Schur algorithms for computing the matrix exponential, *J. Math. Math. Sci.* ID4908185. doi:10.1155/2023/4908185.

Petkov, P. H. and Konstantinov, M. M. (2022). The numerical Jordan form, *Linear Algebra Appl.* **638**, 1–45. doi:10.1016/j.laa.2021.12.014.

Poincaré, H. (1885). Sur l'équilibre d'une masse fluide animé d'un mouvement de rotation, *Acta Math.* **7**, 259–380. doi:10.1007/BF02402204.

Poincaré, H. (1928). Sur les properiétés des fonctions défines par les équations aux différences partielles, in *Oeuvres de Henri Poincaré*, Vol. I (Gauthier-Villars, Paris), pp. L–CXXIX, Thèses présentés à la Faculté des Scinces de Paris in 1879.

Pokrzywa, A. (1986). On perturbations and the equivalence orbit of a matrix pencil, *Linear Algebra Appl.* **82**, 99–121. doi:10.1016/0024-3795(86)90144-8.

Porsching, T. (1966). Diagonal similarity transformations which isolate Gerschgorin disks, *Numer. Math.* **8**, 437–443. doi:10.1007/BF02166669.

Rellich, F. (1969). *Perturbation Theory of Eigenvalue Problem* (Gordon and Breach, New York).

Rice, J. (1966). A theory of condition, *SIAM J. Numer. Anal.* **3**, 287–310. doi:10.1137/0703023.

Ruhe, A. (1970a). An algorithm for numerical determination of the structure of a general matrix, *BIT Numer. Math.* **10**, 196–216. doi:10.1007/BF01936867.

Ruhe, A. (1970b). Perturbation bounds for means of eigenvalues and invariant subspaces, *BIT Numer. Math.* **10**, 343–354. doi:10.1007/BF01934203.

Ruhe, A. (1970c). Properties of a matrix with a very ill-conditioned eigenproblem, *Numer. Math.* **15**, 57–60. doi:10.1007/BF02165660.

Rutherford, D. (1932). On the solution of the matrix equation $ax + xb = c$, *Proc. Kon. Nederl. Akad. Weten. Ser. A* **35**, 53–59.

Rutishauser, H. (1958). Solution of eigenvalue problems with the LR-transformation, *Nat. Bur. Stand. Appl. Math. Ser.* **49**, 47–81.

Saad, Y. (2011). *Numerical Methods for Large Eigenvalue Problems*. Classics in Applied Mathematics (SIAM, Philadelphia, PA). doi:10.1137/1.9781611970739.

Sastre, J., Ibá nez, J., Defez, E., and Ruiz, P. (2015). New scaling-squaring Taylor algorithms for computing the matrix exponential, *SIAM J. Sci. Comput.* **37**, A439–A455. doi:10.1137/090763202.

Schrödinger, E. (1928). *Collected Papers on Wave Mechanics* (Blackie & Son Limited, London).

Schur, I. (1909). Über die charakteristischen Wurzeln einer linearen Substitution mit einer Anwendung auf die Theorie der Integralgleichungen, *Math. Ann.* **66**, 488–510. doi:10.1007/BF01450045.

Segre, C. (1883–84). Sulla teoria e sulla classificazione delle omografie in uno spazio lineare ad un numero qualunque di dimensioni, *Mem. R. Acc. Naz. Lincei* **19**, 127–148, in Opere, a cura della Unione Matematica Italiana, Vol. III, Edizione Cremonese, Roma, 1961, 304–333.

Serra-Capizzano, S. (2005). Jordan canonical form of the Google matrix: A potential contribution to the PageRank computation, *SIAM J. Matrix Anal. Appl.* **27**, 305–312. doi:10.1137/S0895479804441407.

Seyranian, A. P. and Mailybaev, A. A. (2003). *Multiparameter Stability Theory with Mechanical Applications* (World Scientific, Singapore).

Shapiro, H. (1991). A survey of canonical forms and invariants under similarity, *Linear Algebra Appl.* **147**, 101–167. doi:10.1016/0024-3795(91)90232-L.

Shapiro, H. (1999). The Weyr characteristics, *Am. Math. Mon.* **106**, 919–929. doi:10.1080/00029890.1999.12005141.

Shi, X. and Wei, Y. (2012). A sharp version of Bauer–Fike's theorem, *J. Comput. Appl. Math.* **236**, 3218–3227. doi:10.1016/j.cam.2012.02.021.

Simoncini, V. (1996). On the numerical solution of $AX - XB = C$, *BIT Numer. Math.* **36**, 814–830. doi:10.1007/BF01733793.

Simoncini, V. (2007). A new iterative method for solving large-scale Lyapunov equations, *SIAM J. Sci. Comput.* **29**, 1268–1288. doi:10.1137/06066120X.

Simoncini, V. (2016). Computational methods for linear matrix equations, *SIAM Rev.* **58**, 377–441. doi:10.1137/130912839.

Smale, S. (1990). Some remarks on the foundations of numerical analysis, *SIAM Rev.* **32**, 211–220. doi:10.1137/1032043.

Smith, B., Boyle, J., Dongarra, J., Garbow, B., Ikebe, Y., Klema, V., and Moler, C. (1976). *Matrix Eigensystem Routines*-EISPACK *Guide*. Lecture Notes in Computer Science, Vol. 6, 2nd edn. (Springer-Verlag, Berlin).

Smith, R. (1967). The condition numbers of the matrix eigenvalue problem, *Numer. Math.* **10**, 232–240. doi:10.1007/BF02162166.

Solov'ev, V. N. (1984). A generalization of Gershgorin's theorem, *Math. USSR-Izv.* **23**, 545–560. doi:10.1070/IM1984v023n03ABEH001785.

Sridhar, B. and Jordan, D. (1973). An algorithm for calculation of the Jordan canonical form of a matrix, *Comput. Electr. Eng.* **1**, 239–254. doi:10.1016/0045-7906(73)90017-7.

Steer, S. (1982). Formes de Jordan: Application a l'algorithmique numerique, in A. Bensoussan and J. Lions (eds.), *Analysis and Optimization of Systems*. Lecture Notes in Control and Information Sciences, Vol. 44 (Springer Verlag, Berlin), pp. 954–973. doi:10.1007/BFb0044442.

Stewart, G. (1972). On the sensitivity of the eigenvalue problem $Ax = \lambda Bx$, *SIAM J. Numer. Anal.* **9**, 669–686. doi:10.1137/0709056.

Stewart, G. (1973a). Error and perturbation bounds for subspaces associated with certain eigenvalue problems, *SIAM Rev.* **15**, 727–764. doi:10.1137/1015095.

Stewart, G. (1973b). *Introduction to Matrix Computations* (Academic Press, New York).

Stewart, G. (1975). Gershgorin theory for the generalized eigenvalue problem $Ax = \lambda Bx$, *Math. Comput.* **29**, 600–606. doi:10.1090/S0025-5718-1975-0379537-3.

Stewart, G. (1976). Algorithm 506: HQR3 and EXCHNG: Fortran subroutines for calculating and ordering the eigenvalues of a real upper Hessenberg matrix, *ACM Trans. Math. Softw.* **2**, 275–280. doi:10.1145/355694.355700.

Stewart, G. (1977). Perturbation bounds for the QR factorization of a matrix, *SIAM J. Numer. Anal.* **14**, 509–518. doi:10.1137/0714030.

Stewart, G. (1984). Rank degeneracy, *SIAM J. Sci. Stat. Comput.* **5**, 403–413. doi:10.1137/0905030.

Stewart, G. (1990). Stochastic perturbation theory, *SIAM Rev.* **32**, 579–610. doi: 10.1137/1032121.

Stewart, G. (1993). Determining rank in the presense of error, in Moonen, M., Golub, G., and Moor, B. D. (eds.), *Linear Algebra for Large Scale and Real-Time Applications. Proceedings of the NATO Advanced Study Institute*, Leuven, Belgium, August 1992 (Kluwer Academic Publishers, Dordrecht, The Netherlands), pp. 275–291. doi:10.1007/978-94-015-8196-7_16.

Stewart, G. (1994). Perturbation theory for rectangular matrix pencils, *Linear Algebra Appl.* **208–209**, 297–301. doi:DOI:10.1016/0024-3795(94)90445-6.

Stewart, G. (1998). *Matrix Algorithms, Vol. I: Basic Decompositions* (SIAM, Philadelphia, PA).

Stewart, G. (2001). *Matrix Algorithms, Vol. II: Eigensystems* (SIAM, Philadelphia, PA).

Stewart, G. and Sun, J.-G. (1990). *Matrix Perturbation Theory* (Academic Press, San Diego, CA).

Stolovitch, L. (1993). On the computation of a versal family of matrices, *Numer. Algorithms* **4**, 25–46. doi:10.1007/BF02142739.

Strang, G. (2006). *Linear Algebra and its Applications*, 4th edn. (Thomson Brooks/Cole, Pacific Grove, CA).

StratiGraph (2019). StratiGraph and MCS Toolbox, Department of Computing Science, Umeå University, Umeå, Sweden. https://www.umu.se/en/stratigraph-mcs/.

Strutt, J. and Rayleigh, B. (1945). *The Theory of Sound*, 2nd edn. (Dover Publications, New York).

Sun, J. (1992). On condition numbers of a nondefective multiple eigenvalue, *Numer. Math.* **61**, 265–275. doi:10.1007/BF01385508.

Sun, J. (1995a). On perturbation bounds for the QR factorization, *Linear Algebra Appl.* **215**, 95–112. doi:10.1016/0024-3795(93)00077-D.

Sun, J. (1995b). Perturbation bounds for the generalized Schur decomposition, *SIAM J. Matrix Anal. Appl.* **16**, 1328–1340. doi:10.1137/S0895479892242189.

Sun, J. (1998). Stability and accuracy: Perturbation analysis of algebraic eigenproblems, Technical Report 98.07, Department of Computing Science, Umeå University, Umeå, Sweden (Revised version, 2002).

Sylvester, J. (1883). On the equation to the secular inequalities in the planetary theory, *Phylos. Mag. Ser. 5* **16**, 267–269; Collected Mathematical Papers, Vol. 4, pp. 410–411.

Sylvester, J. (1884). Sur l'équation en matrices $px = xq$, *C. R. Acad. Sci. Paris* **99**, 67–71, 15–116.

Symbolic Toolbox (2023). Symbolic Math Toolbox, The MathWorks, Inc., Natick, MA. http://www.mathworks.com/products/symbolic/.

Taussky, O. (1964). On the variation of the characteristic roots of a finite matrix under various changes of its elements, in Schneider, H. (ed.), *Recent Advances in Matrix Theory*, Proceedings of Advanced Seminar, Mathematics Research Center, U.S. Army, University of Wisconsin, Madison, WI, 1963 (University of Wisconsin Press, Madison, WI), pp. 125–138.

Teissier, B. (2020). Foreword, in Molina, J., Lê, D., and Seade, J. (eds.), *Handbook of Geometry and Topology of Singularities*, Vol. I (Springer, Cham, Switzerland), pp. V–VI. doi:10.1007/978-3-030-53061-7.

Thom, R. (1977). Structural stability, catastrophe theory, and applied mathematics, *SIAM Rev.* **19**, 189–201. doi:10.1137/1019036.

Tisseur, F. and Meerbergen, K. (2001). The quadratic eigenvalue problem, *SIAM Rev.* **43**, 235–286. doi:10.1137/S0036144500381988.

Todd, J. (1965). On smallest isolated Gerschgorin disks for eigenvalues, *Numer. Math.* **7**, 171–175. doi:10.1007/BF01397690.

Turnbull, H. and Aitken, A. (1932). *An Introduction to the Theory of Canonical Matrices* (Blackie and Son, London).

Van Dooren, P. (1979). The computation of Kronecker's canonical form of a singular pencil, *Linear Algebra Appl.* **27**, 103–140. doi:10.1016/0024-3795(79)90035-1.

Van Dooren, P. (1981). The generalized eigenstructure problem in linear system theory, *IEEE Trans. Automat. Control* **AC-26**, 111–129. doi:10.1109/TAC.1981.1102559.

Van Dooren, P. (1982). Algorithm 590: DSUBSP and EXCHQZ: Fortran subroutines for computing deflating subspaces with specified spectrum, *ACM Trans. Math. Softw.* **8**, 376–382. doi:10.1145/356012.356017.

Van Dooren, P. (1983). Reducing subspaces: Definitions, properties and algorithms, in Kågström, B. and Ruhe, A. (eds.), *Matrix Pencils.* Lecture Notes in Mathematics, Vol. 973 (Springer, Berlin), pp. 58–73. doi:10.1007/BFb0062094.

Van Loan, C. (1977). The sensitivity of the matrix exponential, *SIAM J. Numer. Anal.* **14**, 971–981. doi:10.1137/0714065.

Van Loan, C. (1978). Computing integrals involving the matrix exponential, *IEEE Trans. Automat. Control* **AC-23**, 395–404. doi:10.1109/TAC.1978.1101743.

Van Loan, C. (1987). On estimating the condition of eigenvalues and eigenvectors, *Linear Algebra Appl.* **88–89**, 715–732. doi:10.1016/0024-3795(87)90131-5.

Van Loan, C. (1997). *Introduction to Scientific Computing: A Matrix-Vector Approach Using MATLAB* (Prentice-Hall, Upper Saddle River, NJ).

Varah, J. (1979). On the separation of two matrices, *SIAM J. Numer. Anal.* **16**, 216–222. doi:10.1137/0716016.

Varga, R. (1965). Minimal Gerschgorin sets, *Pac. J. Math.* **15**, 719–729. doi: 10.2140/pjm.1965.15.719.

Varga, R. (2004). *Geršgorin and His Circles.* Springer Series in Computational Mathematics (Springer-Verlag, Berlin).

Ward, R. C. (1977). Numerical computation of the matrix exponential with accuracy estimate, *SIAM J. Numer. Anal.* **14**, 600–610. doi:10.1137/0714039.

Waterhouse, W. (1984). The codimension of singular matrix pairs, *Linear Algebra Appl.* **57**, 227–245. doi:10.1016/0024-3795(84)90190-3.

Watkins, D. (2010). *Fundamentals of Matrix Computations*, 3rd edn. (John Wiley and Sons, New York).

Wedderburn, J. H. M. (1903). Note on the linear matrix equation, *Proc. Edinb. Math. Soc.* **22**, 49–53. doi:10.1017/S0013091500001954.

Wei, Y. and Li, X. (2003). An improvement on perturbation bounds for the Drazin inverse, *Numer. Linear Algebra Appl.* **10**, 563–575. doi:10.1002/nla.336.

Wei, Y. and Wang, G. (1997). The perturbation theory for the Drazin inverse and its applications, *Linear Algebra Appl.* **258**, 179–186. doi:10.1016/S0024-3795(96)00159-0.

Weierstrass, K. (1868). Zur Theorie der quadratischen und bilinearen formen, *Monatsberichte der Königlich Preussischen Akademie der Wissenschaften zu Berlin*, pp. 310–338; Werke, 2 (1895) (Mayer & Müller, Berlin), pp. 19–44.

Weintraub, S. (2008). *Jordan Canonical Form: Application to Differential Equations* (Morgan & Claypool Publishers, San Rafael, CA).

Weintraub, S. (2009). *Jordan Canonical Form: Theory and Practice* (Morgan & Claypool Publishers, San Rafael, CA).

Weyr, E. (1885). Répartition des matrices en espèces et formation de toutes les espèces, *C. R. Acad. Sci. Paris* **100**, 966–969.

Weyr, E. (1890). Zur theorie der bilinearen formen, *Monatsh. Math. Phys.* **1**, 163–236. doi:10.1007/BF01692475.

Wiggins, S. (2003). *Introduction to Applied Nonlinear Dynamical Systems and Chaos*, 2nd edn. Texts in Applied Mathematics (Springer, New York).

Wilkinson, J. (1965). *The Algebraic Eigenvalue Problem* (Clarendon Press, Oxford, UK).

Wilkinson, J. (1978). Linear differential equations and Kronecker's canonical form, in Boor, C. D. and Golub, G. (eds.), *Recent Advances in Numerical Analysis*, Proceedings of Symposium Conducted by the Mathematics Research Center, University of Wisconsin, Madison, WI (Academic Press, New York), pp. 231–265. doi:10.1016/B978-0-12-208360-0.50017-6.

Wilkinson, J. (1979). Kronecker's canonical form and the QZ algorithm, *Linear Algebra Appl.* **28**, 285–303. doi:10.1016/0024-3795(79)90140-X.

Wilkinson, J. (1984). Sensitivity of eigenvalues, *Util. Math.* **25**, 5–76.

Wilkinson, J. (1986). Sensitivity of eigenvalues, part II, *Util. Math.* **30**, 243–286.

Wilkinson, J. (2023). *Rounding Errors in Algebraic Processes.* Classics in Applied Mathematics (SIAM, Philadelphia, PA), Republication of the edition originally published in 1963.

Wilkinson, J. and Reinsch, C. (1971). *Handbook for Automatic Computation, Vol. 2: Linear Algebra* (Springer-Verlag, Berlin). doi:10.1007/978-3-662-39778-7.

Zeeman, E. (1979). Catastrophe theory, in Güttinger, W. and Eikemeier, H. (eds.), *Structural Stability in Physics.* Springer Series in Synergetics, Vol. 4 (Springer, Berlin), pp. 12–22. doi:10.1007/978-3-642-67363-4_3.

Zeng, Z. (2005). Computing multiple roots of inexact polynomials, *Math. Comput.* **74**, 869–903. doi:10.1090/S0025-5718-04-01692-8.

Zeng, Z. (2016). Sensitivity and computation of a defective eigenvalue, *SIAM J. Matrix Anal. Appl.* **37**, 798–817. doi:10.1137/15M1016266.

Zeng, Z. (2021). Geometric modeling and regularization of algebraic problems, Cornell University Library, ArXiv e-prints in Numerical Analysis [math.NA] 2102.08830v1. http://arxiv.org/abs/2102.08330v1.

Zeng, Z. and Li, T.-Y. (2013). NAClab: A Matlab toolbox for numerical algebraic computation, *ACM Commun. Comput. Algebra* **47**, 170–173. doi:10.1145/2576802.2576829.

Zeng, Z. and Li, T.-Y. (2021). A numerical method for computing the Jordan canonical form, Cornell University Library, ArXiv e-prints in Numerical Analysis [math.NA] 2103.0206v1. http://arxiv.org/abs/2103.02086v1.

Zhang, G., Li, H., and Wei, Y. (2022). Componentwise perturbation analysis for the generalized Schur decomposition, *Calcolo* **59**. doi:10.1007/s10092-022-00463-8.

Index

C

P

Q

R

www.ingramcontent.com/pod-product-compliance
Lightning Source LLC
Chambersburg PA
CBHW060822110426
42492CB00025B/140
9789811286445